13th South-East Asian Congress of Medical Physics 2015 (SEACOMP)

Journal of Physics: Conference Series Volume 694

Yogyakarta, Indonesia
10 – 12 December 2015

Editors:

Freddy Haryanto
Agentte Peralta
Kwan-Hoong Ng
Hidetaka Arimura
Sparisoma Viridi

ISBN: 978-1-5108-2236-8
ISSN: 1742-6588

Printed from e-media with permission by:

Curran Associates, Inc.
57 Morehouse Lane
Red Hook, NY 12571

CURRAN ASSOCIATES INC. proceedings.com

Some format issues inherent in the e-media version may also appear in this print version.

Printed by Curran Associates, Inc. (2016)

Institute of Physics
Dirac House, Temple Back
Bristol BS1 6BE UK

Phone: 44 1 17 929 7481
Fax: 44 1 17 920 0979

techtracking@iop.org

Additional copies of this publication are available from:

Curran Associates, Inc.
57 Morehouse Lane
Red Hook, NY 12571 USA
Phone: 845-758-0400
Fax: 845-758-2633
Email: curran@proceedings.com
Web: www.proceedings.com

Table of contents

Volume 694

13th South-East Asian Congress of Medical Physics 2015 (SEACOMP)10–12 December 2015, Yogyakarta, Indonesia

Accepted papers received: 23 February 2016Published online: 24 March 2016Preface

011001
OPEN ACCESS
The 13th South-East Asian Congress of Medical Physics (SEACOMP 2015)

011002
OPEN ACCESS
Peer review statement

Papers

Therapeutics

012001
OPEN ACCESS
Cell and molecular biology for diagnostic and therapeutic technology

M I Tan pg. 1

012002
OPEN ACCESS
Helical tomotherapy optimized planning parameters for nasopharyngeal cancer

K Yawichai, I Chitapanarux and S Wanwilairat pg. 11

012003
OPEN ACCESS
Verification of Oncentra brachytherapy planning using independent calculation

N A M Safian, N H Abdullah, R Abdullah and C S Chiang pg. 16

012004
OPEN ACCESS
Treatment planning systems for external whole brain radiation therapy: With and without MLC (multi leaf collimator) optimization

T Budiyono, W S Budi and E Hidayanto pg. 21

012005
OPEN ACCESS
The systematic and random errors determination using realtime 3D surface tracking system in breast cancer

J Kanphet, S Suriyapee, N Dumrongkijudom, T Sanghangthum, J Kumkhwao and M Wisetrintong pg. 26

012006
OPEN ACCESS
Comparison of dosimeter response: ionization chamber, TLD, and Gafchromic EBT2 film in 3D-CRT, IMRT, and SBRT techniques for lung cancer

A Fitriandini, W E Wibowo and S A Pawiro pg. 31

012007
OPEN ACCESS
Determination of effective doses in image-guided radiation therapy system

Y Y Pyone, S Suriyapee, T Sanghangthum, S Oonsiri and T Tawonwong pg. 36

012008
OPEN ACCESS
Sensitivity in error detection of patient specific QA tools for IMRT plans

S Z Lat, S Suriyapee and T Sanghangthum pg. 41

012009
OPEN ACCESS
Dosimetric impact of interplay effect in lung IMRT and VMAT treatment using in-house dynamic thorax phantom

Mukhlisin and S A Pawiro   pg. 46

012010
OPEN ACCESS
Dosimetry verification on VMAT and IMRT radiotherapy techniques: In the case of prostate cancer

A Maulana and S A Pawiro   pg. 51

012011
OPEN ACCESS
Evaluation of 3D pre-treatment verification for volumetric modulated arc therapy plan in head region

S Ruangchan, S Oonsiri and S Suriyapee   pg. 56

012012
OPEN ACCESS
RTtxGap: An android radiobiological tool for compensation of radiotherapy treatment interruption

A L Yusoff, M Mohamad, R Abdullah, V M K Bhavaraju and N R Nik Idris   pg. 61

012013
OPEN ACCESS
Focal spot estimation of an Elekta dedicated stereotactic linear accelerator Monte Carlo model

S Herwiningsih and A Fielding   pg. 66

012014
OPEN ACCESS
The implementation in-house dose verification for IMRT and VMAT on breast cancer and NPC cases

A Azzi, F Y Dimitri and S A Pawiro   pg. 71

012015
OPEN ACCESS
Dose profile measurements during respiratory-gated lung stereotactic radiotherapy: A phantom study

W L Jong, J H D Wong, K H Ng and N M Ung pg. 76

012016
OPEN ACCESS
Preliminary experience on the implementation of computed tomography (CT)-based image guided brachytherapy (IGBT) of cervical cancer using high-dose-rate (HDR) Cobalt-60 source in University of Malaya Medical Centre (UMMC)

Z Jamalludin, U N Min, W Z Wan Ishak and R Abdul Malik pg. 81

012017
OPEN ACCESS
Comparison of radiotherapy dosimetry for 3D-CRT, IMRT, and SBRT based on electron density calibration

K Kartutik, W E Wibowo and S A Pawiro pg. 86

012018
OPEN ACCESS
Comparison of vidar dosimetry advantage pro and epson perfection V700 scanner in densitometry of radiochomic EBT2 film in measurement of high dose gradient

W Bura, P Tangboonduangjit and N Damrongkijudom pg. 92

012019
OPEN ACCESS
Determination of output factor for 6 MV small photon beam: comparison between Monte Carlo simulation technique and microDiamond detector

K Krongkietlearts, P Tangboonduangjit and N Paisangittisakul pg. 97

012020
OPEN ACCESS
Neutron contamination of Varian Clinac iX 10 MV photon beam using Monte Carlo simulation

S Yani, R Tursinah, M F Rhani, R C X Soh, F Haryanto and I Arif pg. 101

012021
OPEN ACCESS
Dosimetric investigation of proton therapy on CT-based patient data using Monte Carlo simulation

T Chongsan, T Liamsuwan and P Tangboonduangjit pg. 107

012022
OPEN ACCESS
Comparison of build-up region doses in oblique tangential 6 MV photon beams calculated by AAA and CCC algorithms in breast Rando phantom

P Masunun, P Tangboonduangjit and N Dumrongkijudom pg. 112

012023
OPEN ACCESS
Consistency check of photon beam physical data after recommissioning process

B Kadman, N Chawapun, S Ua-apisitwong, T Asakit, N Chumpu and J Rueansri pg. 117

012024
OPEN ACCESS
Comparison of selected dose calculation algorithms in radiotherapy treatment planning for tissues with inhomogeneities

Y L Woon, S P Heng, J H D Wong and N M Ung pg. 122

012025
OPEN ACCESS
Commissioning and validation of COMPASS system for VMAT patient specific quality assurance

J Pimthong, C Kakanaporn, L Tuntipumiamorn, P Laojunun and P Iampongpaiboon pg. 127

012026
OPEN ACCESS
The dose delivery effect of the different Beam ON interval in FFF SBRT: TrueBEAM

T Tawonwong, S Suriyapee, S Oonsiri, T Sanghangthum and P Oonsiri pg. 132

012027
OPEN ACCESS
Determination of small field output factors in 6 and 10 MV flattening filter free photon beams using various detectors

W Masanga, P Tangboonduangjit, C Khamfongkhruea and C Tannanonta   pg. 136

012028
OPEN ACCESS
$ZnFe_2O_4$ nanoparticles for potential application in radiosensitization

M Hidayatullah, I Nurhasanah and W S Budi   pg. 141

012029
OPEN ACCESS
Validation of in-house treatment planning system software for cobalt-60 teletherapy unit at two radiotherapy installations

I A S Mu'minah, L O H Z Toresano, W E Wibowo, Sugiyantari and S A Pawiro   pg. 146

Diagnostics

012030
OPEN ACCESS
A fully automated calculation of size-specific dose estimates (SSDE) in thoracic and head CT examinations

C Anam, F Haryanto, R Widita, I Arif and G Dougherty   pg. 152

012031
OPEN ACCESS
Assessment of knowledge and awareness among radiology personnel regarding current computed tomography technology and radiation dose

M K A Karim, S Hashim, D A Bradley, N A Bahruddin, W C Ang and N Salehhon   pg. 158

012032
OPEN ACCESS
Occupational radiation doses during interventional procedures

N Nuraeni, E Hiswara, D Kartikasari, A Waris and F Haryanto pg. 164

012033
OPEN ACCESS
Establishment of multi-slice computed tomography (MSCT) reference level in Johor, Malaysia

M K A Karim, S Hashim, K A Bakar, H Muhammad, A Sabarudin, W C Ang and N A Bahruddin pg. 169

012034
OPEN ACCESS
Profile of CT scan output dose in axial and helical modes using convolution

C Anam, F Haryanto, R Widita, I Arif and G Dougherty pg. 174

012035
OPEN ACCESS
Radiation dose to physicians' eye lens during interventional radiology

N A Bahruddin, S Hashim, M K A Karim, A Sabarudin, W C Ang, N Salehhon and K A Bakar pg. 179

012036
OPEN ACCESS
Computer-aided classification of Alzheimer's disease based on support vector machine with combination of cerebral image features in MRI

C Jongkreangkrai, Y Vichianin, C Tocharoenchai, H Arimura for the Alzheimer's Disease Neuroimaging Initiative pg. 184

012037
OPEN ACCESS
Parameter-based estimation of CT dose index and image quality using an in-house android™-based software

S Mubarok, L E Lubis and S A Pawiro pg. 189

012038
OPEN ACCESS
Establishing daily quality control (QC) in screen-film mammography using leeds tor (max) phantom at the breast imaging unit of USTH-Benavides Cancer Institute

K J C Acaba, L D Cinco and J N Melchor   pg. 194

012039
OPEN ACCESS
Effect of filter on average glandular dose and image quality in digital mammography

C Songsaeng, A Krisanachinda and K Theerakul   pg. 198

012040
OPEN ACCESS
Calculation of size specific dose estimates (SSDE) value at cylindrical phantom from CBCT Varian OBI v1.4 X-ray tube EGSnrc Monte Carlo simulation based

M Nasir, D Pratama, C Anam and F Haryanto   pg. 203

012041
OPEN ACCESS
Patient radiation dose from computed tomography angiography and digital subtraction angiography of the brain

Y Netwong and A Krisanachinda   pg. 208

012042
OPEN ACCESS
Three-dimensional image reconstruction of macula from stratus optical coherence tomography (OCT) for diagnosis of macular degeneration

Arinilhaq and R Widita   pg. 214

012043
OPEN ACCESS
Segmentation of solid subregion of high grade gliomas in MRI images based on active contour model (ACM)

P Seow, M T Win, J H D Wong, N A Abdullah and N Ramli   pg. 219

012044
OPEN ACCESS
Comparison of segmentation using fast marching and geodesic active contours methods for bone

A Bilqis and R Widita pg. 224

012045
OPEN ACCESS
Contrast medium usage reduction in abdominal computed tomography by using high-iodinated concentration contrast medium

A Suwannasri, R Kaewlai and S Asavaphatiboon pg. 229

012046
OPEN ACCESS
Measurement of computed tomography dose profile with pitch variation using Gafchromic XR-QA2 and thermoluminescence dosimeter (TLD)

S Purwaningsih, L E Lubis, S A Pawiro and D S Soejoko pg. 233

012047
OPEN ACCESS
Validation of a paediatric thyroid phantom using different multidetector computed tomography models

M Alsabbagh, L Y Ng, A A Tajuddin, M A Manap and R Zainon pg. 238

012048
OPEN ACCESS
Comparison of image segmentation of lungs using methods: connected threshold, neighborhood connected, and threshold level set segmentation

A R Amanda and R Widita pg. 243

012049
OPEN ACCESS
Measuring radiation dose to patients undergoing fluoroscopically-guided interventions

L E Lubis and M K Badawy pg. 248

012050
OPEN ACCESS
The relationships between breast volume, breast dense volume and volumetric breast density with body mass index, body fat mass and ethnicity

N Zakariyah, N B Pathy, N A M Taib, K Rahmat, C W Judy, F Fadzil, S Lau and K H Ng pg. 256

012051
OPEN ACCESS
Analysis of urinary stone based on a spectrum absorption FTIR-ATR

V Asyana, F Haryanto, L A Fitri, T Ridwan, F Anwary and H Soekersi pg. 260

012052
OPEN ACCESS
Using computer aided system to determine the maximum depth of visualization of B-Mode diagnostic ultrasound image

G Maslebu, K Adi and Suryono pg. 265

012053
OPEN ACCESS
Dual energy micro CT SkyScan 1173 for the characterization of urinary stone

L A Fitri, V Asyana, T Ridwan, F Anwary, H Soekersi, F D E Latief and F Haryanto pg. 270

012054
OPEN ACCESS
Eye dose to staff involved in interventional and procedural fluoroscopy

D McLean, D Hadaya and J Tse pg. 275

012055
OPEN ACCESS
Psychophysiological interaction between superior temporal gyrus (STG) and cerebellum: An fMRI study

A N Yusoff, X L Teng, S B Ng, A I A Hamid and S Z M Mukari pg. 280

Nuclear Medicine

012056
OPEN ACCESS
Development of a software tool for an internal dosimetry using MIRD method

A Chaichana and C Tocharoenchai pg. 286

012057
OPEN ACCESS
Assessment of cardiac function using myocardial perfusion imaging technique on SPECT with ^{99m}Tc sestamibi

M R A Gani, F Nazir, S A Pawiro and D S Soejoko pg. 291

012058
OPEN ACCESS
Survival outcome of radioiodine therapy in post thyroidectomy thyroid carcinoma patients: Outcome of long term follow up

F Haque, N Nahar, S Sultana, F Nasreen, Z Jabin and A S M M Alam pg. 296

012059
OPEN ACCESS
Organ doses from hepatic radioembolization with ^{90}Y, ^{153}Sm, ^{166}Ho and ^{177}Lu: A Monte Carlo simulation study using Geant4

N A A Hashikin, C H Yeong, S Guatelli, B J J Abdullah, K H Ng, A Malaroda, A B Rosenfeld and A C Perkins pg. 301

012060
OPEN ACCESS
The standardization methods of radioactive sources (^{125}I, ^{131}I, ^{99m}Tc, and ^{18}F) for calibrating nuclear medicine equipment in Indonesia

G Wurdiyanto and H Candra pg. 306

012061
OPEN ACCESS
Radiation exposure to nuclear medicine staffs during ^{18}F-FDG PET/CT procedures at Ramathibodi Hospital

T Donmoon, W Chamroonrat and M Tuntawiroon pg. 312

012062
OPEN ACCESS
The correlation between effective renal plasma flow (ERPF) and glomerular filtration rate (GFR) with renal scintigraphy ^{99m}Tc-DTPA study

D Ratnasari, F Nazir, L O H Z Toresano, S A Pawiro and D S Soejoko pg. 318

Biomedical Engineering

012063
OPEN ACCESS
Integration of soft tissue model and open haptic device for medical training simulator

G F Akasum, L N Ramdhania, Suprijanto and A Widyotriatmo pg. 324

012064
OPEN ACCESS
Influence of bone morphological properties on a new expandable orthopaedic fastener

M Oldakowski, I Oldakowska, T B Kirk, C T Ford, T B Sercombe, P Hardcastle and R E Day pg. 330

012065
OPEN ACCESS
Image analysis for dental bone quality assessment using CBCT imaging

Suprijanto, L Epsilawati, M S Hajarini, E Juliastuti and H Susanti pg. 336

012066
OPEN ACCESS
Dental panoramic image analysis for enhancement biomarker of mandibular condyle for osteoporosis early detection

Suprijanto, Azhari, E Juliastuti, A Septyvergy and N P P Setyagar pg. 341

012067
OPEN ACCESS
Model-based assessment of probe placement criteria in cancer therapy using RF ablation

E W Tanotogono, G F Akasum, Suprijanto and J J Sudirham pg. 346

012068
OPEN ACCESS
Preliminary study of Alzheimer's Disease diagnosis based on brain electrical signals using wireless EEG

N Handayani, Y Akbar, S N Khotimah, F Haryanto, I Arif and W P Taruno pg. 351

012069
OPEN ACCESS
Two dimensional finite element method simulation to determine the brain capacitance based on ECVT measurement

S H Sirait, W P Taruno, S N Khotimah and F Haryanto pg. 357

012070
OPEN ACCESS
Spectral and brain mapping analysis of EEG based on Pwelch in schizophrenic patients

Y Akbar, S N Khotimah and F Haryanto pg. 362

Biophysics

012071
OPEN ACCESS
Analysis of respirable particulate exposure and its effect to public health around lead smelter and e-waste processing industry in West Java, Indonesia

M Marselina, D Roosmini, I R S Salami, M Ayu A and W Cahyadi pg. 367

012072
OPEN ACCESS
Modelling in conventional electroporation for model cell with organelles using COMSOL Multiphysics

M Y Sulaeman and R Widita pg. 372

012073
OPEN ACCESS
Synthesis and characterization of carbon nanotube from coconut shells activated carbon

A Melati and E Hidayati pg. 376

012074
OPEN ACCESS
Dosimetric impact of interplay effect in lung IMRT and VMAT treatment using in-house dynamic thorax phantom

S Julia, Nurlely and D S Soejoko pg. 381

012075
OPEN ACCESS
Study on the effect of nanoparticle bimetallic coreshell Au-Ag for sensitivity enhancement of biosensor based on surface plasmon resonance

Widayanti and K Abraha pg. 387

012076
OPEN ACCESS
Nonlinear dynamics of specific DNA-protein interactions

D Dwiputra, W Hidayat, R Khairani and F P Zen pg. 392

012077
OPEN ACCESS
Investigation of diffusion process inside the cell using Monte Carlo Cell (MCell)

A Sutresno, F Haryanto, S Viridi and I Arif pg. 397

012078
OPEN ACCESS
Identification and level of organochlorine insecticide contamination in groundwater and iridology analysis for people in Upper Citarum cascade

K Oginawati and M A Pratama pg. 402

012079
OPEN ACCESS
Study of budding yeast colony formation and its characterizations by using circular granular cell

D Aprianti, F Haryanto, A Purqon, S N Khotimah and S Viridi pg. 407

Journal of Physics: Conference Series **694** (2016) 011001 doi:10.1088/1742-6596/694/1/011001

The 13th South-East Asian Congress of Medical Physics (SEACOMP 2015)

13th SOUTH EAST ASIAN CONGRESS OF MEDICAL PHYSICS 2015

UIN Sunan Kalijaga Convention Hall
10-12 December 2015

Improving the Quality of Human Health through Physics

Topics:

- Therapeutics
- Diagnostics
- Nuclear Medicine
- Biomedical Engineering
- Biophysics

Keynote Speakers:

Prof. Kunio Doi, PhD

Prof. Rethy Chhem, MD, PhD

Prof. Kwan-Hoong Ng, PhD, DABMP

Prof. Hidetaka Arimura, PhD

Dr. rer. nat. Marselina Irasonia Tan

Dr. Donald McLean

E-mail: seacomp2015@medphys.fi.itb.ac.id
Website: http://medphys.fi.itb.ac.id/seacomp2015

Organized by:

INSTITUT TEKNOLOGI BANDUNG 1920 · STATE ISLAMIC UNIVERSITY SUNAN KALIJAGA YOGYAKARTA · UNIVERSITAS INDONESIA · HFMBI

13th SEACOMP 2015

Supported by:

SEAFOMP · IOMP

Journal of Physics: Conference Series **694** (2016) 011001 doi:10.1088/1742-6596/694/1/011001

The 13th South-East Asian Congress of Medical Physics (SEACOMP 2015)

UIN Sunan Kalijaga, Yogyakarta, Indonesia
December 10 – 12, 2015

The Editors:

Freddy Haryanto[1], Agnette Peralta[2], Kwan-Hoong Ng[3], Hidetaka Arimura[4], Sparisoma Viridi[1]

[1] Institut Teknologi Bandung, Bandung, Indonesia
[2] Food and Drug Administration, Manila, Philippines
[3] University of Malaya, Kuala Lumpur, Malaysia
[4] Kyushu University, Fukuoka, Japan

Journal of Physics: Conference Series **694** (2016) 011001 doi:10.1088/1742-6596/694/1/011001

WELCOME

These proceedings comprise selected, edited papers that were presented at the 13th SOUTH EAST ASIAN CONGRESS OF MEDICAL PHYSICS with the theme of **"Improving the quality of human health through physics"**, held in Yogyakarta, Indonesia, 10-12 December 2015. It was hosted by Sunan Kalijaga State Islamic University (UIN Sunan Kalijaga), Institut Teknologi Bandung (ITB) and University of Indonesia (UI) in association with the South East Asia Federation of Organizations for Medical Physics (SEAFOMP).

Yogyakarta was the ancient capital of Daerah Istimewa Yogyakarta. It is famous for the UNESCO world heritage site of Borobudur temple. This temple was built in the 9th century during the reign of the Sailendra Dynasty. The temple was designed in the Javanese Buddhist architectural style. The monument was restored by the Indonesian government with UNESCO's help in 1975 and 1982.

The congress attracted about 241 participants from 13 countries. Twelve invited speakers delivered nine plenary lectures as well as four pre-congress workshops. A total of 86 oral and 65 poster presentations were given. A variety of topics were discussed, ranging from the most advanced topics such as proton therapy, image-guided radiotherapy, functional MRI to the more conventional ones such as dose distribution, simulation, dosimetry, quality assurance, etc.

In these proceedings papers are published under five categories, namely therapeutics, diagnostics, nuclear medicine, biomedical engineering and biophysics.

Sincerely,

Freddy Haryanto
Conference Chair

Agnette Peralta
Conference Co-Chair

Kwan-Hoong Ng
Conference Co-Chair

Journal of Physics: Conference Series **694** (2016) 011001 doi:10.1088/1742-6596/694/1/011001

CONFERENCE OFFICIALS

President of Congress
Freddy Haryanto .... *Institut Teknologi Bandung, Indonesia*

Steering Committee
M. Machasin .... *UIN Sunan Kalijaga Yogyakarta, Indonesia*
Maizer Said Nahdi .... *UIN Sunan Kalijaga Yogyakarta, Indonesia*
Djarwani S. Soejoko .... *Universitas Indonesia, Indonesia*
Idam Arif .... *Institut Teknologi Bandung, Indonesia*
Siti Nurul Khotimah .... *Institut Teknologi Bandung, Indonesia*

International Scientific Committee
Kwan-Hoong Ng .... *University of Malaya, Malaysia*
Agnette de Perio Peralta .... *Food and Drug Administration, Philippines*
Hidetaka Arimura .... *Kyushu University, Japan*
James Cheow Lei Lee .... *Nanyang Technological University, Singapore*
Anchali Krisanachinda .... *Chulalongkorn University, Thailand*
Napapong Pongnapang .... *Mahidol University, Thailand*
Nguyen Tan Chau .... *Cho Ray Hospital, Vietnam*
Chai-Hong Yeong .... *University of Malaya, Malaysia*

Local Scientific Committee
Sparisoma Viridi .... *Institut Teknologi Bandung, Indonesia*
Suprianto Ardjo Pawiro .... *Universitas Indonesia, Indonesia*
Gede Bayu Suparta .... *Universitas Gadjah Mada, Indonesia*
Wahyu Setia Budi .... *Universitas Diponegoro, Indonesia*
Johan Noor .... *Universitas Brawijaya, Indonesia*
Rena Widita .... *Institut Teknologi Bandung, Indonesia*
Sihana .... *Universitas Gadjah Mada, Indonesia*
Widayani .... *Institut Teknologi Bandung, Indonesia*

Organizing Committee
Triati Dewi Kencana Wungu .... *Institut Teknologi Bandung, Indonesia*
Sitti Yani .... *Institut Teknologi Bandung, Indonesia*
Adita Sutresno .... *Institut Teknologi Bandung, Indonesia*
Syarif Husein Sirait .... *Institut Teknologi Bandung, Indonesia*
Vepy Asyana .... *Institut Teknologi Bandung, Indonesia*
Nita Handayani .... *Institut Teknologi Bandung, Indonesia*
Asih Melati .... *UIN Sunan Kalijaga Yogyakarta, Indonesia*
Sra Harke Pratama .... *Institut Teknologi Bandung, Indonesia*
Nunung Nuraeni .... *Institut Teknologi Bandung, Indonesia*
Leni Aziyus Fitri .... *Institut Teknologi Bandung, Indonesia*
Budhi Susilawati .... *UIN Sunan Kalijaga Yogyakarta, Indonesia*
Muhammad Nasir .... *Institut Teknologi Bandung, Indonesia*
Yudiansyah Akbar .... *Institut Teknologi Bandung, Indonesia*

Cover photo credit: Andrew JK Tan

Acknowledgements

We gratefully acknowledge, at the time of printing this publication, the financial contributors to the Conference from the following:

Peer review statement

All papers published in this volume of *Journal of Physics: Conference Series* have been peer reviewed through processes administered by the proceedings Editors. Reviews were conducted by expert referees to the professional and scientific standards expected of a proceedings journal published by IOP Publishing.

Journal of Physics: Conference Series **694** (2016) 012001 doi:10.1088/1742-6596/694/1/012001

Cell and molecular biology for diagnostic and therapeutic technology

M I Tan[1]

[1] School of Life Sciences and Technology, Institut Teknologi Bandung, Bandung 40132, Indonesia

E-mail: marsel@sith.itb.ac.id

Abstract. Our body contains 100 trillion cells. However, each cell has certain function and structure. For maintaining their integrity, cells will be collaborating with each other and with extracellular matrix surround them to form a tissue. These interactions effect internally on many networks or pathway such as signalling pathway, metabolic pathway and transport network in the cell. These networks interact with each other to maintain cell survival, cell structure and function and moreover the tissue as well as the organ which the cells built. Therefore, as part of a tissue, genetic and epigenetic abnormality of a cell can also alter these networks, and moreover disturb the function of the tissue itself. Hence, condition of genetic and epigenetic of the cell may affect other conditions in omics level such as transcriptomic, proteomic, metabolomics characteristics which can be differentiated by a particular unique molecular profile from each level, which can be used for diagnostic as well as for targeted therapy.

1. Cells and other components in our body – Complexity in our body

As we know from several decades, that cells are the smallest functional unit in our body [1]. In the human body, there are 100 trillion cells. Each cell in our body has a particular structure and function (figure 1). The cells collaborate with similar cell type or other cell type and acellular components such as connective tissue and construct a particular complex system which is then arranged to develop to specific organs, such as heart, liver, bone.

In keeping their functions and integrity, the cells interacts with cells and other acellular components. Interaction with other cells can be regulated intrinsically or externally through communication with other cells in the vicinity (local interaction) or with cells from a distant place. Interaction with cells in the vicinity could be occurred through interaction between molecules on cell membrane directly with molecules from of neighboring cell membrane (contact-dependent) or through secreted molecules from a cell which influence themselves (autocrine) or other surrounded cell types (paracrine). Distant communications can be happened with the help of blood vessel (endocrine), where the molecule communications delivered through the blood vessels or with the help of nerve cells (synaptic communication) [2] (figure 2).

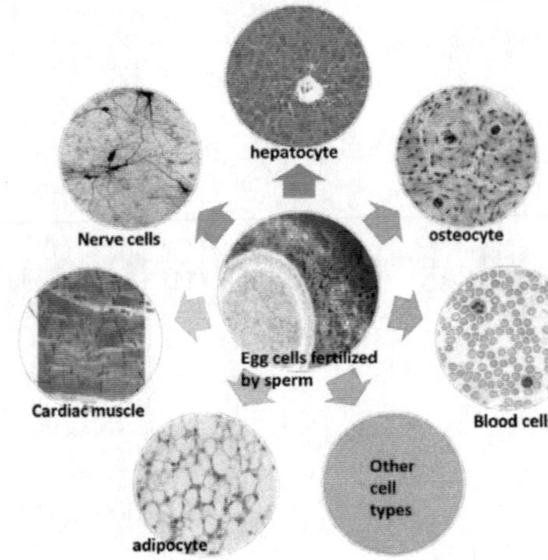

Figure 1. Cells are differentiated into many types of cells

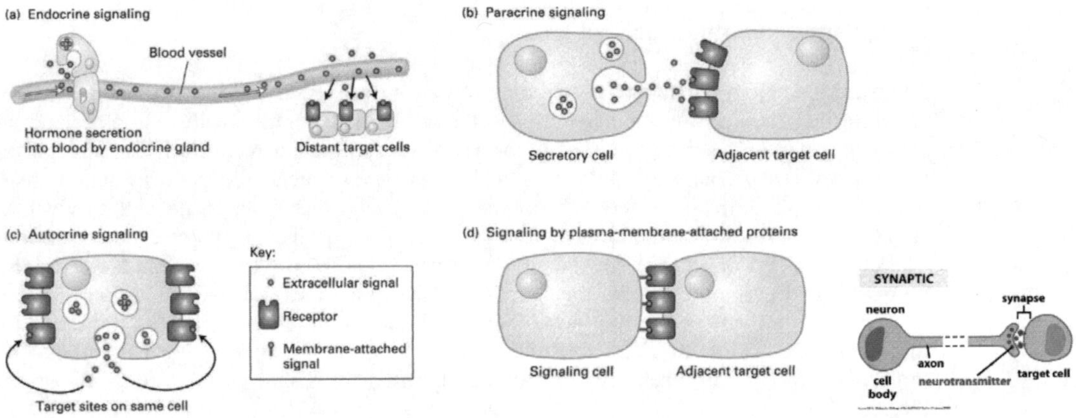

Figure 2. Cell communication type: (a) endocrine signaling; (b) paracrine signaling; (c) autocrine signaling; (d) contact dependent communication; (e) synaptic communication) (adopted from [2]).

Besides interaction with cells, the interaction also occurs with acellular components, extracellular matrix components such as laminin, collagen, etc [3]. Processes that take place in these cells are intrinsically reacted with a variety of cell proteins that work as signaling proteins, or cytoskeletal protein. This component should continue extracellular signals that have been delivered to the cells, and affects the activity of gene expression, metabolism or other cell activity that can further specify the processes that occur in these cells such as cell differentiation, apoptosis, cell proliferation, cell migration (figure 3). The processes that occur in response to a signal received by cell involve genetic and epigenetic signaling pathways related to the intracellular networks dynamics (figure 4).

Journal of Physics: Conference Series **694** (2016) 012001 doi:10.1088/1742-6596/694/1/012001

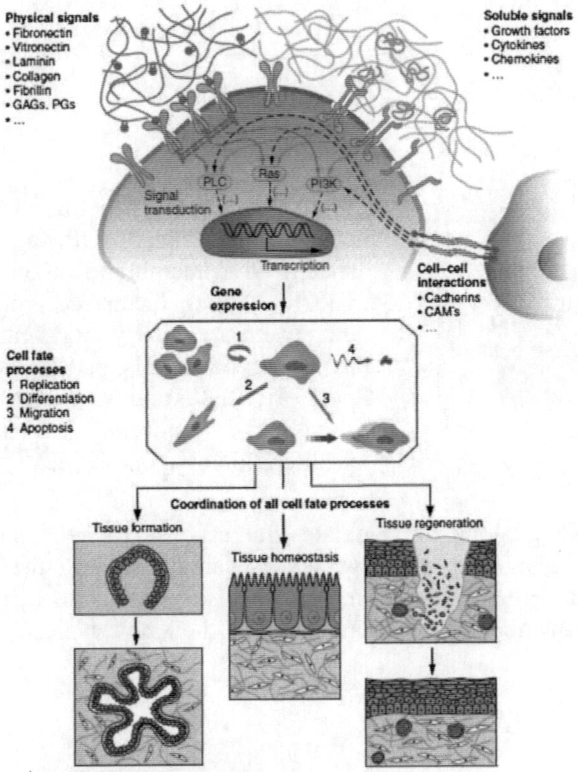

Figure 3. The cell's surrounding is comprised of a highly hydrated atmosphere containing physical and soluble signals, which can control signalling and activation of particular target genes. The activation of these genes will regulate the phenotype of the cell. (adopted from [3])

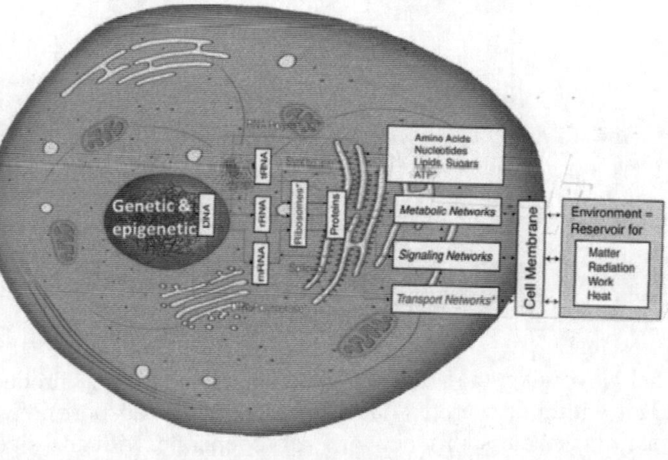

Figure 4. Genetic and epigenetic factors from a cell which are supported by circumstances will regulate metabolic networks, signaling networks, and transport networks (modified from [4]).

There are a variety of intracellular networks participating in the presence of signals from their environment, namely in a metabolic pathway, signaling pathways, transport networks [4]. Overall, the entire pathways collaborate with a very complex network however mutually supporting. Inside the cells, there are different signaling pathways such as MAPK pathway, JAK-Stat pathway, etc., which work together to determine the response of cells as shown in the image below. This signaling pathway will furthermore influence cell response such as cells proliferation or programmed cell death (apoptosis) [5].

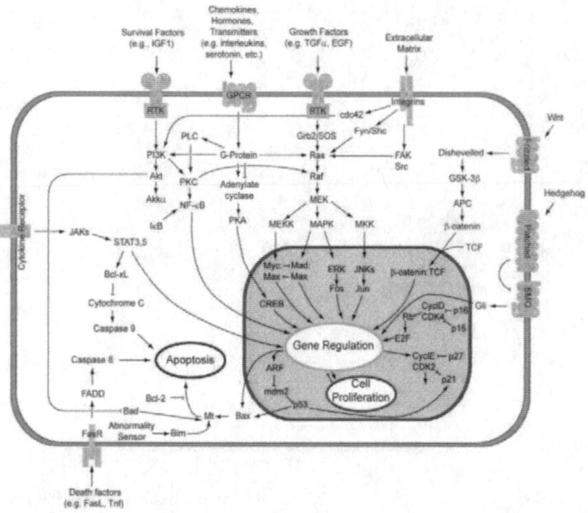

Figure 5. Extracellular molecules activate signaling pathway through specific transmembrane protein such as GPCR, Growth factor receptor, integrin etc. There are variety signaling pathway in eukaryotic cells, such as MAPK pathway, JNK pathway. Activation of signaling pathway cause cell proliferation or apoptosis etc. [5].

This signaling pathway is not operating alone to stimulate the response of cells but also in collaboration with other networks such as metabolic networks. As can be seen in the figure 6, activated myc protein in the MAPK pathway [6], for example, can also play a role in regulating glycolysis, protein metabolism, fatty acid metabolism, or indirectly affect the Krebs cycle.

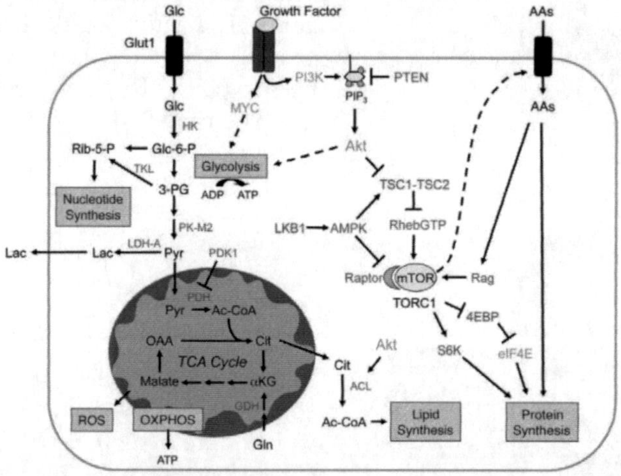

Figure 6. Growth factor-independent activation of the c-Myc pathways stimulates alterations in cellular metabolism to support cell growth and proliferation. c-Myc promote increased rates of glucose uptake and glycolysis (adopted from [6]).

Moreover, transport network, which involving the transport of proteins/other cells products in vesicles that also transferred to its target within or secreted out of the cell. This transport network is not also independent of other systems; nevertheless this network can cooperate with the signaling pathways as show in the picture below [7]. In the image below, some of the components that play a role in the MAPK pathway as Erk1/2, Raf also played a role in the secretory pathway, which furthermore eliciting cell response.

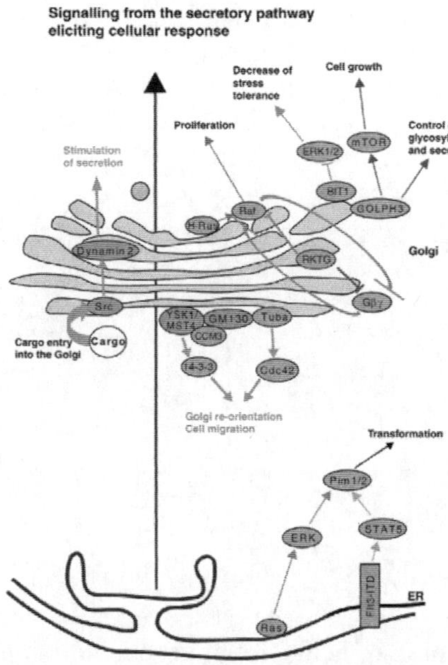

Figure 7. Signalling from the secretory pathway, displays a confirmation on how organelles of the secretory pathway contribute to the compartmentalization of signal transduction to obtain differential cell responses, such as proliferation, transformation and migration (adopted from [7]).

Changes in the secretory or signaling or metabolic pathways will transform cell homeostasis which effect on cell response such as in the figure below. In the figure below as an example, quiescent fibroblast cells can be turned into active proliferative cells when these cells change their metabolic network. When quiescent fibroblast cells transform into an active fibroblast cell in cell division, the cell will increase glucose intake and consequently glycolysis and Krebs cycle, the cells also produce nucleotides and macromolecules, which they need for replication and other activities. Besides that, the active fibroblast cell also activates glutamine and lactate metabolism [8]. This situation shows that alteration in these networks can cause changes in cell activity which could further affect the physiological state of a person. It means that unhealthy cells can cause a disease.

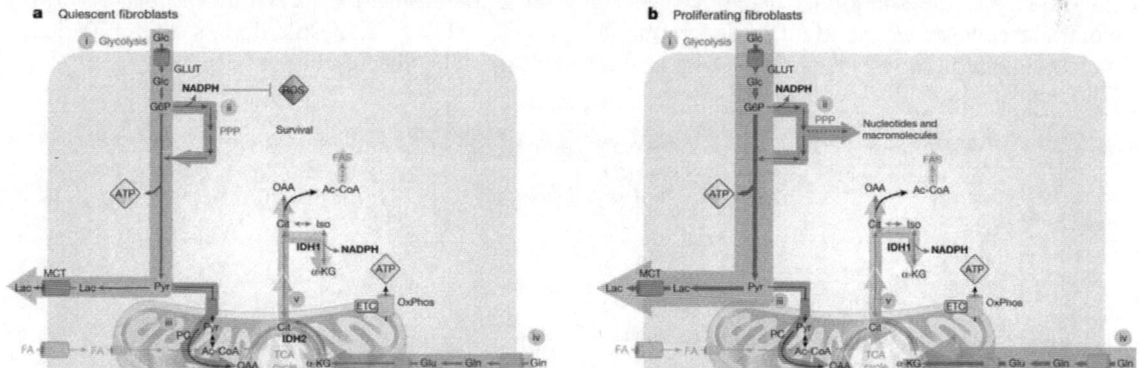

Figure 8. Alteration of fibroblast metabolism could modify proliferation activity of the cells (adopted from [8]).

2. Abnormal cells vs healthy cells

A variety of factors can effect abnormal cells. The prominent causes of the abnormality are genetic and epigenetic alterations. Genetic change might have no impact on the cells when the alteration is only causing silent mutation or is a mutation in the non-coding part of the gene. However, it might also cause severe changes in phenotype, such as sickle cell. In a patient who suffered from sickle cell, there is a mutation in codon 6 of the gene encoding hemoglobin (figure 9). In codon six of hemoglobin gene, nucleotide T is changed to A, which causes a change in amino acid, namely from the glutamic acid into valine, which furthermore resulted in an alteration of hemoglobin function regarding reduced ability for oxygen binding. The blood cell structure furthermore turned into a crescent-shaped and, as a result, people with this disease has some problem in oxygen binding.

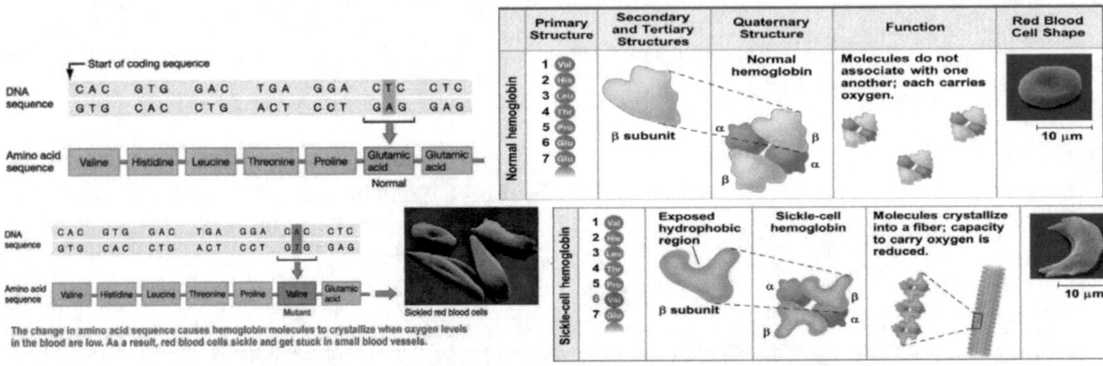

Figure 9. Hemoglobin mutation in sickle cell patient. Alteration of hemoglobin gene in codon 6 from T to A alter amino acid from glutamic acid to valine and furthermore alter structure and function of red blood cell (adopted from [9] and [10]).

As mention before, phenotype alteration can also occur due to epigenetic changes. These epigenetic changes took place in several types of cancer, for example in colorectal cancer. Changes in genetic and epigenetic might further alter structure and organization of cells, or cellular sensing and signaling or secreted products such as the extracellular matrix that are subsequently affecting changes in mechanotransduction signaling that cause a particular disease (figure 10) [11]. Furthermore, these alteration might also change a different profile of molecules on cell surfaces [12]. For example in breast cancer cells, EGFR is found more in membrane cells of breast cancer than in normal cells (figure 11). On the membrane of embryonic stem cells, the sulfated disaccharide is located on the membrane cells and support their differentiation process. However, desulfation of aforementioned disaccharide in these embryonic stem cells could inhibit the cells differentiation [12].

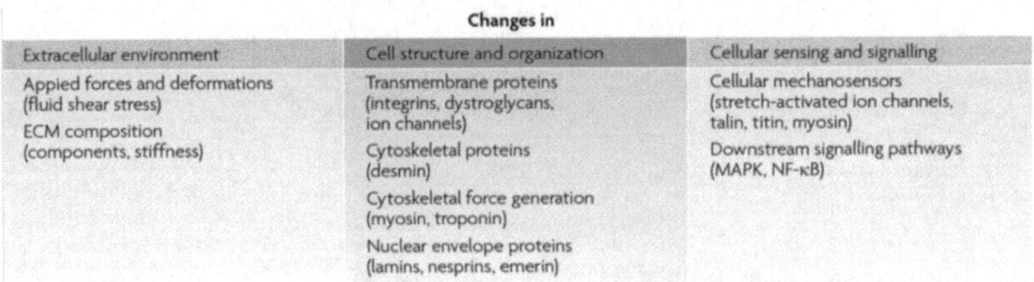

Figure 10. Changes in extracellular environment, cell structure and organization and cellular sensing and signaling are able to alter mechanotransduction signaling and can cause a disease (adopted from [11).

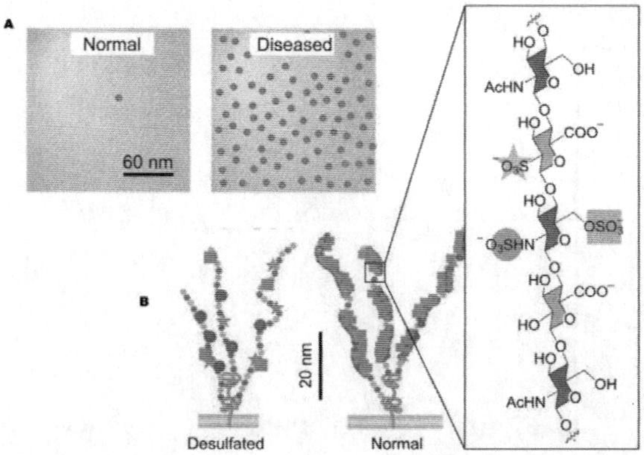

Figure 11. Profile of cell surface features: (A) Alteration of membrane molecule profile on normal and diseased cell (breast cancer cells), where HER2 proteins in breast cancer cells are overexpressed and cover more than 15% of surface cell; (B) Carbohydrate profile on cell surface regulate cell behavior. Desulfation of disaccharide chains on any protein/lipid membrane cell is able to prevent embryonic stem cells from undergoing differentiation (modified from [12]).

As explained before, cell transformation can also be associated with changes in cell metabolism which leads to alterations of extracellular states such as in cancer cells. In a cancer cell, there is any change in metabolism, the process of glycolysis activity is raised so that to maintain cell homeostasis and activity of ion transport, cancer cells increase ions concentration outside the cell (figure 12). It causes changes in potential membrane and cell junctions impairment but at the same time cells become anchorage-independent [13].

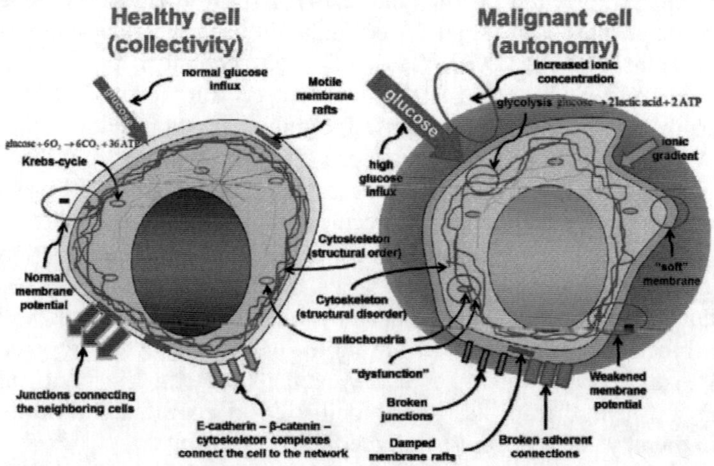

Figure 12. Biophysical differences between the healthy and malignant cells. Malignant cells increase metabolism activity which increase ionic concentration on the surrounding of the cells and disrupt cytoskeleton and junction organization (adopted from [13]).

Abnormalities can also occur in extracellular molecule profiles, for example, it is obvious in cardiac infarction. In myocardial infarction, there are any alterations in the extracellular components, where number and activity of MMP (Matrix metalloproteinase) and cytokine produced by macrophages increase in the extracellular matrix. The activity of macrophages and MMPs in this myocardium extracellular furthermore modifies connective tissue properties and cause moreover myocardial infarction [14].

Journal of Physics: Conference Series **694** (2016) 012001 doi:10.1088/1742-6596/694/1/012001

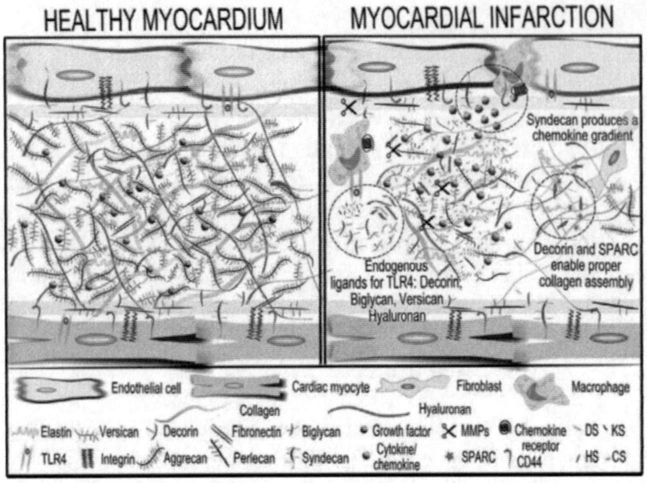

Figure 13. Alteration of architectural arrangement of cardiac ECM (extracellular matrix) in myocardial infarct (Adopted from [14]).

3. Theraphy based on cellular and molecular profile

According to the description above, the condition of genetic and epigenetic of the cell may affect other conditions in omics level such as transcriptomic, proteomic, metabolomic, phenomic characteristics which differentiate by a particular unique molecular profile (figure 14) [15]. In the genomic level we can use specific DNA sequence profile as marker or target for curing a disease such as polymorphism profile (Single nucleotide polymorphism/SNP), LOH (loss of heterozygosity) profile in the chromosomes. In the epigenetic level we can use such methylation or acetylation profile of DNA or histone protein as marker or target for disease therapy. In transcriptomic or proteomic level, specific gene expression profile or miRNA for example can be regulated as targeted therapy or utilized for diagnosis purposes, whereas in metabolomics level, therapy is based on the metabolite level produced in a person. Due to specific molecules profile and certain molecules level in each omics level, it is possible to develop a strategy using these specific molecules for diagnostic and targeted therapy for example by using radiotherapy or chemotherapy.

Cancer treatment can be conducted using chemicals, especially one or more anti-cancer drugs (chemotherapy agents) which can block the activity of certain molecules from any level of omics. Chemotherapy may be applied with therapeutic intent, or it may aim to prolong life or to relieve symptoms (palliative chemotherapy). Chemotherapy is usually delivered in combination with other cancer treatments, such as radiation therapy, surgery, and / or hyperthermia therapy.

Radiation therapy or radiotherapy is a therapy using ionizing radiation, often as part of cancer treatment to control or kill malignant cells. Radiation therapy might be a curative therapy for some types of cancer if they are localized to one area of the body. It can also be applied as part of adjuvant therapy to prevent recurrence of the tumor after surgery to remove the primary malignant tumors (e.g. early stage breast cancer). Radiation therapy is synergistic with chemotherapy and has been used before, during, and after chemotherapy in cancer prone. Radiation therapy is generally administered to the cancerous tumor because of its ability to control cell growth. Ionizing Radiation operates by damaging the DNA of the cancerous tissues that causes cell death.

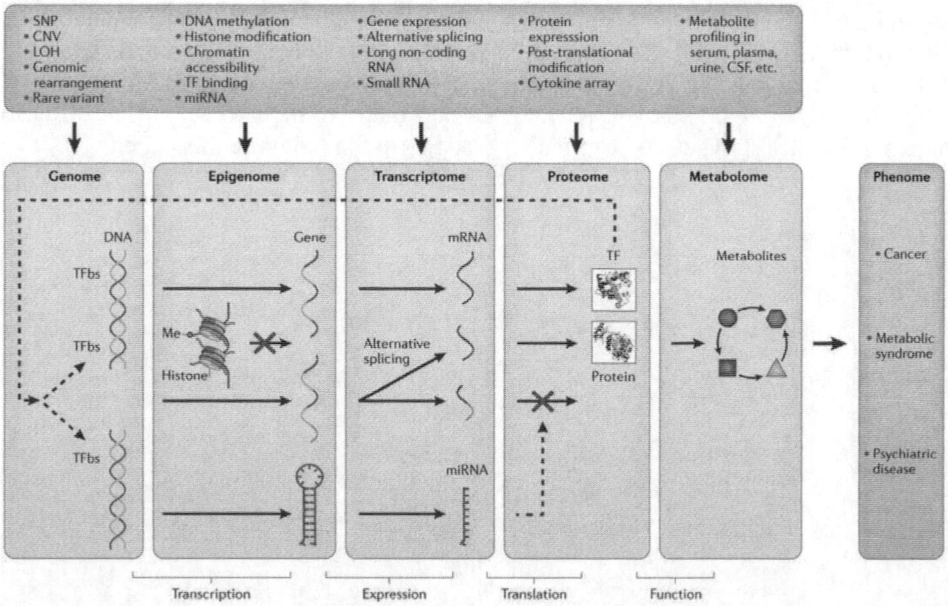

Figure 14. Variety Omics level which can be utilized for diagnostic purposes and therapy namely using specific molecules from each level (adopted from [15]).

References

[1] Sharp L W I 1921 *Introduction to cytology* (New York: McGraw Hill Book Company Inc.)

[2] Lodish H, Berk A, Kaiser C A, Krieger M, Scott M P, Bretscher A, Ploegh H and Matsudaira P 2007 *Molecular cell biology 6th edition* (New York: W. H. Freeman and Company)

[3] Fernandes H, Moroni L, van Blitterswijka C and de Boer J 2009 Extracellular matrix and tissue engineering applications *J Mater Chem.* **19** pp 5474–84

[4] Lipowsky R 2006 The physics of bio-systems: from molecules to networks *Biophys Rev Lett* **1** pp 223–230

[5] https://upload.wikimedia.org/wikipedia/commons/f/fb/Signal_transduction_pathways.png

[6] Jones R G and Thompson C B 2009 Tumor suppressors and cell metabolism: a recipe for cancer growth *Genes Dev* **23** pp 537-548

[7] Farhan H and Rabouille C 2011 Signalling to and from the secretory pathway *J Cell Sci* **124** pp 171-180

[8] Ghesquière B, Wong B W, Kuchnio A and Carmeliet P 2014 Metabolism of stromal and immune cells in health and disease *Nature* **511** pp 167-176

[9] https://thestrangeandspectacularworldofbiochemistry.files.wordpress.com/2013/03/sickle.jpg

[10] https://classconnection.s3.amazonaws.com/19/flashcards/1200019/png/321329795543943.png

[11] Jaalouk D E and Lammerding J 2009 Mechanotransduction gone awry *Nat Rev Mol Cell Biol* **10** pp 63-73

[12] Mager M D, LaPointe V and Stevens M M 2011 Exploring and exploiting chemistry at the cell surface *Nat Chem* **3** pp 582-589

[13] Szasz O 2013 Renewing oncological hyperthermia—Oncothermia *Open J Biophys* **3** pp 245-252

[14] Rienks M, Papageorgiou AP, Frangogiannis N G and Heymans S 2014 Myocardial extracellular matrix: An ever-changing and diverse entity *Circ Res* **114** pp 872-888

[15] Ritchie M D, Holzinger E R, Li R, Pendergrass S A and Kim D 2015 Methods of integrating data to uncover genotype–phenotype interactions *Nat Rev Gen AOP* pp 1-15

Acknowledgments
I would like to thanks Dr. Freddy Haryanto from Department of Physics, Institut Teknologi Bandung and Prof. Dr. Kwan-Hoong Ng from Department of Biomedical Imaging & University of Malaya Research Imaging Centre, who kindly gave me the opportunity to be involved in The 13th South-East Asia Congress of Medical Physics in 2015 and also supported me to write this paper.

Journal of Physics: Conference Series **694** (2016) 012002 doi:10.1088/1742-6596/694/1/012002

Helical tomotherapy optimized planning parameters for nasopharyngeal cancer

K Yawichai[1], I Chitapanarux[1] and S Wanwilairat[1]

[1] Department of Therapeutic Radiology and Oncology, Faculty of Medicine, Chiang Mai University, Thailand

E-mail: somsaksw@gmail.com

Abstract. Helical TomoTherapy(HT) planning depends on optimize parameters including field width (FW), pitch factor (PF) and modulation factor (MF). These optimize parameters are effect to quality of plans and treatment time. The aim of this study was to find the optimized parameters which compromise between plan quality and treatment times. Six nasopharyngeal cancer patients were used. For each patient data set, 18 treatment plans consisted of different optimize parameters combination (FW=5.0, 2.5, 1.0 cm; PF=0.43, 0.287, 0.215; MF2.0, 3.0) were created. The identical optimization procedure followed ICRU83 recommendations. The average D50 of both parotid glands and treatment times per fraction were compared for all plans. The study show treatment plan with FW1.0 cm showed the lowest average D50 of both parotid glands. The treatment time increased inversely to FW. The FW1.0 cm the average treatment time was 4 times longer than FW5.0 cm. PF was very little influence on the average D50 of both parotid glands. Finally, MF increased from 2.0 to 3.0 the average D50 of both parotid glands was slightly decreased. However, the average treatment time was increased 22.28%. For routine nasopharyngeal cancer patients with HT, we suggest the planning optimization parameters consist of FW=5.0 cm, PF=0.43 and MF=2.0.

1. Introduction

The nasopharynx, a cuboidal chamber, is located below the base of skull and behind of the nasal cavity [1]. Radiation therapy oncology group (RTOG 0225) suggested that the intensity modulated radiotherapy (IMRT) combine with chemotherapy is suitable treatment for staging I-IVB squamous cell carcinoma of nasopharynx [2].

The helical tomotheray (HT) is a specialized technique of delivering IMRT by using a 6 MV linear accelerator mount on the ring gantry that rotate around the patient while couch through the gantry. The HT treatment planning parameters including; Field width (FW), pitch factor (PF) and modulation factor (MF).

The FW has three setting values 1.0, 2.5 and 5.0 cm. The chosen FW is compromised between treatment time and dose distribution in superior-inferior direction. Although a smaller FW results in good dose distribution in superior-inferior direction, there are increased treatment times [3].The PF is the couch distance travel for one gantry rotation divided by FW. Kissick *et al.* has recommend the solution to use PF of 0.86/n, where n is an integer. For reduce dose fluctuation due to helical junction of divergent beam in HT as known as "Thread effect" [4].

MF defines as the ratio of maximum leaf open time to the average leaf open time. Increasing MF contributes the larger of beam modulation. However, there are increased treatment times [4].

Journal of Physics: Conference Series **694** (2016) 012002 doi:10.1088/1742-6596/694/1/012002

This study aim to find the optimize parameters of Helical TomoTherapy treatment planning for nasopharyngeal cancer concerning quality and efficiency of the treatment plan.

2. Material and methods

2.1. Patients data
The CT images and structure contour data of six nasopharyngeal cancer patients that had been treated with HT were randomly selected. Each patient data set, three planning target volumes (PTV) were defined as PTV70, PTV59.4 and PTV54. All plans were created using a dose per fraction of 1.8 Gy.

2.2. Treatment planning
For each patient data set, 18 treatment plans were created with different combination of treatment planning parameters FW=1.0, 2.5 and 5.0 cm. PF=0.43, 0.287 and 0.215 MF= 2.0 and 3.0. The initial parameters set used for planning were FW=5.0, PF=0.43 and MF=2.0. These initial planning parameters set was combined with all optimization dose constraints in Table1. as default plan. This plan was 250 iterations optimized; under the dose constraint to PTV followed ICRU83 recommendation. The dose homogeneous in PTV was maintained between V95% less than 2% volumes and V107% less than 2% volumes. The maximum and minimum penalty for PTV was changed during 250 iterations until the dose to PTV fulfilled the ICRU83 recommendations. All plans were optimized base on single dose constraint in PTV.

Table 1. Optimization dose constraints for all the treatment plan.

Name	overlap priority	block	importance	max dose	max dose penalty	DVH volume	DVH dose	min dose penalty
						95	69.5	
PTV 70	1	-	150	70.5	150	50	70	140
						1	70.5	
PTV59.4	4	-	1	70	1	95	59.4	1
						95	59.8	
PTV59.4-ex	2	-	90	69	170	50	60	100
						25	60.5	
						95	53.5	
PTV 54	3	-	100	55.5	60	50	54.5	95
						10	55	

2.3. Plan evaluation
All plans were evaluated using the 50% dose of both parotid glands volume (D50) and treatment times. The statistical result of both parotid glands D50 and treatment times were analyzed using SPSS statistics version 17.0. The student pair t-test was used to compare treatment times and the Wilcoxon test was performed to compare D50 of both parotid glands.

3. Results

3.1 D50 both parotid glands

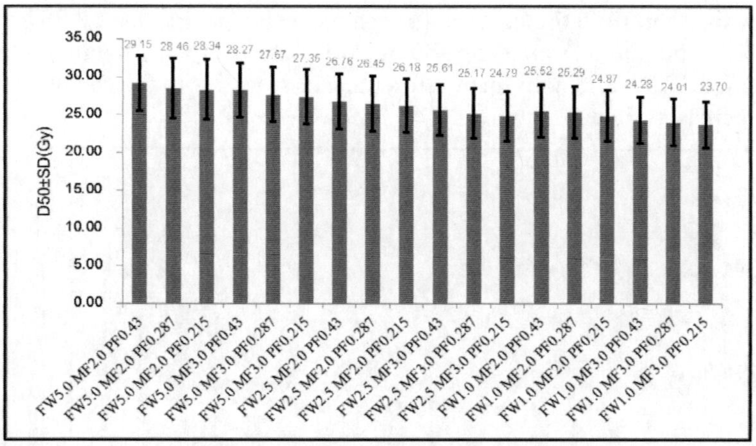

Figure 1. The average of D50 of both parotid glands for all patients data set with different treatment planning parameters combination.

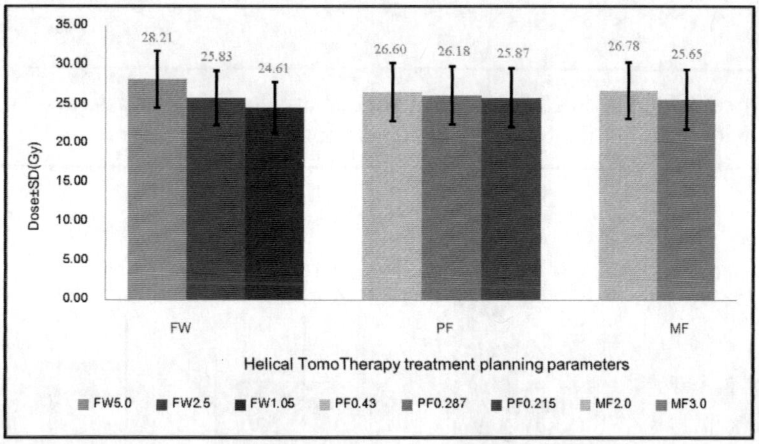

Figure 2. The average D50 of both parotid glands for different optimize parameters.

The figure 1 show average of D50 of both parotid glands for all patients data set with different treatment planning parameters combination, total 108 treatment plans were generated. There were no significantly different of D50 of both parotid glands. But the average of D50 of both parotid glands show the lowest when the FW=1.0 cm was used to combine with other optimize parameters.

The figure 2 show average of D50 of the all patient parotid glands with the different optimize parameter. The average of D50 of all parotid glands using FW 5.0, 2.5 and 1.0 cm. were 28.21, 25.83 and 24.61 Gy, respectively. Treatment plans using PF 0.43, 0.287 and 0.215, the average of D50 of all parotid glands were 26.60, 26.18 and 25.87 Gy, respectively. Treatment plan using MF 2.0, 3.0, the average of D50 of all parotid glands were 26.78 Gy, 25.65 respectively. FW demonstrate the major effect to D50 of the parotid glands compare to PF and MF.

3.2. Treatment time per fraction

The average of treatment time per fraction with 3 combination planning parameter showed in figure 3. Treatment plan with FW=5.0, PF=0.43, MF=2.0 showed the significantly lowest treatment time as 3.23±0.33 minutes when compare with other group except with FW=5.0, PF=0.43, MF=3.0 group.

The average of treatment time per fraction with different planning parameters showed in figure 4. Average treatment time when used FW 5.0, 2.5 and 1.0 cm. showed 4.51, 8.39 and 19.37 minutes, which was 4 times longer than the FW5.0 cm. Average treatment time of PF 0.43, 0.287 and 0.215 showed 9.57, 10.10 and 12.21 minutes. Decreasing PF from 0.43 to 0.215 the average treatment time was significantly increased 18.97%. Finally increasing the MF from 2.0 to 3.0 the average treatment time was significantly increased about 3 minutes.

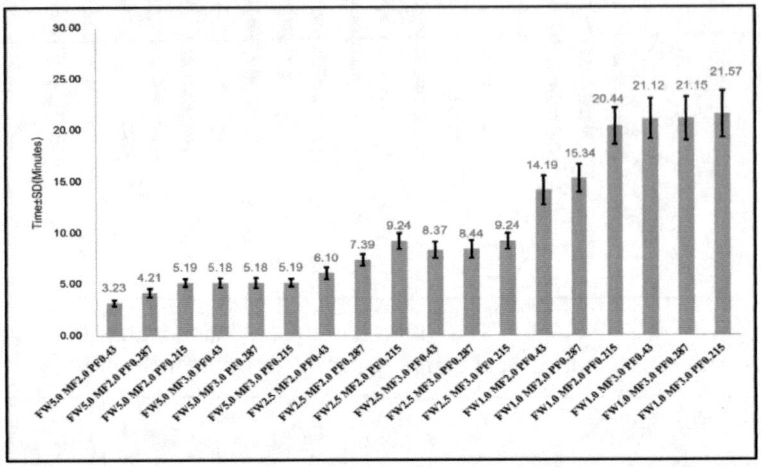

Figure 3. The average treatment times per fraction for all patients data set with different treatment planning parameters combination.

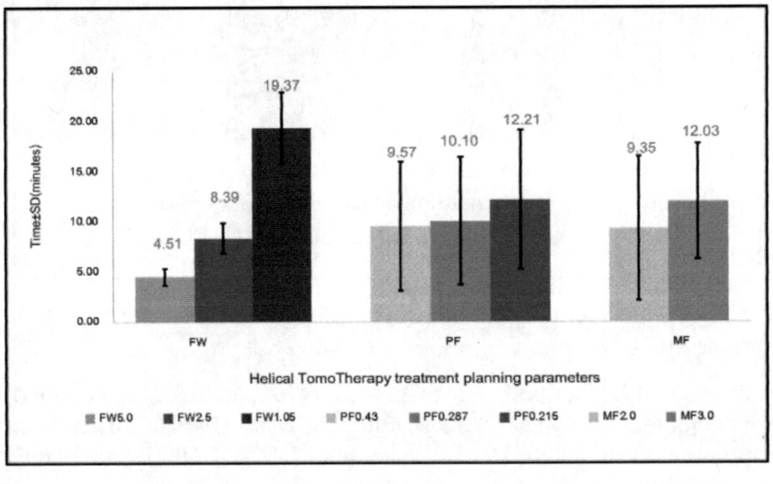

Figure 4. The average treatment times per fraction with different optimize parameters.

4. Discussions

This work investigates the optimized planning parameter in HT for nasopharyngeal cancer. Finding of this work are consistent with Vicent Wu *et al.* that reported the FW effect on the quality of treatment plan in the OAR dose, but not so much on the PTV. Increasing FW was reducing the effectiveness in OAR sparing. When decreasing the PF from 0.43 to 0.215, there was little influence on D50 of both parotid glands [5]. J Lee Roy reported that decreasing PF from 0.43 to 0.287 slight improve dose distribution in PTV, but not significantly reduce the dose to OAR. We increased the MF from 2.0 to 3.0, the average D50 of both parotid glands were slightly decreased 4.2%. Ryczkoski A sustainable recommended the value of MF 1.8-3.0 for head and neck HT planning. However, if it is necessary to reduce the dose in OAR he suggested the MF should not be lower than 2.4 [6].

Our study, the FW had a larger effect on D50 of both parotid glands than PF and MF was consistent with Skorska M *et al.* that they show FW was mainly parameter in dose distribution because FW had effect on dose fall off in longitudinal direction when used the large FW demonstrated the large penumbra. The dose distribution was slightly influence when PF increased from 0.215 to 0.43 [7].

The FW had the major effect on treatment time. Using the FW 1.0 cm. we received the average longest treatment time 19.05±3.58 minutes which was 4 times longer than used the FW 5.0 cm. Therefore, the treatment time is inversely proportional to the FW consistent with previous studied. Monica M. *et al.* reported treatment times ranged from 22-27, 8-12 and 4-6 minutes for the FW1.0, 2.5 and 5.0 respectively [8]. The PF parameter was less effect to treatment time than FW and MF which was consistent with Woch *et al.* that they reported the treatment times decreasing with larger PF value [9]. However, Monica M. *et al.* found that increasing the PF value only marginally influenced treatment times because might be complexity of target volume or which force the HT system to adjust other machine parameter such as gantry period [8]. This effect treatment times to maintain treatment planning object. The last optimize parameter, when increasing the MF from 2.0 to 3.0 the average treatment time was increased 22.28%. Ryczkoski *et al.* reported that the treatment time was approximately linear correlation decrease when reduce the MF range 6.0 to 3.0. They found treatment plan with the MF value below 3.0 the treatment time was not shortened. This was related of the maximum speed of machine gantry [6]. For our HT system version 4.0, the minimum gantry rotation period is 12 sec [10]. This allows the treatment time to be shortened to an MF about 2.5.

5. Conclusions

We had analysed 18 treatment plans with different combination of optimize parameters for nasopharyngeal carcinoma by concerning target dose constraint follow ICRU83 recommendation. Treatment plan consist of FW=5.0 cm, PF=0.43 and MF=2.0 show the optimal compromised between plan quality and efficiency treatment time. These optimize parameters may serve as routinely guideline for nasopharyngeal carcinoma treatment planning with HT system.

References

[1] American Cancer Society 2013 *Nasopharyngeal cancer detailed guide* (Atlanta: America)

[2] RTOG 0225 A Phase II Study of intensity modulated radiation therapy (IMRT) +/- Chemotherapy for Nasopharynx Cancer. [Internet]. 2005 [cited 2014 Mar 23]. Available from: www.rtog.org/members/protocols/0225

[3] Mackie T R, Balog J and Ruchala K 1999 Tomotherapy *Semin Radiat Oncol* **9** 108-117

[4] Kissick M W, Fenwick J, James J A, Jeraj R, Kapatoes J M and Keller 2005 The helical tomotherapy thread effect *J. Med Phys.* **32** 1414-1423

[5] Vincent W C and Mui W L A 2011 A case report on the effect of fan beam thickness in helical tomotherapy of nasopharynx carcinoma *J. Med Dosim.* **36** 57-61

[6] Ryczkowski A and Piotrowski T 2013 Influence of the modulation factor on the treatment plan quality and execution time in tomotherapy in head and neck cancer: In phantom study. *J. Cancer Res Ther.* **9(4)** 618-623

[7] Skorska M and Piotrowski T 2013 Optimization of treatment planning parameters used in tomotherapy for prostate cancer patients J. Physica Medica **29** 285-273

[8] Monica M, Jonas D F, John. P G, Lee T K, Rosen I I and Robert S F 2011 Investigation of pitch and jaw width to decrease delivery time of helical tomotherapy treatment head and neck cancer. *J. Med Dosim* **36** 397-403

[9] Woch K, Soisson E T and Westerley D 2008 Acceleration of tomotherapy treatment delivery by increasing pitch and decreasing modulation *J. Med Phys.* **35** 2636

[10] TomoTherapy 2012 *TomoTherapy® Tomotherapy Treatment Planning System* (Tomo Planning Guide) [CD-ROM] (Accuray Inc.USA)

Verification of Oncentra brachytherapy planning using independent calculation

N A M Safian[1], N H Abdullah[1], R Abdullah[1] and C S Chiang[1]

[1] Medical Radiation Programme, School of Health Sciences, Health Campus, Universiti Sains Malaysia, 16150 Kubang Kerian, Kelantan, Malaysia

Email: aininasafian@gmail.com

Abstract. This study was done to investigate the verification technique of treatment plan quality assurance for brachytherapy. It is aimed to verify the point doses in ^{192}Ir high dose rate (HDR) brachytherapy between Oncentra Masterplan brachytherapy treatment planning system and independent calculation software at a region of rectum, bladder and prescription points for both pair ovoids and full catheter set ups. The Oncentra TPS output text files were automatically loaded into the verification programme that has been developed based on spreadsheets. The output consists of source coordinates, desired calculation point coordinates and the dwell time of a patient plan. The source strength and reference dates were entered into the programme and then dose point calculations were independently performed. The programme shows its results in a comparison of its calculated point doses with the corresponding Oncentra TPS outcome. From the total of 40 clinical cases that consisted of two fractions for 20 patients, the results that were given in term of percentage difference, it shows an agreement between TPS and independent calculation are in the range of 2%. This programme only takes a few minutes to be used is preferably recommended to be implemented as the verification technique in clinical brachytherapy dosimetry.

1. Introduction

Brachytherapy is a treatment method which requires the delivery of radiation directly into or onto the surface of the area to be treated. In order to ensure the optimal treatment of patients, an institution must develop a suitable quality control (QC) programme for brachytherapy sources, equipment and the physical and clinical procedures [1]. A quick calculation check to evaluate the output of a treatment plan allows detection of any possible errors. Seeking a compromise for efficiency both in the time required for the extra calculation and in the level of reliability. The purpose of this study was to verify the point doses in 192Ir HDR brachytherapy between Oncentra brachytherapy treatment planning and independent calculation software. To the specific review, this study compared the values of planned and independent calculated point doses of 192Ir of rectum and bladder in HDR brachytherapy. These studies also compared the values of planned and independent calculated point doses based on prescription points. Lastly, it was the comparison of the values of planned and independent calculated point doses between pair ovoid applicator and full catheter set ups at rectum, bladder and prescription points.

2. Materials and method

2.1. Oncentra masterplan treatment planning system (TPS)

Journal of Physics: Conference Series **694** (2016) 012003 doi:10.1088/1742-6596/694/1/012003

Nucletron Oncentra brachytherapy planning system 4.1 (Elekta, Stockholm, Sweden) was used to perform the clinical brachytherapy planning calculation. The protocol for brachytherapy dose calculations implemented in Oncetra Masterplan TPS was based on dose formalism defined by AAPM task group no. 43. One of the first essential steps to determine the dose distribution in brachytherapy is based on the determination of the source positioning in relation to the target volume and other anatomic characteristics of interest [2].

2.2. Independent calculation workbooks/spreadsheets

Independent calculation is a verification programme that has been developed to verify point dose calculation in brachytherapy treatment planning system. The system works in a workbook formed that represent software in order to calculate point dose independently.

Basically, the working principle of independent calculation programme is based on AAPM TG-43 dose formalism. The TG-43 updated formalism a point source (1D) model and a line source (2D) model to describe the dosimetric behaviour of the seed sources. The orientation of the source is taken into account using the anisotropy function along with the linear geometric factor, the radial dose function and the dose rate constant. In the point dose approximation, the anisotropy factor and the point geometry factor are used.

This verification programme considers the linear nature of the geometry of the sources such as the use of 2D general formalism proposed by TG-43 to perform the calculations. The verification programme has been designed on Microsoft Excel 2002 spreadsheets by a group of medical physicist [3]. All the required information which was included in the text file of the TPS plan was imported directly into the program through a simple interface step. The air-kerma strength with reference dates and times also required for the calculation. Those informations are all independently entered when carried out the verification.

In order to carry out the verification technique of Oncentra TPS, the data of a total of 20 patients with cervix cancer that undergoing brachytherapy were needed. The data consists of two fractions with 10 patients with pair ovoids applicator and 10 patients with a full catheter. Therefore, there were 40 cases were verified using independent calculation. Figure 1 shows the screenshots of independent calculation spreadsheets.

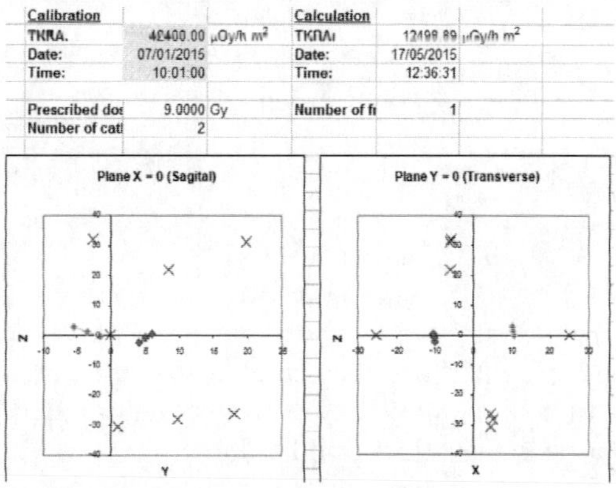

Figure 1. The independent calculation worksheets on sagittal and transverse planes.

3. Results and discussion

Figure 2(a) and (b) show the comparison of absolute doses between Oncentra treatment planning system and independent calculation at rectum and bladder for pair ovoid set up respectively. While figure 3 shows the comparison of absolute doses between Oncentra treatment planning system and independent calculation at the prescription point for pair ovoid set up.

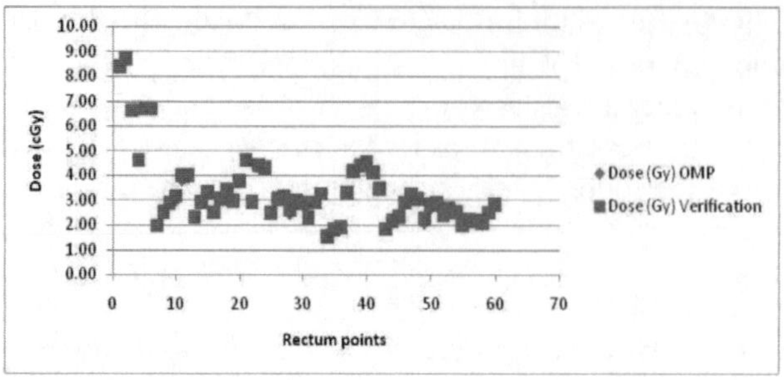

Figure 2(a). Comparison of calculated absolute doses using Oncentra treatment planning system and independent calculation at the rectum.

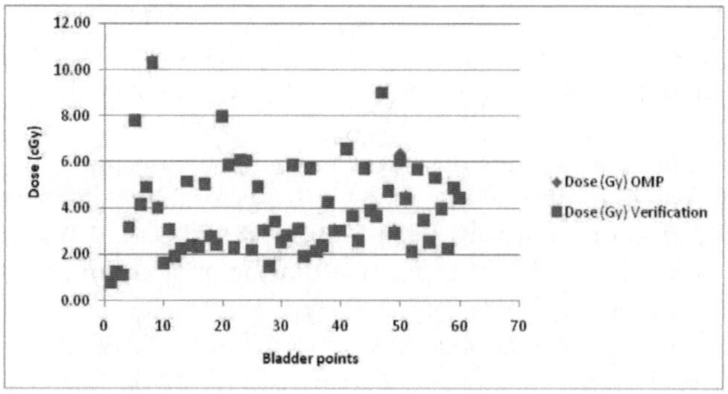

Figure 2(b). Comparison of calculated absolute doses using Oncentra treatment planning system and independent calculation at the bladder.

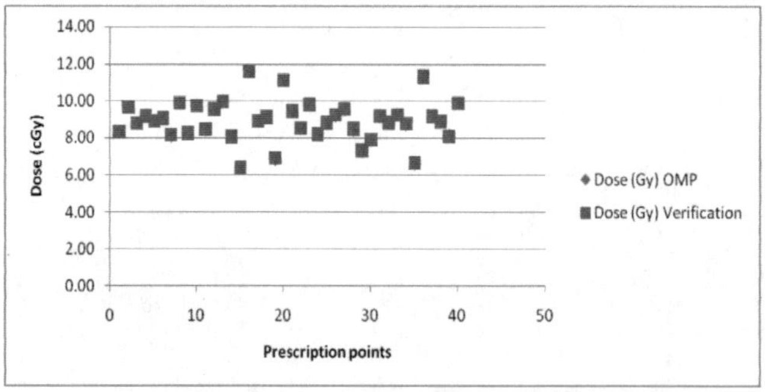

Figure 3. Comparison of calculated absolute doses using Oncentra treatment planning system and independent calculation at the prescription point.

The comparison between the values of the planned and independent calculated point doses of ^{192}Ir in HDR brachytherapy between pair ovoids applicator and full catheter for rectum and bladder were showed in figure 4(a) and (b) respectively. The verification of dose calculation at prescription points in the comparison between pair ovoids and full catheter set up were showed in figure 5.

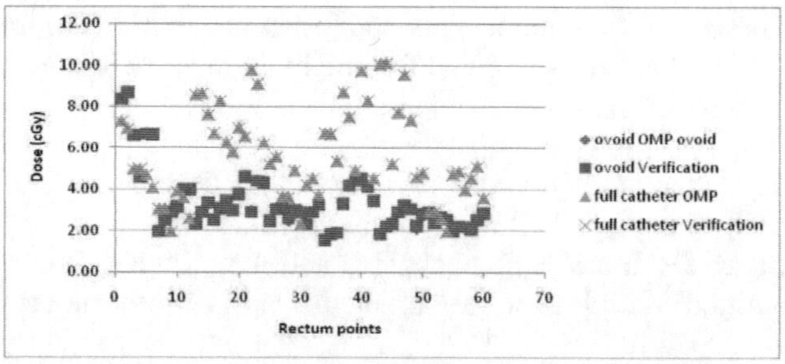

Figure 4(a). Comparison between pair ovoids and full catheter set up for verification of dose calculation at the rectum.

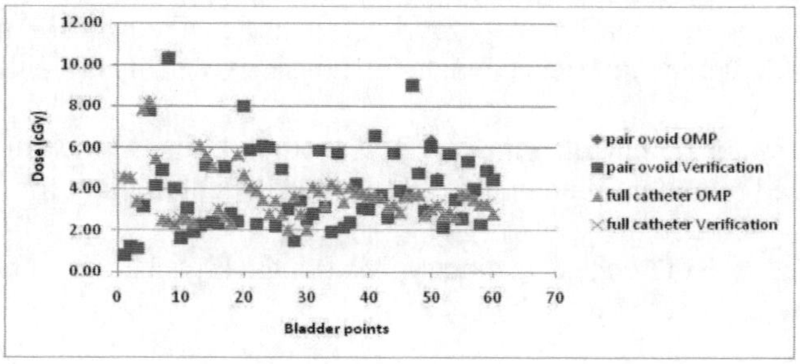

Figure 4(b). Comparison between pair ovoids and full catheter set up for verification of dose calculation at the bladder.

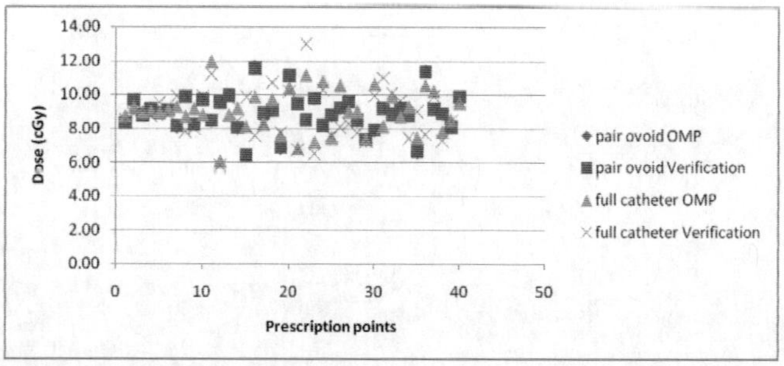

Figure 5. Comparison between pair ovoids and full catheter set up for verification of dose calculation at prescription points.

The reported results showed that for a total sample of 40 cases most of the deviations between the TPS and the independent calculation programme are less than 2%. There were a few cases in ratio of 1:10 that showed deviation slightly greater, but it can be accepted due to small ratio.

4. Conclusions

As the conclusion, the independent calculation programme capable of meeting requirements for independent verification of clinical dosimetry, it can be utilized in all plans for intracavitary brachytherapy. Based on the results that were given in term of percentage difference, it shows a good agreement between TPS and independent calculation in the range of 2%. The absolute dose calculated was evaluated at rectum, bladder and prescription points for 40 clinical cases respectively. Confirmation of the accuracy of optimized calculations with verification evaluation technique is vital in order to assure the accuracy of treatment. This independent verification of computer plan HDR brachytherapy provides a solid base to be applied for brachytherapy treatment.

References

[1] Carlos E A 2011 *The Need for Quality Control in High Dose Rate Brachytherapy* Ed. Ognyan Ivanov, University Campus STeP Ri, InTech, pp 12-25

[2] Jetro P O, Luiz A R R, Delano V S B, Lúcia H B and Arnaldo R C 2009 Rectal dose assessment in patients submitted to high-dose-rate brachytherapy for uterine cervix cancer *Radiol Bras.* **42** 83–88

[3] Vicente C, Jose P C, Françoise L, Jose R, Facundo B, Carmen P C and Venselaar J L M 2010 A program for the independent verification of brachytherapy planning system calculations *J Contemp Brachytherapy* **2** 129-133

[4] Ravinder N, Lowell L A, Gary L, Keith A W, Jeffrey F W and Ali S M 1995 *Dosimetry of Interstitial Brachytherapy Sources* American institute of Physics, American Association of Physicists in Medicine American Center for Physics One Physics Ellipse College Park, MD 20740-3843

[5] Ahmad B, Muhammad A G, Muhammad A R, Naeem A L and Mohammad A 2009 Verification of computerized treatment planning for HDR ^{192}Ir Brachytherapy for Gynaecological Cancer Saeed *J Pak Med Assoc.* **59**

[6] Practical Course in Reference Dosimetry, National Physical Laboratory Jan 2014 HDR Brachytherapy Dosimetry at the NPL **11**

Treatment planning systems for external whole brain radiation therapy: With and without MLC (multi leaf collimator) optimization

T Budiyono[1], W S Budi[1] and E Hidayanto[1]

[1] Department of Physics, Faculty of Science and Mathematics, Diponegoro University, Semarang 50275, Indonesia

E-mail: tris_fis@yahoo.com

Abstract. Radiation therapy for brain malignancy is done by giving a dose of radiation to a whole volume of the brain (WBRT) followed by a booster at the primary tumor with more advanced techniques. Two external radiation fields given from the right and left side. Because the shape of the head, there will be an unavoidable hotspot radiation dose of greater than 107%. This study aims to optimize planning of radiation therapy using field in field multi-leaf collimator technique. A study of 15 WBRT samples with CT slices is done by adding some segments of radiation in each field of radiation and delivering appropriate dose weighting using a TPS precise plan Elekta R 2.15. Results showed that this optimization a more homogeneous radiation on CTV target volume, lower dose in healthy tissue, and reduced hotspots in CTV target volume. Comparison results of field in field multi segmented MLC technique with standard conventional technique for WBRT are: higher average minimum dose (77.25% ± 0:47%) vs (60% ± 3:35%); lower average maximum dose (110.27% ± 0.26%) vs (114.53% ± 1.56%); lower hotspot volume (5.71% vs 27.43%); and lower dose on eye lenses (right eye: 9.52% vs 18.20%); (left eye: 8.60% vs 16.53%).

1. Introduction

Radioactive substances are used in radiation therapy as they have been known to kill malignant or cancerous cells. Radiation therapy is usually a combination of surgery and chemotherapy. In general, more than 50% of patients with malignant cells require radiation therapy [1]. The goal of radiation therapy is to give high doses of radiation that is concentrated in the tumor volume and keep the surrounding healthy tissues and organs from being exposed to those high doses of radiation [2].

O In case of a malignancy in the brain, it is usually subject to whole brain radiation therapy (WBRT) followed by radiation to the primary tumor [4]. WBRT is usually done using two opposing radiation fields from the patient's right and left sides, while blocking radiation on the eyes. A radiation dose 3000 cGy is given in 10 fractions or a radiation dose of 4000 cGy given in 20 fractions, 5 fraction/week and a two-day lag time is allowed for the recovery of healthy tissues [2].

External radiation tele therapy must meet the recommendation of the International Commission on Radiation Units and Measurement (ICRU) which says that the entire volume of the target must obtain a homogeneous radiation tolerance of 95-107% [2]. The recommendation is intended to kill tumors with no complications in the healthy tissue and prevent any recurrence due to lack of radiation dose. Excessive doses will result in side effects and complications, whereas less than adequate will result in leftover malignant cells and recurrence [1]. External whole brain radiation usually results in hot spots on thin heads due to excessive dose. Hence, there is the need to find a solution that will provide homogenous radiation for the whole volume of tumor.

2. Methods

A study of 15 WBRT samples with 1 mm CT slice thickness is done by adding some segments of radiation in each field of radiation and delivering appropriate dose weighting using a TPS precise plan Elekta R 2.15 treatment planning computer.

The procedures conducted in this study are; set up the TPS computer, perform data entry, including delineation of the target volume and critical organs by a radiation oncologist, dosimetry calculations, analysis and evaluation of radiation planning, re-planning using MLC optimization method, and the addition of segmentation, create graphs and tables, and finally compare the results with the currently available standard protocol.

3. Results and discussion

3.1. Calculation results of TPS on whole brain radiation therapy using standard protocol

TPS calculation results show that there is a hotspot on a CT scanner slice as shown in figure 1 below.

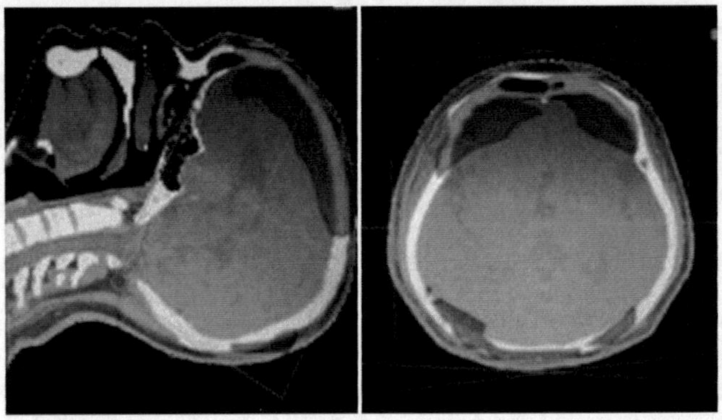

Figure 1. Distribution of whole brain radiation dose using standard protocol.

Red indicates areas receiving a radiation dose ≥ 107% (hot spots), located mainly on the edges of the head that are thinner; top (vertex), front (frontal) and rear (occipital).This sort of radiation planning is not optimum as dose distribution is not even and there is a large hot spot on the target volume. Yet it is still taken as a common practice due to the following reasons:

1. No body compensator or bolus that can be used to modify the curve iso-dosage is available.
2. No MLC on linac machines used for tele therapy is available.
3. No IMRT software that can measure multi-segmented dose is available.
4. The dose given to the whole brain is still low compared to the tolerance limit of brain tissue.
5. The equipment available is not optimally used.

3.2. *TPS results using optimized MLC with radiation field segmentation*

Figure 2 is an example of the same sample that is optimized using a multi-segmented MLC software (Elekta). The hot spot (red/shaded) is significantly narrower, indicating less excessive radiation dose.

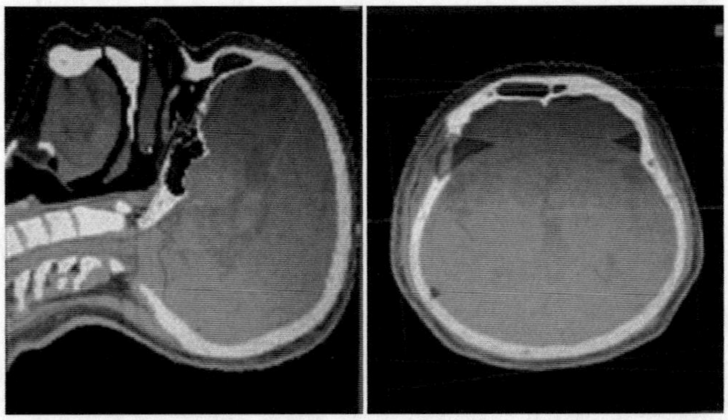

Figure 2. Distribution of whole brain radiation dose using optimized MLC.

3.3. Radiation dose distribution parameters

3.3.1. Homogeneity. Homogeneity of the radiation dose to the target volume is a measure that the radiation given can be concentrated and distributed uniformly on the target tumor. Therefore, all malignant cells are expected to die as radiation doses are sufficient and distributed evenly. The ICRU recommends that homogeneity of radiation dose in an external radiation therapy is in the range of 95% - 107%. Table 1 shows that distribution of radiation dose to the target CTV volume changes with the optimized MLC. D_{min} on PTV increases from 60% ± 3.35% to 77.25% ± 0.47%, D_{max} decreases from 114.53% ± 1.56% to 110.27% ± 0.26%, and D_{mean} decreases closer to the prescribed dose of 100%, from 105.07% ± 1.24% to 102.73% ± 0.26%
Thus, if compared with the standard protocol TPS, the MLC produces smaller variations in optimized dose for PTV and bigger D_{mean} closer to 100%. Hence, PTV dose distribution is more homogeneous.

Table 1. Dose distribution on the brain target volume (CTV).

Pat.	D_{min} (%)		D_{max} target (%)		$D_{average}$ (%)		Hot spot Volume (cc)	
	Standard TPS	Optimized MLC	Standard TPS	Optimized MLC	Standard TPS	Optimized MLC	Standard TPS	Optimized MLC
1	74	76	112	109	105	103	313.23	23.32
2	59	81	115	110	106	102	449.86	34.24
3	53	63	116	110	106	104	443.44	147.45
4	78	87	118	110	106	103	442.78	67.31
5	50	64	115	110	105	102	228.31	97.65
6	51	73	110	109	104	102	202.39	35.58
7	88	90	113	110	104	103	195.81	54.19
8	59	81	115	110	106	102	449.86	70.54
9	46	76	115	111	105	100	348.69	83.64
10	63	84	113	100	104	102	267.42	88.44
11	56	80	112	110	104	103	472.42	67.87
12	86	89	115	111	106	104	458.85	90.43
13	76	84	117	112	106	101	710.29	161.72
14	8	91	114	111	106	104	375.64	111.34
15	75	86	117	113	107	102	404.18	77.89

mean	60	77.25	114.53	110.27	105.07	102.73	384.21	80.77
SD	3.35	0.47	1.56	0.26	1.24	0.26	27.01	7.67

3.3.2. Hot spot. The extent of hot spot could potentially lead to undesirable side effects or complications. More modern equipment is designed to minimize the likelihood of such consequences.

Table 1 shows that the volume of whole brain radiation hotspot for standard protocol is 384.21 cc (27.43% of PTV volume), whereas for optimized MLC, this number shrinks to 80.77 cc (5.71% of PTV volume).

Irradiation using less advanced equipment often results in uneven hot spot distribution, especially on the body surface. This prompts the Radiation Therapy Oncology Group (RTOG) to set a tolerance dose of > 110%, no more than 20% of the dose for the target volume.

3.4. Surrounding healthy tissues/organs

The radiation dose allowed to be taken by the surrounding body tissues/organs is shown in table 2 below.

Table 2. Doses for eyes for both standard protocol and optimized MLC.

Patient No.	Standard TPS		Optimized MLC	
	Right eye (%)	Left eye (%)	Right eye (%)	Left eye (%)
1	9	8	5	5
2	17	17	8	7
3	24	20	9	7
4	29	26	11	10
5	13	9	6	5
6	15	13	7	7
7	22	21	11	14
8	18	15	10	8
9	22	17	15	12
10	17	14	10	9
11	18	22	9	7
12	11	9	10	7
13	18	17	8	9
14	18	23	12	13
15	22	17	7	9
Mean	18.20	16.53	9.52	8.60
SD	1.00	1.11	0.52	0.56

Table 2 shows that radiation doses for eye lenses (organs at risk - OAR) is lower for optimized MLC compared to standard protocol (right eye: 9.52% vs 18.20%); (left eye: 8.60% vs 16.53%).

4. Conclusion

Results of this research show that whole brain radiation therapy using standard protocol cannot prevent the presence of hot spot on the edges of thinner head, as a radiation dose of ≥ 107% is given. They also indicate that compared to standard WBRT radiation, optimized MLC results in smaller hot spot volume. Hence, reducing the risk of both side effects and complications. Furthermore, optimized MLC yields; higher D_{min} (77.25% ± 0:47%) vs (60% ± 3:35%), lower D_{max} (110.27% ± 0:26%) vs (114.53% ± 1:56%), and D_{mean} closer to 100%, (102.73% ± 0:26%) vs (105.07% ± 1:24%). Radiation doses are distributed more evenly or more homogeneous on the target tumor volume. Optimized MLC also produces narrower hot spot on CTV volumes; 80.77 cc (5.71% of CTV volume) vs. 384.21 cc (27.43% CTV of CTV) and finally, optimized MLC gives lower radiation doses on healthy tissues/organs and eye lenses; (right eye: 9.52% vs 18.20%); (left eye: 8.60% vs 16.53%).

References

[1] Podgorsak E B 2005 Radiation Oncology Physics: *A Handbook for Teachers and Students* The 3-rd Edition (Vienna: IAEA) pp 179-191

[2] Parker W and Patrocinio H 2012, *Clinical treatment planning in external photon beam radiotherapy* (Montreal, Canada: McGill University Health Centre)

[3] Khan F M 2003 *The Physics of Radiation Therapy* The 3rd Edition (Philadelphia, USA: Lippincott Williams & Wilkins) pp 136-137,164

[4] Assouline A and Levy A 2011 Whole brain radiotherapy: prognostic factors and results of radiation boost delivered through a conventional accelerator *Journal Radiotherapy and Oncology* **99** 214-217

[5] Oinam A S and Chakraborty S 2010 Shielding in whole brain irradiation in the multileaf collimator era *International Cancer Journal* **6(2)** 152-158

[6] Keall P, Arief I and Shamas S 2008 The development and investigation of prototype three dimensional compensator for whole brain radiation therapy *Phys. Med. Biol.* **53** 2267-76

[7] Chojnacka M and Anna S G 2012 Reirradiation of relapsed brain tumors in children *Report of practical oncology and radiotherapy* **17** 32-37

[8] Dawod T and Omar R 2015 Assesment of brain dose distribution for ARC and conformal radiation therapy (CRT): A comparison study *Journal of Radiation Research and Applied Sciences* **8** 55-60

[9] Fernandez G and Pocinho R 2012 Quality of life and radiotherapy in brain metastasis patients *Report of practical oncology and radiotherapy* **17** 281-287

[10] Sahgal A and Aoyama H 2014 Stereotactic radiosurgery with or without whole brain radiation therapy for 1 to 4 brain metastases *Int J Oncol Biol Phys* **91(4)** 710-717

[11] Olmez I and Donahue B R 2010 Clinical outcome in extracranial tumor site and unusual toxicities with concurrent whole brain radiation (WBRT) and erlotinib treatment in patients with non-small cell cancer *Elsevier Lung Cancer* **70** 174-179

[12] Lippitz B and Lindquist C 2014 *Stereotactic radiosurgery in treatment of brain metastases*: The current evidence, Cancer treatment review 40 48-59

Journal of Physics: Conference Series **694** (2016) 012005 doi:10.1088/1742-6596/694/1/012005

The systematic and random errors determination using real-time 3D surface tracking system in breast cancer

J Kanphet[1,3], S Suriyapee[3], N Dumrongkijudom[2], T Sanghangthum[3], J Kumkhwao[3] and M Wisetrintong[3]

[1] Medical Physics School, Faculty of Medicine, Ramathibodi Hospital, Mahidol University, Bangkok, Thailand
[2] Department of Radiological Technology, Faculty of Medical Technology, Mahidol University, Bangkok, Thailand
[3] Department of Radiology, King Chulalongkorn Memorial Hospital, The Thai Red Cross Society, Bangkok, Thailand

E-mail: jaruek2525@gmail.com

Abstract. The purpose of this study to determine the patient setup uncertainties in deep inspiration breath-hold (DIBH) radiation therapy for left breast cancer patients using real-time 3D surface tracking system. The six breast cancer patients treated by 6 MV photon beams from TrueBeam linear accelerator were selected. The patient setup errors and motion during treatment were observed and calculated for interfraction and intrafraction motions. The systematic and random errors were calculated in vertical, longitudinal and lateral directions. From 180 images tracking before and during treatment, the maximum systematic error of interfraction and intrafraction motions were 0.56 mm and 0.23 mm, the maximum random error of interfraction and intrafraction motions were 1.18 mm and 0.53 mm, respectively. The interfraction was more pronounce than the intrafraction, while the systematic error was less impact than random error. In conclusion the intrafraction motion error from patient setup uncertainty is about half of interfraction motion error, which is less impact due to the stability in organ movement from DIBH. The systematic reproducibility is also half of random error because of the high efficiency of modern linac machine that can reduce the systematic uncertainty effectively, while the random errors is uncontrollable.

1. Introduction

Radiotherapy is a standard treatment for breast cancer patient. The side effect may be occurred by irradiated of radiation dose to critical organ included suddenly and long-term effect. An increased risk of heart disease is associated with patient who treated with left side of breast cancer radiotherapy when compared with the right side [1, 2]. In left sided breast cancer radiotherapy, left ventricle (LV) and left anterior descending artery (LAD) are the organ at risk (OAR) that dose of both organs could be reduced during end-inhale or deep inspiration breath hold (DIBH) [3-6]. The increasing of lung volume is expanded to push heart up and make more distance between heart and chest wall. The larger the distance between heart and chest wall, the lower the received heart dose. The successful of DIBH technique is occurred if the patient is still in the same position during treatment as the position in simulation process. The AlignRT system is the real-time 3D surface tracking system with completely non-invasion, not requires ionizing radiation and easy to use. This system is applied to setup and monitor the breast patient's position during treatment with DIBH technique to improve precision and stability of patient's position and avoid the long-term side effect of heart.

In this study, the systematic error and random error both before treatment (interfraction motion) and during treatment (intrafraction motion) of left sided breast cancer treated with the DIBH technique were determined by using AlignRT system.

2. Material and methods

2.1. Image acquisition system

The new 3D surface base monitoring system, AlignRT (VisionRT, London, UK) was installed in treatment room with TrueBeam linear accelerator as shown in Figure 1. This system consists of 3 camera pods, which 2 of them are ceiling mounted at 30 degree angled with treatment couch and another at the centrally foot end of treatment couch. Each camera pods compose of 2 data cameras, white light flash and speckle projector. The data from every camera are combined to generate 3D surface image. The registration software is designed for patient positioning by aligning the real-time 3D surface image with reference image that can be generated from CT body image (CT_S) or captured surface image of AlignRT (ART_S) at first fraction of treatment. The registration algorithm of AlignRT was used to determine 3D transformations. The six directions of the position shift were reported in vertical, longitudinal, lateral, yaw, pith and rotation that real-time displays at the computer workstation, however this study considered in only first three directions. When the position shifts are larger than setting tolerance limit, the AlignRT will hold radiation beam immediately until the shift back in setting tolerance, the AlignRT will release radiation beam on again.

Figure 1. The 3 camera pods of 3D surface tracking system in TrueBeam linear accelerator.

2.2. Clinical protocol

The 6 left breast cancer patients who treated by DIBH technique aged between 39 to 62 years (mean 49 years) were randomly selected. All patients were simulated by CT scanner (GE Medical System, Waukesha, WI, USA) for acquisition of patient image data with 0.5 cm slice thickness for DIBH and free breathing series. The breast posibord system (CIVCO, Kalona, USA) was used to immobilize patient's positioning with both hand over head on support arm and place knee on knee support. The skin markers were marked with laser system for patient setup.

Patient data were transferred to treatment planning system, Eclipse (Varian Medical System, Plato Alto, CF, USA). Target as tumor and Organs at risk (OAR) were delineated by radiation oncologist. The target volume covers of whole breast tissue with started 5 mm under the skin. The treatment plan using 3D conformal radiotherapy composed of two tangential fields with 200 cGy per fraction in 25 fractions to a total dose of 5000 cGy and the boost at tumor phase was performed with 6 MeV electron beams for 1500 cGy in 5 fractions after photon beam irradiated cover of whole breast tissue

2.3 Image acquisition
At the first fraction of treatment, the patients were setup position by align marker with laser to reproduce the position in CT simulator room. The patients are instructed by intercom to take deep inhale for DIBH technique. During deep inhalation, the patient's position would be matched with CT_S and captured the position in suddenly to generate ART_S. This image was used the reference image for remainder fractions. Before treatment in each fraction, the 3D image was captured, compared with reference, ART_S and reported the error as interfraction motion. During irradiated radiation, the surface images were captured 6 times in each field and the error from ART_S references were reported as intrafraction motion. The setup errors of each patient along the three axis oriented compose along the anterior-posterior, craniocaudal and left-right direction was called vertical, longitudinal and lateral errors, respectively. The mean and standard deviation (SD) were calculated from reported data to illustrate as systematic errors (Σ) and random errors (σ) of each patient along the 3 axis from 3D surface registration (AlignRT). The systematic error of population was represented by the standard deviation of mean of each patient, while the random error of population was defined by the mean error of standard deviation for individual patient. The root mean square was performed to combined the effect of inter-and intrafraction motion to one index as express in equation (1) and (2)

$$\Sigma^2_{tot} = \Sigma^2_{setup} + \Sigma^2_{Patient\ motion} \quad (1)$$

$$\sigma^2_{tot} = \sigma^2_{setup} + \sigma^2_{Patient\ motion} \quad (2)$$

3. Results
From 6 patients of 180 images tracking before and during treatment, the comparison result with the reference image for all setup error were less than 1.3 mm. The maximum systematic error of interfraction and intrafraction motion were 0.56 mm and 0.23 mm, the maximum random error of interfraction and intrafraction motion were 1.18 mm and 0.53 mm, respectively as shown in figure 1. The interfraction was more pronounce than the intrafraction, and the systematic error was less impact than random error.

Table 1. Show the population systematic and random error from inter-and intrafraction of LMT and LLT fields.

	LMT						LLT					
	Inter-Fraction			Intra-Fraction			Inter-Fraction			Intra-Fraction		
	Ver.	Long.	Lat.	Ver.	Long.	Lat.	Ver.	Long.	Lat.	Ver.	Long.	Lat.
Systematic error (mm)	0.46	0.39	0.24	0.23	0.22	0.12	0.48	0.33	0.56	0.21	0.20	0.10
Random error (mm)	1.00	1.18	1.07	0.53	0.50	0.27	0.99	1.13	1.01	0.37	0.32	0.19

Table 2. The systematic and random error after combined effect of inter- and intrafraction motion.

	LMT			LLT		
	Ver.	Long.	Lat.	Ver.	Long.	Lat.
Systematic error (mm)	0.51	0.45	0.27	0.52	0.38	0.57
Random error (mm)	1.13	1.28	1.11	1.06	1.17	1.02

As shown in table 2, the systematic and random errors of left medial tangential field were 0.51 mm and 1.13 mm for vertical, 0.45 mm and 1.28 mm for longitudinal and 0.27 mm and 1.11 mm for lateral, respectively. The systematic and random errors of left lateral tangential field were 0.52 mm and 1.06 mm for vertical 0.38 mm and 1.17 mm for longitudinal and 0.57 mm and 1.02 mm for lateral, respectively.

4. Discussion

There are many studies about 3D surface tracking system (AlignRT) used to verify patient setup uncertainty have been published over the few years ago [7-10]. Then demonstrated that AlignRT could improve the precision of patient's setup position on breast cancer radiotherapy treatment. In this study, the AlignRT was applied to determine the systematic error and random error in a level of sub millimetre that showed the maximum value less than 1.5 mm along the 3 axis.

Breast cancer radiotherapy treatment with DIBH technique was able to treat with more stability and same position of chest wall in each fraction. From the result systematic error was less impact than the random error because the using of modern machine and advance IGRT can reduce the influence from the systematic error while the random error cannot control. The result was agree with the study of Letizia D. *et al.* [11], who investigated the clinical application of a technique for patient set-up verification in breast cancer radiotherapy based on a 3D surface image registration system.

When compared with the Bert *et al.*, [12] study, that investigated breast patient setup uncertainty in free breath using AlignRT, the mean displacement of 1± 1.2 mm was shown. Our result was lesser error than Bert study that may be due to more stable in organ movement from DIBH technique. From this reason, our results exhibited the lesser value of intrafraction motion than interfraction motion.

5. Conclusion

The intrafraction motion error from patient setup uncertainty is about half of interfraction motion error, which is less impact due to the stability in organ movement from DIBH. The systematic reproducibility is also half of random error because of the high efficiency of modern linac machine. It can reduce the systematic uncertainty effectively, while the random errors is uncontrollable.

References

[1] Correa C R, Litt H I, Hwang W T, Ferrari V A, Solin L J and Harris E E 2007 *J Clin Oncol.* **25(21)** pp 3031-7
[2] Gyenes G 2007 *J Clin Oncol.* **25(17)** pp 2489-90
[3] Borst G R, Sonke J J, den Hollander S, Betgen A, Remeijer P and van Giersbergen A 2010 *International journal of radiation oncology, biology, physics* **78(5)** pp 1345-51
[4] Nissen H D and Appelt A L 2013 Radiotherapy and oncology *Journal of the European Society for Therapeutic Radiology and Oncology* **106(1)** pp 28-32
[5] Pedersen A N, Korreman S, Nystrom H and Specht L 2004 *Radiotherapy and oncology: journal of the European Society for Therapeutic Radiology and Oncology* **72(1)** pp 53-60
[6] Remouchamps V M, Vicini F A, Sharpe M B, Kestin L L, Martinez A A and Wong J W 2003 *International journal of radiation oncology, biology, physics* **55(2)** pp 392-406
[7] Bert C, Metheany K G, Doppke K and Chen G T 2005 *Med. Phys.* **32(9)** pp 2753-62
[8] Gierga D P, Riboldi M, Turcotte J C, Sharp G C, Jiang S B and Taghian 2008 *International journal of radiation oncology, biology, physics*. **70(4)** pp 1239-46
[9] Krengli M, Gaiano S, Mones E, Ballare A, Beldi D and Bolchini C 2009 *Radiat Oncol.* **4** pp 9
[10] Schoffel P J, Harms W, Sroka-Perez G, Schlegel W and Karger C P 2007 *Phys Med Biol.* 2007 **52(13)** pp 3949-63
[11] Deantonio L, Masini L, Loi G, Gambaro G, Bolchini C and Krengli M 2011 *Reports of practical oncology and radiotherapy: journal of Greatpoland Cancer Center in Poznan and Polish Society of Radiation Oncology.* **16(3)** pp 77-81
[12] Bert C, Metheany K G, Doppke K P, Taghian A G, Powell S N and Chen G T 2006 *International journal of radiation oncology, biology, physics*. **64(4)** pp 1265-74

Acknowledgements

I am grateful to all medical physicists, radiation technologist and staff at Division of Therapeutic Radiology and Oncology, King Chulalongkorn Memorial Hospital for their kind support of this work

Journal of Physics: Conference Series **694** (2016) 012006 doi:10.1088/1742-6596/694/1/012006

Comparison of dosimeter response: ionization chamber, TLD, and Gafchromic EBT2 film in 3D-CRT, IMRT, and SBRT techniques for lung cancer

A Fitriandini[1]**, W E Wibowo**[2] **and S A Pawiro**[1]

[1] Department of Physics, Faculty of Mathematics and Natural Sciences, University of Indonesia, Depok, Indonesia

[2] Department of Radiotherapy, RSUPN Cipto Mangunkusumo, Jakarta, Indonesia

E-mail: aninda.fitrianiandini@sci.ui.ac.id

Abstract. This research was conducted by measuring point dose in the target area (lungs), heart, and spine using four dosimeters (PTW N30013, Exradin A16, TLD, and the Gafchromic EBT2 film). The measurement was performed in CIRS 002LFC thorax phantom. The main objective of this study was to compare the dosimetry of those different systems. Dose measurements performed only in a single fraction of irradiation. The measurements result shown that TLD has the least accuracy and precision. As the effect of volume averaging, ionization chamber reaches the discrepancy value up to -13.30% in the target area. EBT2 film has discrepancy value of <1% in the 3D-CRT and IMRT techniques. This dosimeter is proposed to be an appropriate alternative dosimeter to be used at point dose verification.

1. Introduction

The role of radiotherapy in the treatment of lung cancer consists of several different techniques, which have been developed to achieve conformal dose distribution in the target volume while minimizing dose to the surrounding normal tissue. Treatment of lung cancer using irradiation needs a high accuracy and precision due to the presence of OARs around the target volume. Research conducted by Brahme in 1984 showed that the dose difference by 5% of target volume resulted in changes in the probability of tumor control by 10-20%; and increase the probability of healthy tissue complication by 20-30% [1].Therefore, the dose verification is needed -especially in small field-, such as IMRT and SBRT- and it can be performed by measuring point dose in several investigated points. Dose verification can be done by measuring the point dose using ionization chamber or in vivo by thermoluminescence dosimeter (TLD) and film [2].

The ionization chamber is a benchmark of dosimeter, and had been known to provides the most accurate and reliable results. One difficulty with ionization chamber is that the measured dose can be perturbed by volume averaging over their relatively large active volume [3]. Practically, there are several things to consider, such as the large volume of active and effective point of measurements (EPOM) [4]. When ionization chamber measurements are impractical, it can be replaced by TLD. The small size of TLD allows it to be inserted in an anthropomorphic phantom. For fractional dose (± 200 cGy) measurement, the corrections for energy and dose response are taking into account [5]. To achieve dose measurement precision on the order of 2-3%, a TLD implementation program requires a rigorous annealing and response measurements protocol, and routine QA of the TLD reader and annealing oven temperature control [6]. Another in vivo dosimeter is radiochromic film, which has high spatial resolution and specifically designed for use in the dose range 0.1-106 Gy. These properties allow the radiochromic film to be used in small field measurements, that difficult to measure by the ionization chamber. In addition, weak energy dependence in a wide range of beam qualities for radiotherapy and near tissue equivalence makes them suitable for dose measurement in radiation fields with high dose gradients [4]. Because this system is not influenced by the effects of volume averaging owned by ionization chamber, the radiochromic film potentially becomes an effective option in point dose measurement. Due to the several types of the dosimeter, it is better to know and understand the characteristic of each dosimeter before deciding which the appropriate one to yield the optimum results.

2. Methodology

2.1. Contouring target and OARs

This research was performed using CIRS thorax phantom 002LFC (CIRS Inc., Norfolk, Virginia). This anthropomorphic phantom consists of four different materials, representing tissues/organs inside the thorax: lungs, spinal cord, and soft tissue (later used for represents heart). All of those different materials became investigated points of target and OARs that will be evaluated.

The target was simulated inside the right lung by a cylindrical form, using an imaginary boundary; which means that there was no real mass, the target was formed only by contouring. In other words, the target has the same density with lungs. The dimensions and volumes of the target were adjusted depend on related techniques. Contouring results are shown in figure 1, where region *A* is the volume of interest (VOI) for 3D-CRT and IMRT techniques; and the smaller one, region *B*, is the VOI for SBRT technique.

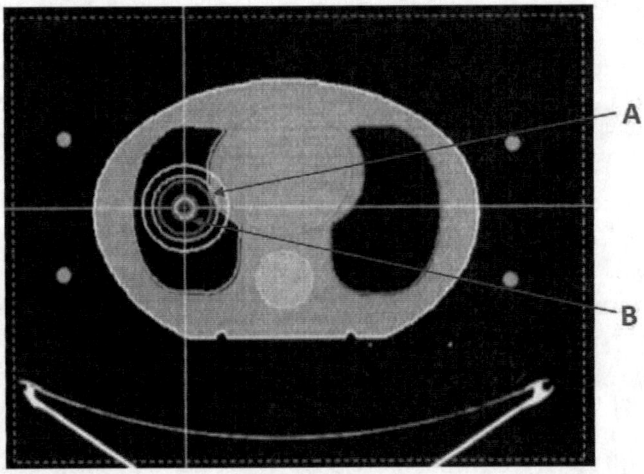

Figure 1. Contouring result: regions A and B are volume of interests (VOIs) for 3D-CRT/IMRT and SBRT techniques, respectively.

2.2. Treatment planning and delivery

Treatment plans were generated in Pinnacle^3® (Philips Radiation Oncology Systems, Fitchburg, WI) TPS, by using superposition-convolution algorithm. A 6 MV photon beam produced by Synergy-S (Elekta AB, Stockholm, Sweden) accelerator has been employed in this study. Treatment plans were generated in three irradiation techniques: 3D-CRT, IMRT, and SBRT. Each technique was planned using two CT-density curves from different phantom (CIRS 062M and CIRS 002LFC). Prescribed dose and number of fields delivered were different for each technique; it was set based on the clinical data. Seven irradiation fields were applied to 3D-CRT and IMRT; and fifteen fields for SBRT. In addition, there were three segments for each field in IMRT plan.

2.3. Dosimeter preparations
The TL dosimetry system consists of LiF TLD cylinders (TLD-100, Harshaw) of 3 × 1 mm diameter and a Harshaw 3500 read-out system. Whereas EBT2 films were cut into small square with a size of 2 × 2 cm^2. Because TLD has a relatively high standard deviation, it is prior to grouping the TLD before being calibrated. TLD and EBT2 film were calibrated in twenty seven-steps dose (50 to 1800 cGy), applied with field sizes of 10 × 10 cm^2 in 5 cm depth.

2.4. Phantom measurements
Generally, the dose measurement was divided into 1) target; and 2) OARs. In target dose measurement, there were four dosimeters employed: PTW N30013, Exradin A16, TLD, and *Gafchromic* EBT2 film. OARs dose were measured by PTW N3013 and Exradin A16 only. Dose measurements using TLD and film were performed using holders which have made from cork. Holders were designed resembling the original interchangeable rod from the thorax phantom.

2.5. Analysis of results
For the evaluation of the TPS calculated dose (D_{calc}) and measured dose (D_{meas}) value, we follow the equation:

$$\delta(\%) = \left(\frac{D_{meas} - D_{calc}}{D_{calc}} \right) \times 100$$

The agreement criterion was determined according to ICRU recommendation, which suggest the accuracy of ±5% in the delivery of an absorbed dose to a target volume [7].

3. Results and discussion

3.1. Target dose measurements
The results of target dose measurements in 3D-CRT, IMRT, and SBRT technique are shown in Figure 2. Generally, the measurement results confirmed that every dosimeter has its own characteristic. The use of different CT-density curves did not provide any significant change in discrepancies; because actually there was no significant difference between the two curves.

In 3D-CRT, EBT2 film provides the smallest discrepancies, -0.71% and -0.18% for CIRS 062M and 002LFC respectively. Other dosimeters gained the similar results, about 6%. This value is similar to results obtained by Watts [8] and Wibowo [9] that perform measurements on the same phantom using 0.6 cm^3 ionization chamber and TLD. It can be inferred that the pattern of results obtained in this study is regarding the medium inhomogeneity contained in the phantom. Watts suggests that irradiation in central axis beam which passes through the lung materials become the largest source of error in the measurement; whereas Wibowo found the irradiation ML/LM (media-lateral/lateral-medial) is the largest contributor of error in the measurement of point dose in lungs.

In IMRT, the measurement results showed that the ionization chamber with smaller active volume yields more accurate results. The dose values obtained by PTW N30013 and Exradin A16 were different enough; the difference was reached 9 cGy for both CT-density conditions. This difference can be caused by several things: 1) the effects of intra-leaf MLC leakage; and 2) the influence of beam intensity modulation. Beam intensity modulation is done by providing multiple segments on the radiation field. There was a possibility that segment variations led to the formation of high dose gradient region, right on the measurement area.

Similar to the results of 3D-CRT, all measurement results in SBRT were lower than planned dose (underestimate). This is caused by lateral electron disequilibrium (LED) [10]. In depths beyond the maximum longitudinal electron range, LED occurs when the lateral range of electrons become equal or greater than the radius of the field. Under this condition, electrons liberated from the beam's central axis will scatter beyond the field edge. They are not replaced and lead to dose reduction along the beam's central axis.

3.2. OARs measurements

The OAR dose measurement results gave the same pattern as the results obtained in the target. Generally, it was found that PTW N30013 yields the lower dose value than Exradin A16 on all variations of irradiations; it indicates volume averaging effect occurred in PTW N30013.

3.3 Dosimetric comparison

PTW N30013 is good to be used in 3D-CRT and IMRT technique; by the condition that the uniform dose area in the target must have a much larger size, incomparable to the ionization chamber cavity dimension. It is important to ensure that the radiation fluence entering the cavity is homogeneous, to obtain the accurate results. The ionization chamber with volume of 0.6 cm3, the commonly used dosimeter, apparently was not suitable for use in the small size and non-standard measurement field due to a lack of accuracy and spatial resolution.

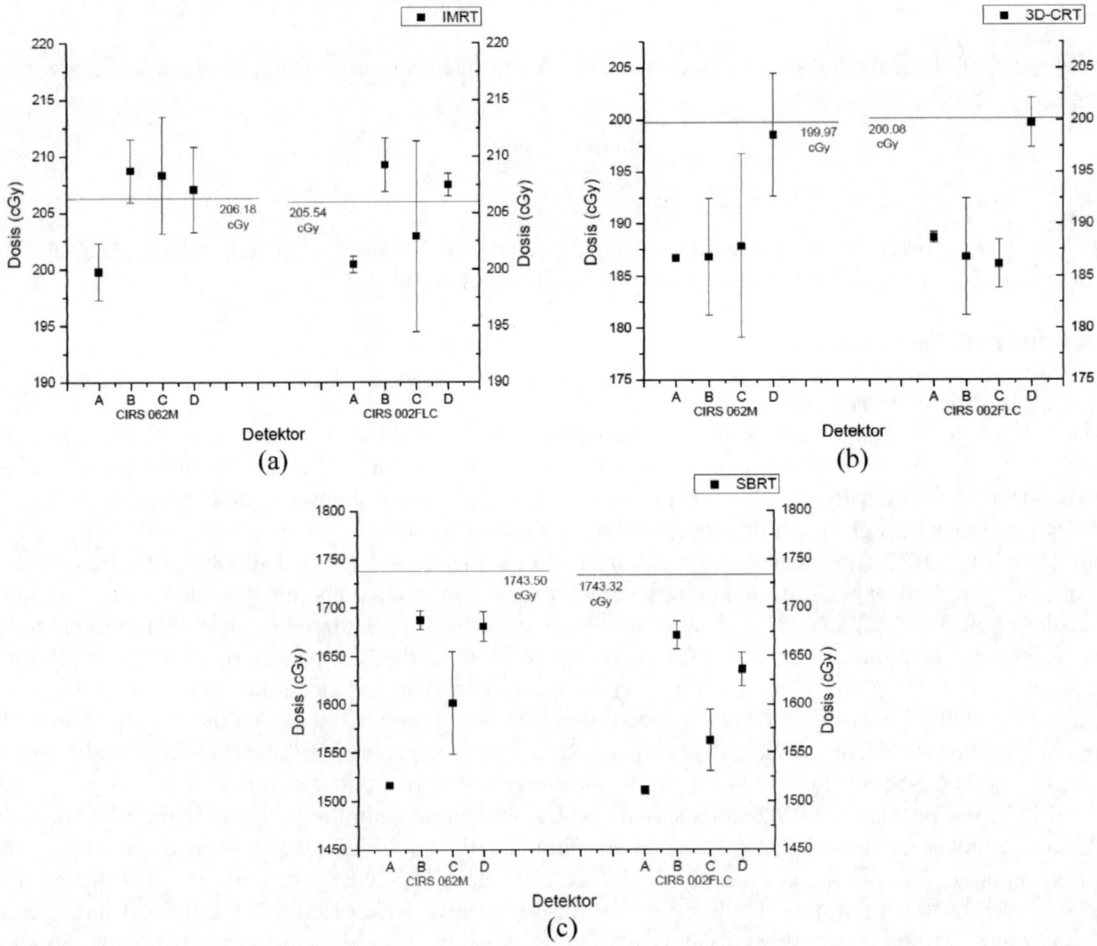

Figure 2. Measurements results for: (a) 3D-CRT; (b) IMRT; (c) SBRT. The horizontal axis represents dosimeter employed: A) PTW N30013; B) Exradin A16; C) TLD-100; D) EBT2 film.

Basically, Exradin A16 with an active volume of 0.007 cm^3 is suitable for use in high-dose measurement in a small field. At low-dose measurement, the signals obtained from the electrometer are probably not much different from the existing noise. As occurred in dose measurement of the spine, the value of SNR (signal-to-noise ratio) is not big enough, so that the results obtained are underestimate (smaller than expected). In SBRT, although it indicates the volume averaging effect, Exradin A16 being the best dosimeter to use in small field dosimetry in this study.

Whereas TLD, showed the lowest accuracy and precision in all measurement result in this study. Although in IMRT the discrepancy of TLD was about 1%, but in the other techniques, TLD did not represents the results well. It proved that the dosimeter with smallest dimension may not necessarily be the best dosimeter to be used in small field measurement. The fluctuation of measurement results makes TLD was considered to being replaced by another dosimeter which more practical in use and have better accuracy and precision.

The EBT2 film gained an average uncertainty value of 1.41%; it confirmed research conducted by Devic *et al.* which declared that EBT2 film has uncertainty value not more than 2% [11]. High accuracy results obtained in 3D-CRT and IMRT, with average discrepancy value of -0.45% and 0.69%, respectively. The results obtained in this study general support the hypothesis: because of its tissue equivalent and high spatial resolution, EBT2 film can be used as an appropriate dosimeter for point dose verification.

4. Conclusion

The selection of dosimeters in dose point verification requires some consideration, such as radiation technique and field size of the investigated case. Generally, Gafchromic EBT2 film yields the better results in all variations, and it is proposed to be an appropriate alternative dosimeter to be used at point dose verification.

References

[1] Brahme A 1984 *Acta. Radiol. Oncol.* **23** 379-391
[2] Nalbant N and Kesen D B 2014 *J. Nucl. Med. Radiat. Ther.* **5** 3
[3] Low D A, Parikh P, Dempsey J F, Wahab S and Huq S 2003 *Med. Phys.* **30** 1705-1711
[4] McEwen M R, Kawrakow I and Ross C K 2008 *Med. Phys.* **35** 950-958
[5] Olko P 2010 *Rad. Meas.* **45** 506-511
[6] Low D A, Moran J M, Dempsey J F, Dong L and Oldham M 2011 *Med. Phys.* **38** 1313-1338
[7] International Atomic Energy Agency 2004 TRS 398-Absorbed Dose Determination In External Beam Radiotherapy: An International Code for Practice for Dosimetry based on Standards of Absorbed Dose to Water (Vienna: IAEA) p 13
[8] Watts R 2001 Method to QA 3-D Treatment Planning System (Utah: AAPM) pp 4-6
[9] Wibowo W E 2008 Perbandingan Dosis pada Simulasi Radioterapi Paru-paru Antara Hasil Kalkulasi TPS dengan Pengukuran (Depok: FMIPA UI) pp 24-31
[10] Disher B 2013 *The impact of lateral electron disequilibrium on stereotactic body radiationtherapy of lung cancer* (Ontario: University of Western Ontario) pp 10-12
[11] Devic S 2009 *Med. Phys.* **39** 429-437

Acknowledgements

The authors would like to express special thanks to the medical physicist of Department of Radiotherapy RS Cipto Mangunkusumo (W. E. Wibowo and N. Nasution) who helped with the measurements and other preparations along this research; also to the Laboratory of Medical Physics, University of Indonesia which provided equipment needed in this research. This study was supported by Hibah Pascasarjana Universitas Indonesia 2015.

Determination of effective doses in image-guided radiation therapy system

Y Y Pyone[1], S Suriyapee[1], T Sanghangthum[1,2], S Oonsiri[2] and T Tawonwong[2]

[1] Medical Imaging, Department of Radiology, Faculty of Medicine, Chulalongkorn University, Bangkok, Thailand

[2] Division of Therapeutic Radiology and Oncology, Department of Radiology, King Chulalongkorn Memorial Hospital, Bangkok, Thailand

E-mail: yinyinpyone71@gmail.com

Abstract. The organ and effective doses in image-guided radiotherapy system are determined in this study. For 2D imaging, incident air kerma (K_i) was measured by 6cc ionization chamber with Accu-Pro dosimeter. The entrance surface air kerma (ESAK) was calculated by multiplying K_i with backscatter factor. The effective dose was calculated by multiplying ESAK with conversion coefficient. For 3D imaging, computed tomography/cone-beam dose index (CTDI/CBDI) measurements were performed by using 100mm pencil ionization chamber with Accu-Pro dosimeter. The dose index in air and in CTDI phantom from planning CT and cone-beam CT were measured. Then, effective dose was calculated by ImPACT software. The effective doses from 2D conventional simulator for anteroposterior and lateral projections were 0.01 and 0.02mSv for head, 0.15 and 0.16mSv for thorax, 0.22 and 0.21mSv for pelvis, respectively. The effective doses from 3D, planning CT and CBCT, were 3.3 and 0.1mSv for head, 13 and 2.4mSv for thorax and 7.2 and 4.9mSv for pelvis, respectively. Based on 30 fractions of treatment course, total effective dose (3D CT, 2D setup verification and 6 times CBCT) of head, thorax and pelvis were 3.93, 27.71 and 37.03mSv, respectively. Therefore, IGRT should be administered with significant parameters to reduce the dose.

1. Introduction

Radiation therapy is one of the most common treatment methods for cancer by delivering the high radiation dose to the tumor while avoiding the normal tissues from receiving doses above their limitation. The precision of radiation dose can be increased by new treatment modalities and advanced delivery techniques that can be applied to a cancer patient. To get the full benefit, accurate positioning of the target volume is needed and the magnitudes of both external and internal motion of the patient are minimized. Therefore, more imaging modalities are used for dose calculation in simulation process and for patient setup verification in the treatment process. However, this can result in additional dose contributions to the integral patient dose. This dose is possible to increase the risk of radiation (stochastic effects) and also the severity of side effects from the cancer therapy treatment.

Image-guided radiation therapy (IGRT) is the combination of imaging procedures in the treatment room and contemporary irradiation techniques. It involves multiple imaging procedures for simulation, planning, setup, and intrafraction monitoring. Common imaging procedures for IGRT are kilovoltage planar imaging, portal imaging, cone-beam computed tomography (CBCT), fan beam computed tomography. With the increasing use of IGRT, the chance for secondary cancer also increases. The objective of this study was to determine the imaging dose of IGRT from the whole process of treatment.

2. Material and methods

The materials used in this study were from the Department of Therapeutic Radiology and Oncology, King Chulalongkorn Memorial Hospital

2.1. *Effective dose calculation from setup imaging*

The Acuity Simulator (Varian Medical System, Palo Alto, CA, USA) supports basic 2D simulation and complex treatment plan verification. For effective dose calculation, a 6 cc ionization chamber (Radcal Corporation, Monrovia, CA) was placed at the isocenter and exposed with the selected exposure parameters according to the protocol and records the dose from dosimeter. Then, Incident Air Kerma (K_i) and Entrance Surface Air Kerma (K_E) were calculated by using equation (1) and (2) [6], below. The effective dose for planar projection radiography is directly obtained from entrance skin dose (mGy) multiply with the conversion coefficient, equation (3) [4].

$$K_i = M_Q \, N_{K,\,Q0} \, k_Q \, k_{TP} \tag{1}$$

$$K_E = K_i * \text{Back Scatter Factor} \tag{2}$$

$$\text{Effective dose} = K_E * \text{conversion coefficient} \tag{3}$$

Where K_i = Incident Air Kerma, K_E = Entrance Surface Air Kerma, M_Q = mean value of the dose meter reading, $N_{k,Q0}$ = dosimeter calibration coefficient, k_Q = radiation quality factor, k_{TP} = temperature, pressure correction factor.

2.2. *Effective dose calculation from planning CT*

Computed tomography dose index (CTDI) measurements were performed with the Philips Brilliance Big Bore CT scanner (Philips Healthcare, Andover, MA, USA). This CT scanner has an 85cm bore and true 60cm scan field of view. This 85 cm bore size can be used to scan patients with immobilization devices. The $CTDI^{air}$ measurement was done by placing the 100 mm ionization chamber in air at the isocenter and recorded the dose from dosimeter. Then, $CTDI^{center}$ and $CTDI^{periphery}$ were measured by placing the chamber at the center, 12, 3, 6, and 9 o'clock positions of the CTDI phantom (16 cm in diameter for head protocol and 32 cm in diameter for chest and pelvis protocol).The CTDI weighted was calculated by using the equation (4), the $CTDI^{air}$, $CTDI^{weighted}$, type of CT unit, total mAs and scan range of each protocol were entered into the ImPACT (Imaging Performance Assessment of CT scanners) software (version 1.0.4) for the organ and effective doses calculation.

$$\text{CTDI/CBDI}_{100}^{\text{weighted}} = \frac{1}{3}\text{CTDI/CBDI}_{100}^{\text{center}} + \text{CTDI/CBDI}_{100}^{\text{periphery}} \tag{4}$$

Where $\text{CTDI/CBDI}_{100}^{\text{weighted}}$, $\text{CTDI/CBDI}_{100}^{\text{center}}$, $\text{CTDI/CBDI}_{100}^{\text{periphery}}$ were CT/Cone-beam Dose Index measured in air, center of phantom and average of the four peripherals position of the phantom, respectively.

2.3. *Effective dose calculation from cone beam CT*

The Cone-beam CT system (Varian Medical System, Palo Alto, CA) attached to the Varian TrueBeam Linear Accelerator (version 2.0), is designed to verify for motion and setup errors in patients undergoing radiation therapy. For Cone-beam CT effective dose calculation, Cone-beam Dose Index (CBDI) measurements were performed as the same as in planning CT, figure 1. As the ImPACT software is designed specifically for the CT scanner, for the cone-beam CT, we used ImPACT factor as shown in equation (4) to match with ImPACT software to select the suitable reference CT unit.

$$\text{ImPACT factor} = a\,\frac{\text{CBDI}_{100}^{\text{center}}}{\text{CBDI}_{100}^{\text{air}}} + b\,\frac{\text{CBDI}_{100}^{\text{periphery}}}{\text{CBDI}_{100}^{\text{air}}} + c \tag{5}$$

Where a, b, c = Empirical factors developed by ImPACT group.

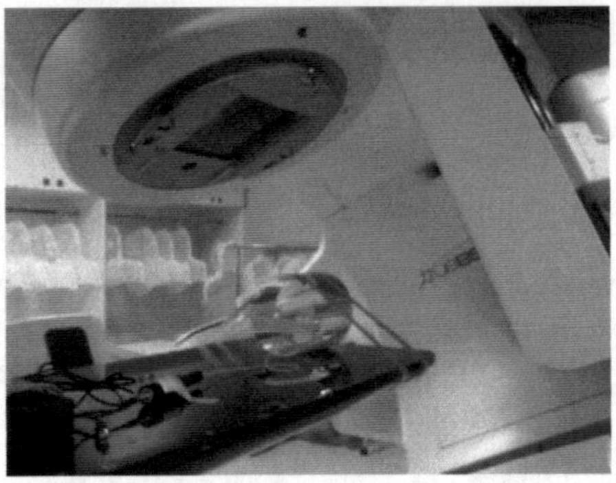

Figure 1. The setup of 100mm pencil ionization chamber in 16 cm diameter CTDI phantom for CBCT head protocol.

3. Results

The effective dose calculation from 2D kilovoltage setup imaging was performed as the AAPM Task Group 75 recommendation [5]. The conversion coefficients were determined according to the exposure parameters and half value layer of the machine. The parameters and total effective doses are shown in table 1.

Table 1. Effective doses and acquisition parameters for each protocol of acuity simulator.

Parameters	Head		Chest		Pelvis	
	AP	Lateral	AP	Lateral	AP	Lateral
kV	85	90	99	100	110	110
mAs	9.97	19.46	10.16	21.58	11.86	20
Field size (cm)	15.3*15.3	15.3*15.3	20.03*20.03	20.03*20.03	20.03*20.03	20.03*20.03
Conversion coefficient	0.01	0.01	0.12	0.05	0.13	0.07
Effective dose (mSv)	0.01	0.02	0.15	0.16	0.22	0.21
Total effective dose (mSv)	**0.03**		**0.31**		**0.43**	

The organ and effective doses from the planning fan beam CT and cone-beam CT are shown in table 2 and 3 for each region. The dose calculations were performed based on the departmental routine of the planning CT and manufacturer protocol of the Varian cone-beam CT system. For effective dose calculation, the organ and tissue weighting factors were used as recommended by ICRP 103 [6].

Table 2. Organ and effective doses from Philips Brilliance Big Bore CT.

Organs	W_T	**Organ Dose, (H_T), (mSv)**			**Effective Dose, ($H_T.W_T$), (mSv)**		
		Head	**Chest**	**Pelvis**	**Head**	**Chest**	**Pelvis**
Gonads	0.08	0.000032	0.056	28	0	0.0045	2.3
Bone Marrow	0.12	2.5	9.3	7.4	0.3	1.1	0.88
Colon	0.12	0.0011	0.4	16	0.00013	0.048	1.9
Lung	0.12	0.88	31	0.04	0.11	3.7	0
Stomach	0.12	0.027	9.2	0.84	0.0032	1.1	0.1
Bladder	0.04	0	0.025	31	0	0.001	1.2
Breast	0.12	0.13	24	0.04	0.016	2.8	0
Liver	0.04	0.043	13	0.51	0.0017	0.54	0.02
Esophagus (Thymus)	0.04	0.69	34	0.01	0.028	1.4	0
Thyroid	0.04	29	18	0	1.2	0.71	0
Skin	0.01	3.1	6.8	6.7	0.031	0.068	0.07
Bone Surface	0.01	9.7	17	11	0.097	0.17	0.11
Brain	0.01	19	0.39	0	0.19	0.0039	0
Salivary Glands (Brain)	0.01	19	0.39	0	0.19	0.0039	0
Remainder	0.12	4.1	12	5	0.49	1.4	0.59
	Total Effective Dose (mSv)				**3.3**	**13**	**7.2**

Table 3. Organ and effective doses from Cone-beam CT attached to the Varian TrueBeam Linear Accelerator.

Organs	W_T	Organ Dose, (H_T), (mSv)			Effective Dose, ($H_T.W_T$), (mSv)		
		Head	Chest	Pelvis	Head	Chest	Pelvis
Gonads	0.08	<0.01	<0.01	19	<0.01	<0.01	1.5
Bone Marrow	0.12	0.22	1.6	4	0.026	0.2	0.48
Colon	0.12	0.000032	0.017	9.3	<0.01	<0.01	1.1
Lung	0.12	0.015	5.7	0.01	0.0018	0.69	0.001
Stomach	0.12	0.00058	0.29	0.2	0.00007	0.034	0.024
Bladder	0.04	0	0.0023	34	0	0.00009	1.3
Breast	0.12	0.0037	6.6	0.02	0.00044	0.79	0.002
Liver	0.04	0.00088	0.44	0.13	0.000035	0.017	0.005
Esophagus (Thymus)	0.04	0.013	10	0	0.0005	0.4	0
Thyroid	0.04	0.42	0.76	0	0.017	0.03	0
Skin	0.01	0.23	1	4.4	0.0023	0.01	0.044
Bone Surface	0.01	0.95	3.1	6.1	0.0095	0.031	0.061
Brain	0.01	2.5	0.039	0	0.025	0.00039	0
Salivary Glands (Brain)	0.01	2.5	0.039	0	0.025	0.00039	0
Remainder	0.12	0.25	1.4	3.1	0.03	0.17	0.37
Total Effective Dose (mSv)					**0.1**	**2.4**	**4.9**

4. Discussion
The effective doses from the Acuity simulator were very small compared with the other modalities. The maximum was observed at pelvis protocol of 0.43 mSv due to the highest exposure parameters. From the planning CT, lung and breast tissue acquired 31 mSv and 24 mSv, gonads and colon acquired 28 mSv and 16 mSv from chest and pelvis protocols. The thyroid, esophagus, and bladder obtained the highest organ dose of 29 mSv, 34 mSv and 31 mSv for the head, chest and pelvis protocols, respectively. The effective dose from planning CT chest protocol was higher than that of the pelvis protocol because the scan length for chest protocol was longer than the pelvis protocol to cover the metastasis area, as shown in table (4). In cone-beam CT imaging, salivary gland, esophagus, and bladder received the highest organ dose of 2.5, 10 and 34 mSv for the standard head, low dose thorax, and pelvis protocols. However, the highest radiation sensitive organ, the lungs acquired 5.7 mSv in low dose thorax, gonads and colon acquired 19 and 9.3 mSv from pelvis protocols. The effective doses from CBCT were lower than that of the planning CT; however, to achieve the precise and accuracy treatment, CBCT has to perform every week to every day which increasing the amount of radiation dose to the patient.

Table 4. Acquisition parameters for planning CT.

Parameter	Head	Chest	Pelvis
kV	120	120	120
mA	325	325	325
Collimation (mm)	16 x 1.5	16 x 1.5	16 x 1.5
Rotation Time (second)	0.75	1	1
Pitch	0.938	0.938	0.938
Scan Length (mm)	357	355	280
$CTDI_{air}$ (mGy)	15.23	15.24	15.24
$^{n}CTDI_{weighted}$ (mGy)	12.41	8.05	8.05
$CTDI_{vol}$ (mGy)	32.2	27.9	27.9
DLP (mGy.cm)	790	990	781

5. Conclusion
In summary, for 30 fractions of the treatment course, at least, one scan of 3D planning CT and 2D setup verification and 6 times CBCT have been performed. Therefore, the total effective doses of the

Journal of Physics: Conference Series **694** (2016) 012007 doi:10.1088/1742-6596/694/1/012007

head, thorax, and pelvis regions are 3.93, 27.71and 37.03 mSv, respectively. This can attribute more radiation dose to the already high treatment dose burden to the patient. Therefore, to reduce the chance of secondary cancer, image-guided radiotherapy should be administered with significant parameters to reduce the dose.

References

[1] Hälg R A, Besserer J and Schneider U 2012 *Med. Phys.* **39** pp 7650- 61
[2] Hyer D E and Hintenlang D E 2010 *Med Phys.* **37** 4620-26
[3] Cheng H C Y, Wu V W C, Liu E S F and Kwong D L W 2011 *Int J Radiat Oncol Biol Phys.* **80** 291-300
[4] Murphy M J *et al*. 2007 Report of the AAPM Task Group 75 *Med Phys.* **34** 4041-63
[5] International Commission on Radiological Protection. ICRP Publication 103, annals of ICRP 2007 **37** pp 1-332
[6] IAEA 2007 *Technical Report Series No. 457* (Vienna, Austria: International Atomic Energy Agency)

Journal of Physics: Conference Series **694** (2016) 012008 doi:10.1088/1742-6596/694/1/012008

Sensitivity in error detection of patient specific QA tools for IMRT plans

S Z Lat[1], S Suriyapee[1] and T Sanghangthum[1,2]

[1] Medical Imaging, Department of Radiology, Faculty of Medicine, Chulalongkorn University, Bangkok, Thailand
[2] Division of Therapeutic Radiology and Oncology, Department of Radiology, King Chulalongkorn Memorial Hospital, Bangkok, Thailand

E-mail: happyzinlay@gmail.com

Abstract. The high complexity of dose calculation in treatment planning and accurate delivery of IMRT plan need high precision of verification method. The purpose of this study is to investigate error detection capability of patient specific QA tools for IMRT plans. The two H&N and two prostate IMRT plans with MapCHECK2 and portal dosimetry QA tools were studied. Measurements were undertaken for original and modified plans with errors introduced. The intentional errors composed of prescribed dose (±2 to ±6%) and position shifting in X-axis and Y-axis (±1 to ±5mm). After measurement, gamma pass between original and modified plans were compared. The average gamma pass for original H&N and prostate plans were 98.3% and 100% for MapCHECK2 and 95.9% and 99.8% for portal dosimetry, respectively. In H&N plan, MapCHECK2 can detect position shift errors starting from 3mm while portal dosimetry can detect errors started from 2mm. Both devices showed similar sensitivity in detection of position shift error in prostate plan. For H&N plan, MapCHECK2 can detect dose errors starting at ±4%, whereas portal dosimetry can detect from ±2%. For prostate plan, both devices can identify dose errors starting from ±4%. Sensitivity of error detection depends on type of errors and plan complexity.

1. Introduction

Intensity modulated radiation therapy (IMRT) technique, represents the advanced form of three-dimensional conformal therapy. It is advantageous for patient treatments because it allows more conformal shaping of the dose to the target and significant gains in the sparing of normal tissues [3]. IMRT achieves desired dose distribution in a complex shaped volume by modulating the intensity map of each treatment field using multileaf collimator (MLC). The complexity in planning and delivery of radiation demands a high level of quality control both in the operation of the equipment and in the treatment delivery for individual patients [2].

Patient-specific QA (quality assurance) for IMRT plans can be done using various kinds of planar dosimeters such as films, diode arrays, ionization chamber arrays and electronic portal imaging devices (EPID). For 2D or 3D QA tools the gamma index combines dose difference and distance to agreement to the single parameter is commonly used for dosimetric evaluation of treatment plans.

According to different designs and configurations of QA tools, the capability of each detector is different. The sensitivity of each QA tool depends on the type and number of detectors, arrangement, and spacing between them. The purpose of this study is to evaluate the error detection capability of patient specific QA tools (MapCHECK 2 system and Portal Dosimetry System) used to evaluate the IMRT plans.

2. Materials and methods

2.1. Creating IMRT plans

Four clinical IMRT plans, two head and neck and two prostate, were selected. The plans were created using Eclipse Treatment Planning System-version 11.0.31 (Varian medical Systems, Palo Alto, CA, USA). The six MV beams were employed with 9 field arrangement for head and neck IMRT cases while 10 MV beams with 7 fields were optimized and calculated for prostate IMRT plans. The head and neck plan represented the complicated plans and prostate as the simple plans.

From the copies of the original plans, the plans with intentional errors were created. The intentional errors were increasing and decreasing prescribed dose (±2, ±4 and ±6%) and position shifting in X-axis and Y-axis (±1, ±2, ±3 and ±5mm).

2.2. Measuring systems

The patient plans were calculated for EPID and MapCHECK 2 and delivered by using ClinaciX (Varian Medical Systems, Palo Alto, CA, USA, Inc). The original and edited plans were measured by EPID and MapCHECK 2 system. The EPID which was fixed with Varian ClinaciX machine was set at 100cm source to detector distance (SDD) and measured both original and modified plans. For the measurement with MapCHECK 2 which was placed on the patient couch and the SDD was set at 95.8cm and the solid water phantom of 3cm was added on its detector surface to acquire the 5cm water equivalent thickness. All measurements were saved in respective software for further analysis.

2.3. Analyzing the measurements

The gamma evaluation tool was available for analysis of the measurement for each QA tool. Firstly the dose distributions of the original plans were compared between those calculated from the TPS and from the measurement. In order to observe the error detection of QA tools, the measurements with intentional errors were compared with the original plan. The gamma index of 3% dose difference and 3mm distance to agreement with 10% threshold were used to set the evaluation criteria. The gamma evaluation pass rate was set at 95% and an error was considered detected when the gamma failure rate was higher than 5%. We then analyzed the gamma result of modified plans and evaluated the error detection capability of each QA tool.

3. Results

The result of 4 original IMRT plans measured with Portal Dosimetry and MapCHECK 2 system are shown in table 1. All the results were analyzed by using 3%/3mm gamma criteria. The pass rates were higher for the prostate plan compared to the head and neck plan. This was attributed to the increased modulation and irregular field shapes in the head and neck plan [5]. Even without error, the pass rate of head and neck was close to the acceptable pass rate of 95% because the beam configuration and tumor shape in head and neck was quite complicated compared with the prostate plan.

Table 1. The gamma results of 4 IMRT plans obtained without intentional error.

	Gamma Passing rate (%)	
Plan	Portal Dosimetry	MapCHECK 2

Plan 1 (H&N)	95.2	98.6
Plan 2 (H&N)	96.6	97.9
Plan 3 (Prostate)	99.6	100
Plan 4 (Prostate)	99.9	100

3.3. Error detection by portal dosimetry system
As the head and neck plan was more complicated, the error detection in this plan was higher than in that of the prostate plan.

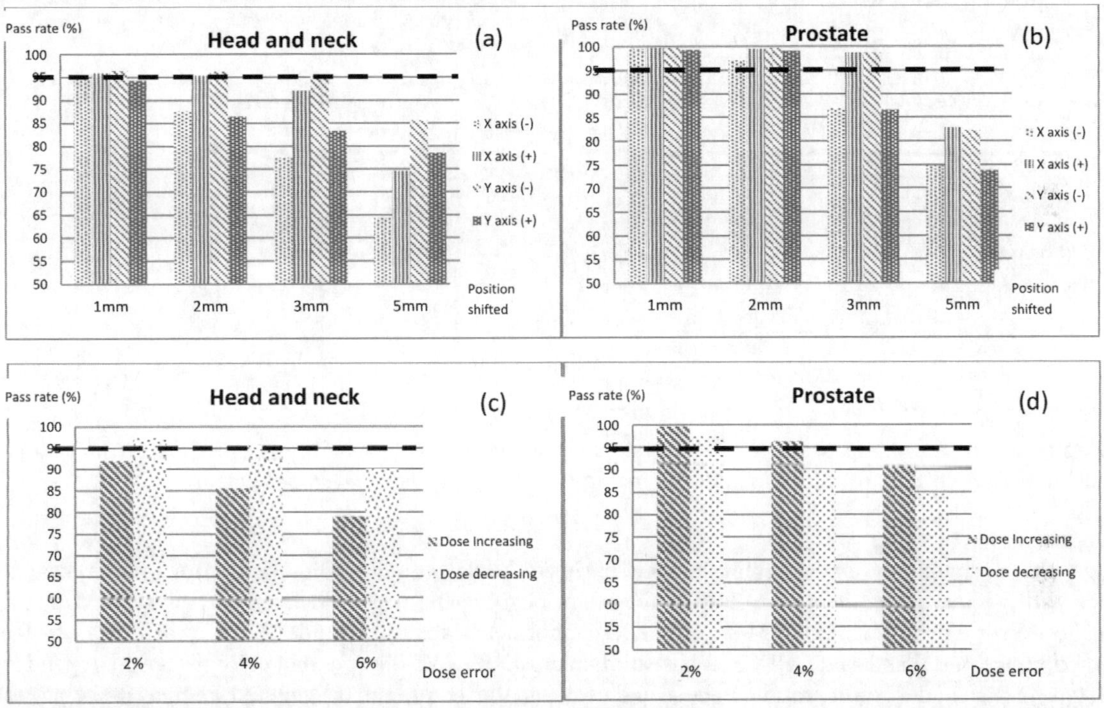

Figure 1. The results for portal dosimetry system for (a) Position shift error in H&N; (b) Position shift error in prostate (c) prescribed dose error in H&N; and (d) prescribed dose error in prostate.

Error of position shifted in X-axis can be detected start from -1mm, +3mm and in Y-axis was -5mm and +1mm. In the prostate plan, the error detection started from ±3mm in both X and Y axis. Prescribed dose error in head neck plan that had a failure rate less than 95% can be observed in the dose increasing error of 2% and decreasing error of 6%. For the prostate plan, the dose started from -4% and +6% errors can be detected. The results for gamma pass measured by portal dosimetry system are shown in figure 1.

3.4. Error detection by MapCHECK 2 system
The MapCHECK2 system can detect error starting from -2mm and +3mm in X axis and +2mm and -3mm in Y axis for the head neck plan while MapCHECK 2 can detect the error starting from ±3mm in both axes for prostate as shown in figure 2. Even small error of 2% dose decreasing can be detected in head and neck plan but in prostate only starting from 4% the system can be detected.

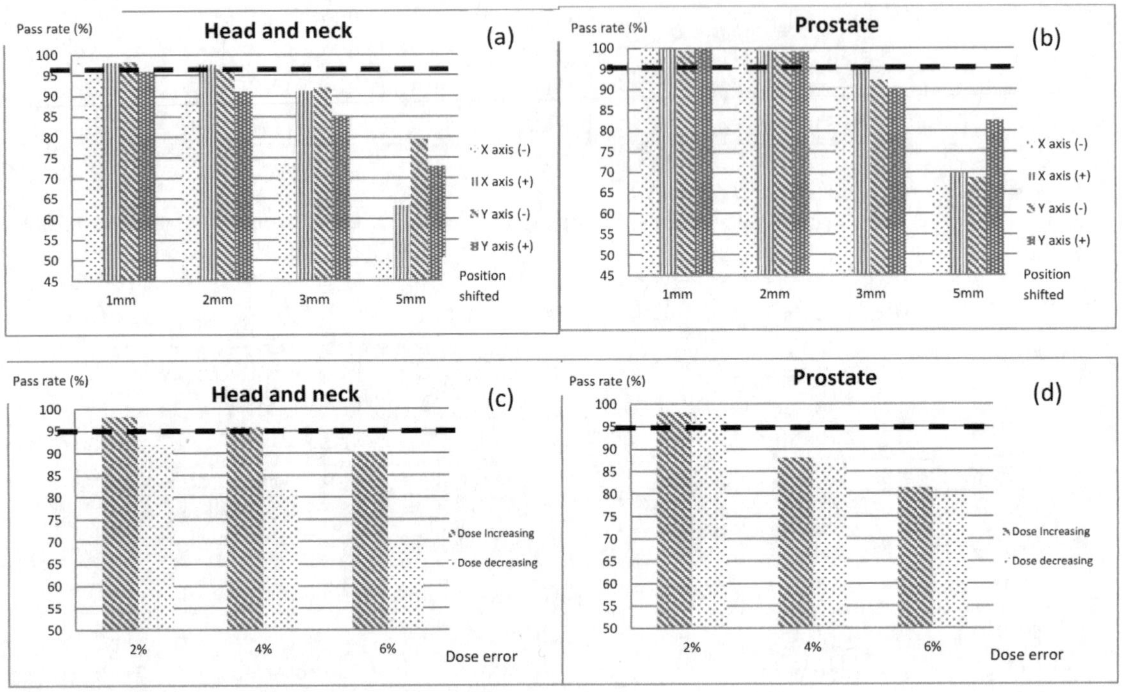

Figure 2. The results for MapCHECK 2 system showing (a) Position shift error in H&N; (b) Position shift error in prostate; (c) prescribed dose error in H&N; and (d) prescribed dose error in prostate

4. Discussion

The portal dosimetry system is slightly higher sensitivity to detect the position shift error in both head and neck and prostate plans than MapCHECK 2 system because the portal dosimetry system can detect the smallest error of 1mm while the MapCHECK 2 cannot detect that magnitude of the error. For dose error detection, the portal and MapCHECK 2 has similar sensitivity. We observe that error detection depends on the type of plan under consideration. In head and neck, smaller errors can be detected whereas in the prostate plan, the error needs to be of a higher magnitude before it can be detected. The sensitivity to the introduced errors depends on the plan and the various systems can detect various errors, and some of the errors cannot be found with these systems [1].

5. Conclusion

Error detection of each QA tools depends on the error type, site, and shape of tumor location. The configurations of QA tools also play an important role in the sensitivity of error detection. The higher the resolution of QA tools, the better the sensitivity. From our results, we can conclude that portal dosimetry system has marginally higher sensitivity than MapCHECK 2 system because it is higher resolution than the MapCHECK 2 system, at least for the plans we considered.

References

[1] Fredh A, Scherman J B and Fog L S 2013 *Med Phys.* **40(3)** 031716-9
[2] Kunnanchath J, Majumdar S D, Tharmarnadar G, Ramachandran M, Joshi R C and Devadason S 2014 *Transworld Medical Journal* **2(4)** 187-192

[3] Stathakis S, Myers P, Esquivel C, Mavroidis P and Papanikolaou N 2013 *Med. Phys.* **40(7)**, 071731-9

[4] Gordon M M, Jonas D F, John P G and Brent C P 2012 *Med. Phys.* **39(7)** 4378-4385

[5] Bawazeer O, Gray A, Arumugam S, Vial P, Thwaites D, Descallar J and Holloway L 2014 *Journal of Physics: Conference Series* 489 doi:10.1088/1742-6596/489/1/012071

Acknowledgments
I am grateful to Assistant Professor Chonlakiet Khorprasert, Department of Radiology, King Chulalongkorn Memorial Hospital for his equipment support and thanks to all medical physicists and staff at the division of Therapeutic Radiology and Oncology, King Chulalongkorn Memorial Hospital for their kind support of this work.

Dosimetric impact of interplay effect in lung IMRT and VMAT treatment using in-house dynamic thorax phantom

Mukhlisin[1,2] and S A Pawiro[1]

[1] Department of Physics, Faculty of Mathematics and Natural Sciences, Universitas Indonesia, Depok, Indonesia
[2] Nuclear Energy Regulatory Agency (BAPETEN), Jakarta, Indonesia

E-mail: r.mukhlisin@bapeten.go.id

Abstract. Tumor motion due to patient's respiratory is a significant problem in radiotherapy treatment of lung cancer. The purpose of this project is to study the interplay effect through dosimetry verification between the calculated and delivered dose, as well as the dosimetric impact of leaf interplay with breathing-induced tumor motion in IMRT and VMAT treatment. In this study, a dynamic thorax phantom was designed and constructed for dosimetry measurement. The phantom had a linear sinusoidal tumor motion toward superior-inferior direction with variation of amplitudes and periods. TLD-100 LiF:Mg,Ti and Gafchromic EBT2 film were used to measure dose in the midpoint target and the spinal cord. The IMRT and VMAT treatment had prescription dose of 200 cGy per fraction. The dosimetric impact due to interplay effect during IMRT and VMAT treatment were resulted in the range of 0.5% to -6.6% and 0.9% to -5.3% of target dose reduction, respectively. Meanwhile, mean dose deviation of spinal cord in IMRT and VMAT treatment were around 1.0% to -6.9% and 0.9% to -6.3%, respectively. The results showed that if respiratory management technique were not implemented, the presence of lung tumor motion during dose delivery in IMRT and VMAT treatment causes dose discrepancies inside tumor volume.

1. Introduction

The utilization of radiotherapy modality in IMRT treatment and the newest modality such as VMAT treatment are widely applied in radiation oncology. Modern linear accelerator is able to modulate radiation beam dynamically and create dose distribution by steep dose gradient that is enable to increase sparing dose on the organ at risk and escalate dose in tumor target [1]. Nowadays, IMRT and VMAT treatment have been widely used in Non-Small Cell Lung Cancer (NSCLC) treatment [2, 3].

Radiotherapy complexity of IMRT and VMAT treatment correlates with several uncertainties [4]. Those can lead to inaccuracy of given dose and has implication in tumor control, morbidity and toxicity treatment [5]. Quality assurance in radiotherapy treatment planning process is very important to ensure the dose calculation is carried out precisely and accurately, as well as minimize the possibility of radiation accident occurrence [6, 7].

The presence of tumor motion driven by patient breathing [8], further limits the maximum dose deliverable to the tumor volume, as an additional margin is generally assigned around the tumor volume to encompass the entire range of tumor motion. This significantly increases the total irradiated volume of normal lung tissues. In addition, the presence of tumor motion during dose delivery can generate unwanted dose discrepancies inside the tumor volume [9].

The aim of this study was to evaluate the dosimetric accuracy between planning dose and measured dose for lung treatment, as well as the possible interplay between MLC motion and tumor motion in IMRT and VMAT treatment using an in-house dynamic thorax phantom.

2. Materials and methods

The experiment was performed using a specially designed dynamic thorax phantom, calibrated TLD-100 LiF:Mg,Ti and Gafchromic EBT2 detectors, calibrated RapidArc Clinac® iX Linac, and Eclipse treatment planning system ver. 11.0. In this study, the following constraints were the tumor target could be moved only in superior-inferior direction and the experiment was performed for point dose measurement at midpoint the tumor target.

2.1. Development of in-house dynamic thorax phantom

A dynamic thorax phantom was constructed to simulate lung tumor motion. Phantom material made from local materials consisting of acrylic (ρ = 1.10 g/cm^3), cork (ρ = 0.22 g/cm^3), teflon (ρ = 1.88 g/cm^3), and polyethylene (ρ = 0.98 g/cm^3) to simulate soft tissue, lung, bone, and baseplate respectively. Phantom was dedicated the human ordinary thorax in term of structure, proportion and composition with oval shaped the phantom size of 19.5 × 30.9 × 20.1 cm^3. A rod equivalent to lung tissue with a diameter of 5.4 cm contains a spherical shaped target with a volume 19.7 cc (diameter of 3.35 cm) is represented tumor target.

Tumor motion was simulated along superior-inferior direction to simulate the respiratory-induced tumor motion. TLDs and EBT2 films can be placed inside the spherical target. The amplitude and period of motion were designed to be adjusted in static mode, 9.3 mm and 2.3s, 20 mm and 3.44s, and 30 mm and 4.22s, which was driven by a DC motor. Using these parameters, the experiment was able to mimic various clinical situations like amplitude and frequency (period) of the tumor motion. The literature stated that the average breathing cycle of lung tumor motion is of amplitude ±2 cm and a frequency of 12-17 cycles/minute (period of 3.5s to 5s). The speed control circuit in the phantom was tuned so that the abovementioned criteria were met.

2.2. IMRT and VMAT dose planning

CT images of the phantom were acquired with a target in static position using Philips Brilliance CT Simulator with a slice thickness of 3 mm. The image series were transferred from the CT control station to the treatment planning system server through the DICOM radiotherapy planning network for planning. After CT data series of the phantom was received by TPS, tumor target, and critical organs were contoured on the acquired CT images.

The CTV was defined to be the GTV plus a 0.5 cm margin as appropriate to account for microscopic tumor extension and then the ITV plus 0.5 cm was added to the ITV to form the CTV. A margin of 0.5 cm was given to simulated IM in the inferior-superior direction and an additional set-up margin of 0.5 cm. Thus, the total PTV includes the CTV plus a total margin of at least 1.5 cm to the superior-inferior dimensions and at least 1.0 cm in the axial plane [10].

IMRT and VMAT plans were created with the Eclipse treatment planning system version 11.0. The plans were generated and delivered based on x-ray photon beam 6 MV from a Varian RapidArc Clinac® iX Linac with a Millennium MLC (spatial resolution of 10 mm at isocenter). All dose calculations with a grid resolution of 2.5 mm using the Eclipse AAA were performed, taking into account heterogeneity correction. TPS dose planning applied dose prescription (95%) for a total dose of 6000 cGy and 30 fractions (200 cGy per fraction). For optimization, seven irradiation fields for IMRT treatment with gantry rotation for 15o each was used, whereas VMAT treatment utilized RapidArc double partial arc with gantry arc from 181o CW 20o and 20o CCW 181o are shown in figure 1. The approved plan was then transferred to the RapidArc treatment linear accelerator unit.

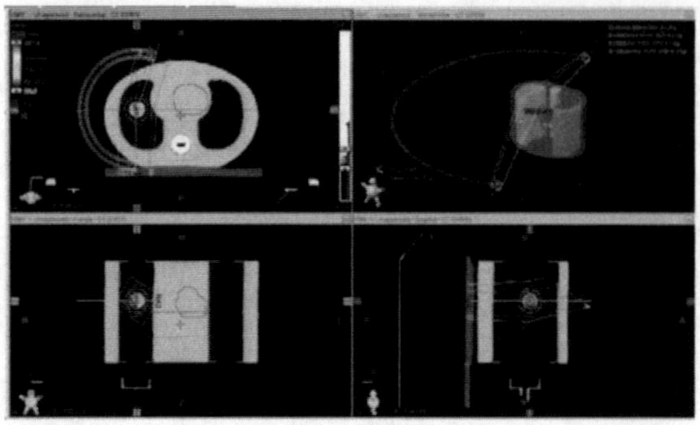

Figure 1. Planning of the dynamic phantom which shows the cross-sectional images of the phantom and field setup.

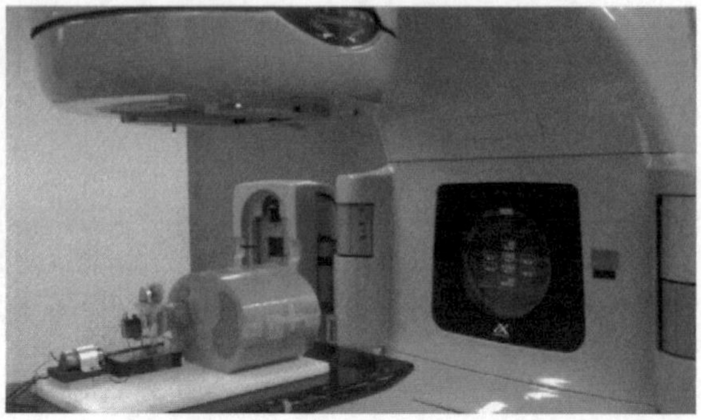

Figure 2. Irradiation technique of the dynamic thorax phantom for both IMRT and VMAT technique.

2.3. Dose measurements

The phantom assembly was positioned on the Varian RapidArc Clinac® iX Linac as if it was positioned during CT image acquisition. TLDs and gafchromic EBT2 films were inserted in tumor target and spinal cord of the thorax phantom. The phantom was irradiated by x-ray photon beam 6 MV using IMRT and VMAT techniques. The dosimetry measurement methods are presented in figure 2.

Gafchromic EBT2 films were scanned at least 48 hours after the irradiation using an Epson V700 flatbed scanner (Seiko Epson Corp., Suwa, Japan) with a resolution 72 DPI and it was stored in TIFF format. Furthermore, the pixel values were measured using FilmQA Pro and ImageJ software in the red channel, whereas TLDs were measured at least 48 hours after the irradiation using Harshaw M3500 TLD reader which integrate with WinREMS software. The dose difference (Δ%) of measured dose (Dmeas) and planned dose (Dplan) was calculated as

$$\left(\left(\boldsymbol{D}_{meas} - \boldsymbol{D}_{plan}\right)/\boldsymbol{D}_{plan}\right) * 100$$

3. Results and discussion

Based on TPS dose calculation with prescribed dose of 200 cGy, PTV dose of IMRT treatment were resulted in the range of 161.4 cGy to 210.2 cGy (mean dose of 200 cGy), while PTV dose of VMAT treatment were in the range of 174.3 cGy to 207.9 cGy (mean dose of 200 cGy). Furthermore, the spinal cord of IMRT treatment were in the range of 0.3 cGy to 51 cGy (mean dose of 8.8 cGy), whereas for VMAT treatment were 0.3 cGy to 49.9 cGy (mean dose of 10.3 cGy).

The point dose measurement in lung tumor target at static and dynamic condition of IMRT and VMAT treatment using TLDs and gafchromic EBT2 film are presented in figure 3, which indicated that mean dose of tumor target decreases with increasing of tumor target amplitude. A decrease dose of tumor volume is possibly caused by interplay effect, where only a fraction of the PTV is irradiated at any given time [11]. According to McCarter and Beckham, the interplay effect is caused by the combination of the intra-fraction target motion and the beam motion which generates variations of the dose in each voxel [12]. On the other hand, the mean dose of the spinal cord is below the dose calculated TPS for both IMRT and VMAT treatment as illustrated in figure 4.

Mean dose deviation of tumor target between TPS calculation and measurement in IMRT treatment by tumor target moves at condition of static, 9.3 mm, 20 mm and 30 mm were 0.3% to 0.5%, -2.7% to -3.0%, -3.7% to -4.6%, and -6.0% to -6.6% respectively while dose deviation in VMAT treatment were 0.2% to 0.9%, -1.6% to -1.9%, -2.9% to -3.1%, and -5.0% to -5.3% respectively. On the other hand, mean dose deviation of the spinal cord in IMRT treatment were -5.6% to -1.0%, -6.8% to -6.9%, -3.7% to -5.9%, and 0.7% to 1.0% respectively and in VMAT treatment were -1.4% to -3.1%, -3.0% to -6.3%, -1.6% to -4.2%, and 0.1% to 0.9% respectively.

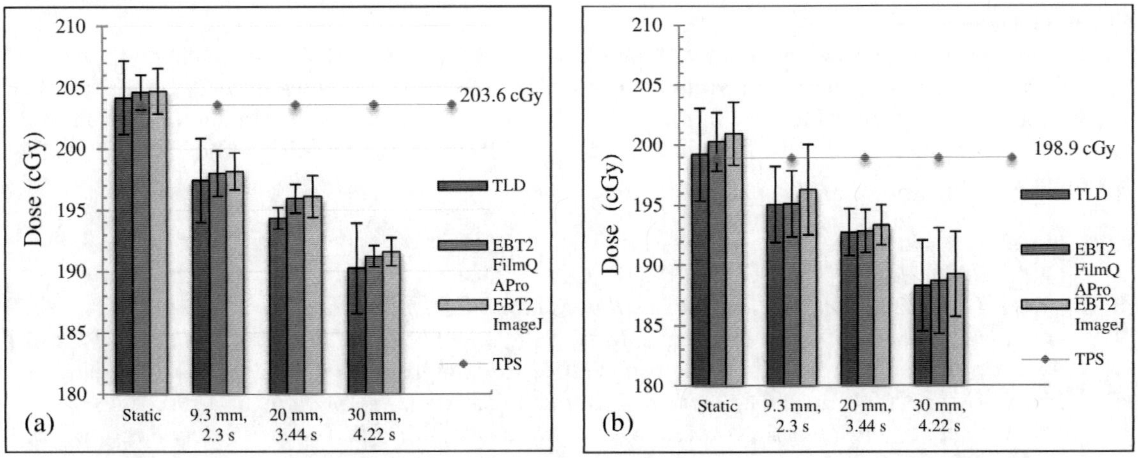

Figure 3. Ratio of mean dose in tumor target (a) IMRT treatment; (b) VMAT treatment.

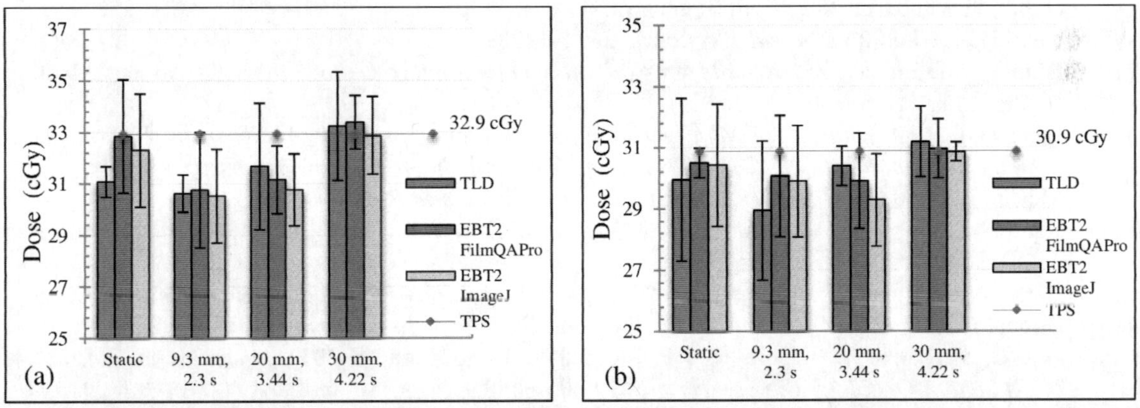

Figure 4. Ratio of mean dose in spinal cord (a) IMRT treatment; (b) VMAT treatment.

According to this experiment result, the interplay effect causes dose discrepancies inside the target, both in IMRT and VMAT treatment. This result is in accordance with previous experimental research by Jiang *et al.*, Berbeco *et al.*, Boopathy *et al.*, Ong *et al.*, and Ceberg C *et al.*, which stated that interplay effect will cause unwanted dosimetry effect [13-17].

The mean dose deviation are relatively higher than previous research of Jiang *et al.*, Berbeco *et al.*, Ong *et al.*, and Ceberg C *et al.*, this might be contributed from material deviation of the phantom, for instance density value and CT number (HU), Linac photon output deviation and the phantom mechanical are required for further development in term of its motion accuracy. Contrary, mean dose deviation percentages are relatively lower than previous research by Boopathy *et al.*, in which its deviation was in the range of 5 to 10%.

Patient respiratory cycle demonstrated the complex pattern, in which amplitude and period tumor motion are altering continuously during irradiation. The motion of tumor position is induced by lung tidal volume alteration as time function [18, 19]. Mathematically, human respiratory is non-stationary circumstance and difficult to be modeled [18].

This investigation can be considered as preliminary research to evaluate lung tumor target motion effect at dose received by tumor target using dynamic thorax phantom by modeling lung tumor motion due to respiratory, as tumor target merely moves along superior-inferior and dose measurement was only performed in the midpoint target. Therefore, a further experiment is required for measurement in the peripheral target.

4. Conclusions

The results indicated that the presence of tumor motion during dose delivery can generate unwanted dose discrepancies inside the tumor volume. The percentage of dose discrepancy for tumor target of tumor motion static, 9.3 mm and 20 mm are closed to tolerance level, while the tumor motion of 30 mm could not meet tolerance level criteria of ICRU (+7% to -5%) and AAPM (±5%) recommendation.

References

[1] Otto K 2008 *Medical Physics* **35** 310-317
[2] Liu H H *et al.* 2004 *International Journal of Radiation Oncology Biology Physics* **58** 1268-1279
[3] Merrow C E, Wang I Z, and Podgorsak M B 2012 *Journal of Applied Clinical Medical Phys.* **14**
[4] Swinnen A, Verstraete J and Huyskens D P 2002 *Radiotherapy and Oncology* **64** 317-326
[5] Ebert M *et al.* 2009 *Journal of Medical Imaging and Radiation Oncology* **53** 119-131
[6] IAEA 2000 *Lessons Learned from Accidental Exposures in Radiotherapy* (Safety Reports Series No. 17)
[7] IAEA 2001 *Investigation of an Accidental Exposure of Radiotherapy Patients in Panama*
[8] Dieterich S and Suh Y 2007 Springer 3-13
[9] Keall P J *et al.* 2006 *Medical Physics* **33** 3874-3900
[10] Bradley J D *et al.* 2013 *Results on Radiation Dose in RTOG 0617* (ASCO Annual Meeting Proceedings) p. 7501
[11] Larsson T 2010 *Accuracy of MLC-tracking for Inversely Optimized Arc Therapy Treatments of Varying Complexity for Two MLCs* (Lund University)
[12] McCarter S and Beckham W 2000 *Physics in Medicine and Biology* **45** 923
[13] Jiang S B, Pope C, Al Jarrah K M, Kung J H, Bortfeld T and Chen G T 2003 *Physics in Medicine and Biology* **48** 1773
[14] Berbeco R I, Pope C J and Jiang S B 2006 *Journal of Applied Clinical Medical Physics* **7**
[15] Boopathy R *et al.* 2010 *Journal of Medical and Biological Engineering* **30** 189-192
[16] Ong C, Verbakel W F, Cuijpers J P, Slotman B J and Senan S 2011 *International Journal of Radiation Oncology Biology Physics* **79** 305-311
[17] Ceberg S, Ceberg C, Falk M, af Rosenschöld P M, and Bäck S Å 2013 *Journal of Physics* (Conference Series) p. 012098
[18] Ozhasoglu C and Murphy M J 2002 *International Journal of Radiation Oncology Biology Physics* **52** 1389-1399
[19] Seppenwoolde Y *et al.* 2002 *International Journal of Radiation Oncology Biology Physics* **53** 822-834

Acknowledgement

This work was partially supported by research grants from Directorate on Research and Community Service, Universitas Indonesia 2015.

Dosimetry verification on VMAT and IMRT radiotherapy techniques: In the case of prostate cancer

A Maulana[1,2] and S A Pawiro[1]

[1] Physic Department, Faculty of Mathematics and Natural Science, Universitas Indonesia
[2] Directorate for Inspection on Radiation Facility and Radioactive Sources, National Nuclear Energy Regulatory Agency (BAPETEN)

E-mail: a.maulana@bapeten.go.id

Abstract. Radiotherapy treatment depends on the accuracy of the dose delivery to patients, the purpose of the study is to verify the dose in IMRT and VMAT technique in prostate cancer cases correspond to TPS dose using phantom base on ICRU No.50. The dose verification of the target and OAR was performed by placing the TLD Rod LiF100 and EBT2 Gafchromic film at slab hole of pelvic part of the Alderson RANDO phantom for prostate cancer simulation. The Exposed TLDs was evaluated using the TLD Reader Harshaw while EBT2 film was scanned using Epson scanner. The point dose measurements were compared between planned dose and measured dose at target volume and OAR. The result is the dose difference at target volume, bladder and rectum for IMRT and VMAT are less than 5%. On the other hand, the dose difference at the Femoral head is more than 5% for both techniques because the location of OAR already in low gradient dose. Furthermore, the difference dose of the target volume for IMRT technique tends to be smaller than VMAT either for TLD and EBT2 film detectors. From the measurement showed that the delivered dose on the phantom simulation match with ICRU No.50 criteria.

1. Introduction

Prostate cancer is a type of cancer that commonly occurs in the prostate gland, and usually affects men over the age of 50 years and above. In Indonesia, the number of patients with prostate cancer is in the third ranks after cervical and breast cancer whose number is estimated around 25.012 people [1]. Common treatment methods and efficiently used for the treatment of prostate cancer is radiotherapy to kill malignant cancer cells or reduce their growth [2].

The success of radiotherapy treatment planning systems depends on the accuracy of the dose delivery to patients, which includes determining the volume of cancer, determination of the dose to be delivered and how the treatment is done. ICRU No. 50 and 62 recommended that the dose uniformity in the target organ must be within uncertainties in the range of + 7% and -5% [3].

Several studies on dosimetry verification of prostate cancer explained that VMAT delivered more efficient technique with the maximum dose of the target cancer and minimal on healthy tissue could be done by (rectum, bladder and femoral head) [4, 5] and also faster with less Monitor Units compare to the IMRT technique [6, 7]. Klein *et al.* conducted a verification dose in the case of prostate cancer on IMRT and VMAT techniques using deformable anthropomorphic phantom and plastic scintillation detector explained that the differences of the measured dose of target volume compared to TPS at about 1.2% and 10.1% for IMRT and VMAT, respectively [8]. In addition Blais *et al.*, did the

Journal of Physics: Conference Series **694** (2016) 012010 doi:10.1088/1742-6596/694/1/012010

experiment using RANDO anthropomorphic phantom and detector TLD and obtained the difference measure dose of the target volume and planned dose around of 1.1% for IMRT techniques [9].

The purpose of the study is to verify the dose in IMRT and VMAT radiotherapy technique in prostate cancer cases correspond to planned dose (TPS) using phantom base on ICRU No.50.

2. Materials and methods

This work was performed at using Linear accelerator RapidArc Varian Clinac with 120 MLC which produce two X-ray photon beams 6 MV and 15 MV. In this study, we only used x-ray photon beams 6 MV during the experiment. The planning was done using Eclipse Treatment Planning System version 11.0 that capable of doing the planning of IMRT and VMAT techniques.

Prior to dosimetry In order to provide the additional data and evaluate the implementation of IMRT and VMAT, we proposed to do the verification, the calibration of TLD for x-ray 6 MV was done at water phantom to obtain a conversion factor using condition of field size 10 cm × 10 cm, 10 cm depth and 100 cm SAD. Prior to the experiment, the annealing process of TLD was carried out for one hour at the temperature of 400°C and 2 hours at the temperature of 100°C. The exposed TLDs with the single dose of 200 cGy were then compared to the readings with the ionization chamber detector under the same conditions.

Additionally, calibration of the Gafchromic EBT2 film was done to determine the pixel value conversion factor to dose. The Films were sandwiched between layers of solid water phantom at 10 cm depth and exposed using x-rays photon 6 MV with 100 cm SAD and field size of 10 × 10 cm^2. The films were irradiated with varied dose as indicated MU variation in the range of 20, 40, 80, 120, 160, 200, 240, 280, 320, 360 and 400. The films were then scanned using Epson scanners V700 with red channel and resolution of 72 dpi. After the image was saved in. TIFF format, the calibration curve was generated as a function between pixel value and dose.

The patient was simulated by RANDO phantom using Philips CT-Scan to get an image and transferred to radiotherapy planning system for IMRT and VMAT technique optimization. In TPS planning, PTV was simulated in the prostate, while OAR of this case are bladder, rectum, right femoral head and left femoral head. The total planned doses for target volume was about 80 Gy and given in 40 fractionation (2 Gy).

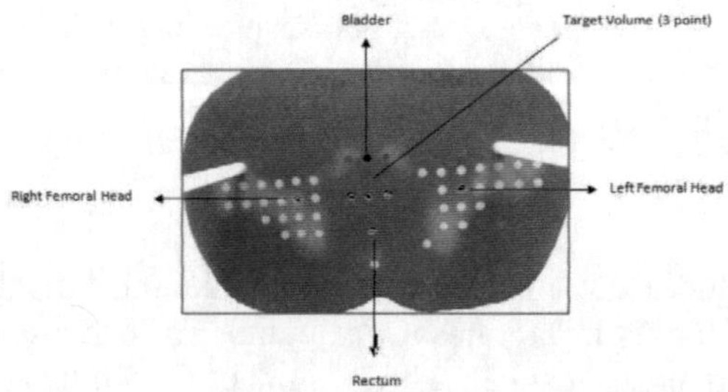

Figure 1. The measurement points in RANDO phantom for prostate cancer simulation.

For dose verification, the point measurements were done using TLDs and EBT2 Gafchromic films. Both detectors were inserted to RANDO phantom at the simulated target volume and organ at risk. We decided to put 3 points of the target and 4 points at OAR for Bladder, Rectum, right Femoral Head and left Femoral Head as illustrated in figure 1.

The orientation of TLDs at RANDO phantom were in coronal position while the EBT2 Gafchromic film will be varied on coronal and axial orientations. The films were cut into 1 × 1 cm^2 for axial orientation, and the size of film of coronal orientation was 1 × 0.5 cm^2.

After irradiation during IMRT and VMAT techniques, TLDs were read using TLD reader Harsaw and the films were scanned using Epson Perfection V700 using transmission mode, the red channel, and resolution of 72 dpi.

In order to evaluate and analyze the data, we used the formula as indicated in AAPM Task Group 119. The dose difference (Δ in %) was calculated using the Equation (1) as follows:

$$\text{Dose Difference (\%)} = \frac{D_{measured} - D_{plan}}{D_{plan}} \times 100\ \% \qquad \text{....(1)}$$

where $D_{measured}$ is measured dose in both detectors, D_{plan} is the planned dose at TPS. Afterward, the results of calculation were evaluated in accordance with ICRU recommendation and criteria.

3. Results and discussion

The point dose measurement of the target volume and organ at risk for IMRT technique are illustrated in figure 2. From the figures 3, the dose difference between point dose at the target between measurement and planning in the range of -0.46 to 6.62%, -2.04 to 6.49%, -1.46 to 6.26% for right peripheral, isocenter and left peripheral, respectively. In addition, the dose difference at OAR in the range of -0.60 to 5.57%, 2.75 to 3.87%, -13.26 to 11.18%, -21.40 to 8.01% for bladder, rectum, right femoral head and left femoral head, respectively.

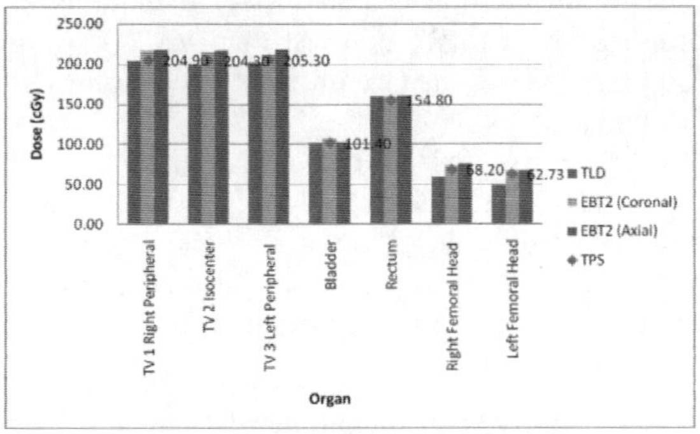

Figure 2. Measured dose of IMRT technique.

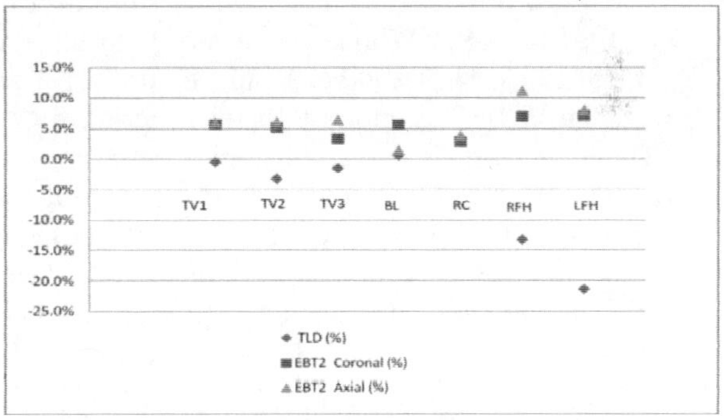

Figure 3. Dose difference of IMRT technique.

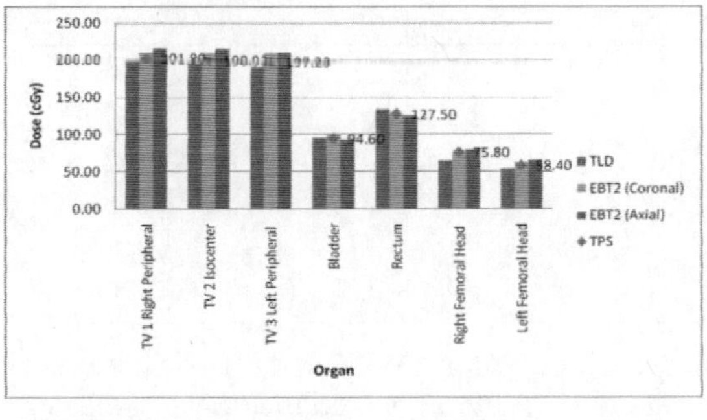

Figure 4. Measured dose of VMAT technique.

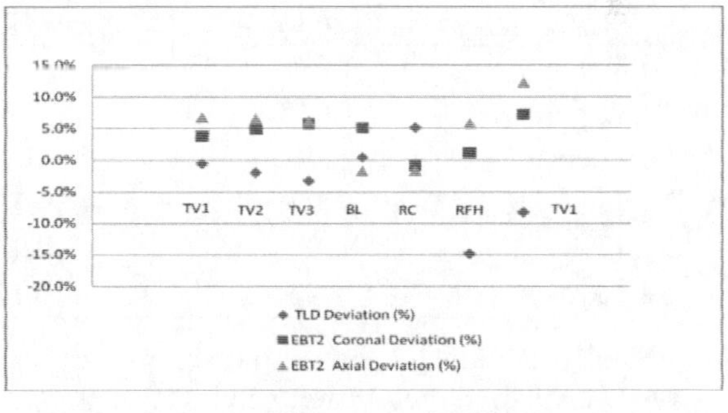

Figure 5. Dose difference of VMAT technique.

On the other hand, the point dose measurement of target and OAR for VMAT can be seen in figure 4. The figure 5 shows that the dose difference of target in the range of -0.58 to 6.68%, -2.03 to 6.46%, -3.31 to 6.13% for right peripheral, isocenter and left peripheral, respectively, while it was found at OAR in the range of -1.72 to 5.12%, -1.68 to 5.04%, -14.89 to 5.72%, -8.26 to 12.18% for bladder, rectum, right femoral head and left femoral head, respectively.

If we compare the results of this work with the previous study, we found that the pattern of dose difference of IMRT technique tends lower than VMAT technique in accordance with Blais works. Blais *et al.* showed the dose discrepancy for IMRT around of 1.1%. Additionally, Klein et. al showed that the difference dose around of 1.2% and 10.1% for IMRT and VMAT techniques, respectively.

From table 1, the dose at the target in VMAT technique has relative error tend to greater than dose deviation of IMRT technique. The hypothesis of this phenomenon is the difference fluence of VMAT happens during the treatment that means the detectors received the difference fluence during the measurement compared to IMRT technique. It means the point of the detector received difference fluence during the measurement, so it can be a source of higher uncertainties of the measurement during VMAT techniques. The use of gafchromic in this case has shown larger deviations than TLD. It can be happen due to the calibration process and angular dependent.

Table 1. Relative errors dose on target volume prostate cancer in IMRT and VMAT techniques.

Research	Detector	IMRT	VMAT
This Work	TLD	-1.69%	-1.99%
	EBT2 Coronal	4.53%	4.84%
	EBT2 Axial	6.03%	6.44%
David Klein [8]	Plastic Scintillation Detector	1.20%	10.10%
Adam Richard Blais [9]	TLD	1.10%	

The doses of both the bladder and rectum in IMRT technique are higher than VMAT technique for TLD, film and TPS. While, the dose of the right femoral head with IMRT is lower than VMAT using TLD and films. On the other hand, the dose of right and left femoral head with IMRT technology is lower than VMAT technology for TLD and film measurement.

The dose difference of bladder and rectum as less than 6% for all detectors, whereas dose difference of femoral head occurred in the range of 8 to 21%, 1 to 7%, 5 to 11% for TLD, EBT2 coronal and EBT2 axial. We may conclude those rectum and bladder doses are around 6%, because it is close to the PTV. However, the dose difference of the femoral head more than 6%, because the location of femoral head occurs in low dose gradient so it makes the higher relative errors on the measured dose compared to planned dose on the TPS.

Tabel 2. Surface dose estimation for IMRT and VMAT techniques.

Organ	TLD IMRT (cGy)	TLD VMAT (cGy)	EBT2 IMRT (cGy)	EBT2 VMAT (cGy)
AP pelvis	18.37	12.50	14.83	19.18
Right pelvis	12.40	17.65	18.17	20.34
Left pelvis	8.76	13.06	21.13	20.42
Tyroid	4.92	0.12	5.32	6.18
Right eye	4.93	0.13	5.67	5.67
Left eye	4.92	0.12	6.69	6.53

Finally, the table 2 shows the comparison of surface dose during IMRT and VMAT treatment for AP pelvis, right pelvis, left pelvis, thyroid, left and right eye. Table 2 showed that the surface dose of IMRT tend to be smaller than VMAT dose if we are using TLDs, but the dose of EBT2 film tends to be similar between IMRT and VMAT techniques.

4. Conclusion

From this study, it can be concluded that the delivered dose on the phantom simulation for the prostate cancer using IMRT and VMAT technique produced the dose difference at target volume for IMRT and VMAT are less than 5% and match with ICRU No.50 criteria.

References

[1] Kementrian Kesehatan RI 2013 *Data riset kesehatan dasar 2013* Balitbangkes
[2] Korhonen L 2009 *Methods for dose calculation and beam characterization in external photon beam radiotherapy* (Finland: Helsinki University of Technology) Dissertations p 197

[3] ICRU 1976 *Determination of absorbed dose in a patient irradiate by beams of X or gamma rays in radiotherapy procedures* ICRU Report 24 Bethesda Maryland

[4] Zhang P, Happersett L, Hunt M, Jackson A, Zelefsky M and Mageras G 2010 Volumetric modulated arc therapy: planning and evaluation for prostate cancer cases *Int J Radiat Oncol BiolPhys* **76** pp1456–62

[5] Davidson M T, Blake S J, Batchelar D L, Cheung P and Mah K 2011 Assessing the role of volumetric modulated arc therapy(VMAT) relative to IMRT and helical tomotherapy in themanagement of localized, locally advanced, andpost-operative prostate cancer *Int J Radiat Oncol Biol Phys* **80** pp 1550–58

[6] Leszczynski W, Slosarek K and Szlag M 2012 Comparison of dose distribution in IMRT and RapidArc technique in prostate radiotherapy *Rep Pract Oncol Radiother* **17** pp 347–51

[7] Hardcastle N, Tomé W A, Foo K, Miller A, Carolan M and Metcalfe P 2011 Comparison of prostate IMRT and VMAT biologically optimized treatment plans *Med Dosim* **36** 292–8.5

[8] Klein D, Briere T M, Kudchadker R, Archambault L, Beaulieu L, Lee A and Beddar S 2012 In-phantom dose verification of prostate IMRT and VMAT deliveries using plastic scintillation detectors *Radiation Measurements* **47** 921-929

[9] Blais A R, Lederer E, Oliver M and Leszczynski K 2012 Static and rotational step-and-shoot IMRT treatment plans for the prostate: A risk comparison study *Medical Physics* **39(2)**

Evaluation of 3D pre-treatment verification for volumetric modulated arc therapy plan in head region

S Ruangchan[1], **S Oonsiri**[1] **and S Suriyapee**[2]

[1] Division of Therapeutic Radiology and Oncology, Department of Radiology, King Chulalongkorn Memorial Hospital The Thai Red Cross Society, Bangkok, Thailand
[2] Department of Radiology, Faculty of Medicine, Chulalongkorn University, Bangkok, Thailand

E-mail: sirinya.bm@gmail.com

Abstract. The development of pre-treatment QA tools contributes to the three dimension (3D) dose verification using the calculation software with the measured planar dose distribution. This research is aimed to evaluate the Sun Nuclear 3DVH software with Thermo luminescence dosimeter (TLD) measurement. The two VMAT patient plans (2.5 arcs) of 6 MV photons with different PTV locations were transferred to the Rando phantom images. The PTV of the first plan located in homogeneous area and vice versa in the second plan. For treatment planning process, the Rando phantom images were employed in optimization and calculation with the PTV, brain stem, lens and TLD position contouring. The verification plans were created, transferred to the ArcCHECK for measurement and calculated the 3D dose using 3DVH software. The range of the percent dose differences in both PTV and organ at risk (OAR) between TLD and 3DVH software of the first and the second plans were -2.09 to 3.87% and -1.39 to 6.88%, respectively. The mean percent dose differences for the PTV were 1.62% and 3.93% for the first and the second plans, respectively. In conclusion, the 3DVH software results show good agreement with TLD when the tumor located in the homogeneous area.

1. Introduction

VMAT is the advance technique that developed from the intensity modulated radiation therapy (IMRT) technique. The VMAT parameters allow for variation of the MLC movement, dose rate and gantry rotation speed. These affect the plan to be more complicated. Therefore, the pre-treatment verification is very important to make sure that the modulated dose is accurate [1].

The pre-treatment verification by dose measurement is used to evaluate the treatment planning. Traditional of devices for doing this process are point dosimetry, planar dose distribution and 3D dose comparison. Previously, the reporting of verification treatment plan illustrated a planar dose distribution in criteria of 95% gamma pass ($\gamma \leq 1$) by using a 2D planar dose distribution. But it was found that the limitation in dose distribution did not relate to the patient dose [2].

Currently, the software can generate the dose measurement to 3D as DVHs to compare with treatment planning system (TPS) such as the 3DVH software. The 3DVH software is commercially available software program and used with the ArcCHECK device (both Sun Nuclear Corp., Melbourne, Florida, USA). The 3DVH software needs patient DICOM files (RT plan, RT dose, RT structures, and CT images) from TPS. The treatment plan is verified by creating a planned dose perturbation (PDP) file from errors in comparison between patient RT dose and ArcCHECK RT dose. The errors are convolved in ArcCHECK measurement data. The evaluation base on estimate dose to patient by overlay the resultant PDP beams on the patient CT images and show clinical impact using a DVHs analysis tool [3].

Since 2013, our previous studies were the evaluation of 3D commercial QA software in patient treatment plans. In some cases, we observed that dose in PTV from 3DVH software were lower than TPS [4]. These issues shall be addressed in this study. The purpose of this work is to evaluate the volume dose from 3DVH software comparing with the TLD measurement in Alderson Rando phantom of VMAT technique.

2. Material and methods

The TLD-100 were calibrated and used for measurement in this study. The two VMAT plans of Rando phantom in the head region were selected and verified by the 3DVH software. The studies were performed by volume dose comparison between measurement in TLD and 3DVH software for pre-treatment verification.

2.1. TLD characteristics

The TLD-100 (LiF:Mg, Ti) with 3.2 x 3.2 x 0.38 mm^3 were employed in this study. The 120 TLD chips were calibrated with 100 cGy of 6 MV photon beam. The 8 TLD chips were selected as a standard and 82 TLD chips were used for measurement in this study.

The dose response of TLD were investigated in 6 MV photon beam. The doses were varied from 0 to 300 cGy (0, 2, 5, 10, 20, 50, 100, 150, 200, 250 and 300 cGy). The parameters of measurement were 400 MU/min dose rate, 12x12 cm^2 field size and 100 cm source to detector distance. The dose linearity was evaluated by regression coefficients. Also, the dose minimum detectability of TLD was reported.

2.2. Treatment planning system

The Alderson Rando phantom was scanned by CT-scanner (LightSpeed RT GE Medical system, Waukesha, WI, USA) with 2.5 mm slice thickness in the head region. The two patient plans of VMAT treatment technique (2.5 arcs) in the head region of our previous study were selected for investigation [5]. The head Rando phantom images were registered with the two patient plans in the PTV, brain stem, lens and TLD position contouring. The head Rando phantom images were optimized and calculated in the Eclipse treatment planning software (Version 11.0.31, Varian Medical System, Palo Alto, CF, USA) by VMAT technique with the 6 MV photon beam, 2.5 arcs rotation and 54 Gy prescription dose in PTV (200 cGy dose per fraction).

2.3. Pre-treatment verification

2.3.1. TLD measurement. The two VMAT plans were transferred to the Varian ClinaciX linear accelerator (Varian Medical systems, Palo Alto, CA, USA) for measurement. The TLD ships were inserted into the Alderson Rando phantom in location of PTV, brain stem and lens. The Alderson Rando phantom was set on treatment couch with laser alignment and irradiated with the parameter from TPS. Also, the standard TLD chips were exposed with 105 MU at 100 cm source to detector distance and 12x12 cm^2 field size for daily output correction. The TLD ships were read by Harshaw TLD 5500 dosimeter reader and WinREMS software program. The mean organs dose in PTV, brain stem and lens were reported.

2.3.2. Arc Check measurement and 3DVH software calculation. The ArcCHECK phantom was placed on treatment couch at 86.65 cm source to detector distance. The calibrated dose of detector were performed before measurement with parameters; 6 MV photon beam, 10x10 cm^2 field size and 200 cGy as a 163 MU dose delivered. The two verification VMAT plans of head region were created in ArcCHECK phantom and transferred to measurement.

For 3DVH software, the Alderson Rando phantom DICOM flies (RT plan, RT dose, RT structures, and CT images) were imported. The DVHs and 3D gamma passed were generated by 3DVH software.

3. Results

3.1. TLD characteristics
The results of TLD dose linearity of 6 MV photon beams over 0 to 300 cGy delivered dose are shown in figure 1. The graphs plotted between collected TLD signal and delivered dose for 6 photon beams are shown with regression coefficients of 0.9994. The dose minimum detectability of TLD was 0.01 cGy.

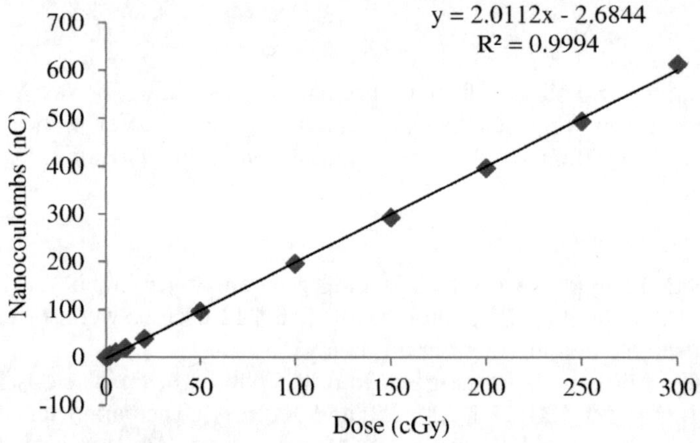

Figure 1. TLD dose response of 6 MV photon beam as function of delivered dose from 0 to 300 cGy.

3.2. Pre-treatment verification
Table 1-2 illustrate the percent dose differences between TLD and 3DVH software calculation, case No.1 and case No.2 were ranged from -2.09 to 3.87% and from -1.39 to 6.88%, respectively. The mean percent dose differences in PTV of case No.1 showed good agreement between TLD and 3DVH software. For case No.2, the mean dose difference in PTV was 3.93%.

Table 3-4 describe the percent dose differences from TLD in OARs which showed the large variation. The percent gamma passed with 3%/3mm criteria of 3DVH was 99.5% and 95.5% for case No. 1 and case No. 2, respectively.

Table 1. The percent dose differences between TLD and 3DVH software calculation in PTV of case No.1.

Organs	TLD No.	TLD ± SD (cGy)	3DVH (cGy)	%Dose diff.
	1	206.14±2.14	201.30	2.35
	2	205.65±2.32	201.70	1.92
PTV	3	205.90±4.85	205.10	0.39
	4	213.35±43.08	206.30	3.30
	5	204.92±5.80	209.20	-2.09

	6	214.55±26.00	207.20	3.43
	7	213.57±34.45	205.30	3.87
	8	205.52±3.29	206.40	-0.43
	Mean	208.70	205.31	1.62

Table 2. The percent dose differences between TLD and 3DVH software calculation in PTV of case No.2.

Organs	TLD No.	TLD ± SD (cGy)	3DVH (cGy)	%Dose diff.
PTV	1	208.18±2.09	198.80	4.51
	2	207.35±6.26	194.90	6.00
	3	204.65±3.62	198.60	2.96
	4	208.34±36.94	194.00	6.88
	5	205.10±4.82	197.10	3.90
	6	213.32±27.57	203.00	4.84
	7	175.27±28.01	177.70	-1.39
	8	200.11±2.35	196.00	2.05
	9	211.02±8.33	201.10	4.70
	Mean	203.70	195.69	3.93

Table 3. The percent dose differences between TLD and 3DVH software calculation in brain stem and lens of case No.1.

Organs	TLD No.	TLD ± SD (cGy)	3DVH (cGy)	%Dose diff.
Brain stem	9	2.57±1.69	5.40	-110.12
	10	5.79±3.36	5.80	-0.17
Lt. lens	11	2.79±1.07	3.70	-32.62
Rt. lens	12	3.05±0.14	4.30	-40.98

Table 4. The percent dose differences between TLD and 3DVH software calculation in brain stem and lens of case No.2

Organs	TLD No.	TLD ± SD (cGy)	3DVH (cGy)	%Dose diff.
Brain stem	10	35.80±20.84	33.70	5.87
Lt. lens	11	34.36±14.47	17.30	49.65
Rt. lens	12	36.55± 1.17	22.70	37.89

4. Discussion

The aim of this study is to evaluate the 3DVH software accurately predict the dose by direct comparing absolute dose to TLD measurement in Alderson Rando phantom. The characteristics of TLD were determined before making the measurement. The TLD signal increase linearly with the dose delivered with the regression coefficients of 0.9994 for 6 MV photon beams and 0.01 cGy dose detectability. The TLD is suitable for measurement in VMAT technique with 200 cGy dose per fraction.

For pre-treatment verification, the results show good agreement between measurement and calculation in PTV of case No.1. The result of case No. 2 shows lower dose about 3.93% from TLD measurement. These agree with our previous study that reported lower dose in 3DVH software of patient plans for tumor located in the inhomogeneity area [4]. The 3DVH software uses the dose error between measured in Arc CHECK and calculated from TPS with applied the PDP algorithm to obtain the 3D dose. The inhomogeneity is included in the dose error, special calculation is not taken. The size and shape of tumor and organ at risks also influence the dose report. For brain stem and lens, the result shows large variation. Some part of OARs is in the PTV resulted in high dose gradient. Also, the OARs have a small volume and locate in low dose region. This can effect to the response of TLD and may cause to misinterpretation.

By comparison of previous works, Olch J evaluated the accuracy of 3DVH software estimates of dose to virtual ion chamber and film that indicated the 3DVH system may accurately substitute for patient-specific IMRT QA [5]. However in VMAT technique, our results illustrating the 3DVH QA software is suitable to calculate in 3D pre-treatment verification as well.

5. Conclusions

The 3DVH software results show good agreement with TLD when tumor is located in the homogeneities area. The 3DVH software is suitable for reporting of dose differences in pre-treatment verification in VMAT technique of head region.

References

[1] Ezzell G, Galvin J, Low D, Palta J, Rosen I, Sharpe M, Xia P, Xiao Y, Xing L and Yu C 2003 *Med. Phys.* **30** 2089-2115
[2] Nelms B, Zhen H and Tome W 2011 *Med. Phys.* **38** 1037-43
[3] Carrasco P, Jornet N, Latorre A, Eudaldo T, Ruiz A and Ribas M 2012 *Med. Phys.* **39** 5040-49
[4] Ruangchan S 2013 *M.Sc. thesis of graduated studies program in Medical Physics* (Bangkok: Mahidol University) pp 87-94
[5] Olch J 2012 *Med. Phys.* **39** 81-86

Acknowledgements

We wish to thank Medical Physicists of Division of Radiation Oncology, Department of Radiology, King Chulalongkorn Memorial Hospital for their kind support and help in carrying out of this work.

Journal of Physics: Conference Series **694** (2016) 012012 doi:10.1088/1742-6596/694/1/012012

RTtxGap: An android radiobiological tool for compensation of radiotherapy treatment interruption

A L Yusoff[1,2], M Mohamad[1,2], R Abdullah[3], V M K Bhavaraju[1,2] and N R Nik Idris[1,2]

[1] Department of Nuclear Medicine, Radiotherapy & Oncology, School of Medical Sciences, Universiti Sains Malaysia, 16150, Kubang Kerian, Kelantan, Malaysia
[2] Hospital Universiti Sains Malaysia, Universiti Sains Malaysia, 16150, Kubang Kerian, Kelantan, Malaysia
[3] School of Health Sciences, Universiti Sains Malaysia, 16150, Kubang Kerian, Kelantan, Malaysia

E-mail: alutfi@usm.my

Abstract. Treatment interruption is not uncommon in radiotherapy. Common reasons for treatment interruption include machine breakdown, holidays and patient severe radiation reactions. Here RTtxGap, an Android application to assist calculations of compensation for treatment gap, is reported. It uses linear quadratic (LQ) model to calculate the biological effective dose (BED) that is used to solve for treatment gap compensations. Solutions are calculated using BED equation, with consideration for tissue proliferation. The accuracy of results has been verified using LQL Equiv software to be accurate within 1%. Five treatment interruption examples were used to illustrate the capability of the software to calculate the treatment compensation schedules. Solving these examples also illustrates the general consensus regarding compensating for unscheduled treatment interruptions, which ultimately involves balancing the BEDs of tumour and organ at risk. In addition to compensation for treatment gap, RTtxGap can also be used to calculate equivalent total dose in 2-Gy fraction (EQD2), to modify treatment schedule and to calculate alternative dose prescriptions having the same isoeffect.

1. Introduction

Treatment interruption is not uncommon in radiotherapy. Common reasons for treatment interruption include machine breakdown, holidays and patient severe radiation reactions. Prolongation of the overall treatment duration due to unscheduled treatment interruption can significantly affect local tumour control. In the case of head and neck tumour, a one day gap can be damaging with a reduction in local tumour control being estimated at 1.4% per day of missed treatment [1]. For recommendations on how to deal with the unscheduled treatment gaps, the reader is referred to The Royal College of Radiologists report [2].

Until sometime in the 90's, the unscheduled treatment interruptions were generally compensated using Orton and Ellis [3] nominal standard dose–time dose factor (NSD-TDF) tables. Following the introduction of the biological effective dose (BED) in 1989, which was originally formulated by Barendsen [4] as extrapolated response dose, the use of NSD-TDF table for treatment compensation has been criticized for being inaccurate. BED, a linear quadratic (LQ)-based formula with the time factor included, is given by

$$BED = nd\left(1 + \frac{d}{\alpha/\beta}\right) - \ln 2\,(T - T_k)/\alpha T_p \qquad (1)$$

Here, BED is related to n fractions of d Gy given in an overall time of T days in which tumour repopulation does not start until day T_k [5].

In the case where unscheduled treatment interruptions occur, equation (1) is used in current practice to calculate the required remaining schedule. However, calculating treatment compensation when many patients are involved can be a very taxing job. Here, RTtxGap, an Android application developed to assist these calculations based on Equation 1, is reported.

2. Methodology

RTtxGap was developed using JavaScript scripting language. It was compiled for Android devices using Intel® XDK. It uses LQ-model to calculate the BED, equivalent total dose in 2-Gy fraction (EQD2), treatment gap correction and alternative fractionation schedules. Solutions are calculated using Equation 1. To offer more flexibility, solutions are searched using iterative method instead of direct mathematical solving of the BED equation.

As stated by [6], the main difficulty in practical applications of the LQ-model is knowing what parameter values to use in individual calculations. However, more data has been available since that statement made in 2001. Many of this data are found compiled within the LQL Equiv software [7], a software developed collaboratively by the CHD Castelluccio radiotherapy unit in Ajac-cio and the University of Corsica. It is a free software released under the GNU license. It was decided that this data is used as it is in RTtxGap and the LQL Equiv software is used to verify results from RTtxGap.

Table 1. Verifications of BED and EQD2 calculation results from RTtxGap using LQL Equiv software. Notice that results calculated by RTtxGap are within 1% for BED values and are within 2% for EQD2. The higher EQD2 errors are due to the coarse iterative steps used in the RTtxGap program to arrive at the EQD2 solutions.

			Tumour BED			Tumour EQD2	
Target/OAR	Prescription	TtxGap	LQL-Equiv	% diff	RTtxGap	LQL-Equiv	% diff
Breast/lung	50Gy/25\#	80.52	80.5	0.02	50.0	50.0	0.00
	42.4Gy/16\#	81.11	81.0	0.14	50.0	50.3	-0.60
Prostate/rectum	70Gy/35\#	113.84	113.8	0.04	70.0	70.0	0.00
	60Gy/20\#	117.73	117.7	0.03	72.0	72.4	-0.55
NPC/std acute	60Gy/30\#	57.00	57.0	0.00	60.0	60.0	0.00
	57.2Gy/26\#	57.56	57.4	0.28	60.0	60.5	-0.83

			OAR BED			OAR EQD2	
Target/OAR	Prescription	TtxGap	LQL-Equiv	% diff	RTtxGap	LQL-Equiv	% diff
Breast/lung	50Gy/25\#	63.36	63.4	-0.06	50.0	50.0	0.00
	2.4Gy/16\#	66.77	66.5	0.41	52.0	52.5	-0.95
Prostate/rectum	70Gy/35\#	91.20	91.2	0.00	70.0	70.0	0.00
	60Gy/20\#	97.75	97.8	-0.05	74.0	75.0	-1.33
NPC/std acute	60Gy/30\#	72.00	72.0	0.00	60.0	60.0	0.00
	7.2Gy/26\#	69.78	69.8	-0.03	58.0	58.2	-0.34

Due to the small telephone screen, separate mini task is performed on separate tabs in RTtxGap. To use the software to calculate possible treatment compensation for unscheduled treatment gap, a user will first enter the tumour type and organ at risk (OAR) so that the LQ parameters, α and β values, for both tumour and OAR can be determined from lists of parameters stored in the program. Then, the original treatment dose prescription is entered. In the next tab, the user enters details for treatments that have already been delivered, including the number of days treatment has been missed. The number of days, instead of the number of fractions, missed should include weekends. This is to allow for greater flexibility of the software to calculate for cases with large treatment gaps. Although large treatment gaps are not generally advised in practice, it can be useful for academic purposes. In the final tab, the maximum number of fractions, the number of fractions per day, the allowed minimum and maximum dose per fraction, and treatment aim are entered. Four options for treatment aims are defined, namely conservative, aggressive, set maximum physical dose, set maximum OAR BED and set minimum tumour BED. For conservative aim, OAR BED is retained in the new compensated schedule, whereas tumour BED is retained in aggressive mode. If one wants to make sure that tumour BED does not fall below a certain percentage of the original planned value, set the minimum tumour BED option has to be used. One might evaluate this if upon choosing the conservative mode, the resultant tumour BED is found to be unacceptably low. By the same token, one can limit how high the OAR BED can be, in percentage of planned value, by using the last option in the aim.

When all needed information is entered, the calculate button at the bottom right of the screen will be enabled. Upon hitting the calculate button, the new remaining treatment schedule will be reported together with the total physical dose, the percent tumour BED and the percent OAR BED, both are with respect to the planned values, and their respective EQD2 values. The new schedule can be recalculated after changing any of the mentioned parameters. This can be used for fine tuning of the results. Results were verified using LQL Equiv software. Note that LQL Equiv calculates BED and EQD2 values for particular given schedules but it does not automatically find solutions for treatment interruptions. To illustrate the usage of RTtxGap in calculating compensations for unscheduled treatment interruptions, examples presented in reference [8] are reworked.

3. Results and discussion

The prescribed BED and EQD2 values for both tumours and organ at risk in three treatment cases, namely breast carcinoma, prostate carcinoma and nasopharyngeal carcinoma, are summarized in table 1. Results calculated by RTtxGap are within 1% for BED values and are within 2%for EQD2. The higher EQD2 errors are due to the coarse iterative steps used in the RTtxGap program to arrive at the EQD2 solutions.

Since table 1 data does not evaluate the capability of RTtxGap in finding solutions for treatment interruptions, results of calculations for the same tumour targets and organ at risks were evaluated. For these cases, seven days of unscheduled treatment gaps, i.e. 5 missed fractions, were introduced after the first ten fractions. The lost five fractions were compensated by having two fractions per day treatment for the first ten fractions following the gaps. Then RTtxGap were used to calculate the remaining fractions. The results of this are summarized in table 2. Again, results calculated by RTtxGap are within 1% for BED values and are within 2% for EQD2.

Table 2. Worked examples for cases where the treatments were missed for the whole third week. Compensations include 10 two-fraction per day treatments followed by regular treatment for the rest of the fractions. The resultant BEDs and EQD2 were confirmed using LQL Equiv software.

		Tumour BED			Tumour EQD2		
Target/OAR	Prescription	RTtxGap	LQL-Equiv	% diff	RTtxGap	LQL-Equiv	% diff
Breast/lung	50Gy/25\#	81.39	81.4	-0.01	50.0	50.6	-1.19
Prostate/rectum	70Gy/35\#	114.65	114.7	-0.05	70.0	70.5	-0.71
NPC/std acute	60Gy/30\#	57.15	57.2	-0.08	60.0	60.2	-0.33
		OAR BED			OAR EQD2		
Target/OAR	Prescription	RTtxGap	LQL-Equiv	% diff	RTtxGap	LQL-Equiv	% diff
Breast/lung	50Gy/25\#	63.43	63.5	-0.11	50.0	50.1	-0.20
Prostate/rectum	70Gy/35\#	95.67	95.7	-0.03	73.0	73.4	-0.54
NPC/std acute	60Gy/30\#	72.25	72.3	-0.07	60.0	60.2	-0.33

Dale *et al.* [8] has worked out five examples to illustrate usage of BED equation to solve for problems with unscheduled treatment gaps in nasopharyngeal carcinoma cases whose original prescriptions are 70 Gy/35 fractions. The first example deals with situation where unscheduled gap occurs for the whole

Journal of Physics: Conference Series **694** (2016) 012012 doi:10.1088/1742-6596/694/1/012012

third week into treatment. The second example deals with the lost of the whole sixth week of treatment whereas the third example deals with the lost of what was supposed to be the final week of treatment. The fourth example deals with lost of the last two treatment weeks. The last examples deal with lost of the final thirteen fractions. The OAR chosen for this case is the standard acute effect, i.e. linear quadratic model without consideration for tissue repopulation. This represents the worst case scenario for acute adverse effect of treatment that might disrupt treatment. The results are summarized in table 3.

Similar to Dale *et al.* [8] findings, the results indicate that for cases where the overall treatment duration can be maintained, it should be done so. Maintaining the total treatment duration make it possible to maintain both tumour and OAR BEDs without even having to actually calculate them. In our case, this is achieved through delivery of two fractions per day. It is also possible to use treatments over the weekend to maintain the total treatment duration, though this is not considered by RTtxGap. When it is no longer possible to maintain the total treatment duration, treatment compensations to maintain tumour BEDs inevitably translate into higher OAR BEDs. The effect can be minimized by minimizing the resulting extra treatment durations, again through multiple treatment fractions per day and/or treatment during weekends. If this is not possible, i.e. treatment can only be continued with single fraction per day as usual, fractionation schedule that results in lower OAR BED can be considered. This can be easily done in RTtxGap by choosing the third treatment aim, and set the maximum allowable values for the OAR BED. This, however, translates into a lower value for tumour BED that can compromise the efficacy of the treatment. A user can try different combinations of requirements to arrive at an optimum solution. A user, for example, can try setting the minimum required tumour BED to numbers slightly below 100% of the prescribed tumour BED and evaluate whether a better OAR BED can be achieved.

Table 3. Worked examples from Dale *et al.* [8] for the case of nasopharyngeal carcinoma. Descriptions of the examples can be found in the text. Here, RTtxGap aims are mostly selected to maintain tumour BED (aggressive aim) and whenever BED3 is mentioned, the aim is to not let OAR BED go beyond the values in the bracket.

			Tumour		OAR	
Example	RTtxGap aim	Compensation	%BED	EQD2	%BED	EQD2
Example 1	aggr(2\#/day)	10+10(2\#/day)+15	100.24	70.0	100.36	70.0
Example 2	aggr(2\#/day)	25+10(2\#/day)	100.24	70.0	100.36	70.0
Example 3	aggr(2\#/day)	30+5(2\#/day)	99.78	68.0	101.87	71.0
	aggr(1\#/day)	30+7	100.00	70.0	105.71	74.0
Example 4	aggr(2\#/day)	25+10(2.25 Gy/\# in 2\#/day)	99.79	71.0	102.43	71.0
	aggr(1\#/day)	25+14	100.00	70.0	111.43	78.0
	BED3 (106\%)	25+11(2.15 Gy/\#)	95.10	65.0	105.64	73.0
Example 5	aggr(2\#/day)	22+13(2.2 Gy/\# in 2\#/day)	100.04	70.0	104.86	73.0
	aggr(1\#/day)	22+16(2.15 Gy/\#)	99.65	68.0	112.61	78.0
	BED3 (107\%)	22+13(2.3 Gy/\#)	96.06	65.0	106.64	74.0

One has to be careful when high dose per fraction is used in treatment compensations. The higher sensitivity of the late-responding normal tissue to changes in dose per fractions, due to lower α-β ratio, means more biological dose proportion to normal tissues are given for higher dose per fraction is used [8].

The results of any BED calculations for radiotherapy treatment gaps should not be taken as an absolute indicator in clinical decision making. Instead, it should be used as a guide to be considered among other factors such as patient's response. One has to be aware that LQ-model, which these calculations are based on, is considered an oversimplification of the biological systems [8]. Moreover the data for variables in Equation 1 are scarce and are currently drawn from only limited number of publications [9]. As such, it is not 100% robust.

4. Conclusion

The values for planned tumour BED and OAR BED reported by RTtxGap are verified to be exactly the same as those obtained from LQL Equiv software. The BED values for the calculated new schedules are within 1% of the values calculated using LQL Equiv. It should be noted that LQL Equiv does not calculate a new schedule. Instead, new schedule is guessed by the user, from which BED values for both tumour target and OAR are calculated. For verification in this work, the guessing job is replaced by new schedule results from RTtxGap.

Worked out examples from nasopharyngeal carcinoma cases illustrates the general consensus regarding compensations for treatment such as the effect of overall time extensions, of the dose per fraction and the needs to balance tumour and OAR BEDs when treatment period has to be prolonged.

In addition to calculating for treatment gap compensation, RTtxGap can also be used for three other purposes, i. e. to calculate EQD2 for a prescribed dose, to modify treatment schedule for reasons such as finishing treatment before a holiday and to prescribe dose fractionation using alternative schedules, example of which includes calculating the hypofractionation dose for treatment of breast carcinoma.

It is important to note that because RTtxGap uses LQ-model for all its calculations, it should not be used when the dose per fraction is large, generally more than 5 Gy per fraction (5 Gy/#).Users should also take extra care when a long treatment gap is involved. Furthermore, it is assumed that complete repairs have taken place between fractions, which means that if two fractions per day is chosen, the fractions have to be separated by at least 6 hours.

References

[1] Fowler J F and Lindstrom M J 1992 *Int J Radiat Oncol Biol Phys* **23** 457
[2] Board of Faculty of Clinical Oncology 2008 *The timely delivery of radical radiotherapy: standards and guidelines for the management of unscheduled treatment interruptions, Third edition* (The Royal College of Radiologists)
[3] Orton G G and Ellis F 1973 *Br J Radiol* **46** 529
[4] Barendsen G W 1982 *Int J Radiat Oncol Biol Phys* **8** 198
[5] Fowler J F 2010 *Br J Radiol* **83** 554
[6] Jones B, Dale R G, Deehan C, Hopkins K I and Morgan D A 2001 *Clinical Onco* **13** 71
[7] Voyant C, Julian D, Roustit R, Biffi K and Lantieri C 2014 *Rep Pract Onco Radiothe* **19** 47
[8] Dale R G, Hendry J H, Jones B, Robertson A G, Deehan C and Sinclair J A 2002 *Clinical Onco* **14** 382
[9] Uzan J and Nahum A E 2012 *Br J Radiol* **85** 1279

Focal spot estimation of an Elekta dedicated stereotactic linear accelerator Monte Carlo model

S Herwiningsih[1, 2] and A Fielding[2]

[1] Physics Department, Faculty of Mathemamatics and Natural Sciences, Brawijaya University, Malang, 65145, Indonesia
[2] Science and Engineering Faculty, Queensland University of Technology, 2 George Street, Brisbane, 4000, Australia

E-mail: sri.herwiningsih@hdr.qut.edu.au

Abstract. The most challenging task in the Monte Carlo modelling of linear accelerators (linacs) is an accurate determination of the electron beam parameters striking the target which are characterised by the mean energy of incident electron beam and the electron beam shape, referred to as the focal spot. This work aims to determine the optimum focal spot size and shape of Elekta Axesse linac equipped with the Beam Modulator. A BEAMnrc Monte-Carlo linac model has been developed to produce a 6 MV photon beam. Different square field sizes of 2.4 cm, 4 cm and 10.4 cm were simulated in a simple water phantom with a source-to-surface distance of 100 cm. The simulation was performed with the incident electron beam energy of 6.2 MeV with the focal spot size varied between 0.1 and 0.3 cm with an increment of 0.05 cm. The field width (50% relative dose) and penumbra width (distance between 80% - 20 % relative dose) of the simulated profiles were compared with the measured profiles. This work found that an elliptical shape of the focal spot results in a better match with the measured data with the size of 0.2 cm in X-axis and 0.3 cm in Y-axis direction.

1. Introduction

There is an increasing trend of adopting stereotactic treatment for extracranial tumors such as early stage non-small cell lung cancer [1]. Its implementation has been supported by the introduction of high-resolution collimation systems which are able to generate high-resolution small fields of radiation [2]. Elekta has developed a Beam Modulator collimation system integrated into an Elekta Axesse linear accelerator which is specifically designed for stereotactic treatments [3].

Stereotactic radiotherapy requires accurate dose calculations for predicting and planning the radiotherapy treatment. The most powerful tool to accomplish this task is Monte Carlo simulation which has the capability to accurately model the particle transport through the linac head and phantom or patient geometry. This technique provides an independent verification, which is less labor intensive compared to a measurement based-technique.

However, the accuracy of the Monte Carlo modelling of the medical linac relies on the accurate determination of the electron source parameters, i.e. the energy of incident electron beam and the electron beam shape, referred to as the focal spot [4]. These parameters significantly influence the dose of the photon beam simulation [5]. It has been shown that the electron radial intensity distribution affects both the lateral dose profiles and the width of the penumbra [6, 7]. Determination of the radiation source parameters for photon beam modelling have been reported by Sheikh-Bagheri *et al.* [4] and Pena *et al.* [8]. In their studies, the focal spot was assumed to have a circular shape, which has been also adopted by Heydarian *et al.* [2] in their Elekta Synergy S linac model. However, other groups reported that an elliptical focal spot in the model produced a better agreement with the measured dosimetry for a 6 MV photon beam produced by the Elekta Synergy linac [6, 7, 9]. It indicates that different linac machines will have different performance as well as a different optimum value of the radiation source parameters although they have the same nominal energy value. Therefore, it is important to use the published radiation source parameters only as an initial approximation.

This work aims to determine the optimum focal spot size and shape of a Monte Carlo model of an Elekta Axesse linac equipped with the Beam Modulator which is designed for stereotactic treatments.

2. Material and methods

The Elekta Axesse linac with a built-in Beam Modulator collimation system was modelled using the BEAMnrc Monte-Carlo code to produce a 6 MV photon beam [10]. The patient-independent components of the linac head model were the same as the previously commissioned Elekta Precise Monte-Carlo linac model [11], however, the collimation system was completely different. The Beam Modulator consists of 40 leaf pairs with a leaf width of 4 mm at the isocenter. The leaf has no tongue-and-groove design with a curved leaf end. As part of the Beam Modulator design, movable jaws have been replaced with a paired fixed inner jaw and a paired fixed outer jaw. Therefore, the treatment field size is only defined by the multi-leaf collimators (MLCs). The linac model has been previously commissioned to determine the optimum electron-beam energy, which has been found to be 6.2 MeV.

The focal spot size was optimized by varying the full width at half maximum (FWHM) of the Gaussian radial intensity distribution from 0.1 to 0.3 cm at a fixed value of the electron beam energy. The optimization process used an initial assumption that the focal shape was circular. Dose profiles and penumbra matching were used to determine whether the optimized focal spot is circular-shaped or elliptical-shaped [12]. Three different square field sizes (2.4 cm, 4 cm and 10.4 cm) were simulated in a water phantom with a dimension of $60 \times 60 \times 60$ cm^3 with a source-to-surface distance of 100 cm. The lateral dose profiles were extracted at 10 cm depth and then compared with the measured dose profiles for the same depth. The dose profiles were normalized to the central axis dose for both simulation and measurement data. The field width was defined as the width between the 50% isodoses of the central dose profiles while the penumbra width was defined as the distance between 20% and 80% isodoses in the central dose profiles.

The Monte Carlo simulation consists of two stages. The first stage was the simulation of the photon beam through the treatment head components of the linac. The phase-space files were saved at a position just below the exit window of the linac, 55 cm from the target. These phase-space files were then used as the input for DOSXYZnrc simulations to produce the dosimetry data in the water phantom. The simulation used electron histories of 1×10^8 in the BEAMnrc simulation and 1-5 $\times 10^8$ particles in DOSXYZnrc simulation. An electron cutoff of 0.7 MeV and a photon cut-off of 0.01 MeV were used in the simulation. To improve the simulation efficiency, the directional Bremsstrahlung splitting was used as well as range rejection with ESAVE of 2 MeV.

3. Results and discussion

The results show that lateral dose profiles of the photon beam are sensitive to the change of the focal spot size. This effect is more obvious for the small field size. As shown in figure 1, increasing the focal spot size results in a broader penumbra shape, and a decrease in the horns of the lateral profile shoulder of 2.4 cm × 2.4 cm field. The FWHM of 0.2 cm gives a better match with the measured dosimetry data for the X-axis profile while the Y-axis profile shows the best match between measured and simulation for the FWHM of 0.3 cm.

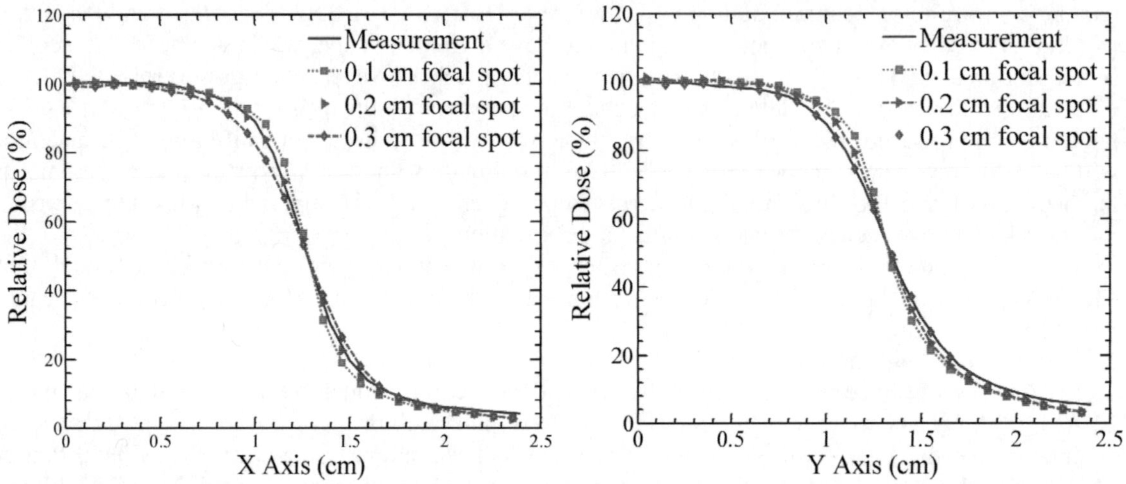

Figure 1. The inline lateral profiles (left) and the crossline lateral profiles (right) of 2.4 × 2.4 cm^2 field showing the sensitivity of the lateral profiles to the change of the focal spot size. The best match was obtained at the focal spot size of 0.2 cm for the inline profile and 0.3 cm for the crossline profile.

The penumbra matching has confirmed that elliptical-shaped focal spot shows a better match with the measurement data rather than a circular-shaped focal spot. Figure 2 shows that the measured penumbra intersects with the simulated penumbra at 0.2 cm for the X-axis and 0.3 cm for the Y-axis.

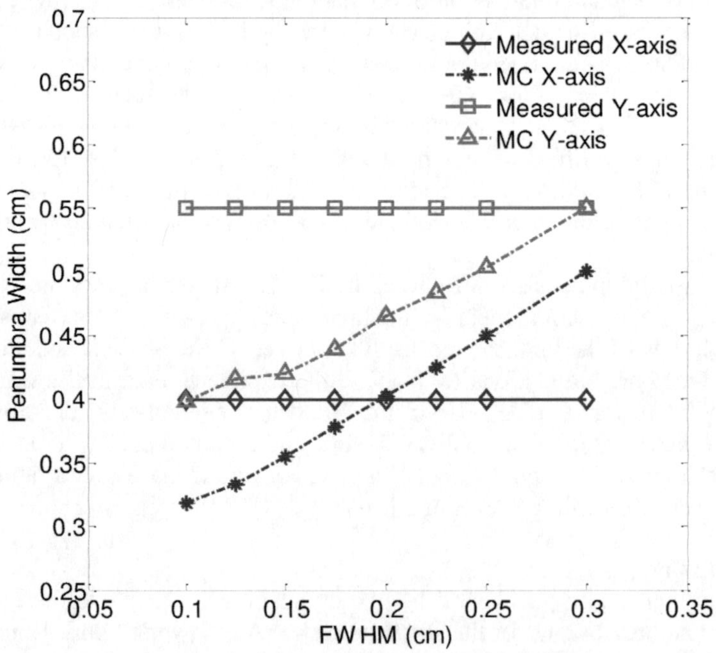

Figure 2. Penumbra matching of the measured (solid lines with markers) and simulated lateral profiles (dashed lines with markers) for a 2.4 × 2.4 cm^2 field with variable electron beam FWHM.

Similar results have been found for 4 × 4 cm^2 field and 10.4 × 10.4 cm^2 field (figure 3), where the elliptic-shaped focal spot yields a better agreement with the measurement data. The optimum value is 0.2 cm in X-axis and 0.3 cm in Y-axis direction. The uncertainty of the Monte Carlo simulation is 0.4

% for the small fields and 0.6% for the large field (10.4 cm × 10.4 cm) over the flat region of the dose profiles.

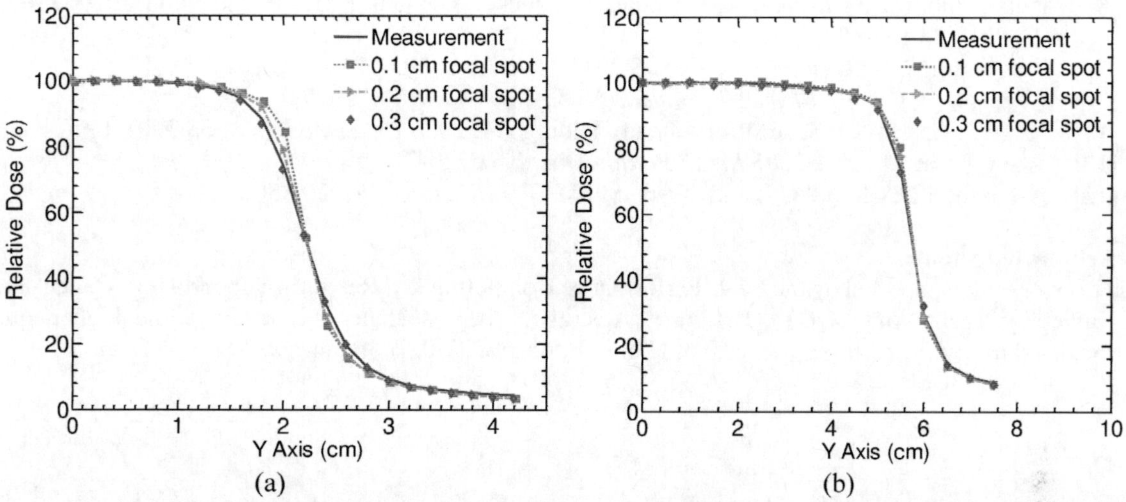

Figure 3. The crossline lateral profiles of 4 cm × 4 cm field (a) and 10.4 cm × 10.4 cm field (b). The focal spot size of 0.3 cm shows better match with the measured data.

The sensitivity of the small field size to the incident beam radial intensity distribution has been also reported by Pena *et al.* [8] and Keall *et al.* [5]. However, their models used an assumption that the electron beam radial intensity has an equal size in both X and Y axes. Similarly, Heydarian *et al.* [2] and Asnaashari *et al.* [13] used a circular-shaped focal spot size with an optimum value of 0.11 cm in their Elekta Synergy S Monte-Carlo model which also has a built-in Beam Modulator collimation system. However, Podder *et al.* [7] found in their study that the penumbra shape of the Elekta Synergy S is different between the leaf-end and the leaf-side penumbra, suggesting a focal spot asymmetry. The elliptical-shaped focal spot approach has been also used by Francescon *et al.* [9] and Almberg *et al.* [6] in their Elekta Synergy S model.

As the Elekta Axesse linac has a similar design as the Elekta Synergy, it is assumed that its radiation source parameters would be the same. We found that an elliptical-shaped focal spot gave the better match with the measurement data for the Elekta Axesse linac, however, the optimum spot size found in this study is slightly larger than that was reported by Franceson *et al.* [9], which is FWHMx = 0.2 cm and FWHMy=0.09 cm.

4. Conclusion

The Monte Carlo simulation has been employed to estimate the focal spot size of the Elekta Axesse linac dedicated for the stereotactic treatment. The study showed that an elliptical-shaped focal spot results in a good agreement with the measurement in both inline and crossline direction. The optimum focal spot size has a size of 0.2 cm in the in-line direction and 0.3 cm in the cross-line direction. The model will be further employed for studying clinical stereotactic treatments involving small field sizes.

References

[1] Tipton K N, Sullivan N, Bruening W, Inamdar R, Launders J, Uhl S and Schoelles K 2011 *Technical Brief No. 6 Stereotactic Body Radiation Therapy* (Rockville: Agency for Healthcare Research and Quality)

[2] Heydarian M, Asnaashari K, Allahverdi M and Jaffray D 2008 *Med. Phys.* **35** 3943-54

[3] Patel I, Glendinning A G and Kirby M C 2005 *Phys. Med. Biol.* **50** 5479-92

[4] Sheikh-Bagheri D and Rogers D W O 2002 *Med. Phys.* **29** 391

[5] Keall P J, Siebers J V, Libby B and Mohan R 2003 *Med. Phys.* **30** 574

[6] Almberg S S, Frengen J, Kylling A and Lindmo T 2012 *Med. Phys.* **39** 40-7
[7] Podder T K, Bednarz G, Yu Y and Galvin J M 2011 *Phys. Medica* **27** 52-7
[8] Pena J, González-Castaño D M, Gómez F, Sánchez-Doblado F and Hartmann G H 2007 *Med. Phys.* **34** 1076-84
[9] Francescon P, Satariano N and Cora S 2011 *Med. Phys.* **38** 6513-27
[10] Rogers D W O 2002 *Physics in Canada* **58** 53-71
[11] Kairn T, Cassidy D, Sandford P M and Fielding A L 2008 *Phys. Med. Biol.* **53** 3903-19
[12] Wang J Z and Li X A 2005 *Int. J. Radiat. Oncol.* **61** 220-7
[13] Asnaashari K, Chow J C L and Heydarian M 2013 *Phys. Medica* **29** 350-6

Acknowledgments
This work was supported by the High-Performance Computing and Research Support Unit, Queensland University Technology (QUT), Brisbane, Australia. The researcher has received the high degree sponsorship from Directorate General of Higher Education (Dikti), Indonesia.

Journal of Physics: Conference Series **694** (2016) 012014 doi:10.1088/1742-6596/694/1/012014

The implementation in-house dose verification for IMRT and VMAT on breast cancer and NPC cases

A Azzi[1], F Y Dimitri[2] and S A Pawiro[1]

[1] Medical Physics and Biophysics Division, Department of Physics, Faculty of Mathematics and Natural Sciences, University of Indonesia, Depok, 16424, West Java, Indonesia
[2] Siloam Hospital TB Simatupang, South Jakarta, 12410, DKI Jakarta, Indonesia

E-mail: akbar.azzi@sci.ui.ac.id

Abstract. IMRT and VMAT have been widely implemented on breast cancer and NPC cases. Dose verification was conducted to ensure the comparison between dose delivered and prescribe dose in TPS to obtain optimal results and the dose delivered. The gafchromic films measurement was done for breast and NPC simulation on RANDO Phantom, irradiating film by 6 MV photon with 200 cGy and scanned with EPSON V700 using 72 dpi. Dose distribution was calculated using the MATLAB-based in-house algorithm. In order to evaluate the gamma index, gafchromic films and MatriXX 2D array were placed in the Multi Cube and irradiated with EPID in same position. Dose distribution differences for IMRT and VMAT with modalities to TPS on film was 5.34% and 6.68% for NPC PTV70; 12.81% and 6.15% for NPC PTV50; 11.14% and 10.80% for breast cancer. Gamma index differences on IMRT with modalities to MatriXX 2D array and EPID was -0.09% and 5.13% for NPC; 0.51% and 2.28% for breast cancer. On the other hand, gamma index difference on VMAT was -1.65% and 1.43% for NPC; 0.19% and 4.38% for breast cancer. This measurement method can be used to verify dose for breast cancer and NPC cases in IMRT and VMAT.

1. Introduction

Radiotherapy is the primary technique for breast cancer and nasopharyngeal carcinoma (NPC) treatment and it could be combined with surgery on breast cancer and chemotherapy on NPC. In NPC planning, there were three CTVs defined, CTV70 for giving a high dose, a CTV60 for reflecting the high risk of local spread in and adjacent to the nasopharynx, and CTV50 for handling nodes around the target. However, the bolus is commonly used in the breast cancer treatment to increase the thickness of the target volume and the usage of high energy photons should be avoided to reduce skin dose [1].

Intensity modulated radiation therapy (IMRT) and volumetric modulated arc therapy (VMAT) are modern techniques in radiotherapy that not only could deliver a high dose to the tumor target precisely but also protected organs and normal tissues nearby the target volume [2]. Dose verification is a series of test to give evidence that the given exposure parameters consistent with the approved plan. Verification on IMRT and VMAT could be performed prior to the treatment (pre-treatment measurements) which are divided into fluence verification and dose verification. Verification is normally done by comparing the dose calculation in irradiated phantom to Treatment Planning System (TPS) [3]. There are two ways of verification, qualitatively and quantitatively. Isodose distribution was the common qualitative evaluation method and gamma index was a quantitative one. Gamma index is a merger of dose difference (DD) and dose to agreement (DTA) in dose comparison. It was using two completion of criteria, usually 3 mm for spatial tolerances and 3% for dose differences [4]. The gamma index concept has been widely used to compare two dose distribution on cancer radiotherapy. The original gamma index of Low *et al.* [4] has been improved for better accuracy and efficiency [5].

In this study, dose distribution verification was done on breast cancer and NPC cases with IMRT and VMAT techniques using gafchromic EBT2 films, TLDs, MatriXX 2D array detector, and EPID. Licensed price of radiotherapy dose verification software for film dosimeter has an expensive price, GUI-MATLAB in-house algorithm intent on optimized the calculation of absorbed dose on gafchromic film and minimized dose verification fee.

2. Method

This work was done at Siloam Hospital TB Simatupang using Linac Varian Trilogy which produce high energy photon beams 6 MV and 15 MV. The methodology was divided into treatment planning system, irradiated RANDO phantom, and Multi Cube, and then measured dose distribution using MATLAB, FilmQA Pro, and ImageJ programs. In this study, the measurement was done using Gafchromic EBT2 films (ISP Technologies), EPID, and MatriXX 2D array (IBA Dosimetry) and TLDs as a comparison detector.

2.1. Treatment planning system

Head and chest of RANDO phantom were scanned with computed tomography (CT) simulator with 3 mm slice thickness to get the images. Additional RANDO breast and chest wall were used for breast cancer case. These images were imported to Eclipse ver.10.0. Planning Target Volume (PTV) was created according to clinical experience of target volume contouring in our hospital. The organ at risks (OARs) delineated included spinal cord, brainstem, optic nerves, eyes, and parotid glands for NPC case[6], while spinal cord, lungs, and heart for breast cancer. The IMRT plan in NPC case was generated with nine irradiation fields with equal-spaced gantry angles of 30°. The irradiation fields in breast cancer were consist of seven fields with half rotation from 305° to 125°. The VMAT double arc plans were same with both of The IMRT plans. The irradiation planning was given 70 Gy at NPC and 50 Gy at breast cancer for total treatment or 200 cGy per fraction [7].

2.2. Calibration

For dose assessment, the gafchromic films and TLDs was calibrated using x-ray 6 MV. The experiment used the slab phantoms with the dimension of 30 cm × 30 cm × 1 cm, Source to Axis Distance (SAD) 100 cm, 5 cm depth, and 10 cm × 10 cm field size. Gafchormic films were cut into 4 cm x 4 cm dimension [8] and irradiated at 14, 35, 69, 104, 139, 173, 208, 242, and 277 cGy. Calibration for TLDs was done using single dose for each TLD at 177 cGy. For the readout of TLDs, a TLD reader model Harshaw 3500 was used with a purified N_2. Readings were taken after 48 hours of irradiation. From TL outputs, the individual calibration factors for TLD were determined in terms of μC/cGy to be used to evaluate absorbed doses from their TL outputs.

2.3. Irradiating RANDO phantom and multi cube

The evaluation of dose distribution for IMRT and VMAT planning has been done with dose measurement at RANDO phantom and Multi Cube. Film gafchromic for dose measurement in NPC case was cut with a dimension of 7 cm × 7 cm, whereas it was cut with a dimension of 1 cm × 1 cm for OARs and 4 mm × 20 mm for in spinal cord area. On the other hand, film gafchromic was used on breast dose assessment with a dimension of 6 cm × 6 cm. it was inserted in between two slices of left RANDO breast phantom. For justification of dose measurement, we also used the TLDs and inserted into the hole of RANDO breast. In order to obtain gamma index analysis, gafchromic films with a size of 20 cm x 25 cm was taken placed in the Multi Cube at 11 cm depth. The film was then irradiated by

Journal of Physics: Conference Series **694** (2016) 012014 doi:10.1088/1742-6596/694/1/012014

using IMRT and VMAT in every case of breast cancer and NPC based on radiotherapy planning for RANDO phantom.

2.4. Dose measurement

Irradiated films were scanned with EPSON V700 with 72 dpi after 24 hours. The dose distribution and gamma index analysis was measured using Matlab in-house algorithm. The algorithm calculated the dose distribution base on red channel calibration curve. Pixel value from the scanned film was converted to dose with interpolation referred to the calibration curve. For justification of dose measurement, we also used the FilmQA Pro, ImageJ programs, and TLDs. Gafchromic films were placed in the same position as MatriXX 2D array and measured with MATLAB in-house algorithm to evaluate the gamma index. Furthermore, we used I'mRT OmniPro and EPID programs to evaluate the gamma index of in-house algorithm MATLAB.

3. Result and discussion

The calibration curve was generated by plotting the pixel value of center irradiated calibration film and delivered dose as shown in figure 1. The dose distribution was measured with in-house algorithm and then cumulative dose histogram of the target volume can be generated after Region of Interest (ROI) was defined as described in figure 2.

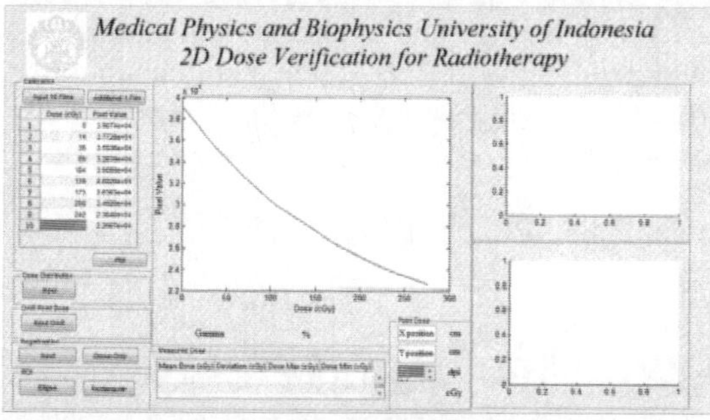

Figure 1. MATLAB in-house calibration curve.

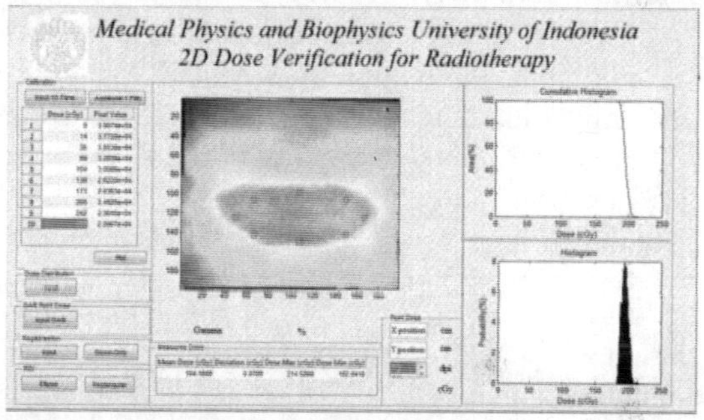

Figure 2. Dose distribution on NPC PTV70 case.

Table 1. Dose differences for IMRT and VMAT with modalities to TPS.

Target	Irradiated Technique	Dose (cGy)					Dose Difference (%)			
		MATLAB	FilmQA Pro	ImageJ	TPS	TLD	MATLAB and TPS	FilmQA Pro and TPS	ImageJ and TPS	TLD and TPS
Breast	IMRT	183.05 ± 7.90	175.50 ± 9.50	185.41 ± 7.53	206.00	191.47 ± 3.36	11.14	14.80	10.00	7.06
	VMAT	185.36 ± 4.86	175.00 ± 6.30	187.63 ± 5.66	207.80	202.12 ± 2.23	10.80	15.78	9.71	3.36
NPC PTV70	IMRT	193.11 ± 4,99	180.40 ± 4.30	196.45 ± 5.09	204.00	-	5.34	11.57	3.70	-
	VMAT	191.31 ± 7.76	176.70 ± 4.50	194.42 ± 5.04	205.00	-	6.68	13.80	5.16	-
NPC PTV50	IMRT	142.49 ± 4.30	135.20 ± 3.00	141.82 ± 4.84	163.43	166.31 ± 2.99	12.81	17.27	13.22	1.76
	VMAT	169.84 ± 3.16	158.00 ± 7.20	149.52 ± 5.26	160.00	162.56 ± 3.62	6.15	1.25	6.55	1.60

From table 1, dose distribution measurement was done with IMRT and VMAT techniques on breast cancer, NPC PTV70, and NPC PTV50. Dose measurement on NPC PTV70 cases can not be done using the TLDs because there is no hole in the RANDO phantom. The average dose difference of in-house algorithm to TPS for all cases was 8.82%. Meanwhile, the dose discrepancy of filmQA Pro, ImageJ, and TLDs to TPS were 12.41%, 8.06%, and 3.45%, respectively. The maximum deviation between the film of in-house algorithm measurement and TLD was found to be 11.05%. Homogeneity in the ROI was affected dose distribution measurement on gafchromic films while the small active region of TLDs was received the appropriate dose of the planning.

Dose differences at OARs around of target volume at NPC for all software to TPS were given in table 2. The average dose difference of in-house algorithm, FilmQA Pro, and ImageJ programs to TPS was 5.98%, 4.27%, and 6.68%, respectively. Tong *et al.* [9] mentioned that the limitation of the dose at spinal cord was 45 Gy for total treatment or 128.57 cGy per fraction. The average absorbed dose for parotid ideally 74.29 cGy per fraction and for eyes was 17.14 cGy per fraction for 70 Gy total dose treatment with IMRT modalities. Based on these references, the dose distribution of the organs at risk was acceptable.

In order to obtain gamma index value, scanned films were compared to planning in DICOM file format by in-house algorithm program as shown in figure 3. Tabel 3 listed gamma index values for each tumor target and radiation technique, and gamma index differences of in-house algorithm with modalities to MatriXX 2D array and EPID.

A positive value on gamma index differences indicated that the analysis of gamma index from the gafcromic film was better than MatriXX and EPID. Generally, the difference in gamma index of gafchromic films had positive values compare to MatriXX 2D array and EPID. The gamma index result of table 3 proved that gafchromic films had a good sensitivity and spatial resolution.

Table 2. Dose differences on OAR in NPC cases.

Irradiated Technique	OAR	Dose (cGy)				Dose Difference (%)		
		MATLAB	FilmQA Pro	ImageJ	TPS	MATLAB and TPS	FilmQA Pro and TPS	ImageJ and TPS
IMRT	Eye	3.06 ± 2.23	3.10 ± 2.30	3.18 ± 0.38	2.98	2.62	4.03	6.71
	R.P[a]	45.27 ±11.48	45.80 ± 8.20	48.74 ± 7.35	45.09	0.40	1.57	8.09
	L.P[b]	48.16 ± 1.87	43.30 ± 2.40	45.98 ± 1.99	44.20	8.96	2.04	4.03
	SC[c]	80.10 ± 6.73	85.20 ± 3.40	83.42 ± 5.41	82.46	2.86	3.32	1.16
VMAT	Eye	2.75 ± 1.30	3.20 ± 1.00	2.61 ± 0.51	3.02	8.94	5.96	13.58
	R.P[a]	45.39 ± 1.55	44.20 ± 4.70	44.13 ± 1.47	48.57	6.55	9.00	9.14
	L.P[b]	52.91 ± 6.87	50.80 ± 9.70	48.13 ±18.25	47.40	11.62	7.17	1.54
	SC[c]	97.33 ±10.79	102.30 ±11.10	93.94 ±10.27	103.42	5.89	1.08	9.17

[a] R.P= Right Parotid.
[b] L.P = Left Parotid.
[c] SC = Spinal Cord.

Table 3. Gamma index evaluation.

Target	Irradiated Technique	Gamma Index (%) Film[a]	MatriXX[b]	EPID	Gamma Index Difference (%) Film and MatriXX	Film and EPID
NPC	IMRT	98.15	98.24	99.80	-0.09	-1.65
	VMAT	99.85	94.72	98.40	5.13	1.45
Breast	IMRT	99.99	99.48	99.80	0.51	0.19
	VMAT	99.58	97.30	95.20	2.28	4.38

[a] Film was calculated using MATLAB in-house algorithm.
[b] MatriXX 2D array was calculated using I'mRT OmniPro program.

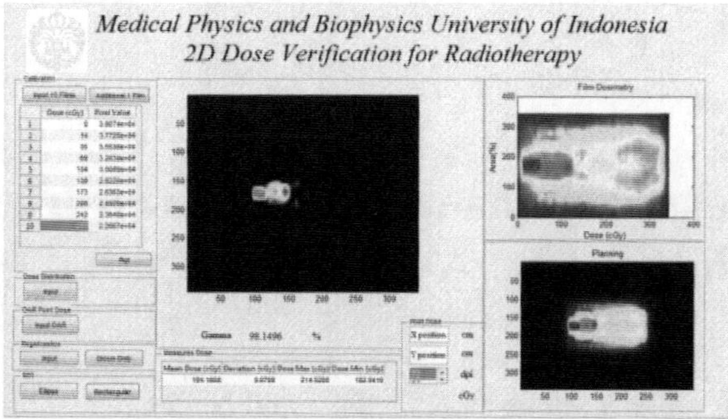

Figure 3. Gamma index calculation on NPC with IMRT technique.

4. Conclusion

This study presents a method in which a gafchromic EBT2 film, TLD, ionization chamber, and EPID detectors were employed to evaluate the dose distribution of LINAC IMRT and VMAT techniques on breast cancer and NPC cases. The MATLAB-based in-house algorithm was implemented for dose verification with differences to TPS of 11.14% and 10.80% for breast cancer; 5.34% and 6.68% for NPC PTV70; 12.81% and 6.15% for NPC PTV70. Gafchromic films can be successfully used for gamma index evaluation with the in-house algorithm. Gamma index differences had a maximum value at 5.13% with modalities to MatriXX and 4.38% for EPID. This measurement method can be used to verify dose for breast cancer and NPC cases in IMRT and VMAT while the constructed GUI assists the researcher in performing the measurement with ease.

References

[1] Barrett A, Dobbs J, Morris S and Roques T 2009 *Practical Radiotherapy Planning* vol 4, ed G Jamieson (Italy: Hodder Arnold) pp 160-278
[2] Lin H, Huang S, Deng X, Zhu J and Chen L 2014 *Radiation Oncology* **9** 71
[3] Nyholm T 2008 Verification of Dose Calculation in Radiotherapy (Sweden: Umeå University) p 25
[4] Low D A, Harms W B, Mutic S and Purdy J A 1998 *Med. Phys.* **25** 656–61
[5] Xuejun G, Xun J and Jiang S B 2011 *Physics in Medicine and Biology* **56** 1431
[6] Zhao N, Yang R, Jiang Y, Tian S, Guo F and Wang J 2015 A hybrid IMRT/VMAT technique for the treatment of nasopharyngeal cancer *BioMed Research International* 2015 940102
[7] Uniyal S C, Naithani U C and Sharma S D 2011 *Report of Practical Oncology and Radiotherapy* **16** 14-20
[8] Setiadi A R 2013 Pengembangan algoritma untuk verifikasi distribusi dosis radioterapi berbasis citra 2D gafchromic (Indonesia: Universitas Indonesia) pp 24-25
[9] Tong W N, Michael C R, Lee C H, Wai M H, Cheuk W C, Kin C L, Oscar S H, Chan F R, Anne W M and Lee F R 2011 *Int. J. Radiation Oncology Biol. Phys.* **79(2)** 420-8

Journal of Physics: Conference Series **694** (2016) 012015 doi:10.1088/1742-6596/694/1/012015

Dose profile measurements during respiratory-gated lung stereotactic radiotherapy: A phantom study

W L Jong[1], J H D Wong[2,3], K H Ng[2,3] and N M Ung[1]

[1] Clinical Oncology Unit, Faculty of Medicine, University of Malaya, Kuala Lumpur, Malaysia
[2] Department of Biomedical Imaging, Faculty of Medicine, University of Malaya, Kuala Lumpur, Malaysia
[3] University of Malaya Research Imaging Centre, Faculty of Medicine, University of Malaya, Kuala Lumpur, Malaysia

E-mail: nmung@ummc.edu.my

Abstract. During stereotactic body radiotherapy, high radiation dose (~60 Gy) is delivered to the tumour in small fractionation regime. In this study, the dosimetric characteristics were studied using radiochromic film during respiratory-gated and non-gated lung stereotactic body radiotherapy (SBRT). Specifically, the effect of respiratory cycle and amplitude, as well as gating window on the dosimetry were studied. In this study, the dose profiles along the irradiated area were measured. The dose profiles for respiratory-gated radiation delivery with different respiratory or tumour motion amplitudes, gating windows and respiratory time per cycle were in agreement with static radiation delivery. The respiratory gating system was able to deliver the radiation dose accurately (±1.05 mm) in the longitudinal direction. Although the treatment time for respiratory-gated SBRT was prolonged, this approach can potentially reduce the margin for internal tumour volume without compromising the tumour coverage. In addition, the normal tissue sparing effect can be improved.

1. Introduction

Stereotactic body radiotherapy (SBRT) refers to hypofraction radiotherapy where high radiation dose (~60Gy) was delivered in small fractionation regime [1]. Treating a lung tumour is a challenging task because of the moving tumour due to the respiratory motion. Erridge *et al.* has reported a mean range of lung tumour motion of 12.5 mm in superior-inferior (S-I) direction [2]. Another study also found that tumour motion of 5.8 mm in S-I direction [3]. According to AAPM Task Group 76, it was recommended that tumour with motion more than 5 mm should be treated with gating or tracking method [4]. This is so that the involvement of normal tissue with high dose can be reduced without compromising the coverage of the moving tumour [5].

In this work, we studied the dose profiles during respiratory-gated SBRT on a gating phantom. Specifically, the effect of respiratory or tumour motion amplitudes, respiratory cycle times and gating window on the dosimetry was studied. In addition, the tracking accuracy of the gating system was assessed.

2. Materials and methods

A linear accelerator (Novalis Tx, Varian Medical System, Palo Alto, CA) equipped with BrainLAB Exactrac gating system and BrainLAB ET Gating Phantom (BrainLAB AG, Heimstetten, Germany) were used in this study. A detailed description of the BrainLAB Exactrac gating system can be found in Tenn *et al.* [6].

A 15 × 15 × 15 cm^3 Perspex phantom embedded with a 7.5 mm diameter spherical fiducial marker at the centre was mounted on the gating phantom [figure 1(a)]. The fiducial marker represented the position of the target tumour during irradiation. The phantom was set to move and mimic normal respiratory pattern. The horizontal motion mimicked the lung tumour motion in the S-I direction. The vertical motion mimicked the chest wall motion due to respiration.

Five infrared (IR) markers were used as surrogates on the phantom to track the respiratory motion for the gated-radiation delivery. In this study, a sine function was used to simulate the respiratory pattern [figure 1(b)], and no offset was set between the vertical and horizontal motion of the phantom [figure 1(a)]. The radiation was delivered only at the exhalation phase within the gating window.

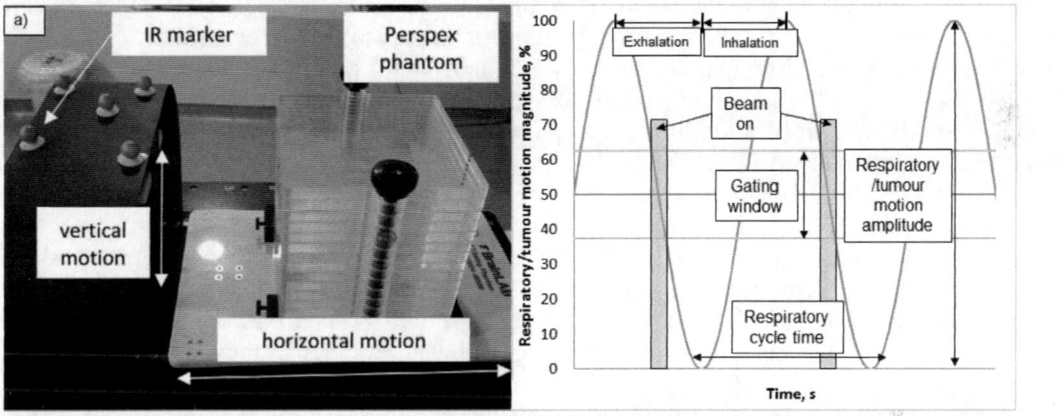

Figure 1. (a) The BrainLAB ET gating phantom and (b) respiratory and tumour motion as a sine function. The dashed line shows the level of the gating and the solid line shows the upper and lower levels (20% gating window) of the gating window (beam on area). The shaded bars show the beam on time during a gated radiation delivery.

2.1. Calibration

Gafchromic EBT2 film (International Specialty Products, Wayne, NJ) was calibrated under a reference condition (1.5 cm depth, 100 cm source-to-surface distance and 10 × 10 cm^2 field size) with 6 MV photon beam and dose rate of 1000 MU min^{-1}.

Gafchromic EBT2 film was cut into strips with dimensions of 10 × 2 cm^2. The films were scanned with Epson 10000XL scanner (Epson America, Inc. Long Beach, CA) 24 hours post-irradiation. Details of the scanning parameters were described by Jong *et al.* [7]. The image were analysed using ImageJ 1.49v software (National Institute of Health, Bethesda, MD).

2.2. Dose profile measurement

A piece of Gafchromic EBT2 film was placed at depth of 8 cm (below the fiducial marker) in the Perspex phantom. The films were irradiated with 300 MU under 6 MV photon beam with a radiation field size of 5×5 cm^2 and 1000 MU min^{-1} at the end of exhalation as in figure 1(b).

The dose profiles for different (i) respiratory amplitudes (0%, 25%, 50%, 75% and 100%, where 100% is the total movement of 25 mm), (ii) respiratory times for a full respiration (3 and 5 seconds) and (iii) gating windows (10% and 20%) were measured. All dose profiles were compared with non-gated delivery (static and moving target).

2.3. *Tracking accuracy*

The accuracy of gated radiation deliveries was studied by assessing the displacement of the projection of the fiducial marker on the film (shown as a 'dip' in the dose profiles). The position of the lower dose of the fiducial marker's projection was considered as the centre of the target. The position of gated radiation deliveries was compared with the static delivery.

3. Results and discussions

The Gafchromic EBT2 film was calibrated from the dose range of 0 cGy to 300 cGy. The centre of the Perspex phantom was marked on the Gafchromic EBT2 films before irradiation to allow better accuracy in comparison. All measurements with Gafchromic EBT2 film were normalised at the maximum dose. All measurements during gated radiation deliveries were benchmarked against static radiation delivery.

3.1. *Dose profiles of gated-lung SBRT*

The dose profiles in the longitudinal direction (moving direction) were measured with Gafchromic EBT2 films. The lung tumour motion has been found to be quite substantial along this direction [3]. The 'dips' near the middle of the radiation beam profiles in figure 2(a)-(d) were due to the presence of the spherical fiducial marker. The 'dip' was also used to indicate the position of the tumour.

Figure 2(a) shows the EBT2 measured dose profiles with different respiratory amplitudes. The respiratory-gated dose profiles for different amplitudes of tumour motion were in good agreement with the static radiation delivery. The mean difference for gated dose profile with 0% tumour motion amplitude was 0.02% (range from -10.0% to 9.33%) while for 100% tumour motion amplitude was 1.16% (range from -8.22% to 11.31%).

Figure 2(b) shows the dose profiles with different gating windows and figure 2(c) showed the dose profiles with different respiratory times per cycle. A previous study reported that the dosimetric error increases with increasing gating window size during respiratory-gated radiotherapy [8]. However, no significant difference was observed with different gating windows in this study. This is due to the small area of radiation delivery [figure 1(b)].

The respiratory time per cycle simulated different respiratory or tumour speed [figure 2(c)]. There were no significant difference in gated radiation delivery for different respiratory times per cycle. The gating window was set based on the respiratory or tumour motion amplitude. This result is consistent with the result obtained by Shiinoki *et al.* [9].

Figure 2(d) shows non-gated radiation dose deliveries while the tumour is moving, simulating free-breathing SBRT without gating. When the tumour motion amplitude increased, the dose profile was degraded [left area of the dose profiles in figure 2(d)] and blurring effect was observed on the dose profile due to respiratory motion [right area of the dose profiles in figure 2(d)]. This indicates that tumour coverage may be compromised whilst more healthy tissues were irradiated unnecessarily. Conventional method of dealing with the inadequacy of tumour coverage involves a larger tumour margin to ensure better tumour coverage but on the expense of higher normal tissue complications. With gating, tumour margin can be reduced for tumour motion [5].

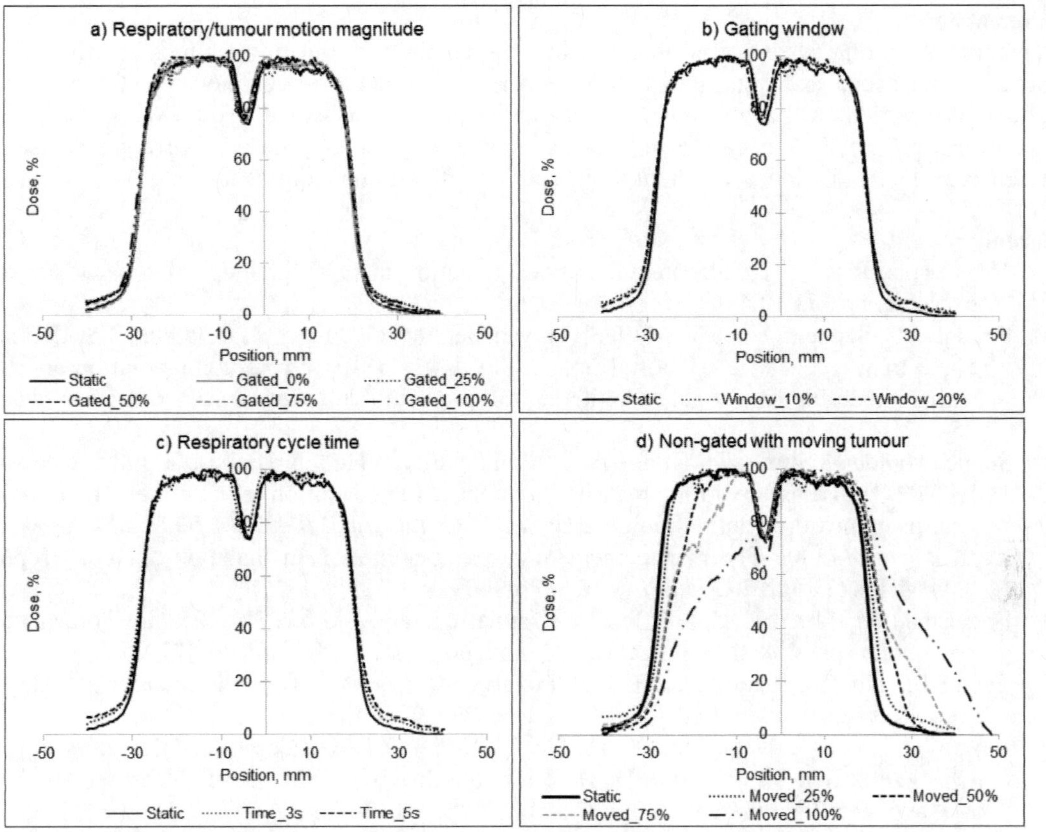

Figure 2. The dose profiles during gated radiation delivery with different (a) respiratory/ target motion amplitude, (b) gating windows, (c) respiratory cycle time, and (d) non-gated radiation delivery with moving target. All measurements were compared with static radiation delivery. The 'dip' in the middle of the dose profiles were caused by the shadowing effect of the fiducial marker.

3.2. *Tracking accuracy*

Table 1 shows the measured displacement of the fiducial marker between gated and static deliveries. Tenn *et al.* has reported a difference of approximately 1.0 mm in position between gated and non-gated radiation delivery at 10% gating level [6]. Willoughby *et al.* has reported the accuracy of 1.7 mm of the Exactrac system on phantom studies [10]. In this study, the maximum displacement of target was -1.05 mm and is in agreement with the previous studies. The result showed that the system is able to deliver the radiation with an accuracy of within -1.05 mm.

Table 1. Measured displacements of fiducial marker during respiratory-gated radiation delivery to the static radiation delivery.

	Displacement (mm)		Displacement (mm)
Amplitude		Time	
0%	-1.05	3 sec	-0.78
25%	0.01	5 sec	0.01
50%	-0.25		
75%	-0.78	Gating windows	
100%	-0.78	10%	0.37
		20%	-0.78

4. Conclusion

The respiratory gating system was able to track the respiratory and tumour motion with different amplitudes and speed and was able to deliver the radiation dose accurately (±1.05 mm) in the longitudinal direction. Although the treatment time for respiratory-gated SBRT was prolonged, this approach can potentially reduce the margin for internal tumour volume without compromising the tumour coverage. In addition, the normal tissue sparing effect can be improved.

References

[1] Timmerman R *et al.* 2010 Stereotactic body radiation therapy for inoperable early stage lung cancer *Jama* **303** 1070-6

[2] Erridge S C, Seppenwoolde Y, Muller S H, van Herk M, De Jaeger K, Belderbos J S, Boersma L J and Lebesque J V 2003 Portal imaging to assess set-up errors, tumor motion and tumor shrinkage during conformal radiotherapy of non-small cell lung cancer *Radiother Oncol* **66** 75-85

[3] Seppenwoolde Y, Shirato H, Kitamura K, Shimizu S, van Herk M, Lebesque J V and Miyasaka K 2002 Precise and real-time measurement of 3D tumor motion in lung due to breathing and heartbeat, measured during radiotherapy *Int J Radiat Oncol Biol Phys* **53** 822-34

[4] Keall P J *et al.* 2006 The management of respiratory motion in radiation oncology report of AAPM Task Group 76 *Med Phys* **33** 3874-900

[5] Engelsman M, Sharp G C, Bortfeld T, Onimaru R and Shirato H 2005 How much margin reduction is possible through gating or breath hold? *Phys Med Biol* **50** 477-90

[6] Tenn S E, Solberg T D and Medin P M 2005 Targeting accuracy of an image guided gating system for stereotactic body radiotherapy *Phys Med Biol* **50** 5443-62

[7] Jong W L, Wong J H D, Ung N M, Ng K H, Ho G F, Cutajar D and Rosenfeld A 2014 Characterization of MOSkin™ Detector for In-vivo Skin Dose Measurement during Megavoltage Radiotherapy *J Appl Clin Med Phys* **15** 120-32

[8] Hugo G D, Agazaryan N and Solberg T D 2002 An evaluation of gating window size, delivery method, and composite field dosimetry of respiratory-gated IMRT *Med Phys* **29** 2517-25

[9] Shiinoki T *et al.* 2014 Quality Assurance for Respiratory-Gated Radiotherapy Using the Real-Time Tumor-Tracking Radiotherapy System *International Journal of Medical Physics, Clinical Engineering and Radiation Oncology* **48** 1175-85

[10] Willoughby T R, Forbes A R, Buchholz D, Langen K M, Wagner T H, Zeidan O A, Kupelian P A and Meeks S L 2006 Evaluation of an infrared camera and X-ray system using implanted fiducials in patients with lung tumors for gated radiation therapy *Int J Radiat Oncol Biol Phys* **66** 568-75

Acknowledgments

We would like to thank the radiographers and medical physicists from the University of Malaya Medical Centre, Malaysia for their assistance. This study was supported by the Postgraduate Research Fund (PPP) PO021-2014B from the University of Malaya, Malaysia.

Preliminary experience on the implementation of computed tomography (CT)-based image guided brachytherapy (IGBT) of cervical cancer using high-dose-rate (HDR) Cobalt-60 source in University of Malaya Medical Centre (UMMC)

Z Jamalludin[1,2], U N Min[1], W Z Wan Ishak[1] and R Abdul Malik[1]

[1] Clinical Oncology Unit, Faculty of Medicine, University of Malaya, Malaysia
[2] Medical Physics Unit, University Malaya Medical Center, Malaysia

E-mail: nmung@ummc.edu.my

Abstract. This study presents our preliminary work of the computed tomography (CT) image guided brachytherapy (IGBT) implementation on cervical cancer patients. We developed a protocol in which patients undergo two Magnetic Resonance Imaging (MRI) examinations; a) prior to external beam radiotherapy (EBRT) and b) prior to intra-cavitary brachytherapy for tumour identification and delineation during IGBT planning and dosimetry. For each fraction, patients were simulated using CT simulator and images were transferred to the treatment planning system. The HR-CTV, IR-CTV, bladder and rectum were delineated on CT-based contouring for cervical cancer. Plans were optimised to achieve HR-CTV and IR-CTV dose (D_{90}) of total EQD_2 80Gy and 60Gy respectively, while limiting the minimum dose to the most irradiated 2cm^3 volume (D_{2cc}) of bladder and rectum to total EQD_2 90Gy and 75Gy respectively. Data from seven insertions were analysed by comparing the volume-based with traditional point-based doses. Based on our data, there were differences between volume and point doses of HR-CTV, bladder and rectum organs. As the number of patients having the CT-based IGBT increases from day to day in our centre, it is expected that the treatment and dosimetry accuracy will be improved with the implementation.

1. Introduction

Intracavitary brachytherapy (ICBT) is a treatment technique of placing radiation source near or in the tumour situated in body cavity. It is known as an integral part of cervical cancer treatment with external beam radiation therapy (EBRT) form the definitive radiation therapy for cervical cancer treatment. Conventional ICBT is based on 2-dimensional (2D) point-based planning determined by International Commissioning Radiation Units (ICRU) definition of prescription and organ at risk (OAR) points on orthogonal x-ray images. This was used for dose calculation irrespective of tumour and OAR location, size and shape, thus proved to results in inadequate target coverage and inaccurate OAR doses. With these limitations, the concept of ICBT prescription and dose evaluation has been developed from point-based planning and evaluation to volume-based technique.

As the new imaging technologies (computed tomography, magnetic resonance and ultrasound) played an important role in radiotherapy, the GEC-ESTRO has published recommendations for volume-based concept of the tumour and OAR with the use 3-dimensional images from magnetic resonance imaging (MRI), thus known as image guided brachytherapy (IGBT)[1]. MRI provides clear tissue discrimination of tumour extension and OARs. However, not all oncology departments have an easy access for MRI, in comparison with computed tomography (CT) which is available in most departments. This give rise to three options of IGBT techniques, depending on departments' convenience and resources; a) MR image guided adaptive brachytherapy – MR images are acquired at each insertion for the purpose of treatment planning, b) MR-CT hybrid planning – MR images are acquired for the first fraction, followed by subsequent CT imaging and c) MR aided CT based planning – CT images are acquired at each insertion for the purposes of treatment planning [2]. We chose the last option for IGBT technique in our centre. This paper gives an overview on our experiences on implementing the IGBT with MR aided CT based planning technique.

2. UMMC - IGBT protocol

In comparison with 2D brachytherapy, IGBT requires: a) additional 3D images from MRI for treatment planning, b) complex CT simulation procedure and c) complex treatment planning techniques and dose evaluation hence more time and personnel were involved in the whole IGBT process. We outline some of the added procedures for IGBT on 2 patients treated in our centre.

2.1. MR images acquisition

A pre-brachytherapy MR image was acquired within 5 days of the first brachytherapy insertion. It was a T2-weighted transverse dataset with para-transverse, para-coronal and para-sagittal imaging orientation with slice thickness of less than 5mm.

2.2. CT simulation images acquisition

Patients underwent bowel preparation as per our institution protocol prior to the brachytherapy procedure. Rectal contrast was injected into the rectum, after the applicators insertion. A radio-opaque liquid consist of 50cc of normal saline plus 5cc of iodine contrast was injected into the bladder. Simulation images were acquired after 10 minutes of contrast injection to give time for the contrast to diffuse homogenously with saline for better image acquisition as shown in figure 1 and 2.

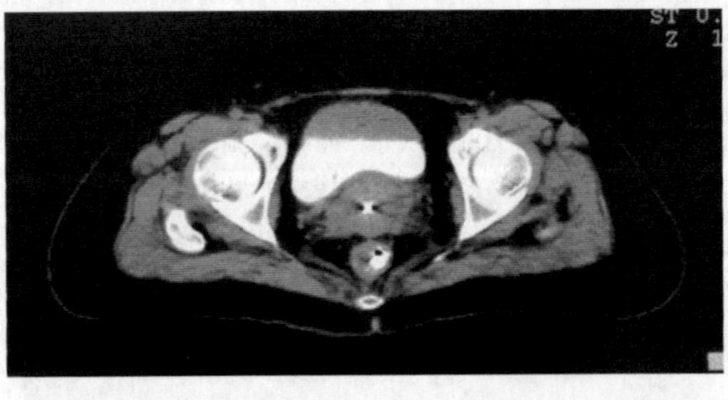

Figure 1. Simulation image acquired after 5 minutes of contrast injection.

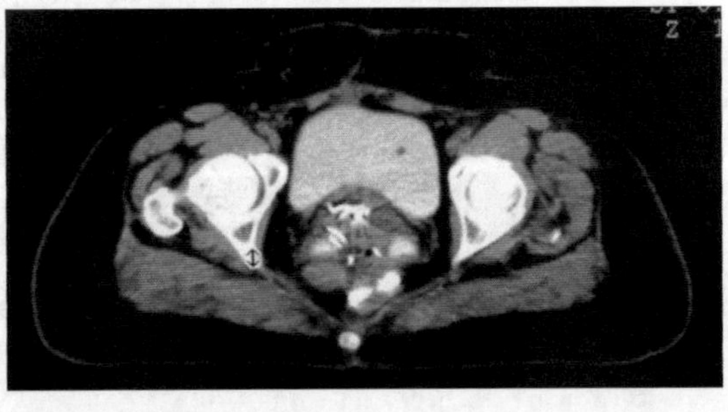

Figure 2. Simulation image acquired after 10 minutes of contrast injection.

The bladders' filling was maintained during the scanning process by clamping the catheter tube and released after the process was completed. Image scanning with 3.0mm slice thickness extending from the iliac crest was performed during each fraction with the applicator in place. A contiguous acquisition of CT slices with ≤ 3.0mm as suggested by GEC ESTRO working group was performed to avoid loss of tumor and OARs volumetric information and to obtain proper applicators reconstruction images [3]. A number of polystyrene base plates with different thickness as shown in figure 3, were placed under the applicator in conditions where: a) the applicators were not secured properly by the gauze packing and b) to push the bladder away from the applicators to reduce the dose to bladder during treatment. The CT images were exported digitally to the HDR Plus 2.6 brachytherapy treatment planning system.

Figure 3. Polystyrene base with variety of plate thicknesses.

2.3. Contouring

While a direct fusion between diagnostic MR images and planning CT datasets was not advised for target and OAR delineation for the MR aided CT based planning technique, a reference high-risk clinical target volume (HR-CTV) may be inferred on the planning CT images from the pre-brachytherapy MR and the clinical exam for the first fraction. The HR-CTV was purely a reference volume to monitor dose in a high-risk region. For subsequent fraction, HR-CTV delineation was also guided by the findings on CT images from the simulation images. The delineation was based on the GEC-ESTRO [1] recommendations.

For OAR delineation, a number of published reports show insignificant differences between OAR volumes and dose to the OARs for CT and MR based treatment planning. Therefore, planning CT images was used to accurately delineate OARs such as rectum and bladder.

2.4. Planning

Treatment planning was performed for each insertion, beginning with standard loading pattern using Manchester system approach. Plans were first normalised 100% to point A in order to achieve a pear shape dose distribution. Further optimisation was performed, by accounting the EBRT dose to result in a total of equivalent dose in 2Gy fractions (EQD_2) dose volume histogram (DVH) constraints of 80-90Gy (α/β=10) for HR-CTV and 60Gy for intermediate risk CTV (IR-CTV), while keeping minimum dose to the most exposed 2cm^3 volume (D_{2cc}) of bladder and rectum/sigmoid to total EQD_2 90Gy (α/β=3) and 75Gy (α/β=3) respectively. If these objectives could not be achieved using the standard loading pattern, we further optimised the plan by adding or removing dwelling positions and adjusting the dwell times manually or using graphical optimisation to improve the dosimetry. Graphical optimisation is a free hand tool of the planning system which can be used to manually shaping the isodose lines. No inverse planning technique was used as this was not recommended by Tanderup [4] which can results in higher deviations from the standard loading shape and higher dose to non-contoured tissue.

For dosimetry evaluation, the following dose parameters were recorded: Point A, ICRU rectal and bladder point, the isodose encompassing 100% and 90% of the tumour target (D_{100} and D_{90}) of HR-CTV and IR-CTV, the percentage of tumour target volume receiving 100% of the prescribed dose (V_{100}), $D_{0.1cc}$, D_{1cc} and D_{2cc} for rectum and bladder.

3. Results and discussion

Table 1 gives an overview of the tumor and OAR doses in total EQD_2 for both patients. Patient A underwent single phase of 3D conformal EBRT (3DCRT) while patient B had two phases of 3DCRT. Based on EQD_2 from EBRT, different IGBT fractionations were applied for both patients. However, the total EQD_2 from both EBRT and IGBT for tumors and OAR was ensured to fulfill the DVH constraints objectives set in our center as shown in table 1.

Table 1. Treatment fractionation schedule and total dose in EQD_2 of tumor and OARs for UMMC IGBT patients.

Patients	EBRT fractionation schedule	IGBT fractionation schedule	Total HR-CTV D_{90} EQD_2 (Gy)	Total IR-CTV D_{90} EQD_2 (Gy)	Total Bladder D_{2cc} EQD_2 (Gy)	Total Rectum D_{2cc} EQD_2 (Gy)
A	Ph1: 1.8Gy x 27	7.0 Gy x 4	80.8	68.6	81.7	73.9
B	Ph1: 2.0Gy x 20 Ph2: 1.8Gy x 3	7.5 Gy x 3	80.4	67.9	82.8	65.9

Table 2 listed out the volume-based and point-based dose for tumor and OARs from each HDR fraction for both patients. It was noted that the D_{90} EQD_2 dose for HR-CTV volume of < 30cc was higher by 32% of mean dose (p-value < 0.05) than ICRU point A dose in both patients and was lower for HR-CTV volume of > 40cc by 26% (p-value > 0.05). Tanderup *et al.* [5] found higher HR-CTV D_{90} by 167% and lower by 36% for tumor of ≤ 31cc and > 31cc respectively. The HR-CTV volume in patient A and B were noted to decrease from the first fraction to last fraction by 52% and 30% respectively.

Table 2. Summary of doses to ICRU point, D_{90} and D_{2cc} EQD_2 of tumor and organ at risks.

Patient	Fraction	Tumor volume (cm^3)	HR-CTV (Gy)		Bladder (Gy)		Rectum (Gy)	
			D_{90} EQD_2	ICRU Point A	D_{2cc} EQD_2	ICRU Point	D_{2cc} EQD_2	ICRU Point
A	1	49.9	6.6	9.0	7.0	5.9	5.2	5.1
	2	30.5	5.0	4.7	4.3	2.9	4.9	5.8
	3	30.9	6.2	5.5	5.2	5.0	4.4	4.8
	4	25.9	6.6	5.4	4.4	3.9	3.4	4.6
B	1	27.5	7.8	5.7	6.4	5.8	4.5	5.1
	2	19.7	7.8	5.6	7.4	7.0	4.6	5.3
	3	19.3	8.0	4.3	5.8	5.0	4.6	5.3

For organ at risks dose analysis, we found that the D_{2cc} EQD_2 for bladder of both patients was higher than ICRU point dose by a mean of 13.5% (p-value=0.018) in comparison with rectum D_{2cc} EQD_2 which was lower than ICRU point dose by a mean of 12.3% (p-value=0.028). The result of dose comparison for bladder organ in this study was in agreement with studies by Onal *et al.* [6] and Madan *et al.* [7]. As the sample size was small (n=7), we expect to strengthen our findings based on future IGBT patients data. The total time spent for IGBT treatment starting from CT simulation to the end of patients' treatment time was in average of 3 hours in comparison of average 2 hours for conventional brachytherapy treatment.

4. Summary

As the international working groups on IGBT are working towards a better and standardised implementation of this technique throughout the world, we chose to perform the IGBT in the most convenient and practical way based on the resources available in our institution. With the implementation of IGBT, we aim to improve the of brachytherapy treatments quality by increasing the tumour dose, thus help in improving local tumour control while reducing the organs at risk doses.

References

[1] Haie-Meder C, Pötter R and Limbergen E V 2005 Recommendations from Gynaecological (GYN) GEC-ESTRO Working Group (I): concepts and terms in 3D image based 3D treatment planning in cervix cancer brachytherapy with emphasis on MRI assessment of GTV and CTV. *Radiother Oncol* **74** 235-245

[2] Cancer Care Ontario 2014 Imaging Strategies for Definitive Intracavitary Brachytherapy of Cervical Cancer: Recommendation Report. June 2014

[3] Pötter R 2002 *Vienna method.* In: Gerbaulet A, Potter R, Haie-Meder C,editors. Cervix cancer. GEC ESTRO handbook of brachytherapy. Brussels: ESTRO p 331–8

[4] Tanderup K 2014 Physics aspects of treatment planning intracavitary +/- interstitial techniques in cervix cancer. ESTRO Teaching Course Image Guided Radiotherapy & Chemotherapy in Gyneacological Cancer. Firenze

[5] Tanderup K, Nielsen S K and Nyvang G B 2010 From point A to the sculpted pear: MR image guidance significantly improves tumour dose and sparing of organs at risk in brachytherapy of cervical cancer. 2010 *Radiother Oncol* **94** pp 173-180

[6] Onal C, Arslan G and Topkan E 2009 Comparison of conventional and CT-based planning for intracavitary brachytherapy for cervical cancer: Target volume coverage and organs at risk doses *J. Exp & Clinical Cancer Research* **28** 95

[7] Madan R, Pathy S and Subramani V 2014 Comparative Evaluation of Two-dimensional Radiography and Three Dimensional Computed Tomography Based Dose-volume Parameters for High-dose-rate Intravitary Brachytherapy of Cervical Cancer: A Prospective Study *Asian Pacific J. Cancer Prevention* **15**:4717-21

Comparison of radiotherapy dosimetry for 3D-CRT, IMRT, and SBRT based on electron density calibration

K Kartutik[1], **W E Wibowo**[2] **and S A Pawiro**[1]

[1] Department of Physics, Faculty of Mathematics and Natural Sciences, Universitas Indonesia, Depok, Indonesia
[2] Department of Radiotherapy, National General Hospital Dr. Cipto Mangunkusumo, Jakarta, Indonesia

E-mail: kartutik@sci.ui.ac.id

Abstract. Accurate calculation of dose distribution affected by inhomogeneity tissue is required in radiotherapy planning. This study was performed to determine the ratio between radiotherapy planning using 3D-CRT, IMRT, and SBRT based on a calibrated curve of CT-number in the lung for different target's shape in 3D-CRT, IMRT, and spinal cord for SBRT. Calibration curves of CT-number were generated under measurement basis and introduced into TPS, then planning was performed for 3D-CRT, IMRT, and SBRT with 7, and 15 radiation fields. Afterwards, planning evaluation was performed by comparing the DVH curve, HI, and CI. 3D-CRT and IMRT produced the lowest HI at calibration curve of CIRS 002LFC with the value 0.24 and 0.10. Whereas SBRT produced the lowest HI on a linear calibration curve with a value of 0.361. The highest CI in IMRT and SBRT technique achieved using a linear calibration curve was 0.97 and 1.77 respectively. For 3D-CRT, the highest CI was obtained by using calibration curve of CIRS 062M with the value of 0.45. From the results of CI and HI, it is concluded that the calibration curve of CT-number does not significantly differ with Schneider's calibrated curve, and inverse planning gives a better result than forward planning.

1. Introduction

The human body is composed of various tissues and cavities with different physical and radiological properties. The most important from the perspective of radiotherapy dosimetry is that there are physical differences between the tissues and cavities found in the body, such as lung, oral cavity, teeth, nose, sinuses and bones. In some cases, in the human body, there are even foreign objects deliberately planted, such as metal prostheses. It becomes important that the absorbed dose is given to the entire irradiated tissues must be corrected using tissues homogeneity as predicted [1].

One of the most important steps in radiotherapy planning process is the acquisition of patient data needed to define the target volume and an accurate calculation of the dose distribution [2, 3]. This data is mostly derived from the patient's radiology images which are usually obtained from a scanning using computed tomography (CT), which contains all the important information, for example in order to see the density distribution of organs and tissues. This information is then used for an accurate dose calculation, especially when the radiation beam passes through non-homogeneous tissues such as lungs and bones where the attenuation coefficient is very different with soft tissue's attenuation coefficient [3, 4].

Images from computed tomography (CT) have largely become the standard for radiotherapy treatment planning due to its ideal structure (transverse images with high-resolution images and its relation to Hounsfield Units as well as the density of electrons). Inhomogeneity corrections can be made on the basis of pixel based on accurate considerations of the density from different tissues in the body of the patient.

Radiotherapy planning is performed to get the optimal therapeutic ratio that provide sufficient dose distribution in target and minimal dose to normal tissue [5]. CT-simulator can generate 3D images, then the purpose of radiotherapy will be easier to be achieved. In lung cancer, radiotherapy treatment becomes more complicated due to anatomic structures that have different densities between lung, bone and soft tissue. Hounsfield Unit (HU) that were very different between those tissues caused homogeneity of the radiation dose on the medium will be difficult to be achieved.

Research conducted by Klein, *et al.* showed quite different results between planning using heterogeneity correction and without heterogeneity correction in three-dimensional treatment planning (3DTP). Klein, *et al.* conducted a study using a homogeneous medium (water) and a heterogeneous medium, and comparing the results. The results showed that 6 to 18% doses on isocenter would be greater in the planning without heterogeneity correction [6]. Because of different results that have been gotten previously, this study was performed to compare radiotherapy planning between 3D-CRT and IMRT for lung and to compare radiotherapy planning using different targets that have different mass density (lung and spinal cord) for SBRT based on calibrated and not calibrated of electron density curve by calculating homogeneity index (HI), conformity index (CI) and also dose volume histogram (DVH).

2. Materials and methods

This study was conducted at the Department of Radiotherapy Cipto Mangunkusumo Hospital. Electron Density Phantom CIRS Model 062M and Thorax Phantom CIRS Model 002LFC was used. CIRS Phantom was scanned on CT-simulator GE Bright Speed. Target volume and normal tissue cross-sections were contoured in Oncentra Masterplan. Radiotherapy planning was conducted in Pinnacle^3 Treatment Planning System using energy 6 MV.

Calibrating CT number curves was the early step in this research. Initially, the calibration curve of CT number was obtained by scanning phantom using CT-simulator and then transferred to the virtual simulator. Afterwards, CT density calibration curve were obtained by entering density (ρ) of each plug as the y-axis, and the number of CT as the x-axis. This radiotherapy planning used 3 calibration curves of electron density are calibration curve of CT number from Electron Density Phantom CIRS Model 062M, Thorax CIRS Model 002LFC and linear calibration curved which is obtained from research conducted by Uwe Schneider in 1995 [7].

Furthermore, Thorax CIRS Model 002LFC was scanned and then transferred to the virtual simulator. The image was contoured to determine the target organs and organ at risk. Afterwards, target organ were formed in three shapes in order to make it as well as the real target in the patient for instance cylinder, half cylinder and concave. Details of the target volume in different shapes were shown in table 1. Then the contour was transferred to TPS Pinnacle^3. Radiotherapy planning technique performed in this study was forward planning for 3D-CRT and SBRT, whereas inverse planning was for IMRT.

Seven radiation fields which were 0° (AP), 51.4°, 113°, 164.4°, 215.8°, 275°, and 331° were used for 3D-CRT and IMRT technique. Although 5 radiation fields were used clinically, but 7 radiation fields gave optimal result according to Arianty. [8] On the other hand 15 radiation fields which were 0° (AP), 51.4°, 113°, 164.4°, 215.8°, 275°, 331°, 98°, 70°, 60°, 257°, 195°, 130°, non-coplanar 30°, and non-coplanar 330° were used for SBRT technique. Radiation field used in SBRT technique was much more than 3D-CRT and IMRT in order to achieve isodose line that can cover small target volume.

In order to evaluate the planning, the calculation of homogeneity index (HI) [9] and conformity index (CI) [10] can be done follow equation (1)-(2). The evaluation of radiotherapy planning was also performed at the organ at risk by referring to the Quantitative Analysis of Normal Tissue Effects in the Clinic (Quantec). Afterwards, the comparison of radiotherapy planning could be made.

$$HI = (D_{2\%} - D_{98\%}) / D_{50\%} \quad (1)$$

$$CI = V_{95} / V_{PTV} \tag{2}$$

Table 1. Target volume in different shapes; cylinder, half-cylinder, concave.

	Cylindrical target				Half-cylindrical target	Concave target
	3D-CRT and IMRT		SBRT			
Organ	Diameter target (cm)	Volume (cm^3)	Diameter target (cm)	Volume (cm^3)	Volume (cm^3)	Volume (cm^3)
GTV	4.6	37.37	2.1	8.43	18.68	14.97
CTV	5.72	59.72	2.1	8.43	27.74	28.02
PTV	7.34	117.39	2.1	8.43	56,45	69.55

3. Results and discussions

The comparison between 3D-CRT and IMRT technique can be seen in Table 2. IMRT technique formed higher coverage isodose line on GTV and PTV compared to 3D-CRT. The higher coverage isodose line was on PTV, it means doses could cover target organ and the more cancer cell would be killed. Both in 3D-CRT and IMRT, the organ at risk received doses that can be tolerated according to QUANTEC, for example, 20 Gy dose received by lung was still below 30% total volume of the lung. 30 Gy dose received by heart was still below 46% total volume of the heart, and D_{max} of the spinal cord was less than 46 Gy. Homogeneity index (HI) differences from 3D-CRT and IMRT could be seen distinctly in figure 1(a). HI obtained from 3D-CRT and IMRT using calibration curve of Phantom CIRS Model 062M, Thorax Phantom CIRS Model 002LFC, and Schneider's calibration curve (linear) were 0.25, 0.243, 0.248, and 0.103, 0.101, 0.102 respectively. The HI of IMRT planning was closer to ideal value 0 according to ICRU No. 83. IMRT result was more homogeneous because optimization was performed in TPS. Different calibration of electron density curves would make the different planning results because electron density curve affected dose calculation. A phantom that had more mass density would deliver dose calculation accurately than phantom that had less mass density.

Table 2. Comparison of radiotherapy planning result between 3D-CRT and IMRT

Parameter	3D-CRT	IMRT
Coverage isodose line	86%	95%
MLD (Mean Lung Dose)	7.23 Gy	8.47 Gy
V_{20} (Lung)	25%	28%
V_{30} (Heart)	10%	5%
D_{max} (Spinal cord)	31.67 Gy	32.27 Gy

Moreover, conformity index (CI) using 3D-CRT technique was far below CI using IMRT that was shown in figure 1(b). CI obtained from 3D-CRT and IMRT using calibration curve of Phantom CIRS Model 062M, Thorax Phantom CIRS Model 002LFC, and Schneider's calibration curve (linear) were 0.45, 0.43, 0.41, and 0.97, 0.91, 0.97 respectively. The CI of IMRT was closer to ideal value 1 which means that prescription isodose line was overlapping with PTV. High conformity on IMRT was occurred because there were dose constraints that had been given in the planning so that PTV would receive dose according to the prescription. IMRT also delivered non-uniform dose so that radiation beam would be modulated. The weighting of each segment would be different so that prescription isodose line would follow the form of planned targets. Both HI and CI using 3D-CRT and IMRT had a quite different results from Ganesh [11]. The differences might be caused of different target volume used because the HI was extremely effected by target volume [8].

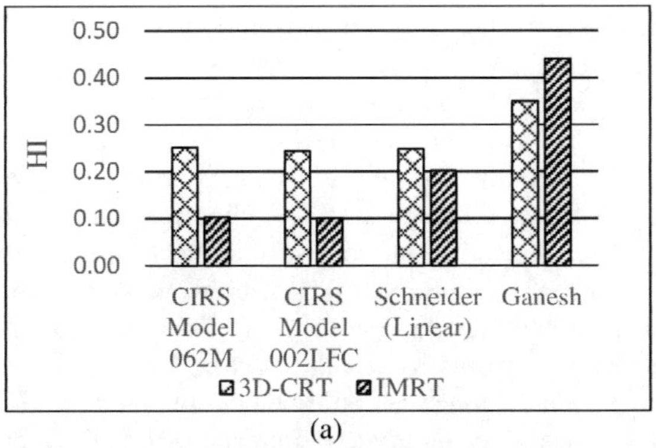

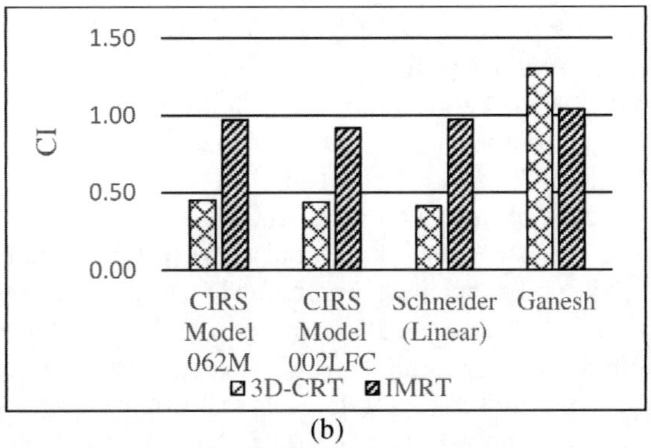

Figure 1. Comparison of (a) HI and (b) CI from radiotherapy planning result using different calibration curve of CT number between 3D-CRT and IMRT.

The comparison of target organs which were lung and spinal cord was only performed in SBRT. The HI value for lung and spinal cord using calibration curve Phantom Model 062M CIRS, CIRS Model 002LFC, and Schneider's were respectively 0.366, 03.65, 0.361 and 0.173, 0.174, 0.158. Spinal cord had better HI than lung shown in figure 2(a) because it has a greater density (ρ) so that particle fluence in spinal cord would be greater. The dose would also spread evenly through PTV, and gave HI closer to the ideal. CI of the spinal cord was closer to ideal than lung shown in figure 2(b). Both lung and the spinal cord had The CI slightly different from ideal because target volume used was very small which is about 8 cc, and the prescription dose was not 100% on isocenter for SBRT. According to the RTOG protocol 0813, the prescription dose in SBRT is conducted by calculating the dose that covers 95% PTV then divided by the total dose. With this calculation, the goal of radiotherapy planning in SBRT can be achieved that deliver the high dose in a small fraction, but isodose line of PTV will be slightly widened.

The isodose line of different target shapes for 3D-CRT and IMRT were shown in figure 3.a and 3.b. All radiotherapy planning with different shapes had a high dose in the boundary area between the lung and soft tissue which is caused by the algorithm in TPS. It used adaptive convolution algorithm which required only a short time in dose calculation. Adaptive convolution algorithm did not take into account the presence of inhomogeneity tissue according to Todd McNut. It gave the high dose in the boundary of lung and soft tissue because lung has a mass density (ρ) which is smaller than the soft tissue. Hotspot area in the boundary could be happened because there was a re-built-up PDD from the lung to soft tissue.

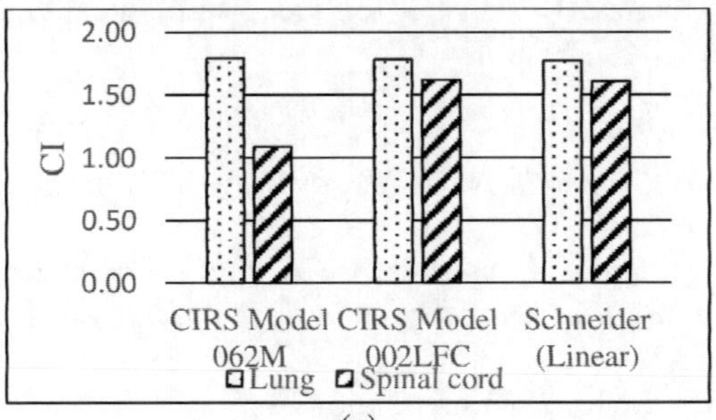

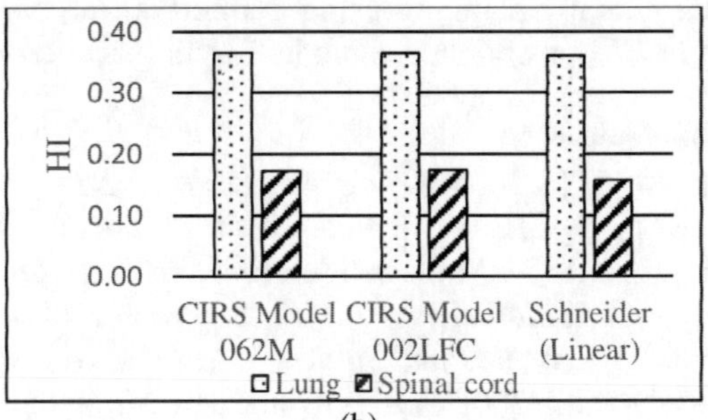

Figure 2. Comparison of (a) HI and (b) CI from radiotherapy planning result using different target organ; lung and spinal cord for SBRT.

The HI obtained using different electron density curves from different shapes of target gave the same trend which produced lowest HI for the cylindrical target, concave, and half-cylindrical target respectively. Cylindrical target had a regular shape so that dose uniformity could be easily achieved. Half-cylinder and concave had more complex shape than cylinder so that uniformity dose was more difficult to be achieved using same weighting and beam angle. The HI in IMRT was better than 3D-

CRT for all different target. It did not have the same trend for the different calibration curve. The HI from calibration curve of Phantom CIRS Thorax Model 002LFC gave best result because it was phantom that was used at irradiation so that this calibration curve gave dose distribution more uniform compared with calibration curve of Phantom CIRS Model 062M and Schneider's calibration curve (linear).

Ideal CI was difficult to be achieved using 3D-CRT especially for targets that have more complex shapes such as a concave. Half-cylindrical target had a better CI compared to cylindrical target and concave because it had smaller volume so that 95% of prescription dose would be more evenly distributed. Concave target had a lower CI because of its irregular shape so that prescription isodose line could not follow target shape easily. More beams were given from above to avoid OAR such as the spinal cord. It would cause dose distribution did not evenly distributed and the isodose line would not overlap with target volume. More complex target shape such as concave could be covered using IMRT because it delivered non-uniform dose distribution. It means that each radiation beam angle had different weighting depending on the target shape so that isodose line would overlap with target planned. The CI in IMRT planning was higher than 3D-CRT even for the concave target. The average conformity index for different target shape was 1,033. This value was ideal according to the RTOG 0813.

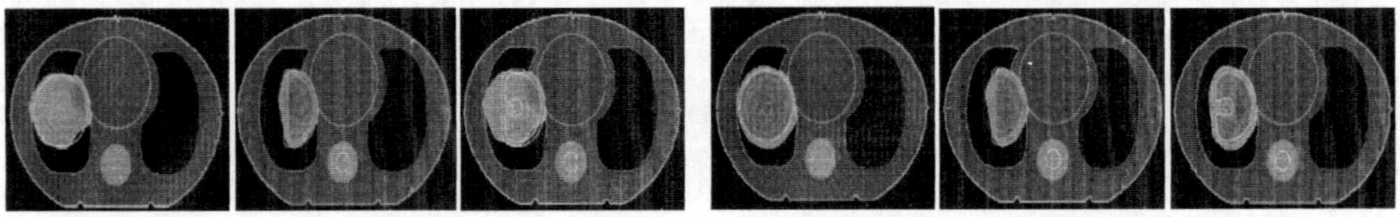

(a) (b)

Figure 3. Comparison of radiotherapy planning result from different target shapes; cylinder (left), half-cylinder (center), concave (right) using (a) 3D-CRT and (b) IMRT.

4. Conclusion

Homogeneity index and conformity index obtained from planning of radiotherapy using calibration curve of Phantom CIRS Model 062M, and CIRS Model 002LFC did not significantly differ with available literature (Schneider's calibrated curve). IMRT could cover dose was around 95%, while 3D-CRT only 88%. The results obtained could support IMRT as a standard technique used for radiotherapy treatment today. Spinal cord's radiotherapy planning gave better DVH, HI, and CI than lung. Radiotherapy planning using cylindrical target gave a better HI than half-cylinder and concave, and better CI was obtained using half-cylindrical target for 3D-CRT and IMRT.

References

[1] AAPM 2004 *Report No 85: Tissue Inhomogenity Corrections for Megavoltage Photon Beams* (Madison: Medical Physics Publishing)

[2] Parker W and Patrocinio H 2005 *Clinical Treatment Planning in External Photon Beam Radiotherapy. In: E. B. Podgorsak (Ed.). Radiation Oncology Physics: A Handbook for Teachers and Students* (Austria: IAEA) p. 229-273

[3] Neal A 2007 *Patient Data Acquisition* (Inglaterra: Taylor & Francis) p.647-657

A. Barret, Dobbs J, Morris S and Roques T 2009 *Practical Radiotherapy Planning*

[4] Khan F M 2010 *Physics of Radiation Therapy, The 4th Edition* (Lippincot williams & Wilkins)

[5] Klein E, Astrid M, James A, Marry V, Graham and John M 1997 *Int. J. Radiation Oncology Biol. Phys.* **37** 1163-1170

[6] Schneider U, Pedroni E and Lomax A 1995 *Phys. Med. Biol.* **41** 111-124.

[7] Arianty and Dessy 2010 *Tesis: Optimasi Jumlah Lapangan Radiasi pada Perencanaan IMRT* Universitas Indonesia

[8] ICRU Report 83. (2010). *Prescribing, Recording and Reporting Photon Beam Intensity Modulated Radiation Therapy* (*IMRT*). The International Commision on radiation Units and Measurements.

[9] Vaezzadeh S, Allahverdi M 2012 *Iran. J. Radiat* **10** 145-150.

[10] Ganesh. N, Drew G, Diana B, Ying L, Alonso G, Niko P, Sotirios S 2015 *Clinical Engineering and Radiation Oncology* **4** 104-112.

Comparison of vidar dosimetry advantage pro and epson perfection V700 scanner in densitometry of radiochomic EBT2 film in measurement of high dose gradient

W Bura[1], P Tangboonduangjit[1] and N Damrongkijudom[2]

[1] Department of Radiation Therapy, Ramathibodi Hospital, Mahidol University, 270 Rama 6 Road, Phayathai, Bangkok 10400, Thailand.
[2] Faculty of Medical Technology, Mahidol University, Salaya Campus, 999 Phuttamonthon 4, Salaya Phuttamonthon, Nakhon Pathom 73170, Thailand.

E-mail: wthn_nuk@hotmail.com, puangpen@gmail.com

Abstract. Nowadays the radiochromic film is widely used to obtain dose distribution in two dimensions with high spatial resolution, less energy dependence and near tissue equivalent. It can be a commissioning tool to verify high dose gradient of dose distribution for IMRT and VMAT techniques. However, the film scanner could affect the accuracy of dose distribution if lack of precaution. In this study, the comparison between Epson perfection V700 and Vidar Dosimetry Pro Advantage (RED) is evaluated in terms of the capability to verify the 2D dose distribution for conventional and VMAT techniques. The Gafchromic® EBT2 films were read from two types of scanners (Epson perfection V700 and Vidar Dosimetry Pro Advantage) for volumetric modulated radiation therapy (VMAT) dosimetry. The software for analyzing the results of Epson perfection V700 and Vidar Dosimetry Pro Advantage are SNC Patient software and Omnipro' IMRT software, respectively. Comparisons between measured and calculated dose distributions are reported as %passing rate and the gamma index for tolerance parameters of 3% and 3mm. The study found that the %passing rate obtained from Vidar scanner and Epson V700 scanner compared with Eclipse treatment planning system is more than 98% with the criteria of (3%/3mm).

1. Introduction

Treating patients with radiation is aimed towards the projection of the highest amount of radiation upon the tumor, with least amount of radiation affecting the surrounding tissues in order to cut back the side-effects resulting from radiation itself. Currently, there are numerous developments upon radiation technology and various techniques which mainly focus on the ability to project high amount of radiation on the tumor. Therefore, intensity modulated radiation therapy (IMRT) and volumetric modulated arc therapy (VMAT) are the two impacting techniques that are widely used due to the capabilities to control the projection of radiation upon objects similar to or, those near a to tumor (conformal). Furthermore, the accurate amount of radiation required for treatment plans could be obtained through the standards of measurement along with the suitable equipment for a conclusive result with accuracy. However, areas around the edges of the penumbra and intense radiation exposure techniques are the areas with a high amount of fluctuating radiations (High dose gradient). The examination of the latter technique requires radiation equipment with high spatial resolution and dose distribution (2D dose distribution) [2, 3].

Radiation detectors with ability to measure radiations flat or in a 2D form are often designed through the spread of radiations out to different points, often referred to as "Detector Array". The amount of detectors will determine the resolution of radiation measurement. Universal 2D radiation detectors (2D detector) include those like MapCheck and Matrixx, for instance. These equipment and their resolution upon radiation measurement depend upon the amount of detectors while the film is one of the radiation detectors that can measure radiations flat dimensionally and at the same instance, own higher resolution compared to other detectors. [4] Due to its usage limitations in which requires chemical processes to create visible images and must be processed only in dark rooms without light, the popularity of films gradually fades and led to the production of films that are usable in lighted rooms called "Radiochromic Films" [5, 6].

Among the Radiochromic films, the Gafchromic EBT2 film is the most common film in radiation treatment which is developed from its previous version (EBT film) to lessen the affects resulting from energy dependence, UV sensitivity, scratches from cuts, and further levels and balances the film's texture. In addition to the latter statements regarding the Gafchromic EBT2 film, its components are very similar to the tissues (Z_{eff} value of EBT2 = 6.84, Z_{eff} value of tissues = 7.42) [1] with a high spatial resolution appropriate for areas with high dose gradient of dose distribution for the IMRT and VMAT techniques. Nevertheless, a scanner is required and essential for accurate readings or to analyze results, as it will aid in the restriction of film resolution. Advice from major producers in reading EBT2 value will be needed during red light radiation activation and the scanner as Epson 10000XL, which is high in price. Along with radiotherapy in Thailand, either the "Roller Base Scanner" or the "Vidar Scanner" is used.

This research aims to compare the readings of EBT2 film values with the two scanners: Epson perfection V700 and Vidar Dosimetry Pro Advantage (RED), specifically high dose gradient and penumbra in volumetric modulated arc therapy (VMAT) according to Ramathibodi Hospital's Radiotherapy processes.

2. Materials and methods

Gafchromic EBT2 film (ISP Corporation, Wayne, NJ) owns an non-symmetrical structure with an active layer for UV and visible light protection, where its atomic composition is made up of H (40.85%), C (42.37%), O(16.59%), N(0.01%), Li(0.1%), K(0.01%), Br(0.01%), and Cl(0.04%) [7]. EBT2 film that has not been radiated will give off a yellowish hue; once radiated, areas of radiation will be affected as the initial color yellow changes into blue or navy blue in which the intensity of colors relate to the amount of radiation received. Within this research conducted, the Conventional technique and the advance technique are used along with Varian's Clinac iX linear accelerator: LINAC of 6 MV and treatment planning equipment Eclipse V.8.9.21.

Throughout this research, the EBT2 film was used from the same batch (lot A04251303) in which was cut down into pieces. Each piece has the size of 3×3 cm^2, where the uniformity was measured by placing the film pieces onto a 10 cm-thick model (solid water phantom: RW3, PTW, Freiburg, Germany) and a 5 cm-thick solid water phantom was placed on top of the films. Then the amount of radiation was measured with photon energy of 6 MV, field size of 10×10 cm^2, SSD 95 cm, depth of 5cm, and 200 cGy of radiation. After 24 hours of radiation, the films were placed at the center of the Epson scanner V700 which used for reading the pixel values through a program called "Image J (ROI 1×1 cm^2). Each film will be measured 3 times.

Calibration curve was created as the well-cut EBT2 films which were radiated with a photon energy of 6 MV, field size of 10×10 cm^2, SSD 95 cm, depth of 5 cm, and the amount of radiation as follows: 0, 10, 20, 30, 40, 50, 100, 150, 200, 250, 300, 350, 400, 450, 500, 600, 700, 800 cGy. The setup was similar to that of the uniformity measurement of the films, where each amount of radiation was given onto 3 pieces of film. After 24 hours of radiation, an Epson scanner V700 and Vidar Dosimetry Pro Advantage (RED) was used to read the films in order to create a calibration curve through SNC patient software and Omnipro I'mRT as for the Epson scanner V700 and for Vidar Dosimetry Pro Advantage (RED), accordingly.

In terms of the conventional technique, this technique consists of both open fields and wedge fields with sizes of 60° and a penumbra of 5 × 5 and 10 × 10 cm^2, treatment planning equipment Eclipse TPS, photon energy of 6 MV, SSD 95 cm, depth of 5 cm, and the amount of radiation at 200 cGy. Then, within each treatment plan, export dose plan was performed to compare the EBT2 film that measures with monitoring units resulting from the calculation and that has been setup similar to that of the field's uniformity measurement.

Treatment planning for volumetric modulated arc therapy (VMAT) runs its treatment procedures according to the provisions of TG-119, specifically the mock head and neck, and mock prostate with its CT structure from test suited by Task Group 119 which photon energy of 6 MV of 2 arcs and a half (2Arc - half) by Eclipse treatment planning with AAA algorithm version 8.9.17. For mock prostate, the structures of prostate CTV is roughly ellipsoidal, with the posterior concavity, with right-left, anterior-posterior, and superior-inferior dimensionsof 4.0, 2.6, and 6.5 cm, respectively. The prostate PTV is expanded 0.6 cm around the CTV. The rectum is a cylinder with a diameter of 1.5 cm that abuts the indented posterior aspect of the prostate. The PTV includes about one per three of the rectal volume on the widest PTVslice. The bladder is roughly ellipsoidal with right-left, anterior-posterior, and superior-inferior dimensions of 5.0, 4.0, and 5.0 cm, respectively, and is centered on the superior aspect of the prostate. As part of mock head and neck, The PTV includes all anterior volume from the base of the skull to the upper neck, including the posterior neck nodes. The PTV is retracted from the skin by 0.6 cm. There is a gap of about 1.5 cm between the cord and the PTV. Dose goals used for planning and accepted treatment plan when dose distribution at 95 percent isodose is cover the PTV.[8] Then, a verification plan will be created in a phantom of 30 × 30 × 10 cm^3 and a 5 cm deep export dose plan within each treatment plan in order to compare with EBT2 film that will measure with monitor units obtained from the verification plan. The EBT2 film uses the Isocentric Mounting Fixture (IMF) in which stretches across the head of the gantry, then inserts and tightly locks Mapcheck2 into IMF. After the latter process, place EBT2 film with a size of 8 × 10 cm^2 on the center of the CAX, then a 5 cm-thick solid water phantom on the EBT2 film, and expose radiation according to the treatment plan. Film reading will occur after 24 hours radiation of red light (636 nm) through the Epson scanner V700 (SNC patient software) and Vidar Dosimetry Pro Advantage (RED) (Omnipro I'mRT) with a resolution of 75 and 72 dpi respectively. The comparison between the calculation of dose distribution and values resulting from measurement with EBT2 film will be done with the usage of a % pass rate and Gamma index with a tolerance of 3% and 3 mm.

3. Results

The results have found that scanning with Epson scanner is affected strongly by light source uniformity more than that of Vidar Dosimetry Pro Advantage (RED), especially areas around the edges of the scanned parts which are different from the values measured at the center of the scanning area over 5%. With uniformity measured at the center of the scan area, the results have shown that at each area of the film has a percentage difference within 2%.

Figure 1 and 2 show the calibration curves of Epson Perfection V700 and Vidar Dosimetry Pro Advantage (RED), respectively. Due to different software analysis, Dose (cGy) and Scanner response go on the X and Y axis respectively for the calibration curve of Epson Perfection V700 however Optical Density (OD) and Dose (cGy) are the function in X and Y axis, respectively for the calibration curve of Vidar Dosimetry Pro Advantage (RED).

Once the percentage of the passing rate (% passing rate) obtained from the Eclipse treatment planning is compared to film reading with Epson perfection V700 and Vidar Dosimetry Pro Advantage (RED) through criteria of 3% and 3 mm, it is found that both scanners give off over 98% of the % passing rate in open fields and wedge fields in all areas of penumbra as illustrated in table 1.

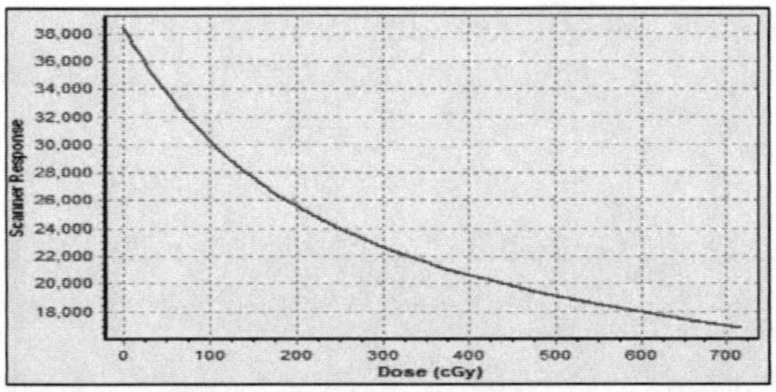

Figure 1. Show the calibration curve of Epson Perfection V700.

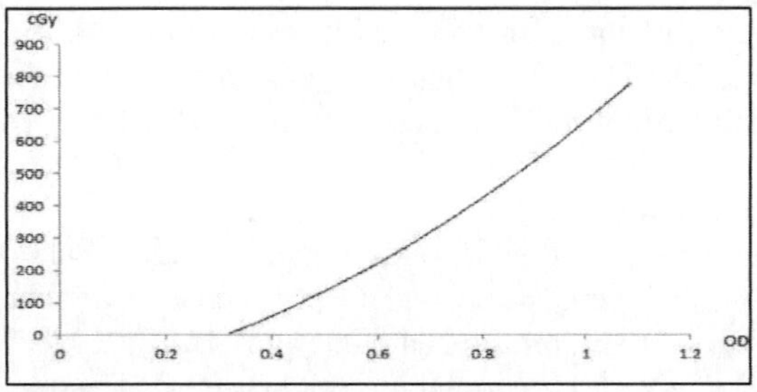

Figure 2. Show the calibration curve of Vidar Dosimetry Pro Advantage (RED).

Table 1. % passing rate in open fields and wedge fields in all areas of penumbra.

	Field Size (cm^2)	Vidar Dosimetry Pro Advantage (RED)	Epson Perfection V700
Open field	5 × 5	99.6	99.8
	10 × 10	99.3	99.1
Wedge Field	5 × 5	98.5	98.5
	10 × 10	98.2	98.1

On the other hand, treatment plans within the mock head and neck, which were done with the VMAT technique, has a % passing rate of 96.5% in which it is slightly more than that of those scanned with Epson perfection V700 (94%). However, a treatment plan in mock prostate gives a % passing rate of more than 96%, scanned by both scanners: Epson perfection V700 and Vidar Dosimetry Pro Advantage (RED). The results are shown in table 2.

Table 2. % passing rate in mock head and neck and mock prostate.

	Vidar Dosimetry Pro Advantage (RED)	Epson Perfection V700
Mock head and neck	96.5	94.0
Mock Prostate	97.6	96.7

4. Conclusion

This research has no means to correct uniformity (uniformity correction) for Epson scanners (which are flatbed scanners) as there is an aim to compare the abilities of the scanners in the appliance to realistic procedures. Within this conducted research, it is crucial to use the software in order to compare and contrast between Epson perfection V700 and Vidar Dosimetry Pro Advantage (RED). As a result, there is a difference in the specification of spatial resolution of 75 dpi for Epson perfection V700 and 72 dpi for Vidar Dosimetry Pro Advantage (RED). With % passing rates of both open fields and wedge fields obtained from the comparison between treatment plan equipment and scanning with Epson perfection V700 in a penumbra of 5 × 5 cm^2 are considered, the outcome is more than that of the Vidar Dosimetry Pro Advantage (RED) which is also measured in the same penumbra. However, when the penumbra area increases within the VMAT, the results have showed that the % passing rate of Vidar Dosimetry Pro Advantage (RED) is slightly more than that of Epson perfection V700. The calibration curve of Epson perfection V700 and Vidar Dosimetry Pro Advantage (RED) are used in different software As a result, the calibration curve is different in plotting. However when analysis by gamma index it gives nearly the same result, so the calibration curve from different software does not affect the evaluation when using the gamma index. For Vidar Dosimetry Pro Advantage (RED), it was difficult for controlling the position of film to be the same position. In contrast Epson perfection V700 will not found this issue. But from preliminary tests about the intensity of the light source, we found that the flatbed scanner (Epson perfection V700) have significant difference of the intensity of light compared to that at the center of the scan area. With results in hand, it is most probable to learn that there is a difference in procedure between light source uniformity and the software used in comparing, which then leads to the main aim of this conducted research in order to encourage future research and studies.

References

[1] Arjomandy B, Tailor R, A Anand, Sahoo N, Gillin M, Prado K and Vicic M 2010 Energy dependence and dose response of Gafchromic EBT2 film over a wide range of photon, electron and proton beam energies *Med. Phys.* **37** 1942–7

[2] Dong L, Antolak J, Salehpour M, Forster K, O'Neill L, Kendall R and Rosen I 2003 Patient-specific point dose measurement for IMRT monitor unit verification *Int. J. Radiat. Oncol. Biol. Phys.* **56** 867–77

[3] Low D A, Mutic S, Dempsey J F, Gerber R L, Bosch W R, Perez C A and Purdy J A 1998 Quantitative dosimetric verification of an IMRT planning and delivery system *Radiother. Oncol.* **49** 305–16

[4] Khan F M and Gibbons J P 2014 *The Physics of Radiation Therapy 5th edition* (Baltimore and Philadelphia: Lippincott Williams & Wilkins)

[5] Niroomand-Rad A, Blackwell C R, Coursey B M, Gall K P, Galvin J M, McLaughlin W L, Meigooni A S, Nath R, Rodgers and Soares C G 1998 Radiochromic film dosimetry: Recommendations of AAPM Radiation Therapy Committee Task Group 55 *Med. Phys.* **25** 2093–115

[6] Pai S, Das I J and Dempsey J F 2007 AAPM TG-69: Radiographic film for megavoltage beam dosimetry *Med. Phys.* **34** 2228–58

[7] Lewis D F 2010 *Radiochromic Film* Wayne New Jersey International Specialty Product

[8] Ezzell G A, Burmeister J W and Dogan N 2009 IMRT commissioning: Multiple institution planning and dosimetry comparisons, a report from AAPM Task Group 119 *Med. Phys.* **36** 5359-79

Journal of Physics: Conference Series **694** (2016) 012019 doi:10.1088/1742-6596/694/1/012019

Determination of output factor for 6 MV small photon beam: comparison between Monte Carlo simulation technique and microDiamond detector

K Krongkietlearts[1], P Tangboonduangjit[1] and N Paisangittisakul[2]

[1] Department of Radiation Therapy, Ramathibodi Hospital, Mahidol University, 270 Rama 6 Road, Phayathai, Bangkok, Thailand, 10400
[2] Department of Physics, Faculty of Sciences, Chulalongkorn University, 254 Phayathai Road, Patumwan, Bangkok, Thailand, 10330

E-mail: komrt39@hotmail.com, puangpen@gmail.com

Abstract. In order to improve the life's quality for a cancer patient, the radiation techniques are constantly evolving. Especially, the two modern techniques which are intensity modulated radiation therapy (IMRT) and volumetric modulated arc therapy (VMAT) are quite promising. They comprise of many small beam sizes (beamlets) with various intensities to achieve the intended radiation dose to the tumor and minimal dose to the nearby normal tissue. The study investigates whether the microDiamond detector (PTW manufacturer), a synthetic single crystal diamond detector, is suitable for small field output factor measurement. The results were compared with those measured by the stereotactic field detector (SFD) and the Monte Carlo simulation (EGSnrc/BEAMnrc/DOSXYZ). The calibration of Monte Carlo simulation was done using the percentage depth dose and dose profile measured by the photon field detector (PFD) of the 10×10 cm^2 field size with 100 cm SSD. Comparison of the values obtained from the calculations and measurements are consistent, no more than 1% difference. The output factors obtained from the microDiamond detector have been compared with those of SFD and Monte Carlo simulation, the results demonstrate the percentage difference of less than 2%.

1. Introduction

Radiotherapy for treatment with cancer involve numerous radiation techniques including the conventional technique (2D), 3DCRT, the development of a computerized radiation technique: "Multileaf Collimator: MLC" that enables radiation to be sheltered and shielded through a computerized program, introducing new techniques like the Intensity Modulated Radiation Therapy (IMRT) and the Volumetric Modulated Arc Therapy (VMAT), following the increasing demands to treat patients [1, 2]. Aimed towards the projection of high radiation upon the tumor and the tissues surrounding the tumor least affected, the projection of high radiation towards the tumor is limited by the Tolerance Dose or the amount of radiation the surrounding tissues received and are affected [3]. However, with radiotherapies like those of the IMRT and the VMAT, treatment extends the range from the ability to control radiation intensity according to the tumor, conformal radiation technique, enabling the affected tissues in the surrounding area to be controlled whilst the radiation upon the tumor can be increased. Therefore, these two techniques are widely accepted as treatments for patients.

Through the continuous movement of the MLC, radiation intensity can be controlled in both IMRT and VMAT. As radiation is performed dynamically to fully cover the tumor cell according to its figure, "beamlets" are produced where its sizes depend upon the sizes of MLC's leaf and its movement [4, 5]. Therefore, it is observed that IMRT and VMAT have smaller fields compared to that of the traditional radiotherapy. Treatment involves accuracy, that is, the amount of radiation the tumor receives must be as stated in the treatment plan. Thus, the study aims to reduce errors resulting from radiation in which involves an experiment to determine the specific beam characteristic and is a first step to controlling the errors of radiation and beam data collection. As beams are used as a standard in the calculation, accurate percentage depth dose (PDD), beam profile, and output factor will determine patient's treatment plans.

To accurately measure a small field area, it is crucial to understand the characteristics of the field and the measuring process. The measurement of radiation within a small field may result in errors in measurement due to the lack of charged particle equilibrium (Non-CPE) and the errors in the measurement of field width that overlaps with the edges, or the penumbra, and perturbation that results in volume averaging measurement [6, 7]. Thenceforth, the measurement of radiation in a small field requires a suitable detector which is dose rate independent, energy independent, angular independent, high in resolution and sensitivity, active in linearity, small and less in perturbation, and lastly, tissue equivalent [8].

Radiation detectors used in small fields are those of ionization chambers (IC) and semiconductors. Although the development of smaller ionization chambers is in progress, the detectors are not small enough to reduce perturbation [9], which then leads to volume averaging that results in underestimation. As the detectors become smaller in size, it also leads to lower sensitivity. In contrast to the latter, semiconductors like diode have higher sensitivity than IC. It is smaller in size, and higher in measurement resolution. However, the disadvantage of this radiation detector is its sensitivity that depends upon the dose rate (dose rate dependent), is angular dependent, and is non-water equivalent, which leads to the possibility of an increase in a secondary electron that then results in overestimation [10]. As a result, there is yet a suitable detector to measure the amount of radiation projected over a small field without any disadvantages. However, Ramathibodi hospital uses the stereotactic field detector (Scanditronix IBA SFD) with a diameter of 0.6 mm and sensitive volume of 0.017 mm^3 in the radiotherapy process to measure small fields. Currently, detectors are still in development to suit small fields, shifting from PTW's diamond detector to microDiamond detector (PTW-60019) with a diameter of 2.2 mm and a sensitive volume of 0.004 mm^3, which is 4 times smaller than that of SFD's. Moreover, the sensitivity of energy dependence and dose rate dependence are lowered along with the characteristics of being tissue equivalent, marking it a beneficial detector towards small field, but it is a costly tool. Furthermore, as the measurement of small fields may result in errors and lacks the standard conditions in measurement compared to large field, the evaluation of c measurement correct processes and controls are as well difficult. Monte Carlo simulation can be used to track the interactions between particle and media in random or stochastic process. With random numbers, probability distribution, the fundamentals of physics, and the absorbing of center photon radiation as interaction forecasting factors, the study of complex problems can be simpler and will further lead to standard acceptance [11].

This study aims to compare the relative output factor obtained from the measurement of the stereotactic field detector (Scanditronix IBA SFD), which is considered as a standard detector for small fields at Ramathibodi hospital, and microDiamond detector (PTW Freiburg, Germany, PTW-60019), with the outcome from Monte Carlo simulation.

2. Materials and method

2.1. Monte Carlo simulation

Monte Carlo Simulation models a linear accelerator (LINAC) Clinac iX through EGSnrc which is divided into the physical part model in linear accelerator coded: BEAMnrc, using data and information obtained from (Monte Carlo Data Package: High Energy Accelerator DWG NO. 100040466-02) and calculation of radiation done in virtual mediums coded with DOSXYZnrc.

The simulation of radiation projectors includes several unknown parameters as a model. Here, the energy of initial electrons beam and its radial intensity distribution width are in our consideration. To find these parameters percentage depth dose and the dose profiles at 5 cm and 10 cm depth, for the field size of 10 × 10 cm^2 in water obtained from the simulation are used for the comparison with the measured data. The comparison of these factors involves systematic changes in the values of both parameters. The set of parameters yielding the lowest Chi-square value and the difference in comparison between measurement and simulation of less than ± 1% is selected for the model of the linac. Then, the radiation-in-water for the field sizes of 1 × 1, 5×5 and 10 × 10 cm^2 at 5 cm and 10 cm depths are simulated with the DOSXYZnrc code. The relative outputs are then evaluated and compared with the measured values obtained from SFD and microDiamond detector (PTW Freiburg, Germany, PTW-60019).

2.2. Measurement

Radiation measurement (dosimetry) is divided into two sectors: (i) common beam data for simulation examination and (ii) relative output factor for the comparison with the simulation.

Common beam data requires the SFD in the measurement of depth dose, dose profiles with SSD 100 cm, F.S 10 × 10 cm^2, and depths of 5 and 10 cm of dose profile with IBA Blue Phantom2 through a step-by-step scan with a resolution of 1 mm. They are the database to find the energy of initial electron beam and radial intensity distribution width used in beam simulation. The microDiamond detector (PTW Freiburg, Germany, PTW-60019) is a synthetic single crystal diamond detector with a small sensitive volume of 0.004 mm^3 (smaller than that of SFD's which is: 0.017 mm^3) and is tissue equivalent. It has been used to measure the absorbed dose across each field size at 5 and 10 cm depth five times. The average dose value at the beam center is required to compute the relative output factor which is later compared with those obtained from the Monte Carlo simulation.

3. Results

For our MC simulation, the best energy of initial electron beam and the full width at half-maximum (FWHM) of the radial intensity is 6.2 MeV and 0.6 mm, respectively. These parameters give the difference between measurement and simulation of less than 1%. Consequently, the percentage depth dose and dose profile, at depth 5 and 10 cm, for the square field sizes between 1 × 1 and 5 × 5 cm^2 have been measured and compared with the values obtained from Monte Carlo simulation. The results from the measurement match those of simulation well where the difference does not exceed over 2%.

Table 1. Comparison of between the relative output factors obtained from the measurement using SFD and microDiamond detector and from the calculation using Monte Carlo simulation (MC). The percentage difference from the MC value is given in parenthesis.

Type	Relative Output Factor (%diff)							
	Depth 5 cm				Depth 10 cm			
	1×1 cm^2	2×2 cm^2	3×3 cm^2	5×5 cm^2	1×1 cm^2	2×2 cm^2	3×3 cm^2	5×5 cm^2
SFD	0.7574 (1.6)	0.8786 (1.3)	0.9140 (1.4)	0.9373 (1.2)	0.6995 (1.5)	0.8243 (1.8)	0.8669 (1.6)	0.9052 (1.3)
microDiamond	0.7586 (1.7)	0.8799 (1.4)	0.9115 (1.1)	0.9381 (1.2)	0.7028 (1.9)	0.8113 (1.7)	0.8645 (1.3)	0.8920 (1.3)
Calculation (MC)	0.7457	0.8673	0.9018	0.9266	0.6893	0.8099	0.8533	0.9037

In the case of the relative output factor, the measurement results from both the SFD and microDiamond detector have been compared with the simulated results where the differences are found to be less than 2% as illustrated in table 1.

4. Conclusion

This study has found that microDiamond detector (PTW Freiburg, Germany, PTW-60019 owns appropriate characteristics in the measurement of the output factor in small fields although it's high price. The results of this study have revealed that the measurement of output factor in small fields results in same values obtained from SFD, with a difference not exceeding 2%. At any rate, this study aims to compare the values of the output factor obtained from the measurement of both detectors with the calculation. Moreover, this study does not involve the study of specific characteristics of the micrDiamond detector, therefore, the information obtained from this study may not be enough to point the advantages and disadvantages, the necessity upon purchases, replacement, or encourage promotional agencies the detectors in use.

References

[1] Nutting C 2003 Intensity-modulated radiotherapy (IMRT): the most important advance in radiotherapy since the linear accelerator? *Br. J. Radiol.* **7** 673

[2] Van Dyk J (Ed) 1999 *The Modern Technology of Radiation Oncology: A Compendium for Medical Physicists and radiation Oncologists* (Madison (WI): Medical Physics Publishing)

[3] Lee N Y and Terezakis S A 2008 Intensity-modulated radiation therapy *J. Surg. Oncol.* **97** 691-6

[4] Weeb S 2003 The Physical basis of IMRT and inverse planning *Br J. Radiol.* **76** 678-89

[5] Arno J M and John C R 2005 *Intensity modulated radiation therapy: A clinical perspective* (London: Hamilton London BC Decker Inc.)

[6] Das I J, Ding G X and Ahnesjo A 2008 Small fields: Nonequilibirium radiation dosimetry *Med. Phys.* **35** 206-15

[7] Laub W and Wong T 2003 The Volume effect of detectors in the dosimetry of small fields used in IMRT *Med. Phys*. **30** 341–7

[8] Low D A, Moran J M, Dempsey J F, Dong L and Oldham M 2011 Dosimetry tools and techniques for IMRT *Med. Phys*. **38** 1313–38

[9] Sauer OA, Wilbert J 2007 Measurement of output factors for small photon beams *Med. Phys.* **34** 1983-88

[10] Eklund K and Ahnesjo A 2010 Modeling silicon diode dose response factors for small photon fields *Phys. Med. Biol.* **55** 7411-23

[11] Rogers D W O, Faddegon B A, Ding G X, Ma C M, We J and Mackie T R 1995 BEAM: A Monte Carlo code to simulate radiotherapy treatment units *Med. Phys.* **22** 503-24

[12] Hendee W 2013 *Monte Carlo Techniques in Radiation Therapy* Seco J and Verhaegen F Editor Boca Raton CRC Press Taylor & Francis Group

Neutron contamination of Varian Clinac iX 10 MV photon beam using Monte Carlo simulation

S Yani[1,2], R Tursinah[3], M F Rhani[4], R C X Soh[5], F Haryanto[1] and I Arif[1]

[1] Department of Physics, Faculty of Mathematics and Natural Sciences, Institut Teknologi Bandung, Jalan Ganesa 10, Bandung 40132, Indonesia
[2] Akademi Kebidanan Pelita Ibu Kendari, Jalan Syech Yusuf 12 Kendari, Indonesia
[3] National Nuclear Energy Agency, Jalan Tamansari 71 Bandung, Indonesia
[4] Department of Radiation Oncology, National University Cancer Institute Singapore, Singapore/Singapore
[5] School of Physical and Mathematical Sciences, Nanyang Technological University, Singapore

E-mail: sitti.yani@s.itb.ac.id

Abstract. High energy medical accelerators are commonly used in radiotherapy to increase the effectiveness of treatments. As we know neutrons can be emitted from a medical accelerator if there is an incident of X-ray that hits any of its materials. This issue becomes a point of view of many researchers. The neutron contamination has caused many problems such as image resolution and radiation protection for patients and radio oncologists. This study concerns the simulation of neutron contamination emitted from Varian Clinac iX 10 MV using Monte Carlo code system. As neutron production process is very complex, Monte Carlo simulation with MCNPX code system was carried out to study this contamination. The design of this medical accelerator was modelled based on the actual materials and geometry. The maximum energy of photons and neutron in the scoring plane was 10.5 and 2.239 MeV, respectively. The number and energy of the particles produced depend on the depth and distance from beam axis. From these results, it is pointed out that the neutron produced by linac 10 MV photon beam in a typical treatment is not negligible.

1. Introduction

Nowadays, radiotherapy has become the most powerful tool to treat cancer in many countries. Some new technologies are developed to increase the effectiveness of treatments, both of photon and electron beam. For example, intensity modulated radiation therapy (IMRT) and volumetric modulated arc therapy (VMAT) technique use high energy medical accelerator photon beam by controlling the multileaf collimator (MLC) movement during the treatment to kill cancer without any harming in the healthy tissue surrounding the cancer [1,2]. Generally, the linear accelerator (linac) used for IMRT and VMAT is high energy linac. Neutrons can be emitted from medical accelerator if incident X-ray hits any its materials. This unwanted neutrons are mostly created by photonuclear reactions. High energy photons created in the high energy photon mode of the linac interact with the nuclei of high Z materials inside the accelerator and liberate neutrons [3]. These neutrons scatter throughout the treatment room and reach the patient. As neutrons have high relative biological effectiveness (RBE), even small neutron doses may be harmful to the patient [4, 5] and medical staff [6]. In this case, special radiation protection methods have to be implemented in order to prevent patients and medical staff from exposure to it. This study concerns the simulation of neutron contamination emitted from Varian Clinac iX 10 MV photon beam using the Monte Carlo code system. The neutron production process is so complex, and therefore Monte Carlo simulation with MCNPX and PHITS code system was carried out to study this contamination. Many studies have been performed to characterize the secondary neutrons from high-energy linacs using Monte Carlo simulation [7, 8].

The aim of this study is to investigate the neutron contamination in Varian Clinac iX 10 MV photon beam.

2. Methods

Monte Carlo calculation MCNPX version 2.4.0 and PHITS 2.81 Monte Carlo code were used to model photon beams from the Varian Clinac iX 10 MV Photon beam. MCNPX is a well-known general purpose of Monte Carlo code developed at Los Alamos National Laboratories. It extends the transport capability of Monte Carlo program to include 34 particles over a more complete energy range. This range (<150 MeV) of energy can be used to simulate radiation transport of these particles in radiation therapy applications. Figure 1 shows the setup of the simulated geometry for the Varian Clinac iX 10 MV modelled by PHITS.

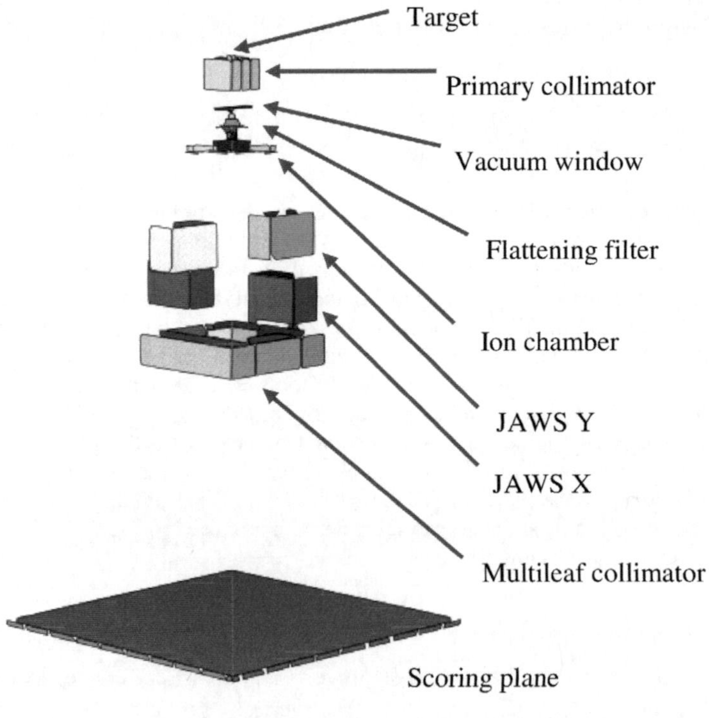

Figure 1. Sketch Varian Clinac iX 10 MV photon beam.

The design of this medical accelerator was modelled based on on the materials and geometry of its actual. In this simulation, the optimal initial electron beam parameters were 10.5 MeV in energy. The electron source with a point source was put before the target. The simulated model included the bremsstrahlung target, the primary collimator, the vacuum window, the flattening filter, the monitor ion chamber, the mirror, and the upper and lower jaws. A beam monitoring chamber and flattening filter were accurately modelled due to the fact that they are the main sources of contaminating electrons. This detailed description of the geometry required for the accurate simulation was provided by the manufacturer. The electron and photon cut-off energies were set to 0.5 MeV and 0.01 MeV, respectively. Table 1 represents the material of each linac component and its density used in simulation.

Table 1. Details of CMs parameters.

CMs name	Distance from reference plane (cm)	Materials	Density (g/cm^3)
Target	0	Copper	8.96
Primary Collimator	5.5	Tungsten	19.25
Exit windows	8.66	Beryllium	1.85
Flattening filter	9.2499	Copper	8.96
Ion Chamber	14.79	Kapton and Air	1.42 and 0.0012
JAWS Y	27.88	Blok Tungsten	19.25
JAWS X	36.63	Blok Tungsten	19.25
MLC	48.185	Tungsten	19.25
Scoring plane	100	Water	1

The MLC are used to create a square field size (10 × 10 cm^2) at the isocenter. Up to this point, the simulations included geometries for the MLC and upper jaws and lower jaws that were not quite representative for the Varian machine (the tip of MLC and JAWS not rounded shape). The number of history used in this simulation was 2×10^9 particles. The F4 and *F4 tallies were used to calculate the fluence profiles, the energy spectra, and the mean energy profiles at the phantom surface. The maximum statistical uncertainty of our results was about 7% for the neutron part and 0.4% for the electron and photon part in the scoring plane. The importance of electron, photon and neutron was set to 1 for each linac component (IMP: E 1, IMP: P 1, IMP: N 1).

3. Results

Figure 2 shows the photon, positron, electron, and neutron spectra calculated by PHITS code system in the target and in 50 × 50 cm^2 the scoring plane (100 cm from target).

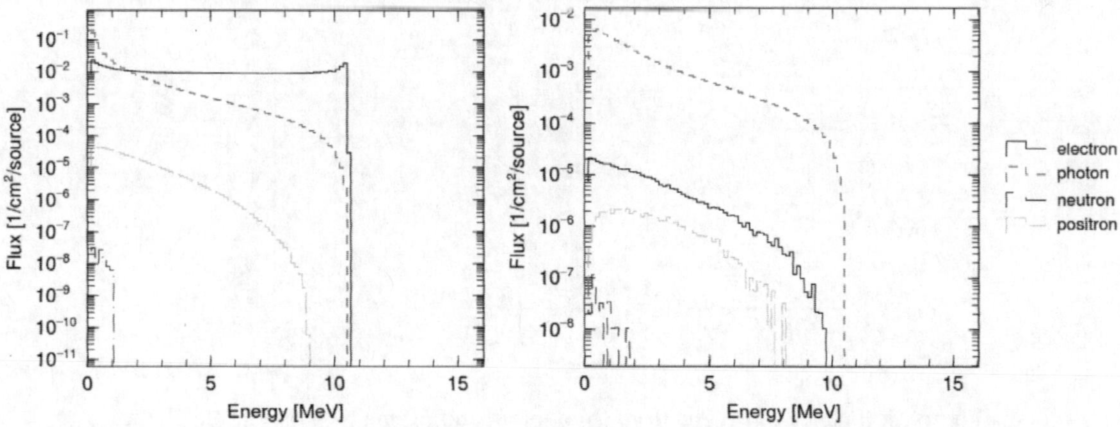

Figure 2. Photon (red), positron (green), electron (black), and neutron (blue) spectra after target and in the depth 100 cm generated by 10 MV Varian Clinac iX 10 MV photon beam.

The neutron energy produced in the target is less than 1 MeV, The neutron with energy more than 1 MeV was produced in the depth 100 cm from target. The electron dominated the particle in the target component module with a maximum energy of 10.5 MeV.

Journal of Physics: Conference Series **694** (2016) 012020 doi:10.1088/1742-6596/694/1/012020

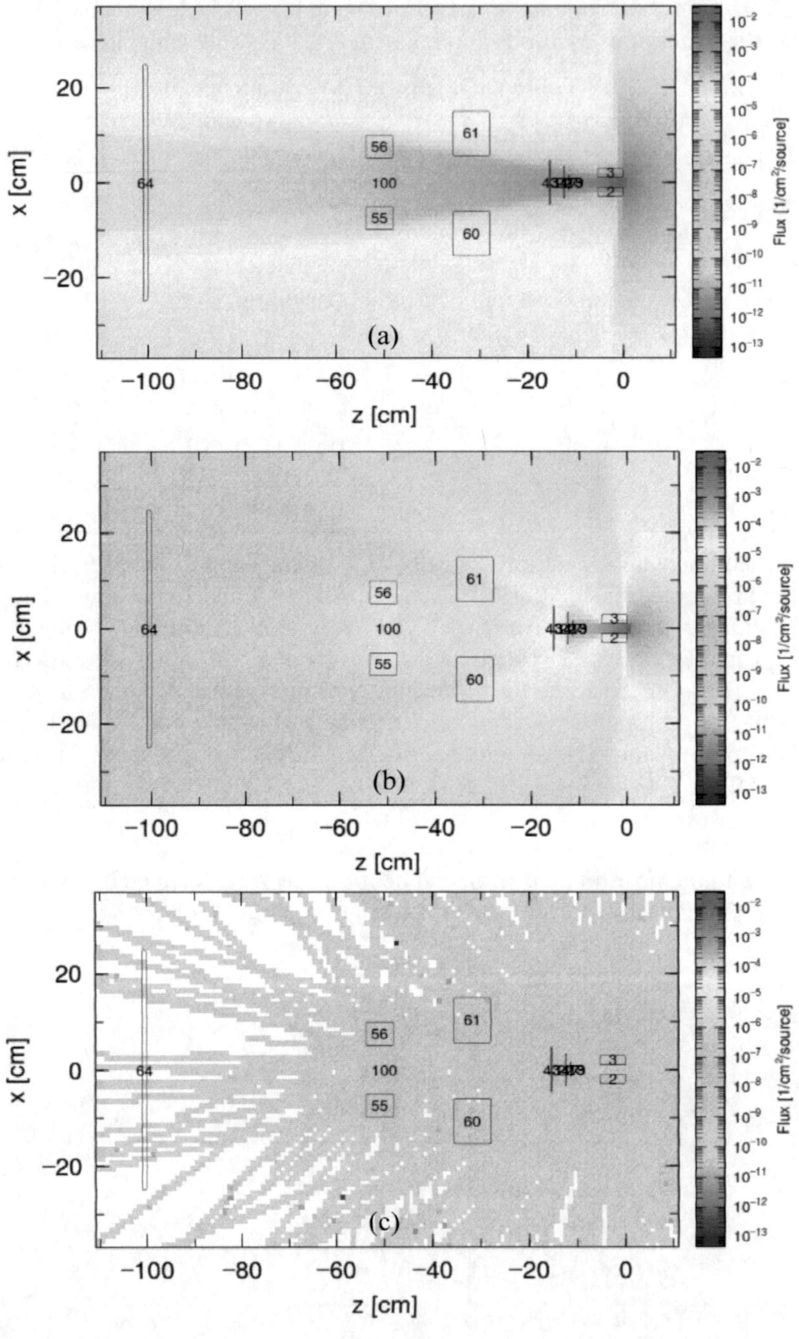

Figure 3. Fluence of (a) electron, (b) photon, and (c) neutron of Varian Clinac iX 10 MV photon beam.

Figure 3 shows the fluence of electron, photon and neutron in x-z plane. The unwanted particles (electron and neutron) give contribution to this photon beam but most of them are scattered to the treatment room. The statistical uncertainties in the simulation depend upon the number of particle histories used to determine the results. In this case the uncertainties are +7% due to the lack of neutron in the scoring plane. Another uncertainty was caused by uncertainties in the photoneutron cross sections. This systematic uncertainty has resulted in the overall uncertainty in the calculated results of both fluence and energy in the range from *25% to +35%.

Table 2 shows the maximum and minimum energy of neutron produced in some component modules of linac Varian Clinac iX 10 MV photon beam.

Table 2. The maximum and minimum energy of neutron in some components of linac.

Energy (MeV)	Target	Primary Collimator	JAWS Y	MLC	Scoring plane
Maximum	1.778	2.818	1.41	1.512	2.239
Minimum	0.005	0.0028	0.0199	0.045	0.01

4. Discussion

The linac incident particle was electron with a maximum energy of 10 MeV. In Figure 2, it can be seen that the electron and photon dominated the particles in the target and scoring plane, respectively. The maximum energy of photons in the scoring plane was 10.5 MeV. The number of photons, electrons and neutrons decreased gradually with the depth. The bremsstrahlung process dominated the reaction in the target (Cu with the thickness of 0.508 cm) that produced gamma. The photon passed the primary collimator, vacuum window, flattening filter, ion chamber, JAWS Y and JAWS X and multileaf collimator. The bremsstrahlung radiation through vacuum window was modified by a flattening filter. The energy fluence flattens after the flattening filter (Figure 3 (a)).

Figure 3(b) and (c) show the unwanted electron and neutron produced by the linear medical accelerator. The electrons and photons produced in high energy linacs may undergo electroneutron (e,e'n) interactions and photoneutron (γ,n) interactions, respectively. Both types of interactions produce neutrons as by-products. Electroneutron production is 100 times less probable and therefore neglected in the simulations. Photoneutron production is governed by the neutron separation energy and by photoneutron cross-sections. Figure 3(c) shows the neutron fluence in x-z plane. Most of the neutron is scattered in the treatment room and this neutron effect was not good for medical staff.

Table 2 below shows that the energy of neutron changed depending on the distance between linac component, the target, and the distance to the beam axis. This results have a good agreement with another study that the energy and amount of neutron decreases gradually from the center of the field [9].

5. Conclusion

From these results it is pointed out that the neutron produced by linac 10 MV photon beam in a typical treatment is not negligible. Therefore, this new system, consisting of Monte Carlo simulation and spectral measurements, represents a reliable approach to evaluate the leakage neutron field and to optimize the treatment plan.

References

[1] Zelefsky M J *et al.* 2000 *Radiother. Oncol.* **55(3)** pp 241-9
[2] Wolff D *et al.* 2009 *Radiotherapy and Oncology* **93** pp 226–33
[3] National Council on Radiation Protection and Measurements 1984 *Neutron Contamination from Medical Electron Accelerators* (Madison, WI: Medical Physics Publishing)
[4] Wakabayashi G, Fukunaga J, Hosono M, Nohtomi A, Umezu Y, Itoh T, Yahiro E, Nakamura Y, Fujibuchi T and Nakamura K 2014 *Japanese Society of Radiological Technology and Japan Society of Medical Physics*
[5] Mao X S, Kase K R, Liu J C, Nelson W R, Kleck J H and Johnsen S 1997 *Health Phys.* **72** 524-529
[6] McGinley P H 1998 *Shielding Techniques for Radiation Oncology Facilities* (Madison, WI: Medical Physics Pub. Corp)
[7] Barquero R, Mendez R and del Castillo A 2004 Collateral Patient Doses in the Varian 21iX Radiotherapy Linac. Madrid, Spanyol

[8] Barquero R, Mendez R, Vega-Carrillo H R, Iniguez M P and Edwards T M 2006 *Health Physics Society*
[9] Hashemi S M, Malayeri B H, Raisali G, Shokrani P and Sharafi A A 2007 *Nukleonika* **52(1)** pp 39–43

Acknowledgements
The authors thank Varian Medical Systems for providing technical data of linear accelerator to be modelled with Monte Carlo. This study was fully supported by Hibah Disertasi Doktor 2016 and Kerjasama Luar Negeri Dikti, Ministry of Education Indonesia 2016.

Journal of Physics: Conference Series **694** (2016) 012021 doi:10.1088/1742-6596/694/1/012021

Dosimetric investigation of proton therapy on CT-based patient data using Monte Carlo simulation

T Chongsan[1]**, T Liamsuwan**[2] **and P Tangboonduangjit**[1]

[1] The School of Medical Physics, Faculty of Medicine Ramathibodi Hospital, Mahidol University, Bangkok, Thailand

[2] Nuclear Research and Development Division, Thailand Institute of Nuclear Technology (Public Organization), Ongkharak, Nakorn Nayok, Thailand

E-mail: Thanaphat.cho@student.mahidol.ac.th

Abstract. The aim of radiotherapy is to deliver high radiation dose to the tumor with low radiation dose to healthy tissues. Protons have Bragg peaks that give high radiation dose to the tumor but low exit dose or dose tail. Therefore, proton therapy is promising for treating deep-seated tumors and tumors locating close to organs at risk. Moreover, the physical characteristic of protons is suitable for treating cancer in pediatric patients. This work developed a computational platform for calculating proton dose distribution using the Monte Carlo (MC) technique and patient's anatomical data. The studied case is a pediatric patient with a primary brain tumor. PHITS will be used for MC simulation. Therefore, patient-specific CT-DICOM files were converted to the PHITS input. A MATLAB optimization program was developed to create a beam delivery control file for this study. The optimization program requires the proton beam data. All these data were calculated in this work using analytical formulas and the calculation accuracy was tested, before the beam delivery control file is used for MC simulation. This study will be useful for researchers aiming to investigate proton dose distribution in patients but do not have access to proton therapy machines.

1. Introduction

The aim of radiotherapy is to delivery high radiation dose to the tumor with low radiation dose to healthy tissue, resulting in the increased therapeutic ratio [1]. Conventional radiotherapy has limitation because photons have exit dose to healthy tissue behind the tumor which can cause side effects such as secondary malignancy. In contrast, protons have Bragg peaks with high radiation dose to the tumor but low exit dose or dose tail. Therefore, proton therapy is promising for treating deep-seated tumors and tumors locating close to organs at risk [2]. Compared to current radiotherapy technique such as intensity modulated radiation therapy (IMRT) and volumetric modulated arc therapy (VMAT) using photons, proton therapy can decrease secondary malignancy risk [3]. In particular, the pencil beam scanning technique is considered to be advantageous for decreasing risks of developing secondary cancer after proton therapy compared to the passive scattering technique [4]. The physical characteristic of protons was shown to be suitable for treating cancer in pediatric patients [5], especially primary brain tumors such as high-grade glioma (HGG) that is aggressive and has high recurrent rates in pediatric patients [6].

This work is a part of the development of a computational approach for calculating proton dose distribution in patients treated with the pencil beam scanning method. The computational platform consists of a MATLAB (MathWorks, Inc.) optimization program for generating a beam delivery control file and the Particle and Heavy Ion Transport Code System (PHITS) for Monte Carlo simulations to calculate proton dose distribution. The Monte Carlo simulation will be use the beam configuration as described by the beam delivery control file and patient-specific CT data. The conversion of patient-specific CT-DICOM data to the Monte Carlo input and the calculation of monoenergetic proton beam data for the development of the optimization program will be described in this paper.

2. Material and method

The studied case is a pediatric patient with a primary brain tumor (HGG). The selected patient was originally treated by photon therapy at Radiation Therapy and Oncology Section, Radiology Division, Faculty of Medicine Ramathibodi Hospital, Mahidol University Bangkok Thailand. The CT-DICOM files of this patient were used for the calculation of the patient-specific proton dose distribution.

In the first step, the CT-DICOM files were converted to the PHITS input file using the DICOM2PHITS program, which is a part of PHITS. The conversion accuracy was tested against known values of electron densities in the Tissue Characterization Phantom (TCP) Gammex^{TM} 467. The TCP rods are displayed in figure 1(a). The verification process of the DICOM2PHITS program followed these steps:

- The TCP rods were scanned by a CT simulator (radiotherapy simulated exposure technique ; GE Optima 580, 120 kV, 255 mA, 1.25 mm slice thickness, 512 x 512 pixels).
- The TCP rod CT-DICOM files were sent to the DICOM2PHITS program for conversion to the Monte Carlo input file.
- The Monte Carlo input file was read by PHITS Version 2.76 for the simulation.
- The electron density value on the rod label was compared with the Monte Carlo output for each rod.
- Percentage differences between the calculated electron densities and the references should be less than 3%. However, if the percentage difference was more than 3% but less than 4%, it was considered acceptable. The selected tolerance was modified from [7].

A MATLAB optimization program was developed to create the beam delivery control file for the patient-specific Monte Carlo simulation. Apart from the patient's anatomical data, the optimization program required the depth dose distributions of monoenergetic protons, the proton range-energy relationship, the stopping power ratios to water for protons in different materials and the scattering characteristics of protons in different materials. All these data were calculated using analytical functions, as follow:

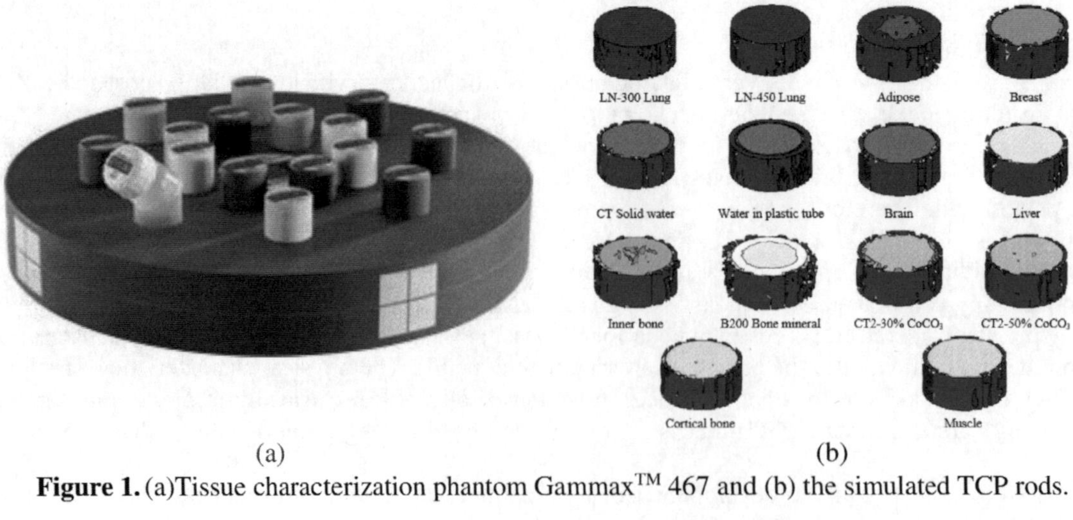

(a) (b)

Figure 1. (a)Tissue characterization phantom Gammax^{TM} 467 and (b) the simulated TCP rods.

- Depth dose of monoenergetic proton [8]

$$D(z) = \Phi_0 \frac{e^{-\varsigma \frac{2}{4}\sigma \frac{1}{p}}\Gamma(1/p)}{\sqrt{2\pi\rho p\alpha^{1/p}(1+\beta R_0)}} \times \left[\frac{1}{\sigma} D_{-1/p}(-\varsigma) + \left(\frac{\beta}{p} + \gamma\beta\right) D_{-1/p}(-\varsigma)\right] \quad (1)$$

where Φ_0 is the primary fluence, $\Gamma(x)$ is the gamma function, β is the slope parameter of the fluence reduction relation and γ is the fraction of locally absorbed energy from nuclear interactions. The parameters α and p stem from the range-energy relationship, Eq. (2).

- Proton range-energy relationship [9]

$$R_0 = \alpha E_0^p \quad (2)$$

where α is 0.002579 that is approximately proportional to the square root of the effective atomic mass of the absorbing medium, E_0 is initial energy of protons and p is 1.736 for energies between 10 and 250 MeV (Geiger's rule). The α and p were fitted from Monte Carlo simulation.

- Stopping power ratio to water for protons in different material (Bragg's rule) [10, 11]

$$\left(\frac{S}{\rho}\right) = \sum W_i \left(\frac{S}{\rho}\right)_i \quad (3)$$

where w is the fraction by weight and $\left(\frac{S}{\rho}\right)$ is the mass stopping power of element i.

- In-patient scattering characteristics of proton in different materials [12]

$$\theta_0 = 14.1\left[1 + \frac{1}{9}\log_{10}\left(\frac{t}{L_R}\right)\right]\left[\int_{-\infty}^{t}\left(\frac{1}{pv}\right)^2 \frac{dt\prime}{L_R}\right]^{1/2} \quad (4)$$

$$y_0(t') = 14.1\left[1 + \frac{1}{9}\log_{10}\left(\frac{t\prime}{L_R}\right)\right]\left[\int_{-\infty}^{t\prime}\left(\frac{t'-z}{pv}\right)^2 \frac{\rho}{L_R} dz\right]^{1/2} \quad (5)$$

where θ_0 is the angular distribution of proton, y_0 is radial distribution of proton, t in grams per square centimetre is the areal density of the material, L_R in grams per square centimetre is the radiation length of the material, pv in MeV is the product of momentum and the velocity of the protons at depth t' and ρ is the density of the material in grams per cubic centimeter.

The accuracy of these functions was tested by comparing the calculation results with literature data or Monte Carlo simulations of monoenergetic protons, before the delivery control file is used for simulating patient-specific proton dose distribution. The results of equations (1) and (2) were compared with the Monte Carlo simulations of monoenergetic protons of different energies, while the results of equation (3) were compared with the material data of PSTAR (Stopping Power and Range Tables for Protons) [11] and the material data implemented in PHITS, and the results of equations (4) and (5) were compared with available literature data [12]. In the final step, outside the scope of this paper, the patient-specific proton dose distribution from the PHITS Monte Carlo simulation and the dose distribution to be achieved according to the optimization program will be compared using the Gamma evaluation index of 3%, 3 mm (%dose difference and DTA (Distance to Agreement)), with the passing criterion of more than 95% (the similar criterion was used in Grevillot *et al.* [13]).

3. Result and discussion

To test the accuracy of the DICOM2PHITS program, the converted and referenced electron densities were compared for 14 different materials. The simulated rod images are displayed in figure 1(b). The electron densities were derived from the CT number with the conversion table that has been constructed with the best accuracy [14]. The conversion was accurate for 8 material rod values (differences were less than 3%), was moderately accurate for 4 material rod values (differences were more than 3% but less than 4%) which may be caused by the broad ranges of CT numbers per material of the DICOM2PHITS database, and was insufficiently accurate for 2 material rod values representing lung equivalent materials LN-300 and LN-450 due to the low electron densities of these materials.

Regarding the proton beam data to be used in the MATLAB optimization program, the depth dose distributions of monoenergetic protons, the range-energy relationship and the stopping power ratios to water for protons in different materials were calculated based on equations (1)-(3). These results are in good consistency with the literature data or Monte Carlo simulations of monoenergetic proton beams. Some of the depth dose calculation results are displayed in figure 2. Investigations of in-patient scattering characteristics of protons and the Monte Carlo simulation of patient-specific proton dose distribution are works in progress.

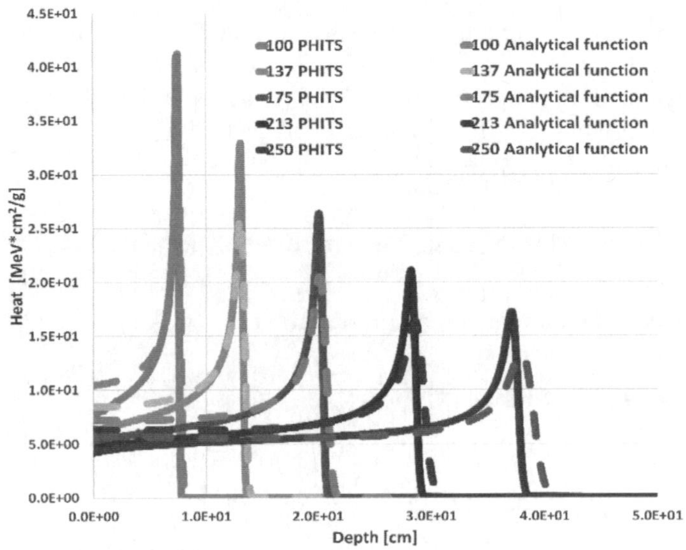

Figure 2. Proton depth dose comparison between PHITS simulation vs. analytical function in equation (1)

4. Conclusion

The paper describes the data preparation for the Monte Carlo simulation of patient-specific proton dose distribution. The described works include the conversion of CT-DICOM files to the input file of the Monte Carlo simulation and the generation of proton beam data to be used in a beam optimization program for creating a beam delivery control file for the Monte Carlo simulation. The DICOM2PHITS program used in this work was found to be suitable for conversion of CT-DICOM data to the Monte Carlo input file for most of the materials relevant to radiation therapy. The analytical functions used for calculating the proton beam data for the beam optimization program were found to be sufficiently accurate compared to literature data and Monte Carlo simulations of monoenergetic protons. This work is a part of a pilot project for development of a computational platform for dosimetric investigation of proton therapy using Monte Carlo simulation and patient CT data. The outcome of this project will be interesting for researchers aiming to investigate patient-specific proton dose distribution but do not have access to proton therapy machines and related equipment.

References

[1] Podgoršak, E B and International Atomic Energy Agency 2005 *Radiation oncology physics: A handbook for teachers and students* (Vienna: International Atomic Energy Agency)
[2] Khan F M 2010 *The Physics of Radiation Therapy 4th edition* (Lippincott Williams & Wilkins)
[3] Paganetti H, Athar B S, Moteabbed M, Adams J A, Schneider U and Yock TI 2012 *J. Phys. Med. Biol.* **57** 6047
[4] Moteabbed M, Yock TI and Paganetti H 2014 *J. Phys. Med. Biol.* **59** 2883

[5] Zhang R, Howell R M, Giebeler A, Taddei P J, Mahajan A and Newhauser W D 2013 *J. Phys. Med. Biol.* **58** 807
[6] Fangusaro J 2012 *J. Front Oncol.* **2** 1
[7] Schneider W, Bortfeld T and Schlegel W 2000 *J. Phys. Med. Biol.* **45** 459
[8] Grassberger C, Lomax A and Paganetti H 2015 *J. Phys. Med. Biol.* **60** 633
[9] Bortfeld T 2014 *J. Med. Phys.* **24** 2024
[10] Thwaites D I 1992 *J. NIMB.* **69** 53
[11] National Institute of Standards and Technology (NIST), *http://physics.nist.gov/PhysRefData/Star/Text/PSTAR.html*
[12] Hong L 1996 *J. Phys. Med. Biol.* **41** 1305
[13] Grevillot L, Bertrand D, Dessy F, Freud N and Sarrut D 2011 *J. Phys Med Biol.* **56** 5203
[14] Kanematsu N, Inaniwa T and Koba Y 2012 *J. Med. Phys.* **39** 1016

Acknowledgments
The authors would like to acknowledge National e-Science Infrastructure Consortium for providing computing resources that have contributed to the research result reported in this paper (http://www.e-science.in.th). This research is partially supported by Graduate Studies of Mahidol University Alumni Association. TC would like to thank all medical physicists at Radiation Therapy and Oncology Section, Radiology Division, Faculty of Medicine Ramathibodi Hospital, Mahidol University for providing the experimental instruments and Thailand Institute of Nuclear Technology (Public Organization) for the beam optimization program used in this study.

Comparison of build-up region doses in oblique tangential 6 MV photon beams calculated by AAA and CCC algorithms in breast Rando phantom

P Masunun[1], P Tangboonduangjit[1] and N Dumrongkijudom[2]

[1] Medical Physics School, Faculty of Medicine Ramathibodi Hospital, Mahidol University, Bangkok, Thailand
[2] Department of Radiological Technology, Faculty of Medical Technology, Mahidol University, Bangkok, Thailand

E-mail: newzoozwen@gmail.com

Abstract. The purpose of this study is to compare the build-up region doses on breast Rando phantom surface with the bolus covered, the doses in breast Rando phantom and also the doses in a lung that is the heterogeneous region by two algorithms. The AAA in Eclipse TPS and the collapsed cone convolution algorithm in Pinnacle treatment planning system were used to plan in tangential field technique with 6 MV photon beam at 200 cGy total doses in Breast Rando phantom with bolus covered (5 mm and 10 mm). TLDs were calibrated with Cobalt-60 and used to measure the doses in irradiation process. The results in treatment planning show that the doses in build-up region and the doses in breast phantom were closely matched in both algorithms which are less than 2% differences. However, overestimate of doses in a lung (L2) were found in AAA with 13.78% and 6.06% differences at 5 mm and 10 mm bolus thickness, respectively when compared with CCC algorithm. The TLD measurements show the underestimate in build-up region and in breast phantom but the doses in a lung (L2) were overestimated when compared with the doses in the two plannings at both thicknesses of the bolus.

1. Introduction

Radiation is currently important in the medical field that used for diagnostic and treating the patients, especially in cancer patients. The giving radiation doses to the patients must be precise, accurate and appropriate for the local of any type of cancer. The evaluation of delivered doses on surface area, in build-up region or in the heterogeneous medium is quite uncertain due to the algorithm calculation that may cause the effect directly to the patients such as erythema and hyperpigmentation. Too low doses from the uncertain calculation are insufficient to damage the gross tumor or too high doses are over limited for curing cancer and affected to the result of treatment.

Nowadays, new technologies of treatment machines are developed due to the complex of planning and treatment. Dose calculation algorithms are influential and controlled the accuracy of radiation doses in treatment planning process. In all superposition/convolution methods, the analytical anisotropic algorithm (AAA) and the collapsed cone convolution algorithm (CCC) are popular because of the accuracy in heterogeneous media calculation, especially in the lung. However, both algorithms cannot well predict the surface doses and build-up region doses in solid water phantom shown by the study of James, Runqing and Micheal [1]. On the other hand, the study by Hasenbalg *et al.* [2] have shown that

CCC algorithm performed overall better than AAA algorithm compared to Monte Carlo in clinical cases of lung and breast.

Mostly irradiation treatment in breast cancer case is oblique tangential photon beam technique. The prediction of surface doses variation for the tangential like photon beams are extremely caution in radiation therapy. James and Grigor [3] have shown that the relative dose profile in the phantom skin did not change with oblique beam using full phantom geometry. Nevertheless, the surface relative depth dose was increased for half phantom geometry (phantom-air interface) like the breast. Above all, in this study is to compare the build-up region doses on Breast Rando phantom surface with the bolus covered, the doses in Breast Rando phantom and also the doses in a lung that is the heterogeneous region by two algorithms (AAA and CCC). Thermoluminescent dosimeters (TLDs) were calibrated with Cobalt-60 and used to measure the doses in irradiation process.

2. Materials and methods

2.1. AAA and CCC algorithms

The analytical anisotropic algorithm (AAA) was developed by the research group of Varian Medical Systems. The calculation of dose distribution for photon beams consists of two parts (the configuration module and the dose calculation module) [4]. The purpose of configuration module is to characterize the phase space of a photon beam. The phase space is approximated using a multiple source model: a point source radiation from the target, a finite source for extra focal radiation, and a third source to model the electron contamination. The dose calculation module is calculated as a superposition of the dose deposited by the primary and secondary photons and the contamination electron for every beamlet. The collapsed cone convolution algorithm (CCC) determines the distribution of total energy released per unit mass (TERMA) in a 3D matrix of the irradiated volume, based on the ray-tracing technique [5].

The analytical anisotropic algorithm (AAA) version 8.9 available in Eclipse treatment planning system and the collapsed cone convolution algorithm (CCC) version 9.2 available in Pinnacle treatment planning system were used for planning and calculation in this study.

2.2. TLDs preparation and calibration

Cobalt-60 teletherapeutic machine (Theratron Elite 80 unit) was used to irradiate radiation dose for TLDs. Harshaw TLD-100 rods with 1mm×5mm size were used. Firstly, all of 200 rods TLD were irradiated and annealed 4 times. TLDs were calibrated at dose 200 cGy at the depth of maximum dose 0.5 cm (d_{max} = 0.5 cm), field size 15×15 cm^2 and SAD technique 80 cm with bolus 0.5 cm. TLDs selected depend on their responsiveness and uniformity. Finally, linearity test of TLDs was done and prepared for measuring the radiation dose in irradiation process.

2.3. Dose calculation in treatment planning

TLDs were put in the rod holders and put into totally 13 holds in Rando phantom (see figure 1 and 2) follow these points

- 4 points in the right upper Breast Rando phantom slab. There was a superior point (B1), a medial point (B2), an inferior point (B3) and a lateral point (B4).
- 4 points in the right lower Breast Rando phantom slab. There was a superior point (B5), a medial point (B6), an inferior point (B7) and a lateral point (B8).
- 2 points in Lung Rando phantom slab that close to the breast phantom. There was a lateral point (L1) and a medial point (L2).

And other 3 points of TLDs with holders free were placed on Breast Rando phantom surface under bolus covered (5 and 10 mm). There was a lateral point (S1), a center point (S2) and a medial point (S3).

After that, CT scan was used to scan the Breast Rando phantom with 5 and 10 mm bolus covered and obtained the slice images of the Breast Rando phantom. Images were sent to treatment planning system both Eclipse and Pinnacle. The planning was done by SAD technique with Varian ClinaciX 6 MV photon beams and prescribed 200 cGy per fraction to 100% of point dose at point B6. Opposing tangential field technique with no wedge was planned with 69° in medial gantry angle and 240° in lateral gantry angle. Field sizes were 5 × 7.9 cm^2 (X collimators), 8 × 8 cm^2 (Y collimators) in medial fields and 7.9 × 5 cm^2 (X collimators), 8 × 8 cm^2 (Y collimators) in Lateral fields. Both Eclipse and Pinnacle were planned by the same technique and calculated with dose contribution method at a prescription point. The comparisons of all 13 points between Eclipse (AAA) and Pinnacles (CCC) were analyzed.

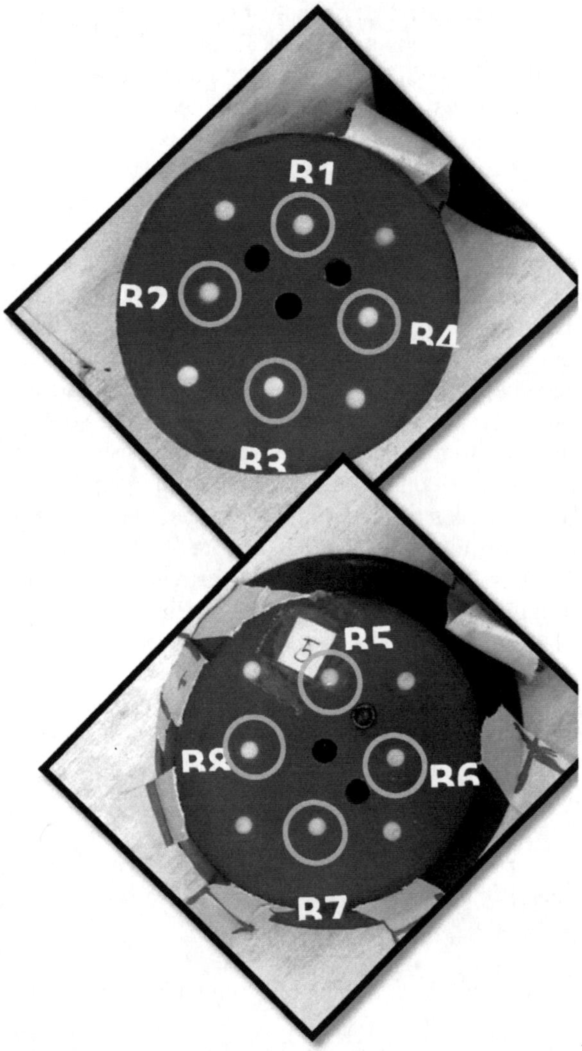

Figure 1. The points of TLD in Breast Rando phantom slab.

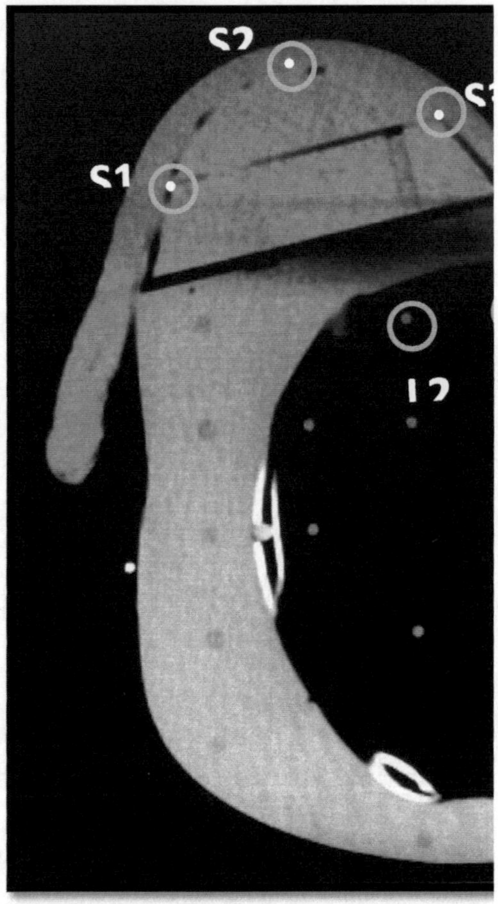

Figure 2. The points of TLD in lung and on the surface with bolus covered.

2.4. *Irradiation and measurement*

TLDs were put into the points and placed on the surface points followed 2.3 of all 13 points. Varian ClinaciX was used to follow the planning both AAA and CCC algorithms. The irradiation was done 3 times for each algorithm. TLDs were read by Harshaw TLD reader and converted to average doses for each point. The comparisons of all 13 points between both algorithms planning and TLD measurements were found.

3. Results

3.1. TLD preparation

TLDs were selected and separated into 3 groups followed their uniformity (±10%) to give the dose of 200 cGy. The linearity test was found by a linear relation of $R^2 = 0.996$ which shows a good performance for measurement.

3.2. Radiation dose in treatment planning

The results show in Table 1 that the doses on Breast Rando phantom surface with bolus covered (point S1, S2, and S3) were closely matched in both algorithms and both thicknesses of the bolus. The differences were between -0.21 to 1.33%. The doses in Breast Rando phantom (Point B1 to B6) were also well matched in both algorithms and both thicknesses of the bolus in which the differences were -0.2 to 1.16%. The lung does in point L1 were somewhat similar with 0.95% and 0.42% differences at 5mm and 10 mm bolus thicknesses. Nevertheless, overestimate doses in a lung (point L2) were found in AAA with 13.78% and 6.06% differences at 5 mm and 10 mm bolus thicknesses, respectively when compared with CCC algorithm.

Table 1. A comparison of point doses in treatment planning by AAA algorithm (Eclipse) versus CCC algorithm (Pinnacle) in Breast Rando phantom with 5 mm and 10 mm bolus covered.

POINTS	Bolus 5 mm		Bolus 10 mm	
	AAA (cGy)	CCC (cGy)	AAA (cGy)	CCC (cGy)
S1	189.9	187.4	199.2	199
S2	202.5	202.1	208.6	208.5
S3	186.2	186.6	195.3	195.3
B1	209.3	206.9	207.9	207.1
B2	203.2	203.5	204.2	203.2
B3	209.4	207.6	209.9	207.9
B4	206	204.9	206.1	204.8
B5	201.8	202.1	200.8	200.4
B6	200.1	200.5	200	200.2
B7	203.6	203.5	202.2	202.1
B8	200.9	201.2	200	200.3
L1	191.9	190.1	188.8	188
L2	89.2	78.4	89.2	84.1

3.3. Irradiated doses versus planning doses

The irradiated TLDs were read in charge particle and converted to average doses shown in table 2. The irradiated doses on Breast Rando phantom surface with bolus covered (point S1, S2 and S3) and the irradiated doses in Breast Rando phantom (Point B1 to B6) were underestimated in both algorithms and both thicknesses of bolus when compared with planning dose which the differences were ranged between -10.98% to -3.00% and -15.35% to -4.26%, respectively. However, the irradiated doses in a lung (point L2) were overestimated in both algorithms at 10 mm bolus thicknesses when compared with planning dose. Although the irradiated doses in a lung (point L2) in CCC algorithm were overestimated at 5 mm bolus thicknesses, they were underestimated in AAA algorithm. The CCC algorithm performed better in TLD measurements when compared with AAA algorithm.

Table 2. A comparison of point doses from TLD measurements versus treatment planning doses in Breast Rando phantom with 5 mm and 10 mm bolus covered in both algorithms.

POINTS	Bolus 5 mm				Bolus 10 mm			
	AAA		CCC		AAA		CCC	
	Irradiated dose (cGy)	Planning dose (cGy)	Irradiated dose (cGy)	Planning dose (cGy)	Irradiated dose (cGy)	Planning dose (cGy)	Irradiated dose (cGy)	Planning dose (cGy)
S1	178.9	189.9	176.4	187.4	186.1	199.2	191.2	199
S2	185.3	202.5	192.3	202.1	194.0	208.6	194.3	208.5
S3	165.7	186.2	169.2	186.6	180.3	195.3	189.4	195.3
B1	177.1	209.3	184.9	206.9	194.4	207.9	192.9	207.1
B2	179.2	203.2	176.1	203.5	189.6	204.2	189.8	203.2
B3	186.8	209.4	183.8	207.6	195.7	209.9	197.6	207.9
B4	178.1	206	185.5	204.9	186.9	206.1	187.8	204.8
B5	176.6	201.8	176.0	202.1	186.2	200.8	190.6	200.4
B6	181.4	200.1	180.7	200.5	185.0	200	189.7	200.2
B7	182.8	203.6	182.9	203.5	187.2	202.2	189.9	202.1
B8	180.1	200.9	174.5	201.2	177.0	200	191.7	200.3
L1	168.0	191.9	164.8	190.1	174.2	188.8	176.2	188
L2	76.9	89.2	94.7	78.4	105.6	89.2	105.5	84.1

4. Conclusions

In this study, both algorithms were not difference in treatment planning of build-up region and breast Rando phantom in which the differences doses were less than 2%. However in the heterogeneous region, lung (point L2), it was overestimated in AAA algorithm. The TLD measurements in build-up region and Breast Rando phantom were underestimated except in lung (point L2) that overestimated in both thicknesses of the bolus. However, the TLD measurements show that CCC algorithm performed better when compared with AAA algorithm at both thicknesses of the bolus. The uncertainties in measurement may cause through the process of placing bolus over the curve of Breast Rando phantom that could make the air gap between bolus and surface of the phantom. Further study might be required in other detectors or other techniques to have more evidence to illustrate the accuracy of dose calculation in build-up region and in lung tissue from the algorithms.

References

[1] James C L C, Runqing J and Micheal K K L 2011 Dosimetry of oblique tangential photon beams calculation by superposition/convolution algorithms *J. Appl. Clin. Med. Phys.* **12** 108-21

[2] Hasenbalg F, Neuenschwander H, Mini R and Born E J 2007 Collapsed cone and analytical anisotropic algorithm dose calculation compared to VMC++ Monte Carlo simulations in clinical cases *J. Phys.: Conf. Series* **74** 012007

[3] James C L C and Grigor N G 2008 Surface dosimetry for oblique tangential photon beams *Med. Phys.* **35** 70-6

[4] Ann V E, Laura T, Jukka P, Mikko T, Hannu H, Sami S, Jyrki A, Marta P, Mauro L and Dominique P H 2006 Testing of the analytical anisotropic algorithm for photon dose calculation *Med. Phys.* **33** 4130-48

[5] Anders A 1989 Collapsed cone convolution of radiant energy for photon dose calculation in heterogeneous media *Med. Phys.* **16** 577-92

Consistency check of photon beam physical data after recommissioning process

B Kadman[1], N Chawapun[1], S Ua-apisitwong[1], T Asakit[2], N Chumpu[2] and J Rueansri[2]

[1] Department of Radiology, Faculty of Medicine, Chiang Mai University, Chiang Mai, Thailand
[2] Lampang Cancer Hospital, Lampang, Thailand

E-mail: double_bo@hotmail.com

Abstract. In radiotherapy, medical linear accelerator (Linac) is the key system used for radiation treatments delivery. Although, recommissioning was recommended after major modification of the machine by AAPM TG53, but it might not be practical in radiotherapy center with heavy workloads. The main purpose of this study was to compare photon beam physical data between initial commissioning and recommissioning of 6MV Elekta Precise linac. The parameters for comparing were the percentage depth dose (PDD) and beam profiles. The clinical commissioning test cases followed IAEA-TECDOC-1583 were planned on REF 91230 IMRT Dose Verification Phantom by Philips' Pinnacle treatment planning system. The Delta[4PT] was used for dose distribution verification with 90% passing criteria of the gamma index (3%/3mm). Our results revealed that the PDDs and beam profiles agreed within a tolerance limit recommended by TRS430. Most of the point doses and dose distribution verification passed the acceptance criteria. This study showed the consistency of photon beam physical data after recommissioning process. There was a good agreement between initial commissioning and recommissioning within a tolerance limit, demonstrated that the full recommissioning process might not be required. However, in the complex treatment planning geometry, the initial data should be applied with great caution.

1. Introduction

Radiotherapy treatment is a high risk procedure since it is complicated, involves multiple data and interaction of many professional groups [1]. The treatment planning and treatment delivery are two major sub-processes in this procedure which may be further sub-divided into several tasks [2]. Therefore, the quality assurance (QA) process is necessary to ensure the correct patient delivery dose [3]. The commissioning management is an important part of QA program for both the radiotherapy treatment planning system (RTPS) and the planning process [4]. As stated by IAEA-TECDOC-1583, the commissioning involves testing of system functions, documentation of the different capabilities and verification of the ability of the dose calculation algorithms to reproduce measured dose calculations. In radiation treatment delivery, medical linear accelerator (Linac) is the key system used in this process. Recommissioning was recommended by AAPM TG53 [5] after major repairs, tuning, changing of beam parameters or machine's parts. The amount of work involved can vary from a few hours to several days works or even longer for a complex 3D dose calculation algorithm. Full recommissioning process is a laborious task that might not be practical in heavy workloads radiotherapy center.

The main purpose of this study was to compare photon beam physical data between initial commissioning and recommissioning in terms of PDD data, beam profiles, and verify by point dose and dose distribution treatment planning.

2. Materials and methods

2.1. The basic beam data sets comparison

The two sets of basic measured beam data from initial commissioning and full recommissioning process for 6MV Elekta Precise linac machine at the Lampang Cancer Hospital were imported to the Pinnacle[3] version 9.8 treatment planning system for photon beam modelling. Measurements of the dose profiles and the depth dose were performed at the source to surface distance of 100 cm in water with a Wellhofer dosimetric system. In this study, the measured open field depth doses and profiles for the 2×2, 3×3, 4×4, 5×5, 10×10, 20×20, 30×30 and 40×40 cm^2 field size were compared. The difference values between commissioning and recommissioning data were expressed as %δ, a percentage of the dose deviation calculated by equation 1 and δ, difference distance at point of interested with normalization dose at d_{max} (D_{max}) calculated by equation 2 modified from TRS430.

$$\%\,\delta = 100 \times (D_{recom} - D_{com}) / D_{com} \quad (1)$$

where % δ is a deviation of percentage difference, D_{recom} is the recommissioning dose at a particular point and D_{com} is the commissioning dose of the same point.

$$\delta = d_{recom} - d_{com} \quad (2)$$

where δ is a deviation of distance, d_{recom} is the recommissioning position at a particular dose level and d_{com} is the commissioning position of the same dose.

Graphical examples in figure 1 showed the analytical criteria of depth dose curves and beam profiles. The tolerance limits were shown in table 1.

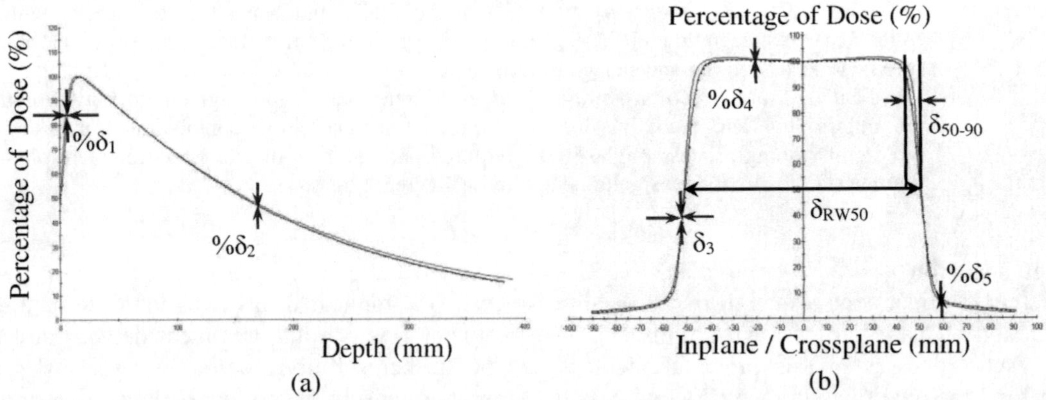

Figure 1. Regions of validity of the criteria $\%\delta_1$ and $\%\ \delta_2$ to compare commissioning data and recommissioning data for depth-dose (PDD) curves (a) and δ_3, $\%\delta_4$, $\%\delta_5$, $\delta_{50\text{-}90}$ and δ_{RW50} for beam profiles(b).

Table 1. The tolerance limits for the δ and %δ[1].

photon beam physical data	parameter	location	tolerance limit
PDD	$\%\delta_1$	build up region : surface and the depth of the 90% isodose surface	10%
	$\%\delta_2$	central beam axis : the high dose and small dose gradient region	2%
beam profile	δ_3	penumbra region of profiles : high dose and large dose gradient regions	2 mm
	$\%\delta_4$	outside central beam axis : high dose and small dose gradient region	3%
	$\%\delta_5$	outside beam edges: data points off the geometrical beam edges and below shielding blocks	30%
	$\delta_{50\text{-}90}$	beam fringe : distance between the 50% and the 90% point	2 mm
	δ_{RW50}	radiological width : width of a profile measured at half its height compared to the value at the beam axis	2 mm

2.2. Point dose verification
Test cases in this study were based on guidelines described in the IAEA-TECDOC-1583 representing the typical techniques in radiotherapy clinical practices. The test cases were planned on REF 91230 IMRT Dose Verification Phantom by Philips' Pinnacle treatment planning system and the defined point doses were measured by 0.6 cm^3 PTW-Freiburg TW30013 cylindrical ionization chamber and PTW-Freiburg T10009 UNIDOS E electrometer. The measured doses were compared with corresponding calculated points from Philips' Pinnacle treatment planning system that used photon beam data from both initial commissioning and recommissioning sets. In brief, the purpose of each test case was as followed:

Test case 1: Testing for reference
Test case 2: Oblique incidence, lack of scattering and tangential fields
Test case 3: Significant blocking of the field corners
Test case 4: Four field box
Test case 5: Automatic expansion and customized blocking
Test case 6: Oblique incidence with irregular field and blocking the center of the field
Test case 7: Three fields, two wedge paired, asymmetric collimation
Test case 8: Non coplanar beams with couch and collimator rotation

2.3 Dose distribution verification
In this study, dose distribution of all test cases on the REF 91230 IMRT Dose Verification phantom were measured by ScandiDos Delta^{4PT} pre-treatment verification system. The passing criteria of the dose distribution was 90% of a gamma index (3% of dose difference and 3 mm distance to agreement). The measurement was performed only for test cases 1-7 and test case 8 was omitted as the crucial electronic part of Delta^{4PT} was in beam direction.

2.4 Statistical analysis
For %δ (deviation of percentage difference) and δ (deviation of distance), this difference can be demonstrated in a comparison of the criteria of TRS 430 and analyzed using SPSS version 17.0 with 95% confidence interval. Mean and standard deviation of three measurements were used for point dose and dose distribution verification.

3. Results

As shown in table 2, according to the tolerance limit recommended by TRS430, the $\%\delta_1$ of PDDs for all field sizes were within tolerance limit of 10%. The PDDs in build-up region of initial commissioning and recommissioning were not statistical significantly difference (p-value = 0.261). However, the $\%\delta_2$ for field sizes of 20×20 cm^2 and larger were out of the tolerance limit of 2%. The PDDs along central axis of initial commissioning and recommissioning were not statistically significantly different (p-value = 0.053).

For beam profiles, at the penumbra region of profiles, δ_3 for both inplane and crossplane were analyzed, only the crossplane was statistically significantly different (p-value = 0.003) but both were within tolerance limit of 2 mm. Outside central beam axis region, there were no statistically significantly difference of $\%\delta_4$ in both planes and the $\%\delta_4$ value were all in tolerance limit of 3%, except for field size 2×2 cm^2. Outside beam edges, there were no statistically significantly difference of $\%\delta_5$ in both planes and the $\%\delta_5$ value were all in tolerance limit of 30%, except for field size 20×20 cm^2 and larger. At beam fringe, there were statistically significantly difference of $\delta_{50\text{-}90}$ in both planes (p-value = 0.000), but both were within tolerance limit of 2 mm. For radiological width, there were no statistically significantly difference of δ_{RW50} in both planes and the δ_{RW50} value were all in tolerance limit of 2 mm, except the large field size as 40×40 cm^2 in crossplane.

For point doses verification, the measured doses for all test cases agreed well within a tolerance limit compared to the corresponding calculated point using both initial commissioning and recommissioning data sets. The percentage difference of measured dose using both data sets for monitor unit calculations were within ±2%.

Dose distribution verified by Delta^{4PT} passed the acceptance criteria 88.9% of the case except in test case 6.

Table 2. The result of commissioning and recommissioning physical beam data comparison.

Field sizes (cm^2)	PDD		Beam profiles									
	$\%\delta_1$	$\%\delta_2$	δ_3		$\%\delta_4$		$\%\delta_5$		$\delta_{50\text{-}90}$		δ_{RW50}	
			CR.	IN.	CR.	IN.	CR.	IN.	CR.	IN.	CR.	IN.
2×2	-2.2	-0.4	-0.6	-0.6	-2.5	-3.1	2.7	22.3	1.0	1.5	0.1	0.1
3×3	4.6	-1.6	-1.0	-0.7	-2.9	-0.7	20.5	28.5	1.5	1.8	0.0	-0.4
4×4	4.1	-1.8	-1.0	-1.2	-0.6	1.2	-8.3	12.7	1.3	1.7	0.3	0.3
5×5	-4.1	-0.8	-1.1	-0.7	-0.1	-0.6	-13.2	8.0	0.8	1.8	0.2	-0.4
10×10	-8.1	-1.0	-0.9	0.1	-1.2	-0.9	-24.5	-27.0	0.5	1.2	0.0	0.2
20×20	-8.1	-5.0	-0.1	0.0	-0.5	-0.6	-65.7	-61.7	0.8	1.6	-0.4	-0.9
30×30	-4.2	-13.6	-0.1	-0.1	-1.7	0.0	-54.4	-50.9	0.7	1.5	-0.5	-0.4
40×40	1.3	-20.6	-1.3	0.7	-1.1	0.1	-72.9	-50.6	1.1	2.0	2.3	-2.0

4. Discussion

For PDDs, in the build-up region, no difference of the $\%\delta_1$ for all field sizes was observed, but at higher depth, the $\%\delta_2$ for field sizes larger than 20×20 cm^2 were out of tolerance limit. This was shown in the PDDs curves, the commissioning dose was higher than recommissioning dose at the same depth, indicated that the photon energy was slightly lower in the latter case. As energy of incident photon decreases, the contribution of the scattered radiation to the absorbed dose is obviously seen at larger field sizes.

The effect of decreasing photon energy was also pronounced for beam profiles. Most of the analysed parameters were within a tolerance limit, except for the $\%\delta_5$ of field sizes larger than 20×20 cm^2. For data points out off the geometrical beam edges in large field sizes, the doses of recommissioning data were lower than that of commissioning. The lower fraction of the central axis energy fluence is due to the lower scattered of flattening filter when the collimator jaws are wide opened [6]. The Gaussian height, Gaussian width, and jaw transmission parameters affect the dose in the low-dose regions of the off-axis profile [7].

All of the point doses verification in the designed test cases agreed well within a tolerance limit for both data sets.

In dose distribution verification, only test case 6, the irregular central blocked fields was not met the acceptable criteria for both data sets. The accuracy of a leaf leakage calculation in treatment planning system depends on the accuracy of an input data while the Delta^{4PT} actually responds to both scattered and leakage radiations. The other test cases representing the typical techniques in radiotherapy clinical practices passed the acceptable criteria, deduced that the isodose plan from initial commissioning and recommissioning data were compatible.

5. Conclusion

In our study, the physical beam data from initial commissioning and recommissioning process were compatible confirmed by point dose and dose distribution verification, demonstrated that the full process of recommissioning might not be required. However, in some circumstance where the treatment depth exceeds 10 cm or in the complex treatment planning geometry, the initial data should be applied with caution.

References

[1] International Atomic Energy Agency 2004 Technical Reports Series no.430: Commissioning and quality *assurance of computerized planning systems for radiation treatment of cancer* (Vienna: International Atomic Energy Agency)

[2] Booth J T 2002 *Modelling the impact of treatment uncertainties in radiotherapy* (Adelaide: South Australia University of Adelaide) p 133

[3] International Atomic Energy Agency 2007 *IAEA-TECDOC-1540: Commissioning of radiotherapy, Specification and Acceptance Testing of Radiotherapy Treatment Planning Systems* (Vienna: International Atomic Energy Agency)

[4] International Atomic Energy Agency 2008 2007 *IAEA-TECDOC-1583*: *Commissioning of radiotherapy treatment planning systems* (Vienna: International Atomic Energy Agency Frass BA, Doppke K, Kutcher G, Starkschall G, Stern R and Van DJ 1998 American association of physicists in medicine radiation therapy committee task group 53: Quality assurance for clinical radiotherapy treatment planning *Med Phys.* **25**(10) 1773-1829

[5] Phillips Medical Systems 2006 *Pinnacle^3 Physics reference guide version 8.0* (Fitchburg, WI: Philips Healthcare)

[6] Bedford J L, Childs P J, Hansen V N, Mosleh-Shirazi M A, Verhaegen F and Warrington A P 2003 Commissioning and quality assurance of the Pinnacle^3 radiotherapy treatment planning system for external beam photons *Br J Radiol.* **76** 163-176

Acknowledgement

I am greatly thankful for Lampang Cancer Hospital for providing the instruments in this project.

Journal of Physics: Conference Series **694** (2016) 012024 doi:10.1088/1742-6596/694/1/012024

Comparison of selected dose calculation algorithms in radiotherapy treatment planning for tissues with inhomogeneities

Y L Woon[1, 2], S P Heng[3], J H D Wong[1,4] and N M Ung[5]

[1] Department of Biomedical Imaging, Faculty of Medicine, University of Malaya, Kuala Lumpur, Malaysia
[2] Department of Radiotherapy and Oncology, Sarawak General Hospital, Malaysia
[3] Radiotherapy Unit, Pantai Hospital Kuala Lumpur, Kuala Lumpur, Malaysia
[4] University of Malaya Research Imaging Centre, Faculty of Medicine, University of Malaya, Kuala Lumpur, Malaysia
[5] Clinical Oncology Unit, Faculty of Medicine, University of Malaya, Kuala Lumpur, Malaysia

E-mail: nmung@ummc.edu.my

Abstract. Inhomogeneity correction is recommended for accurate dose calculation in radiotherapy treatment planning since human body are highly inhomogeneous with the presence of bones and air cavities. However, each dose calculation algorithm has its own limitations. This study is to assess the accuracy of five algorithms that are currently implemented for treatment planning, including pencil beam convolution (PBC), superposition (SP), anisotropic analytical algorithm (AAA), Monte Carlo (MC) and Acuros XB (AXB). The calculated dose was compared with the measured dose using radiochromic film (Gafchromic EBT2) in inhomogeneous phantoms. In addition, the dosimetric impact of different algorithms on intensity modulated radiotherapy (IMRT) was studied for head and neck region. MC had the best agreement with the measured percentage depth dose (PDD) within the inhomogeneous region. This was followed by AXB, AAA, SP and PBC. For IMRT planning, MC algorithm is recommended for treatment planning in preference to PBC and SP. The MC and AXB algorithms were found to have better accuracy in terms of inhomogeneity correction and should be used for tumour volume within the proximity of inhomogeneous structures.

1. Introduction

Today, various dose calculation algorithms for tissue inhomogeneity correction have been developed by commercial Treatment Planning Systems (TPSs). The choice of dose calculation algorithm is particularly important especially for the radiotherapy of tissues with large inhomogeneities, including lungs, oral cavities, teeth, nasal passages, sinuses and bones that have different physical and radiological properties [1]. Even though inhomogeneity correction is considered in the model-based algorithm [2], each model has its own limitations [3].

The purpose of this study is to assess the accuracy of five dose calculation algorithms in tissue inhomogeneity correction, including Pencil Beam Convolution (PBC), Superposition (SP), Anisotropic

Analytical Algorithm (AAA), Monte Carlo (MC) and Acuros XB (AXB). In addition, special attention was given to an advanced treatment planning technique, intensity modulated radiotherapy (IMRT).

2. Methodology

Figure 1 shows the setup for percentage depth dose (PDD) measurement. Expanded polystyrene (EPS) was used to simulate the air cavity. It is made of pre-expanded polystyrene beads. Plaster of Paris (POP) was chosen as bone equivalent material since its main composition is calcium sulphate ($CaSO_4$), which is similar to the bone mineral, calcium phosphates [4].

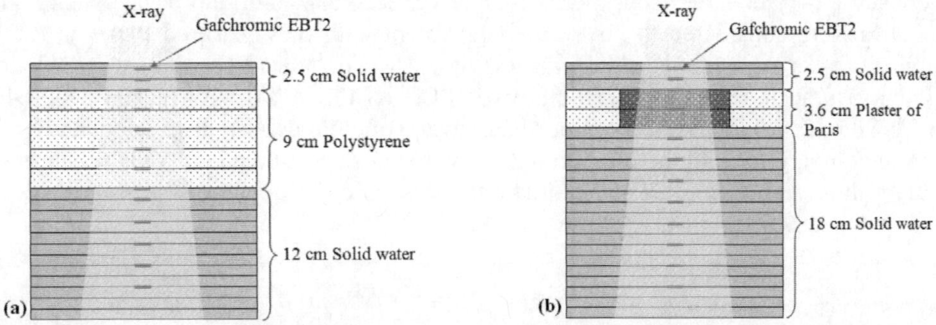

Figure 1. Setup for phantom consisting air cavity (a) and bone (b) equivalent materials.

Five dose calculation algorithms were chosen from three TPSs. They were PBC and SP from XiO TPS (version 4.70, Elekta AB, Stockholm, Sweden), MC from Monaco TPS (version 3.20.01, Elekta AB, Stockholm, Sweden) as well as PBC, AAA and AXB from Eclipse TPS (version 10.0, Varian Medical Systems, California, USA). Irradiations were delivered using the linacs linked with the TPSs used for dose calculation. The XiO and Monaco TPSs were commissioned for Elekta Synergy linac (Elekta AB, Stockholm, Sweden), while the Eclipse TPS was commissioned for Varian Novalis TX (Varian Medical Systems, California, USA).

The dose measurements were carried out using Gafchromic External Beam Therapy 2 (EBT2) (International Specialty Products, New Jersey, USA). Three cut films (2×2 cm^2) were centered along the beam central axis at different depth within the phantom. The EBT2 films were scanned using Epson Flatbed Scanner, Expression 10000XL (Epson America, Inc., California, USA) 24 hours after irradiation. This waiting time is for optimum post-irradiation darkening to take place [5].

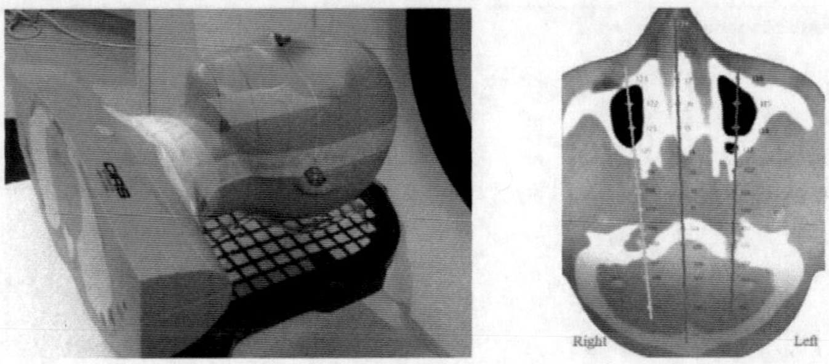

Figure 2. (a) Phantom setup during irradiation, (b) Line profiles selected.

Besides PDD, IMRT for nasopharyngeal carcinoma (NPC) were planned and compared using three different algorithms, which were the PBC, SP and MC. Head and neck region was chosen because this region consists of sinuses, air cavity and bone tissues which are highly inhomogeneous. Dose verifications were carried out using anthropomorphic phantom under actual treatment condition, as shown in figure 2(a). Figure 2(b) illustrates three line profiles chosen across the areas with high inhomogeneities and were named center, left and right line profiles respectively.

3. Results and discussion

3.1. Percentage depth dose

Figure 3 shows the PDDs obtained from measurements and calculations by different algorithms in the phantom with air cavity. The grey dotted line labeled 'water' indicates the calculated PDD of 6MV photon in a homogeneous water phantom as reference PDD for the corresponding linac. The measurement points for EBT2 were interpolated with dotted line.

The measured PDDs in the air cavity region were found to fall rapidly until the minimum as they reached the distal air-water interface. After the interface, secondary buildup began to take place. All calculated PDDs were considered to have good agreement with the measured PDDs at the distance beyond the air cavity, except for SP which was less by 7.2%. The other algorithms were able to predict the dose within 5% accuracy. The deviations were PBC (XiO) 3.2%, AAA 2.7%, MC 2.4%, PBC (Eclipse) 1.9% and AXB 0.4%. At the center of air cavity (or 6 cm depth), dose calculated by MC was less by 2.3% in comparison with the measured dose, while dose calculated by AXB was less by 9.0%. The other algorithms were found to overestimate the dose and the deviations were AAA 24.0%, SP 28.1%, PBC (XiO) 35.6% and PBC (Eclipse) 40.3%.

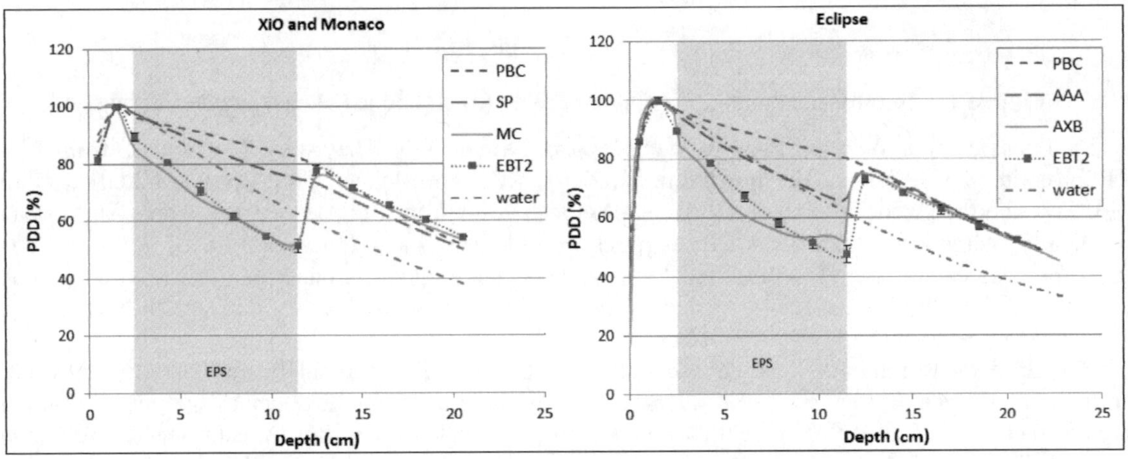

Figure 3. PDDs in phantom with air cavity equivalent material.

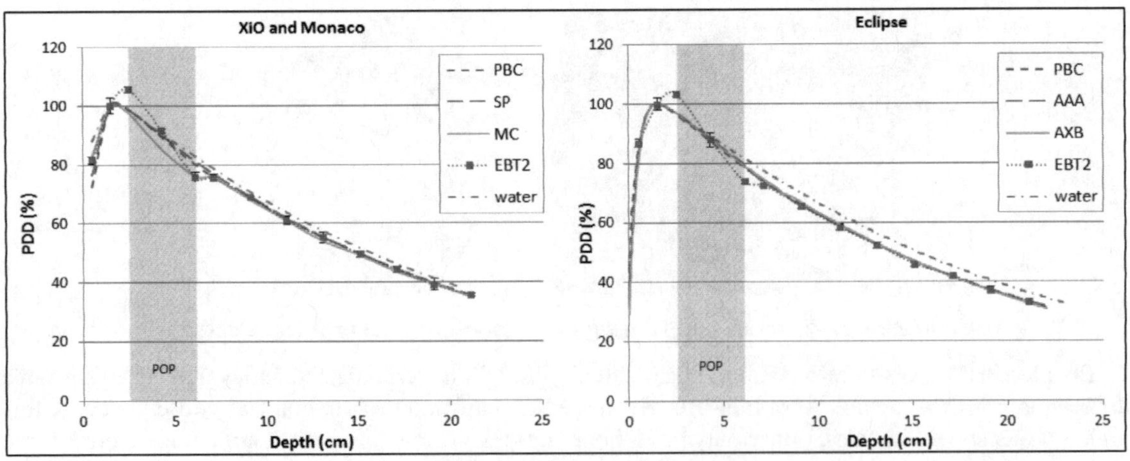

Figure 4. PDDs in phantom with bone equivalent material.

Figure 4 shows the PDDs obtained from measurements and calculations by different algorithms in the phantom with bone equivalent material. All the dose calculation algorithms agreed with each other reasonably well at the distance beyond the high density bone region. The measured PDDs were found to have maximum dose at the proximal water-bone interface. This was followed by high gradient drop off within the bone region. After the distal bone-water interface, the measured PDDs remained constant for a short distance of about 1 cm and then continued to decrease gradually. For calculated PDDs, the deviations were found to range from 6.2% to 7.6% with the measurements at the water-bone interface. Within the bone region, only MC showed the rapid fall off and gave accurate dose prediction at the distal bone-water interface. Generally, all algorithms could calculate the dose accurately after the interface, with deviations of less than 2%.

From the results, MC and AXB were found to have better accuracy in inhomogeneity correction since they simulate the actual electron transport based on the material compositions [6, 7]. Under the effect of air cavity, they managed to show the dose build down and secondary buildup considerably well. In addition, MC could predict the rapid fall off within the bone region and gave the correct dose at the distal bone-water interface. The overall accuracy was followed by AAA, SP and PBC, in terms of inhomogeneity corrections. Overestimation of dose by PBC was the highest in air cavity. However, their predictions at the distance beyond the inhomogeneous regions were still reasonably acceptable.

3.2. *Intensity modulated radiotherapy*

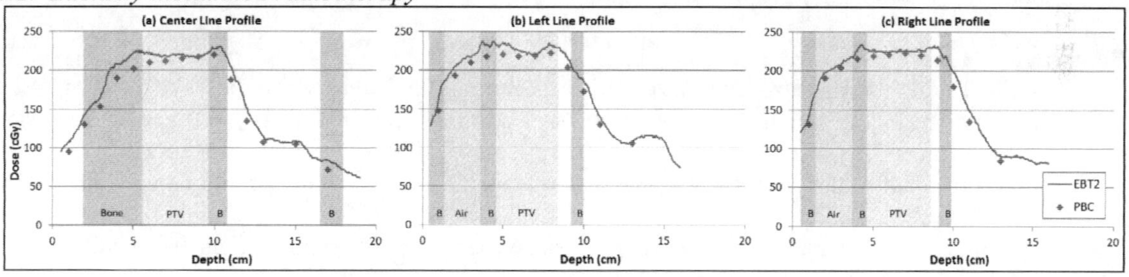

Figure 5. IMRT plan generated using PBC algorithm.

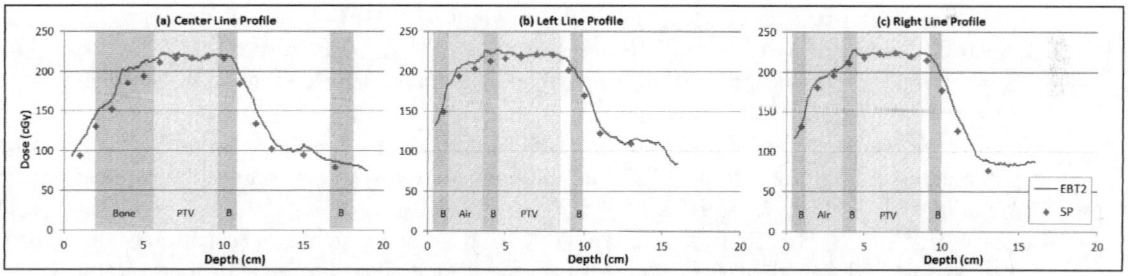

Figure 6. IMRT plan generated using SP algorithm.

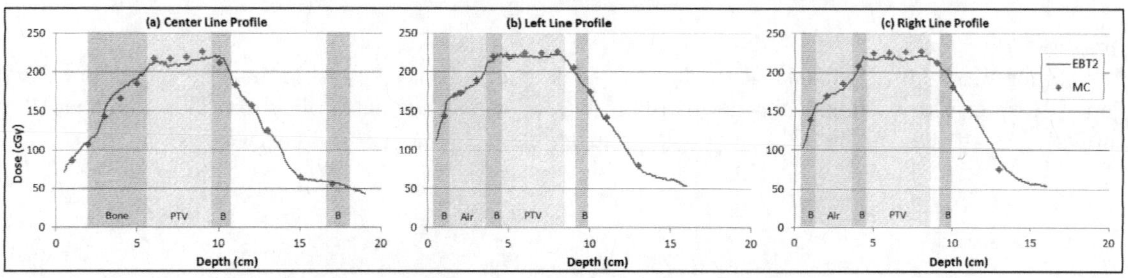

Figure 7. IMRT plan generated using MC algorithm.

Figure 5, 6 and 7 show the results on the agreement between calculations using PBC, SP and MC algorithms and measurements of IMRT plans using EBT2 films respectively. For PBC, the deviation within the planning target volume (PTV) had exceeded 5% at two points and most of the dose points outside PTV were underestimated by PBC and the mean underestimation was 7.1 ± 2.9%. The discrepancies were more predominant in the high density regions such as bones. For SP, the delivered dose to PTV was within 5% but the normal tissues were received higher dose than the calculation by 7.3 ± 3.6% averagely. MC was able to calculate the dose more accurately than PBC and SP, all points were within 5% tolerance except for three points which were located at bones and area of high dose gradient.

When planning IMRT for NPC cases, MC algorithm is recommended since MC could provide accurate dose prediction not only within the PTV, but also to the surrounding tissues. Generally, PBC and SP underestimated the dose in IMRT plan and the surrounding normal tissues received higher dose than the calculation. If the planned margin was too tight, overdosing might cause the tissue tolerance dose of critical organs to be exceeded.

4. Conclusion

MC had the best agreement with the measured PDD within the inhomogeneous region. This was followed by AXB, AAA, SP and PBC. Furthermore, MC algorithm is more accurate for IMRT treatment planning compared to PBC and SP. In conclusion, MC and AXB algorithms were found to have better accuracy in terms of inhomogeneity correction and are suggested to be used for tumour volume within the proximity of inhomogeneous structures.

References

[1] AAPM Radiation Therapy Committee Task Group no. 65 2004 *AAPM Report No. 85: Tissue inhomogeneity corrections for megalovoltage photon beams* (Madison, WI: Medical Physics Publishing)

[2] Robinson D 2008 Inhomogeneity correction and the analytic anisotropic algorithm *J Appl Clin Med Phys* **9** 2786

[3] Animesh 2005 Advantages of multiple algorithm support in treatment planning system for external beam dose calculations *J Cancer Res Ther* **1** 12–20

[4] Rey C, Combes C, Drouet C and Glimcher M J 2009 Bone mineral: update on chemical composition and structure *Osteoporosis International* **20** 1013–21

[5] Ganapathy K, Kurup P G G, Murali V, Muthukumaran M and Velmurugan J 2013 A study on comparison of Gafchromic EBT2 film response under single and cumulative exposure conditions *J Med Phys* **38** 173–7

[6] Schneider W, Bortfeld T and Schlegel W 2000 Correlation between CT numbers and tissue parameters needed for Monte Carlo simulations of clinical dose distributions *Physics in Medicine and Biology* **45** 459–78

[7] Kan M W K, Leung L H T, So R W K and Yu P K N 2013 Experimental verification of the Acuros XB and AAA dose calculation adjacent to heterogeneous media for IMRT and RapidArc of nasopharygeal carcinoma *Medical Physics* **40** 031714

Acknowledgements

The authors would like to thank the medical physicists from University of Malaya Medical Centre and Pantai Hospital Kuala Lumpur for their help in film calibration and data collection. This study has been supported by the University of Malaya postgraduate research grant (PPP: PO016-2014B).

Commissioning and validation of COMPASS system for VMAT patient specific quality assurance

J Pimthong[1], C Kakanaporn[2], L Tuntipumiamorn[2], P Laojunun[2] and P Iampongpaiboon[2]

[1] Medical Physics School, Faculty of Medicine, Ramathibodi Hospital, Mahidol University, Bangkok, Thailand
[2] Division of Radiation Oncology, Department of Medicine, Siriraj Hospital, Mahidol University, Bangkok, Thailand

E-mail: iamjeerawat@hotmail.com

Abstract. Pre-treatment patient specific quality assurance (QA) of advanced treatment techniques such as volumetric modulated arc therapy (VMAT) is one of important QA in radiotherapy. The fast and reliable dosimetric device is required. The objective of this study is to commission and validate the performance of COMPASS system for dose verification of VMAT technique. The COMPASS system is composed of an array of ionization detectors (MatriXX) mounted to the gantry using a custom holder and software for the analysis and visualization of QA results. We validated the COMPASS software for basic and advanced clinical application. For the basic clinical study, the simple open field in various field sizes were validated in homogeneous phantom. And the advanced clinical application, the fifteen prostate and fifteen nasopharyngeal cancers VMAT plans were chosen to study. The treatment plans were measured by the MatriXX. The doses and dose-volume histograms (DVHs) reconstructed from the fluence measurements were compared to the TPS calculated plans. And also, the doses and DVHs computed using collapsed cone convolution (CCC) Algorithm were compared with Eclipse TPS calculated plans using Analytical Anisotropic Algorithm (AAA) that according to dose specified in ICRU 83 for PTV.

1. Introduction

The goal of radiotherapy is to deliver as much dose to the tumor while sparing normal tissue. In advanced radiotherapy techniques, volumetric modulated arc therapy (VMAT) is one of the techniques have helped achieve this goal. The VMAT produces highly conformal dose distribution while simultaneously sparing the organs at risk (OAR), for several indications potentially with a further shortening of treatment times [1, 2]. With this technology, it is possible to deliver an intensity modulated treatment in the form of arcs by dynamically controlling various parameters such as position and speed of the gantry and multi –leaf collimator (MLCs) as well as dose rate [3]. However, with the complexities of treatment deliveries, a comprehensive pre-treatment patient specific quality assurance (QA) procedure is required. Pre-treatment patient specific QA in radiotherapy is the method used to ensure that the correct amount of radiation is being delivered to the correct location. It is a comparison between calculated radiation dose in treatment planning system and radiation dose from measuring. Advanced treatment techniques such as VMAT require appropriate pre-treatment QA to verify 3D dose distribution [4]. The fast and reliable dosimetric device also is required.

A traditionally routine QA method for treatment plan verification is performed in a phantom using small volume ion chambers for absolute dosimetry [5] or radiographic film for relative dosimetry [6] are time consuming. Recently two-dimensional (2D) array and electronic portal imaging device (EPID) show the dose distribution immediately after the treatment delivery. Currently, our department has used commercially two-dimensional (2D) detector arrays, MapCHECK diode array (Sun nuclear, Melbourne, Florida, USA) for the quality assurance of patient-specific IMRT and VMAT treatment plans. To evaluate the treatment plan, gamma index passing rate is usually used in criteria of 95% gamma pass ($\gamma \leq 1$). Due to the lack of information about the correlations between the verification measurement results and the patient anatomical structures. And also the resulting lack of information about the actual irradiation doses to the different target volume and organ at risk. Recently some 3D verification tools that provide patient anatomical structure information were applied clinically. The COMPASS QA system (IBA Dosimetry, Germany) is one such technique which uses MatriXX detector (IBA Dosimetry, Schwarzenbruck, Germany) along with gantry angle sensor. It consists of two parts; COMPASS software and an associated measuring device [7]. COMPASS software is able to recalculate and to reconstruct a 3D dose distribution on the CT data of the patient based on measured fluence from MatriXX detector of the patient based on measured fluence from MatriXX detector and collapsed cone convolution/superposition algorithm. Therefore, the aim of the present study is to validate the accuracy of the COMPASS system by comparing computation dose and reconstruction dose in each VMAT plan of prostate cancer and nasopharyngeal cancer radiation dose from the TPS. The doses and DVHs computed using Collapsed Cone Convolution (CCC) Algorithm were compared with Eclipse TPS calculated plans using Analytical Anisotropic Algorithm (AAA) that according to dose specified in ICRU 83 for PTV.

2. Material and methods

To commission the COMPASS system, reformatting 6 and 10 MV photon beam data from a Varian Clinac iX linear accelerator (Varian Medical Systems, Palo Alto) into a software acceptable beam model was required. The COMPASS system is composed of an array of ionization detectors (MatriXX) mounted to the gantry using a custom holder and software for the analysis and visualization of QA results. Furthermore, the validation of COMPASS software was studied in two aspects; the basic application and advanced application studies. To study the basic application, the homogeneous phantom was employed. For the advanced application, fifteen prostate cancer and nasopharyngeal cancer VMAT plans using 10 MV and 6 MV photons respectively were chosen to investigate. The MatriXX detector was performed in warm-up phase for 15 minutes, then was pre-irradiated at dose 10 Gy at the beginning with an open field size of 25 × 25 cm^2 for SDD = 100 cm. Next, performed the geometry calibration step with an open field size of 10 × 10 cm^2, with 100 MU delivered. The absolute calibration (cGy/MU) was measured with an open field size of 10×10 cm^2, with 100 MU delivered. Lastly, performed calibration gantry angle sensor and then the gantry angle sensor was tested if the calibration of the sensor is still inside accuracy criteria. To generate the dose calculation from COMPASS software, the DICOM files of each treatment plan (RT plans, RT doses, RT structures and CT images) was exported to the COMPASS software (version 3.0a, IBA Dosimetry GmbH, Schwarzenbruck, Germany) with an associated measuring device, the MatriXX detector. The MatriXX was attached on the gantry with gantry holder. The solid water phantoms (Gammex RMI GmbH, Germany) were placed on a surface of MatriXX detector as build up material 2.0 cm and 5.0 cm thickness for 6 and 10 MV photon beam, respectively.

2.1. Basic clinical application study

The aim of the basic clinical study was to verify the accuracy of the COMPASS® software; the simple open field in various sizes were validated in homogeneous phantom. Firstly, the homogeneous water phantom of size 30 × 30 cm^3 was created and then a 3.5 × 3.5 cm^3 PTV1, 4.0 × 4.0 cm^3 PTV2 and 4.5

× 4.5 cm^3 PTV3 were created at the center of the homogeneous water phantom. The fluence for open fields in various sizes of 5 × 5, 5 × 10, 10 × 10, 15 × 15, 20 × 20, 20 × 5 and 25 × 25 cm^2 at 100 cGy Prescribed dose for 6 MV and 10 MV were acquired by the COMPASS® software on a homogeneous phantom that was created in Eclipse TPS. Afterward, the dose calculated and measured by COMPASS QA system and Eclipse TPS respectively were compared. The doses and DVHs computed using collapsed cone convolution (CCC) Algorithm were compared with Eclipse TPS calculated plans using analytical anisotropic algorithm (AAA) that according to dose specified in ICRU 83 for PTV.

2.2 Advanced clinical application study
According to the aim of this study is to clinically implement the COMPASS® software for pre-treatment QA of VMAT treatment plans. The fifteen prostate cancer and nasopharynx cancer VMAT plans were chosen to investigate. To generate DVHs using COMPASS software, the fluence from MatriXX measurement were transferred to the software. The output of the COMPASS software gave 3D dose reconstruction on patient CT images using beam modeling, detector measurement and TPS. The DVHs were compared between the COMPASS® dose reconstruction and Eclipse TPS dose calculation according to dose specified in ICRU 83 for PTV and OARs. The DVHs analysis of Planning Target Volume (PTV).

3. Results

3.1. Basic clinical application study
For basic clinical application study, to verify the accuracy in dose computation of COMPASS software which an independent secondary treatment planning check to verify conventional TPS calculation. The doses were analyzed by the definition of $D_{98\%}$, $D_{95\%}$, $D_{50\%}$, $D_{2\%}$ and average dose in each PTV. The results showed that the calculated dose difference between COMPASS software and TPS was less than 2%. In addition, the measured dose and reconstruction dose from COMPASS was differed from TPS dose less than 3%.

3.2 Advanced clinical application study
The advanced clinical application study for the prostate cancer and nasopharyngeal cancer VMAT plans, the comparison of DVHs between TPS calculated dose and computed dose from COMPASS software was evaluated in term of $D_{98\%}$, $D_{95\%}$, $D_{50\%}$, $D_{2\%}$ and average dose of PTV at tolerance level of 3%, the results showed that the mean of percentage difference was 0.19%, 0.13%, 0.11%, 0.16% and 0.11%, respectively are shown in table 1(a). For the nasopharyngeal cancer VMAT plans, the results showed that the mean of percentage difference was 1.85%, 1.87%, 1.61%, 0.76% and 1.55%, respectively are shown in table 2(a).

For the prostate cancer VMAT plans, the comparison of DVHs between TPS and reconstructed dose from COMPASS software was also evaluated. The results showed that the mean of percentage difference was 2.55%, 1.65%, 0.18%, 0.73% and 0.25%, respectively are shown in Table 1(b). For the nasopharyngeal cancer VMAT plans, the results showed that the mean of percentage difference was 2.87%, 2.35%, 1.23%, 0.87% and 1.17 %, respectively are shown in table 2(b).

Table 1. (a) Comparison of COMPASS computed and Eclipse calculated for prostate cancer (10 MV) (b) Comparison of COMPASS measured and Eclipse calculated for prostate cancer (10 MV).

(a)

Site	$D_{98\%}$	$D_{95\%}$	$D_{50\%}$	$D_{2\%}$	Average
	Absolute Relative Difference (%)				
Prostate1	0.10	0.00	0.10	0.00	0.10
Prostate2	0.10	0.20	0.00	0.20	0.10
Prostate3	0.20	0.00	0.00	0.10	0.00
Prostate4	0.40	0.30	0.10	0.20	0.20

(b)

Site	$D_{98\%}$	$D_{95\%}$	$D_{50\%}$	$D_{2\%}$	Average
	Absolute Relative Difference (%)				
Prostate1	2.30	1.50	0.00	0.80	0.10
Prostate2	2.80	1.40	0.60	1.40	0.40
Prostate3	3.40	2.20	0.10	0.60	0.30
Prostate4	2.50	1.80	0.00	0.70	0.20

Prostate5	0.40	0.30	0.20	0.30	0.20	Prostate5	3.00	2.20	0.10	0.40	0.40
Prostate6	0.20	0.20	0.30	0.10	0.20	Prostate6	2.50	1.60	0.30	0.30	0.40
Prostate7	0.00	0.10	0.10	0.10	0.00	Prostate7	3.00	1.80	0.10	0.60	0.10
Prostate8	0.60	0.40	0.20	0.20	0.20	Prostate8	2.60	1.90	0.10	0.70	0.30
Prostate9	0.10	0.10	0.00	0.10	0.00	Prostate9	2.80	1.70	0.30	0.30	0.50
Prostate10	0.10	0.10	0.00	0.10	0.00	Prostate10	1.40	0.90	0.10	0.70	0.10
Prostate11	0.20	0.10	0.30	0.30	0.30	Prostate11	2.30	1.40	0.10	0.80	0.10
Prostate12	0.30	0.00	0.00	0.50	0.00	Prostate12	3.20	2.40	0.50	0.40	0.60
Prostate13	0.00	0.00	0.10	0.20	0.10	Prostate13	1.70	1.20	0.10	1.20	0.10
Prostate14	0.00	0.00	0.20	0.00	0.10	Prostate14	2.40	1.30	0.10	0.90	0.10
Prostate15	0.20	0.20	0.10	0.00	0.10	Prostate15	2.40	1.50	0.20	1.20	0.00
Average	0.19	0.13	0.11	0.16	0.11	Average	2.55	1.65	0.18	0.73	0.25

Table 2. (a) Comparison of COMPASS computed and Eclipse calculated for NPC (6 MV) (b) Comparison of COMPASS measured and Eclipse calculated for NPC (6 MV)

(a)

Site	$D_{98\%}$	$D_{95\%}$	$D_{50\%}$	$D_{2\%}$	Average
	Absolute Relative Difference (%)				
NPC1	1.00	1.20	1.60	1.70	1.60
NPC2	1.70	1.60	1.20	0.70	1.30
NPC3	1.70	1.90	1.50	1.50	1.60
NPC4	1.50	1.40	1.50	0.40	1.30
NPC5	1.80	1.70	1.40	1.10	1.40
NPC6	1.40	1.30	1.30	0.40	1.20
NPC7	1.90	1.70	1.60	1.00	1.50
NPC8	3.00	2.40	1.40	0.40	1.60
NPC9	1.90	2.00	1.50	0.70	1.40
NPC10	2.10	2.20	2.10	0.30	1.90
NPC11	2.00	2.30	1.60	0.00	1.50
NPC12	1.60	1.90	1.90	1.10	1.70
NPC13	2.70	2.50	2.20	0.80	2.10
NPC14	2.00	1.90	1.80	0.60	1.70
NPC15	1.40	2.00	1.50	0.70	1.50
Average	1.85	1.87	1.61	0.76	1.55

(b)

Site	$D_{98\%}$	$D_{95\%}$	$D_{50\%}$	$D_{2\%}$	Average
	Absolute Relative Difference (%)				
NPC1	3.10	2.60	1.70	0.80	1.70
NPC2	3.20	2.30	1.50	0.10	1.50
NPC3	2.80	2.40	1.20	0.60	1.40
NPC4	3.90	2.50	1.00	1.20	0.90
NPC5	3.30	2.60	1.50	0.70	1.60
NPC6	2.00	1.40	0.90	1.40	0.60
NPC7	2.50	2.00	1.00	0.60	0.90
NPC8	3.50	2.70	1.00	1.40	1.10
NPC9	1.80	1.50	0.10	0.80	0.20
NPC10	3.80	3.40	1.90	0.80	1.70
NPC11	2.60	2.50	1.30	0.50	1.20
NPC12	2.00	1.90	1.10	0.70	1.00
NPC13	2.60	2.10	1.30	2.10	1.00
NPC14	3.10	2.50	1.20	0.60	1.20
NPC15	2.80	2.80	1.70	0.70	1.50
Average	2.87	2.35	1.23	0.87	1.17

4. Discussion

In this study, the COMPASS computed DVHs with Eclipse TPS calculated DVHs for fifteen nasopharyngeal cancers as well as fifteen prostate cancers were compared and shown good agreement with the comparison between the COMPASS system and Eclipse TPS, the results showed that the percent difference between COMPASS computed and Eclipse TPS calculated DVHs was less than 2%. And the percent difference between the COMPASS measured and Eclipse TPS calculated dose was less than 3%. Such difference may be due to the difference in beam modeling and inhomogeneity corrections between two different calculation algorithms, Collapsed Cone Convolution in COMPASS system and AAA in Eclipse TPS. Another reason it may be due to the difference resolution in MatriXX 2D detector array and Eclipse TPS grid calculation. Although, the dose measured from COMPASS and TPS expected dose distribution differed, it showed less than 3% which is acceptable. Therefore, it is recommended to use other traditional pretreatment QA such as film and absolute dose measurement by ionization chamber accompanied with the COMPASS software to confirm the results.

5. Conclusion

In conclusion, the COMPASS QA system is able to perform a full three-dimensional Collapsed Cone Convolution/Superposition algorithm, whereas the Eclipse TPS uses Analytical Anisotropic Algorithm (AAA) to generate calculated dose. The results in this study showed that the percentage difference of the dose between COMPASS software and TPS calculation was less than 2% in order to be analyzed by the definition of $D_{98\%}$, $D_{95\%}$, $D_{50\%}$, $D_{2\%}$ and average dose in each PTV. Moreover, the dose measured from COMPASS and TPS expected dose distribution was differed less than 3%. Taken all together, it indicated that using COMPASS QA system along with Matrix Detector (IBA dosimetry) can be an effective tool for 3D dose pre-treatment verification of VMAT plans in the patient anatomy when compared to traditional TPS.

References

[1] Wolff D *et al.* 2009 Volumetric modulated arc therapy (VMAT) vs. serial tomotherapy, step-and-shoot IMRT and 3D-conformal RT for treatment of prostate cancer *Radiother. Oncol.* **93** 226-33

[2] Palma D, Vollans E, James K, Nakano S, Moiseenko V, Shaffer R, McKenzie M, Morris J and Otto K 2008 Volumetric modulated arc therapy for delivery of prostate radiotherapy: comparison with intensity modulated radiotherapy and three dimensional conformal radiotherapy *Int. J. Radiat. Oncol. Biol. Phys.* **15** 996-1001

[3] Yu C X 1995 Intensity-modulated arc therapy with dynamic multi-leaf collimation: an alternative to tomotherapy *Phys. Med. Biol.* **40** 1435-49

[4] Boggula R, Jahnke L, Wertz H, Lohr F and Wenz F. 2011 Patient-specific 3D pre-treatment and potential 3D online dose verification of Monte Carlo-calculated IMRT prostate treatment plans *Int. J. Radiat Oncol. Biol. Phys.* **81** 1168-75

[5] Yoda K, Nakagawa K, Shiraishi K, Okano Y, Ohtomo K and Pellegrini R G 2009 Dose verification of intensity modulated arc therapy using an ERGO treatment planning system and Elekta internal multi-leaf collimators for prostate cancer treatment *Br. J. Radiol.* **82** 328–31

[6] Ezzell G A, Galvin J M, Low D, Palta J R, Rosen I, Sharpe M B, Xia P, Xiao Y, Xing L and Yu C X 2003 Guidance document on delivery, treatment planning and clinical implementation of IMRT: report of the IMRT Subcommittee of the AAPM Radiation Therapy Committee *Med. Phys.* **30** 2089-115

[7] IBA Dosimetry 2011 *COMPASS Manual* Germany GmbH

Journal of Physics: Conference Series **694** (2016) 012026 doi:10.1088/1742-6596/694/1/012026

The dose delivery effect of the different Beam ON interval in FFF SBRT: TrueBEAM

T Tawonwong[1], S Suriyapee[2], S Oonsiri[1], T Sanghangthum[1] and P Oonsiri[1]

[1] Department of Radiology, King Chulalongkorn Memorial Hospital, BKK, Thailand
[2] Department of Radiology, Faculty of Medicine, Chulalongkorn University, BKK, Thailand

E-mail: ultrakji@hotmail.com

Abstract. The purpose of this study is to determine the dose delivery effect of the different Beam ON interval in Flattening Filter Free Stereotactic Body Radiation Therapy (FFF-SBRT). The three 10MV-FFF SBRT plans (2 half rotating Rapid Arc, 9 to10 Gray/Fraction) were selected and irradiated in three different intervals (100%, 50% and 25%) using the RPM gating system. The plan verification was performed by the ArcCHECK for gamma analysis and the ionization chamber for point dose measurement. The dose delivery time of each interval were observed. For gamma analysis (2%&2mm criteria), the average percent pass of all plans for 100%, 50% and 25% intervals were 86.1±3.3%, 86.0±3.0% and 86.1±3.3%, respectively. For point dose measurement, the average ratios of each interval to the treatment planning were 1.012±0.015, 1.011±0.014 and 1.011±0.013 for 100%, 50% and 25% interval, respectively. The average dose delivery time was increasing from 74.3±5.0 second for 100% interval to 154.3±12.6 and 347.9±20.3 second for 50% and 25% interval, respectively. The same quality of the dose delivery from different Beam ON intervals in FFF-SBRT by TrueBEAM was illustrated. While the 100% interval represents the breath-hold treatment technique, the differences for the free-breathing using RPM gating system can be treated confidently.

1. Introduction

The stereotactic body radiation therapy (SBRT) was introduced after the stereotactic radiosurgery (SRS) and the stereotactic radiation therapy (SRT) for the extra-cranial lesion. The delivery treatment time is prolonging for the dose of 6-30 Gray per fraction (Gy/Fx) [1-3]. The using of the Flattening Filter Free (FFF) beam for SBRT (FFF-SBRT) with the Linac-Based SBRT becomes more practicable. The flattening filter was removed out of the beam then the dose rate is increasing from 600 monitor unit per minute (MU/min) for the flattened beam to 1400 and 2400 MU/min for 6MV-FFF and 10MV-FFF beams, respectively. The total dose delivery time is shortening at least 2 times for FFF SBRT [4].

The margin added to the Gross Tumor Volume (GTV) due to the respiratory is essential. The breath hold technique allows the SBRT to be irradiated with the smallest added up margin. The repeatability of the same breath hold level may require using some from many systems used to monitor the patient's respiratory [5, 6].

The patient specific QA is very important to make sure that the LINAC machine can deliver the dose correctly. The gamma analysis with different criteria was investigated in many studies [7-9]. The accepted percent pass of the gamma analysis is depending on the instruments used. The point dose measurement is also performed. The purpose of this study is to determine the effect of the difference Beam

ON interval in Flattening Filter Free Stereotactic Body Radiation Therapy (FFF-SBRT) according to the phase selection using RPM gating system.

2. Materials and methods

The TrueBEAM LINAC, the Eclipse treatment planning version 11.0.31 and the Real-time Position Management (RPM) respiratory gating system (Varian Medical Systems, Palo Alto, CA) were employed in this study. The dose distribution was measured using the ArcCHECK 2 (Sun Nuclear Corp., FL). The CC13 ionization chamber (IBA dosimetry, Barlett, TN) was inserted into the center of ArcCHECK2 to measure the point dose at the isocenter. The setup of the instruments was illustrated in figure 1.

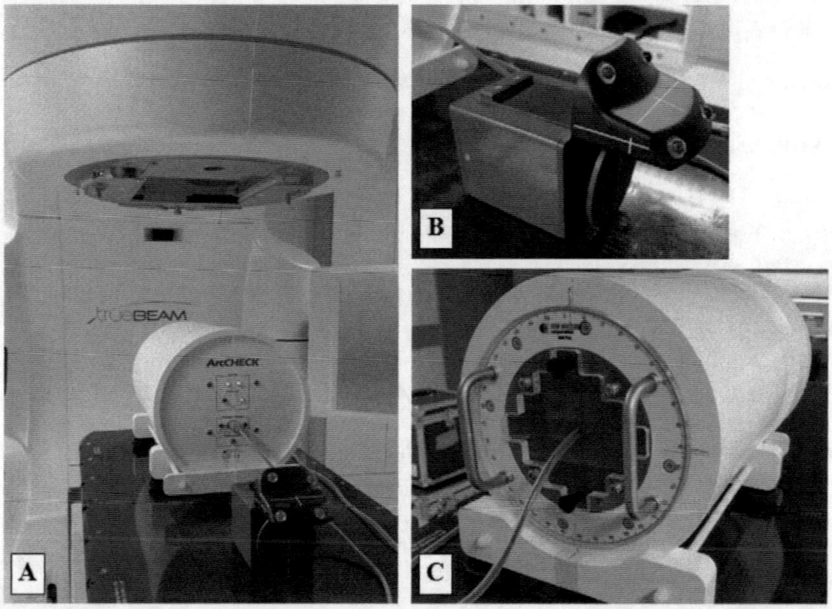

Figure 1. (A) The ArcCHECK2 and RPM system, (B) The Reflector box and the Breathing cycle simulator and (C) The ionization chamber inserted into ArcCHECK2

The three 10MV-FFF SBRT plans (2 half rotating Rapid Arc, 9 and 10 Gy/Fx) were selected and irradiated in three different Beam ON interval (100%, 50% and 25%). The RPM Gating System was used as the trigger of the BEAM ON interval in SBRT. The 100% Beam ON interval has represented the treatment of breath hold technique with intentional hold the beam by the therapist. For 50% and 25% intervals, the gating phase was selected as 25%-75% and 37%-62%, respectively. As shown in Figure 2, the yellow strip represents the Beam ON interval.

The ArcCHECK2 was designed for higher rep rate detectable which compatible with FFF mode. At the center of the phantom, the ionization chambers and inhomogeneity materials can be inserted as the option [8].

The measured dose distribution from ArcCHECK2 was compared with the calculated dose from the TPS. The 2% dose different and 2 mm distance-to-agreement threshold was used for gamma analysis. The integral point dose was calculated from the charge which corrected for pressure and temperature. The dose delivery time of each interval were observed.

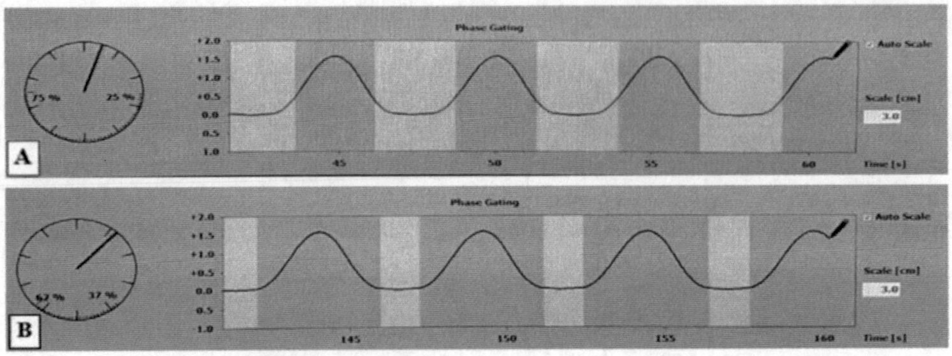

Figure 2. The gating phase selection for different Beam ON interval (A)50% and (B)25%

3. Results and discussion

3.1 Gamma analysis

For the 2% dose different and 2 mm distance-to-agreement threshold, the average percent pass of all plans for 100%, 50% and 25% interval were 86.1±3.3%, 86.0±3.0% and 86.1±3.3%, respectively. The demonstrated percent pass was representing the similarity of the different Beam ON interval. The results were shown in Table 1.

In the real clinical situation, we still observed for the accepted percent pass and the suitable criteria. The percent pass was 94.3±0.5%, 94.4±0.6% and 94.7±0.8% in the same order as above when using the 3% dose different and 3 mm distance-to-agreement threshold.

Table 1. The results of the 3 Rapid arc cases from the different Beam ON interval

Case No	Percent pass (2%, 2mm)			Point dose ratio (Measured/TPS)			Delivery time (sec)		
	100%	50%	25%	100%	50%	25%	100%	50%	25%
1	88.3	87.1	86.2	1.0065	1.0062	1.0059	69	141	325
2	87.6	88.3	89.4	1.0283	1.0265	1.0252	75	156	351
3	82.3	82.6	82.8	1.0007	1.0007	1.0007	79	166	365
Average	86.1	86.0	86.1	1.012	1.011	1.011	74.3	154.3	347
SD	3.3	3.0	3.3	0.015	0.014	0.013	5.0	12.6	20.3

3.2 Point dose

For point dose measurement, the mean dose for assumed ionization chamber volume at the isocenter of the ArcCHECK2 from the TPS was used to compare with the measurement. The average ratios of each interval to the TPS were 1.012±0.015, 1.011±0.014 and 1.011±0.013 for 100%, 50% and 25% interval, respectively. The deviation of the integral dose at the isocenter was about 1% from interval to interval.

3.3 Delivery time

The average dose delivery time was increasing from 74.3±5.0 second for 100% interval to 154.3±12.6 and 347.9±20.3 second for 50% and 25% interval, respectively. The delivery time was depending on the dose per fraction and the size of the target. The relative ratio was 2 and 4 times for 50% and 25% compared to 100% interval, respectively.

Even though the percent gamma passed result of our study was lower than the other studies [7, 8], the maximum ratios of point dose were within 2-3%. The higher dose rate of 2400 MU/min and the higher dose gradient in delivering the dose of 9-10 Gy in two half rotating arcs may be the major factors which effected on the gamma analysis outcome. The more number of the SBRT patient specific QA have to be investigated to establish the optimal percent pass acceptable and the suitable gamma criteria.

4. Conclusion

The performance of gating SBRT in TrueBEAM was investigated. The same quality of the dose delivery from different Beam ON interval in FFF-SBRT was illustrated. The increasing number of the patients can be confidently treated by free breathing and using the RPM gating system. The Breath hold technique is the best selection to minimize the side effect from the treatment. While not all patients can hold their breath for a long period, free breathing plus the RPM Gating System may be used to increase the opportunity for these people. The increasing treatment time may be negotiated by the benefit of receiving SBRT with the optimal margin.

References

[1] Benedict S H *et al.* 2010 Stereotactic body radiation therapy: The report of TG101

[2] Amedola B E, Amendola M, Perez N, Iglesias A and Wu X 2013 Volumetric-modulated arc therapy with RapidArc®: An evaluation of treatment delivery efficiency *reports of practical oncology and radiotherapy* **18** 383–6

[3] Roa D E ,Schiffner D C, Zhang J, Dietrich S N, Kuo JV , Wong J, Ramsinghani N S and Al-Ghazi M S. 2012 The use of RapidArc®volumetric-modulated arc therapy to deliver stererreotactic radiosurgery and stereotactic body radiotherapy tointracranial and extracranial targets. *Med Dosim* 257–64

[4] Thomas E M, Popple R A, Prendergast B M, Clark G M, Dobelbower M C and Fiveash J B 2013 Effects of flattening filter-free and volumetric-modulated arc therapy delivery on treatment efficiency *Journal of Applied Clinical Medical Physics* **14** 155-66

[5] Wang P M, Chung N N, Hsu W C, Chang F L, Jang C J and Scorsetti M 2014 Feasibility of stereotactic body radiation therapy with volumetric modulated arc therapy and high intensity photon beams for hepatocellular carcinoma patients *Radiation Oncology* 9:18

[6] Korreman S 2005 Breathing adapted radiotherapy for breast cancer: comparison of free breathing gating with the breath hold technique *Radiother. Oncol.* **76** 311–8

[7] Nicolini G, Vanetti E, Clivio A, Fogliata A and Cozzi L 2010 Pre-clinical evaluation of respiratory-gated delivery of volumetric modulated arc therapy with RapidArc Phys. Med. Biol. **55** 347-57

[8] Fakir H, Gaede S, Mulligan M and Chen J Z 2012 Development of a novel ArcCHECK™ insert for routine quality assurance of VMAT delivery including dose calculation with inhomogeneities *Medical Physics* **39** 4203-8

[9] Noto K, Ueda S, Kojima H, Isomura N, Takemura A, Takamatsu S, Kumano T and Takanaka T 2014 Accuracy of Dose Delivery in Multiple Breath-Hold Segmented Volumetric Modulated Arc Therapy: A Static Phantom Study *Journal of Radiotherapy* Article ID 743150

Acknowledgement

The authors would like to express our thanks to all staffs and facilities in the division of Therapeutic Radiation and Oncology, Department of Radiology, King Chulalongkorn Memorial Hospital – The Thai Red Cross Society for all support.

13th South-East Asian Congress of Medical Physics 2015 (SEACOMP) IOP Publishing
Journal of Physics: Conference Series **694** (2016) 012027 doi:10.1088/1742-6596/694/1/012027

Determination of small field output factors in 6 and 10 MV flattening filter free photon beams using various detectors

W Masanga[1,3], P Tangboonduangjit[2], C Khamfongkhruea[3] and C Tannanonta[3]

[1] Medical Physics School, Faculty of Medicine Ramathibodi Hospital, Mahidol University, Bangkok, Thailand
[2] Department of Diagnostic and Therapeutic Radiology, Faculty of Medicine Ramathibodi Hospital, Mahidol University, Bangkok, Thailand
[3] Department of Radiation Oncology, Chulabhorn Hospital, Bangkok, Thailand

E-mail: wm.namaoi@gmail.com

Abstract. The study aimed to determine appropriate detectors for output factor measurement of small fields in 6 and 10 MV flattening filter free photon beams using five different detectors. Field sizes were varied between 0.6×0.6 and 4.0×4.0 cm^2. An indirect method (daisy-chaining) was applied to normalize the output factors. For the smallest field size, the variations of output factors compared among the detectors were 13%. Exradin A16 had the lowest output factor and increasing in sequence with CC01, microDiamond, microLion and EDGE detectors, respectively, for both energies. The similarity between CC01 and microDiamond output factor values were within 1.6% and 1% for all field sizes of 6 and 10 MV FFF, respectively. EDGE and microLion presented the highest values while ExradinA16 gave lowest values. In conclusion, IBACC01, Exradin A16, microLion, microDiamond and EDGE detectors seem to be the detectors of choices for small field output factor measurement of FFF beams down to 1.6×1.6 cm^2. However, we could not guarantee which detector is the most suitable for output factor measurement in small field less than 1.6×1.6 cm^2 of FFF beams. Further studies are required to provide reference information for validation purposes.

1. Introduction

Radiotherapy has been greatly developed for both treatment machines and delivery techniques. High energy photon linear accelerator can be divided in three major subtypes, namely helical linac, robotic linac and standard linac. The former two are without a flattening filter while the standard linac, typically comes with flattening filter (FF) to produce uniform dose for certain depths, could also be used without flattening filter,(flattening filter free; FFF). Several studies used the FFF photon beams for clinical treatment particularly for stereotactic radiotherapy, stereotactic body radiotherapy, intensity-modulated radiotherapy (IMRT) and volumemetric modulated arc therapy (VMAT) [1-3]. The main advantages of FFF technology are increasing in dose rate and reducing in head scatter which provided a short delivery time with reducing out of field dose. For IMRT and VMAT techniques, they are based on delivery of non-uniform fluence distribution; thus flattened beam is not required [4].

Generally these advanced techniques are used in small field or segment deliveries which are non-equilibrium conditions. A concise definition of small field is still unclear. Typically, it is considered from the field dimension which is less than the lateral range of charged particle which leads to the occlusion of source from collimating device [5]. Dosimetry of small fields remains a challenging topic and difficult to achieve with accurate measurements because of loss of charged particle equilibrium, source occlusion and finite size of detectors [6]. Detectors with high spatial resolution, good tissue equivalence, and high precision are preferred. Several studies examined measurement of output factor with various detectors in small field. Dieterich and Sherouse assessed performance of various commercially available diodes by measuring SRS cone factors [7]. They demonstrated the "daisy-chaining" method to limit biasing effects of spectral differences between field sizes. Small field output factor of CyberKnife® and linear accelerator (Novalis) with microMLC and circular cone were determined and compared with different types of detectors by Bassinet *et al.* Their results showed significant differences among output factor values measured by several detectors, particularly in the smallest field size, was more than 20% [5].Wolfgang *et al.* studied output factor measurements of FF and FFF beams at field size of 10 $\times$10 cm^2down to 0.6 $\times$ 0.6 cm^2 by using various detectors [8]. They reported that the dose response ratio (ratio of the relative output measured with online detectors to alanine) had no significant difference between FF and FFF beams and concluded that microDiamond and unshielded diode (SFD and EFD) were suitable candidates for small field dosimetry in FF and FFF beams.The purpose of this study was to determine appropriate detector types for output factor measurement of small fields in the 6 and 10 MV FFF mode of a standard linac.

2. Materials and methods

2.1. Linear accelerator

A TrueBeam linear accelerator (Varian) at Chulabhorn Hospital (Bangkok, Thailand) is a new C-series machine designed to deliver both flattened photon beams and flattening filter free (FFF) photon beams. In this study, FFF beam energy of 6 and 10 MV at maximum dose rate of 1400 and 2400 MU/min, respectively, were used for measurement of output factor of collimator jaw-shaped square field sizes of 0.6 to 4.0 cm at 10 cm depth in water.

2.2. Detectors

Detectors which have small size of active volume including the followings: air filled ionization chambers (IBA CC01 and Exradin A16), liquid filled ionization chambers (microLion PTW 31018), Synthetic single crystaldiamond detector (PTW 60019 microDiamond), and Sun Nuclear EDGE detector were included in the study, details as shown in table 1.

Table 1. Characteristics of detectors used in this study.

Detector	Detector type	Sensitive volume (mm)	Volume	Material	Effective point
PTW 60019 microDiamond	Synthetic diamond	Thickness 0.001 Diameter 2.2	0.004 mm^3	Carbon Polystyrene	1 mm
Sun Nuclear Edge detector	Shieldeddiode	Thickness 0.03 Diameter 0.8	0.0019 mm^3	Silicon Brass	0.5 mm
IBA CC01	Airfilled IC	Cavity lehgth3.6 Diameter 2	0.01 cm^3	Air Wall: C-522 Electrode: Steel	0.6r
Exradin A16	Airfilled IC	Diameter 2.4	0.007 cm^3	Air Wall: C-522	0.6r
PTW 31018 microLion	Liquid filled IC	Thickness 0.35 Diameter 2.5	0.002 cm^3	Liquid isooctane	0.975 mm

All detectors were positioned inside Blue Phantom1 System (ScanditronixWellhofer GmbH, Schwarzenbruck, Germany). For CC01, Exradin A16 and Edge detectors, their stems were placed perpendicular to the beam axis. In contrast, microDiamond and microLion were placed with their stem parallel to the beam axis. In the first step of measurement, the effective point of measurement of all detectors was placed at water surface. And then, it was moved to depth of 10 cm with a source axis distance of 100 cm. In order to align the center of the detectors to the maximum dose, beam profiles were scanned for field size of 1.0 × 1.0 cm^2 in step mode with OMNIPRO-ACCEPT software.

2.3. Measurement conditions

Each detector measured the output factor by opening collimator jaw-shaped for field sizes of 0.6 × 0.6, 1.0 × 1.0, 1.6 × 0.6, 2.0 × 2.0, 2.6 × 2.6, 3.0 × 3.0, 4.0 × 4.0 and 10 × 10 cm^2 at 10 cm depth with a source axis distance of 100 cm. The measurements were performed using 150 MU. According to the manufacturer's specifications, the nominal voltages for CC01, Exradin A16 and microLion are +400V, +300V, and +800V, respectively. Edge and microDiamond have no bias voltage.

Practically, the output factors were normalized to their corresponding values at calibration reference field of 10 × 10cm^2. In this study, relative output factor for small volume detectors was calculated by daisy-chaining method which is the ratio of output reading for small field size to intermediate 4.0 x 4.0 cm^2 field size of small volume detector and renormalized by applying the ratio of output reading of suitable ionization chamber at intermediate field of 4.0 x 4.0 cm^{2}to reference field of 10 x 10 cm^2 according to equation (1).

$$Outputfactor = \frac{M_{SD(S_{\text{small}})}}{M_{SD(S_{\text{Intermediat}})}} \cdot \frac{M_{IC(S_{\text{Intermediat}})}}{M_{IC(S_{\text{ref}})}} \quad (1)$$

where *SD* is small field detector, *IC* refers to a suitable ionization chamber for intermediate field (CC13), $S_{\text{intermediate}}$ refers to the intermediate field 4.0 x 4.0 cm^2 and S_{small} refers to small defined fields.

3. Results

Table 2 and figure 1 showed the result of output factors measured with each of small detectors in 6 MV FFF photon beams with the maximum dose rate of 1400 MU/min. The largest variations of output factor values compared among detectors were found to be13% in the field size of 0.6 x 0.6 cm^2. It is noteworthy that output factor of all detectors obtained good proximity within 2% except the field size equal or less than 1.0 x 1.0 cm^2. For the smallest field size, output factors measured by Exradin A16 was much smaller than those of the other detectors while output factors measured with microLion and EDGE were the largest.

Table 2. Output factors for 6MV FFF (1400 MU/min) measured with various detectors at depth of 10 cm of SAD 100 cm.

Field size(cm^2)	Detectors				
	Exradin A16	CC01	microDiamond	microLion	EDGE
4.0 ×4.0	0.880	0.880	0.880	0.880	0.880
3.0 ×3.0	0.846	0.844	0.849	0.852	0.844
2.6×2.6	0.832	0.830	0.834	0.839	0.830
2.0 ×2.0	0.805	0.804	0.810	0.816	0.807
1.6×1.6	0.778	0.783	0.786	0.794	0.786
1.0 ×1.0	0.689	0.701	0.713	0.724	0.725
0.6×0.6	0.501	0.531	0.537	0.566	0.569

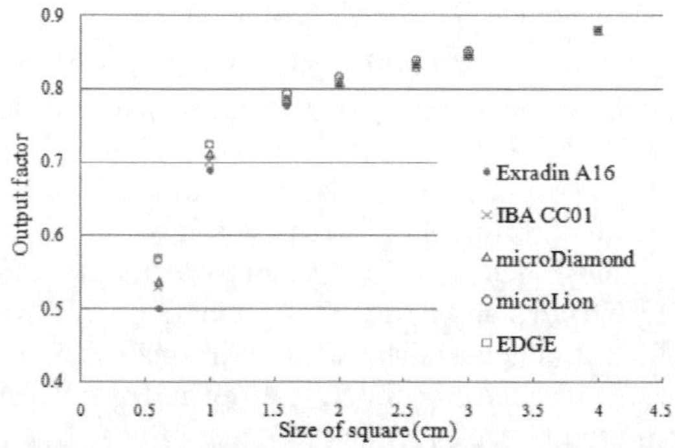

Figure 1. Output factors for 6MV FFF (1400 MU/min) measured with various detectors.

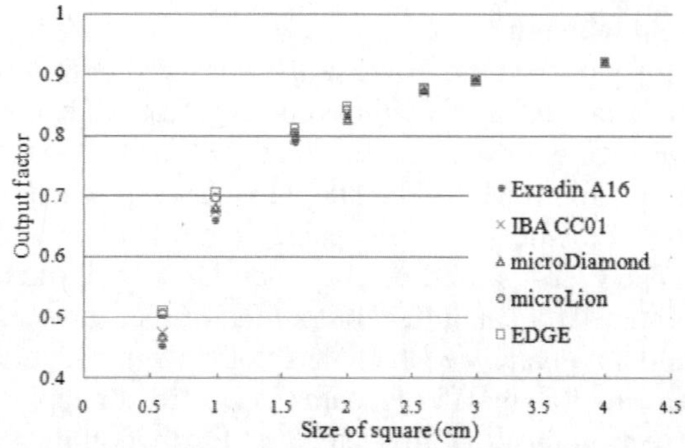

Figure 2. Output factors for 10MV FFF (2400 MU/min) measured with various detectors.

Table 3 and figure 2 present the output factors of 10 MV FFF photon beam with the maximum dose rate 2400 MU/min. At the smallest field size, the output factors significantly differed with the largest deviation was at 12.9%. The range of the output factors was within 3% for all field sizes larger than 1.6 × 1.6 cm^2. For microLion and EDGE detectors, the output factors are mutually within1.5% and higher than the other detectors while Exradin A16 obtains the lowest value. For CC01 and microDiamond, the output factors agree within1.2% for all field sizes and within the range ofthe otherdetectors.

Table 3. Output factors for 10MV FFF (2400 MU/min) measured with various detectors at depth of 10 cm of SAD 100 cm.

Field size(cm^2)	Detectors				
	Exradin A16	CC01	microDiamond	microLion	EDGE
4.0 × 4.0	0.920	0.920	0.920	0.920	0.920
3.0 × 3.0	0.889	0.892	0.891	0.893	0.893
2.6× 2.6	0.872	0.871	0.876	0.878	0.879
2.0 ×2.0	0.833	0.834	0.827	0.843	0.847
1.6 × 1.6	0.789	0.794	0.798	0.805	0.812
1.0 × 1.0	0.661	0.675	0.681	0.696	0.707
0.6 × 0.6	0.452	0.475	0.469	0.506	0.511

4. Discussion

For field size equals or less than 1.6 × 1.6 cm^2, the air filled ionization chambers including Exradin A16 and CC01 showed the lowest values with decreasing field sizes may be caused by the volume averaging effect. In fact the larger the volume of detector gives the smaller output factor. In this work, the output factor measured by Exradin A16, which has the smallest volume, shows the lowest values instead of CC01, which has the largest volume size, owing to its electrode made by air equivalent plastic while CC01 made by steel which causes more scatter to the output factor. Although, CC01 provided less values, but it agrees with alanine detector within 2% over the range of fields shown by Lechner W [6] and within 3.5% for field size of 0.6 × 0.6 cm^2 demonstrated by Godfrey A [9].

For microDiamond, made of tissue equivalent material, its output factor value was intermediate of other detectors. The agreement between CC01 and microDiamond output factor values are within 1.6% and 1% for all field sizes of 6 and 10 MV FFF, respectively.

Most of all filed sizes, microLion shows mostly higher value of output factor and it was only one to use highest voltage and concern about ion recombination effect in FFF beams as shown by Stephanie L [10] and Eunah C [11]. In addition, the output factor values were closed to the values of EDGE for all field sizes down to 1.0×1.0 cm^2 within 1.5% which agrees with the result of Shin H J [12].

For EDGE, is made of silicon chip and brass shielded material which might increase the scatter dose.

5. Conclusion

In this work, we experimentally measured output factors for 6 and 10 MV FFF photon beams of Varian TrueBeam using five different detectors. The output factors obtained by different detectors differ significantly in smallest field size of 0.6×0.6 cm^2 for 6MV FFF and field size equals or less than 1.6×1.6 cm^2 for 10MV FFF. The variation can reach to 12.9% for smallest field size for both energies and within 2% and 3% for field sizes 1.6×1.6 cm^2 and above of 6 and 10 MVFFF, respectively.

The CC01, Exradin A16, microDiamond, microLion, and EDGE detectors seem to be the detector of choices for small field output factor measurement of FFF beams down to 1.6×1.6 cm^2. However, we could not guarantee which detector is the most suitable for output factor measurement in small field less than 1.6×1.6 cm^{2}of FFF beams as we did not applying a reference detector for comparing and calculating correction factor of the detectors. Further studies are required to seek for the reference information to be compared with.

References

[1] Sharma S D 2011 Unflattened photon beams from the standard flattening filter free accelerators for radiotherapy: Advantages, limitations and challenges *J Med Phys*. **36** 123–5

[2] Kragl G, Wetterstedt S, Knausl B, Lind M, Georg D *et al.* 2009 Dosimetric characteristics of 6 and 10 MV unflattened photon beams *Radiother. Oncol*. **93** 141-6

[3] Hrbacek J, Lang S and Klöck S 2011Commissioning of photon beams of a flattening filter-free linear accelerator and the accuracy of beam modeling using an anisotropic analytical algorithm *Int. J. Radiat.Oncol. Biol. Phys*. **80** 1228-37

[4] Georg D, Knöös T and McClean B 2011 Current status and future perspective of flattening filter free photon beams *Med. Phys.* **38** 1280-93

[5] Bassinet C *et al.* 2013 Small fields output factors measurements and correction factors determination for several detectors for a CyberKnife® and linear accelerators equipped with microMLC and circular cones *Med. Phys.* **40** 071725

[6] Indra J D, X Ding G and Ahnesjö A 2008 Small fields: nonequilibrium radiation dosimetry *Med. Phys.* **35** 206-15

[7] Dieterich S and Sherouse G W 2011 Experimental comparison of seven commercial dosimetry diodes for measurement ofstereotactic radiosurgery cone factors *Med. Phys.* **38** 4166

[8] Lechner W, Palmans H, Sölkner L, Grochowska P and Georg D 2013 Detector comparison for small field output factor measurements in flattening filter free photon beams *Radiotherapy and Oncology*. **109** 356-60

[9] Azangwe G *et al.* 2014 Detector to detector corrections: a comprehensive experimental study of detectorspecific correction factors for beam output measurements for small radiotherapybeams *Medical Physics* **41** 072103

[10] Lang S, Hrbacek J, Leong A and Klöck S 2012 Experimenyal analysis of general ion recombination in a liquid-filled ionization chambers in high dose rate flattenning-filter-free photon beams *Phys. Med. Biol.* **57** 2819-27

[11] Chung E, Davis S and Seuntjens 2013 Experimental of general ion recombination in a liquid-filled ionization chamber in high energy photon beams *Med. Phys.* **40** 062104

[12] Shin H J, Khang Y and Jung J Y. 2012 Evalation of the EDGE Detector in Small-filed Dosimeter *Journal of the Korean Physical Society* **63** 128-134

Acknowledgements

The authors gratefully appreciate the Department of Radiology, Ramathibodi Hospitalfor loaning of the microDiamond andExradin A16 detector and thank Dr. Danupon Nantajit at Chulabhorn Hospital, for grammar review of the manuscript.

Journal of Physics: Conference Series **694** (2016) 012028 doi:10.1088/1742-6596/694/1/012028

$ZnFe_2O_4$ nanoparticles for potential application in radiosensitization

M Hidayatullah[1], I Nurhasanah[1] and W S Budi[1]

[1] Departement of Physics, Faculty of Mathematics and Natural Sciences, Diponegoro University Jl. Prof. Soedarto, SH, Semarang, Indonesia

E-mail: hidayat24@st.fisika.undip.ac.id

Abstract. Radiosensitizer is a material that can increase the effects of radiation in radiotherapy application. Various materials with high effective atomic number have been developed as a radiosensitizer, such as metal, iron oxide and quantum dot. In this study, $ZnFe_2O_4$ nanoparticles are included in iron oxide class were synthesized by precipitation method from the solution of zinc nitrate and ferrite nitrate and followed by calcination at 700°C for 3 hours. The XRD pattern shows that most of the observed peaks can be indexed to the cubic phase of $ZnFe_2O_4$ with a lattice parameter of 8.424 Å. SEM image reveals that nanoparticles are the sphere-like shape with size in the range 84–107 nm. The ability of $ZnFe_2O_4$ nanoparticles as radiosensitizer was examined by loading those nanoparticles into *Escherichia coli* cell culture which irradiated with photon energy of 6 MV at a dose of 2 Gy. $ZnFe_2O_4$ nanoparticles showed ability to increase the absorbed dose by 0.5 to 1.0 cGy/g. In addition, the presence of 1 g/L $ZnFe_2O_4$ nanoparticles resulted in an increase radiation effect by 6.3% higher than if exposed to radiation only. These results indicated that $ZnFe_2O_4$ nanoparticles can be used as the radiosensitizer for increasing radiation effect in radiotherapy.

1. Introduction

Radiotherapy has been long used in cancer treatment by application of X-ray and high-energy particles radiation. The problem in radiotherapy is the optimization of the cancer cell killing and preserving healthy tissues. One strategy to enhance the cell killing is the use radiosensitizers. Recently, research focused on the development of nanomaterials for radiosensitizer under irradiation with X-ray and high-energy particle. Several nanomaterials have been developed as radiosensitizer ie heavy metal with a high atomic number, iron oxide, and quantum dot [1]. Materials with high atomic numbers can increase the number of secondary electrons formation which can increase the radiation absorbed dose. Gold and gadolinium nanoparticles are heavy metal which commonly used as radiosensitizer [2, 3]. Another radiosensitizer proposed for radiotherapy is an iron oxide which has the ability to increase free radicals formation. Free radicals such as superoxide played the main role in the death of cancer cells indirectly [4, 5].

$ZnFe_2O_4$ nanoparticles with low toxicity have been intensively studied for medical applications such as hyperthermia treatment [6], magnetic resonance imaging (MRI) [7] and drug delivery [8]. $ZnFe_2O_4$ nanoparticles were explored as radiosensitizer because it's effective atomic number higher than the effective atomic number of the body's tissues and included in the class of iron oxide [9]. In this study, $ZnFe_2O_4$ nanoparticles were synthesized by precipitation method. These method has been used for production $ZnFe_2O_4$ [10-13]. $ZnFe_2O_4$ nanoparticles were evaluated as a radiosensitizer by measuring the ability to increase the absorbed dose and loading into *E. coli* culture as the living system to determine the enhancing radiation effect.

2. Experimental

2.1. Preparation of $ZnFe_2O_4$ nanoparticle

$ZnFe_2O_4$ nanoparticles were synthesized by precipitation method from the mixed solution of iron nitrate [$Fe_3(NO)_3.9H_2O$] and zinc nitrate [$Zn(NO_3)_2.4H_2O$]. A stoichiometric amount of iron nitrate and zinc nitrate with molar ratio of 1:2 were dissolved in distilled water. Ammonium hydroxide was used as precipitant and mixed with nitrate solution under magnetic stirring at 80°C and pH until 10. The precipitate was obtained then aged for 2 hours. The solid was filtered and washed with distilled water, dried at 100°C for 2 hours in an oven. The reddish brown solid was calcined at 700°C for 3 hours and grinded in a mortar to obtain fine powder $ZnFe_2O_4$ nanoparticles.

2.2. Characterization of $ZnFe_2O_4$ nanoparticles

The structure and phase of $ZnFe_2O_4$ nanoparticles were characterized by X-ray diffraction (XRD). XRD pattern was obtained using PW 1710 diffractometer. The crystallite size was determined using Scherer formula. Scanning electron microscope (SEM) JEOL JSM-6360 was used to obtain the morphological image of $ZnFe_2O_4$ nanoparticles.

Absorbed dose measurement was carried out using linac with energy radiation of 6 MV, 2 Gy of dose, SSD of 100 cm, the field (10 x10) cm^2, the position of the detector located at a depth of 1 cm below phantom surface. Furthermore, the absorbed dose was measured without and with $ZnFe_2O_4$ nanoparticles for the different mass of 1 g, 2 g, 3 g, 4 g and 5 g. $ZnFe_2O_4$ nanoparticles were placed on the phantom surface and above the sensitive area detector. The ability of $ZnFe_2O_4$ nanoparticles as radiosensitizer was investigated by loading those nanoparticles into *E. coli* culture and irradiated using the energy of 6 MV with 2 Gy of dose. A number of bacteria survival was calculated using total plate count method.

3. Results and discussion

3.1. Structure and morphology analysis

The XRD pattern of nanoparticles is shown in figure 1. A pattern shows that nanoparticles contain three phases ie most strong peaks can be indexed as high-quality crystalline of $ZnFe_2O_4$ which is agreement with JCPDS 22-1012. Several peaks can be ascribed from low crystalline of ZnO and Fe_2O_3 phases indicating the obtained nanoparticles is still contained small another phase. However, the obtained value of lattice parameter is 8.424 Å comparable to the value of 8.427 Å [14] and 8.429 Å [15] and close to the value reported in JCPDS 8.441 Å. Therefore, the major phase of nanoparticles in this study can be referred as spinel cubic structure of $ZnFe_2O_4$. Crystallite sizes in the range 11.85 nm to 22.63 nm were obtained by using full width at half maximum (FWHM) of the dominant peaks (220), (311), (511), (440), and (400) to Scherer formula.

The morphology of the $ZnFe_2O_4$ nanoparticles was observed using SEM image as presented in figure 2. It was found that the prepared nanoparticles consist of spherical clusters with sizes in the range 84.44 nm-106.67 nm.

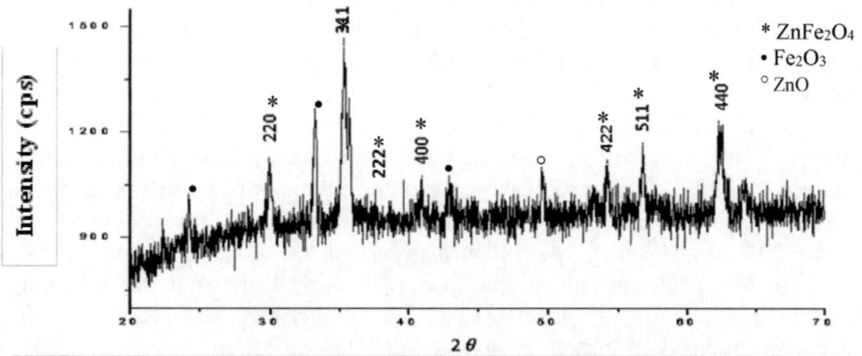

Figure 1. The XRD graphs of $ZnFe_2O_4$ nanoparticles.

Figure 2. SEM image of $ZnFe_2O_4$ nanoparticles.

3.2. Radiosensitizer ability study

The increase in absorbed dose by radiosensitizer can enhance radiation effect to the living system. Potential $ZnFe_2O_4$ nanoparticles as a radiosensitizer can be evaluated from the ability those nanoparticles to increase absorbed dose. The increase in absorbed dose was determined by subtracting the measured absorbed dose without $ZnFe_2O_4$ nanoparticles to the measured absorbed dose $ZnFe_2O_4$ nanoparticles for various mass. Figure 3 shows the increase in absorbed dose for the various mass of $ZnFe_2O_4$ nanoparticles.

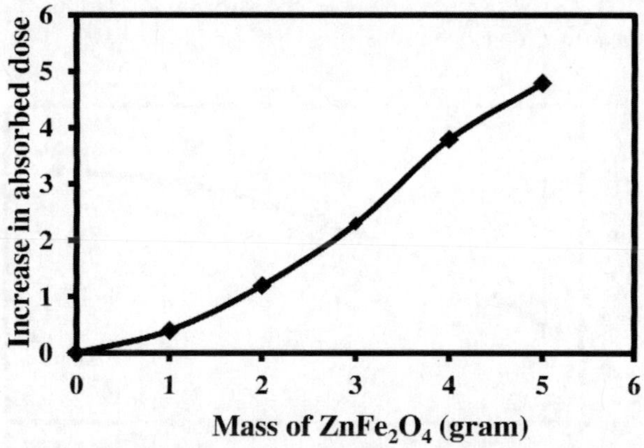

Figure 3. The increase in absorbed dose for various mass of $ZnFe_2O_4$ nanoparticles.

The increase of absorbed dose increases with increasing mass of nanoparticles $ZnFe_2O_4$. It was found that $ZnFe_2O_4$ nanoparticles exhibited the ability to increase of absorbed dose by 0.5 to 1 cGy/g. 1 Gy absorbed dose can produce 2×10^5 ionization, causing damage to 1000 single-strand DNA breaks and 40 double-strand breaks [16, 17]. This means that the increase in absorbed dose of 1 cGy is equivalent to an increase in ionization of 2×10^3, 24 single-strand breaks and 1 double-strand breaks. From this view point can be suggested that $ZnFe_2O_4$ nanoparticles have potential as radiosensitizer.

The influence of $ZnFe_2O_4$ nanoparticles on the cell killing was investigated by loading those nanoparticles into *E. coli* cultures with various concentrations. The cultures were irradiated with the photon energy of 6 MV at a dose of 2 Gy. It was observed that *E. coli* survival decreases with the addition of $ZnFe_2O_4$ nanoparticles as can be seen in figure 4. The decrease in the percentage of bacteria survival indicates the addition of $ZnFe_2O_4$ nanoparticles enhances the radiation effect. Therefore, the radiation effect enhancement determined by subtracting the percentage of bacteria survival with the addition of $ZnFe_2O_4$ nanoparticles to the percentage of bacteria survival without $ZnFe_2O_4$ nanoparticles. Figure 5 reveals the radiation effect enhancement for various $ZnFe_2O_4$ nanoparticles concentrations. The radiation effect enhancement increases with increasing $ZnFe_2O_4$ nanoparticles concentrations. The addition of 1 g/L $ZnFe_2O_4$ nanoparticles enhances radiation effect of about 6.3% greater than if the culture exposed to radiation only without an addition of $ZnFe_2O_4$ nanoparticles. The radiation effect enhancement is occurred due to an increase in the number of free radicals formation because of $ZnFe_2O_4$ nanoparticles addition. These results suggest that $ZnFe_2O_4$ nanoparticles can be used as a radiosensitizer.

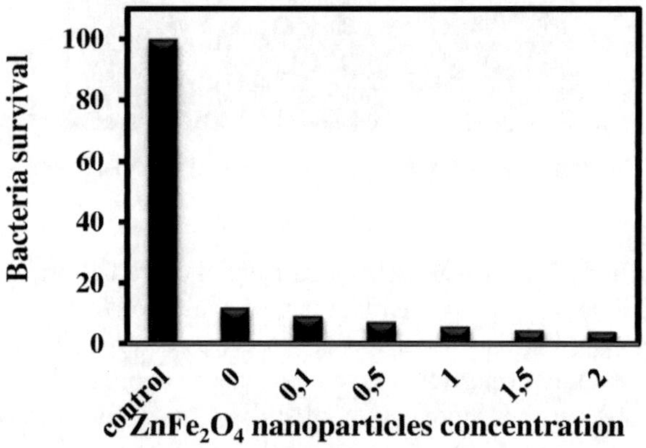

Figure 4. The percentage of bacteria survival for various $ZnFe_2O_4$ nanoparticles concentrations.

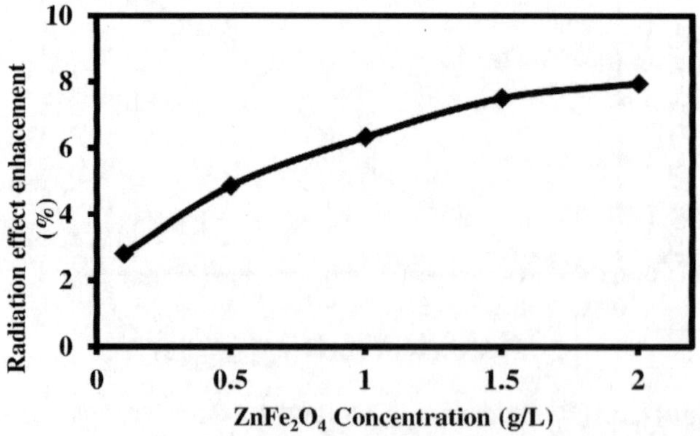

Figure 5. The radiation effect enhancement for various $ZnFe_2O_4$ nanoparticles concentrations.

4. Conclusions

$ZnFe_2O_4$ nanoparticle was synthesized by precipitation method. It has been found that nanoparticle is spinel structure with crystallite size in the range of 11.85 to 22.63 nm and spherical in shape. $ZnFe_2O_4$ nanoparticle was used as radiosensitizer which exhibited improve absorbed dose ability of 0,5–1 cGy/g. The use of $ZnFe_2O_4$ nanoparticle with concentration 1 g/L resulted in an increase in the effect of radiation by 6.3%. Therefore, the nanoparticles $ZnFe_2O_4$ exhibited potential as a radiosensitizer for radiotherapy application.

References

[1] Kwatra D, Venugopal A and Anant S 2013 *Transl. Cancer Res.* **2** 330
[2] Ngwa W, Korideck H, Kassis A, Kumar R, Sridhar S, Makrigiorgos G and Cormack R 2013 *Nanomedicine* **9** 25
[3] Duc G *et al.* 2011 *ASC Nano* **5** 9566
[4] Huang G, Chen H, Dong Y, Luo X, Yu H, Moore Z, Bey E, Bootman D and Gao J 2013 *Theranostics* **3** 116
[5] Choi G, Seo S, Kim K, Kim H, Park S, Lim J and Kim J 2012 *Radiat. Oncol.* **7** 184
[6] Meidanchi A, Akhavan O, Khoei S, Shokri A, Hajikarimi Z and Khansari N 2015 *Material Science and Engineering C* **46** 394
[7] Ichiyanagi Y, Shigeoka D, Hiroki T, Mashino T, Kimura S, Tomitaka A, Ueda K and Takemura Y 2012 *Thermochimica Acta* **532** 123
[8] Zhang S, Liu X, Zhou L and Peng W 2012 *Materials Letters* **68** 243
[9] Rana S, Gallo A, Srivastava R and Misra R 2007 *Acta Biomaterialia* **3** 233
[10] Valenzuela M A, Bosch P, Jimenez-Becerrill J, Quiroz O and Paez A L 2002 *J. Photochemistry and Photobiology* **148** 182
[11] Maletin M, Moshopuolou E, Kontos A G, Devlin E, Delimitis A, Zaspalis V T, Nalbandian L and Srdic V V 2007 *J. European Ceramic Society* **27** 4391
[12] Upadhyay C, Verma H.C, Sathe V and Pimpale A.V 2007 *J. Magnetism and Magnetic Materials* **312** 271
[13] Li Q, Bo C and Wang W 2010 *Materials Chem. And Phys.* **124** 891
[14] Lemine O.M, Bouodina M, Sajieddine M, Al-Saie A M, Shafi M, Khatab A, Al-Hilali M and Henini M 2011 *Physica Cond. Matt. B* **406** 1989
[15] Zhou Z, Liu J, Wang C, Sun P, Hu X, Li X, Shimanoe K, Yamazoe N and Lu G 2015 *Sensor and Actuators* B **206** 577
[16] Lehnert S 2007 *Biomolecular Action of Ionizing Radiation* (New York, London : Taylor and Francis Group) p157
[17] Mayles P, Nahum A and Rosenwald J 2007 *Handbook of Radiotherapy Physics* (New York, London : Taylor and Francis Group) p152

Validation of in-house treatment planning system software for cobalt-60 teletherapy unit at two radiotherapy installations

I A S Mu'minah[1], L O H Z Toresano[1], W E Wibowo[2], Sugiyantari[3] and S A Pawiro[1]

[1] Department of Physics, Faculty of Mathematics and Natural Sciences, Universitas Indonesia, Depok 16424, Indonesia
[2] Department of Radiotherapy, Cipto Mangunkusumo National Center Hospital, Jakarta, Indonesia
[3] Department of Radiotherapy, Persahabatan Center Hospital, Jakarta, Indonesia

E-mail: intan.apriliani@sci.ui.ac.id

Abstract. DSSuperDose v.1.0 is an in-house treatment planning system (TPS) developed by Medical Physics and Biophysics Laboratory (LFMB) Universitas Indonesia as a treatment planning software for Cobalt-60 teletherapy unit. The main objective of this study was the validation of in-house TPS calculation as an essential part in quality assurance (QA) of radiotherapy. Validation of an in-house TPS was performed with two Cobalt-60 teletherapy units by comparison between in-house TPS and ISIS TPS and by measurements of absorbed dose. Mean dose deviations between in-house TPS and measurement were (1.97 ± 2.42)% for open field, (1.32 ± 1.30)% for tray field, and (2.91 ± 2.36)% for wedge field treatments. In-house TPS provide optimal planning for open and tray beam conditions with depth fewer than 10 cm (≤ 10 cm) and field sizes up to 20×20 cm^2, while for wedge beam conditions with field sizes fewer than the physical size of the wedge. Comparison of in-house TPS and ISIS TPS demonstrated a good match of 96%. From the results, it is concluded that DSSuperDose v.1.0 is adequately accurate for treatment planning of radiotherapy.

1. Introduction

Radiation therapy (both internal and external radiation) is a complex and potentially a high risk procedure. It is very important that radiation therapy is accurately planned and properly delivered to the patients. In external beam radiation therapy, either treatment time of teletherapy unit or monitor unit (MU) of linear accelerator becomes the important parameter in ensuring the accuracy of prescribed dose to patients [1]. Efforts to reduce errors and uncertainties of the calculation in treatment planning allow important role in achieving the optimal therapy. Improving the accuracy of the calculation, computational tools should be applied in the treatment planning process. Such as in-house treatment planning system (TPS) was developed in beginning of 2015 by Medical Physics and Biophysics Laboratory (LFMB) Universitas Indonesia as the treatment planning software for cobalt-60 teletherapy unit.

Quality assurance (QA) in treatment planning process of the radiation therapy is essential for minimizing the possibility of accidental exposure. The accuracy of either treatment time or MU calculation is one of important parameter in QA TPS [2]. Validation of MU calculation as part of the commissioning process is one way to identify errors on TPS algorithms, before it should be used clinically [3]. This validation is done by comparison of dose calculation and measurement. Some researchers have described several criteria of dose deviation or acceptable tolerances [2-4]. Another verification methods can also be reached to strengthen confidence in the accuracy of an in-house TPS. Verification calculations are traditionally performed by applying empirical algorithms in a manual calculation procedure [2, 5]. Moreover, comparison of in-house TPS with independent MU calculation software can also be an alternative that is sufficient to identify significant errors and uncertainties in clinical dosimetry [6, 7]. Another method is by comparing the in-house TPS with others commercial software that has been implemented clinically, this method could be a way to identify systematic errors of the algorithm implemented in the TPS [3].

The main objective of this study was to validate in-house TPS of LFMB UI as an essential part in QA of computerized planning systems for radiotherapy. This study was conducted to identify errors on in-house TPS algorithm for the calculation of treatment time for varied beam conditions. Finally, the results of this study can be a used to evaluate the accuracy of the algorithm, before it should be implemented clinically in the patient therapy planning.

2. Materials and methods

In-house TPS was tested to two well-establish radiotherapy department where were Cipto Mangunkusumo National Center Hospital and Persahabatan Center Hospital. Both teletherapy units were FCC 8000 type Cobalt-60 teletherapy of Shandong Xinhua Medical Co. Ltd, China which source activity of each teletherapy unit were 231.25 TBq at October 7th, 2005 and 319.7 TBq at Mart, 2007.

2.1. Treatment time calculation of DSSuperDose v.1.0

DSSuperDose v.1.0 is in-house TPS developed by Medical Physics and Biophysics Laboratory (LFMB) Universitas Indonesia as a treatment planning software for Cobalt-60 teletherapy unit. The semi-empirical method was implemented in the Java™ algorithm. The general formula used to calculate each technique of treatment were allowed in equation (1), is used in fixed SSD (Source to Surface Distance) technique, and equation (2), is used for fixed SAD (Source to Axis Distance) technique [2]. However, several correction factor such inhomogeneities and off-axis correction weren't considered in DSSuperDose v.1.0. The general flow chart of DSSuperDose calculation was displayed by figure 1.

$$time = \frac{PresribedDose}{\dot{D}(d_{ref}, A_{ref}) \times RDF(d_{ref}, A) \times \frac{PDD(d,A)}{100} \times WF \times TF} \quad (1)$$

$$time = \frac{PresribedDose}{\dot{D}(d_{ref}, A_{ref}) \times RDF(d_{ref}, A) x TMR(d,A) \times ISF \times WF \times TF} \quad (2)$$

Data of dosimetry quantities was imported into SQL database system of in-house TPS, to be used for basic beam data of treatment time calculation. Data of percentage depth dose (PDD) and tissue-maximum ratio (TMR) were implemented based on table data of Cobalt-60 beams in British Journal of Radiology Supplement 25 [8], while data of dose rate, relative dose factor (RDF), wedge factor (WF), and tray factor (TF) were implemented by calibration certificate of each teletherapy unit. Linear interpolation was used to calculate numbers of *PDD(d,A)*, *TMR(d,A)* and *RDF(A)*, while equivalent square field (*Aeq*) was determined by 4 Area/Periphery (Day's rule) [2,8]. Calculation of dose rate decay of cobalt-60 source was determined by exponential extrapolation of radioactive decay which half-time ($t_{1/2}$) of cobalt-60 decay is 5.27 years or 1919.9 days [2,8].

2.2. Treatment planning

Test plans were created by both DSSuperDose v.1.0 and ISIS TPS to determine quantities of treatment time in delivering prescribed dose of 50 cGy or 100 cGy for each teletherapy units. Results were acquired for 471 treatment plans of patient-related data including varied treatment techniques, both fixed SSD and fixed SAD technique, for varied beam conditions.

2.3. Measurement of absorbed dose

Absorbed dose were measured at the central beam axis which SSD technique was implemented. Treatment times for each measurement were taken by calculation of DSSuperDose v.1.0 which were acquired for 119 treatments including varied beam conditions as open field, tray field, and wedge field with varied of field sizes and depths. Absorbed dose was calculated based on IAEA TRS 398 [9]. Dose deviations were evaluated with respect to beam conditions, as follow [2]:

$$\delta(\%) = \frac{D_{planned} - D_{measured}}{D_{measured}} 100\% \tag{3}$$

Measurements were performed with 0.6 cc ionization chamber in a homogeneous condition. At Cipto Mangunkusumo National Center Hospital, type chamber (PTW Farmer® type N30013) which was connected to PTW Unidos electrometer with the calibration factor of the detector is (53.99 ± 0.67) cGy/nC was used in a scanning water phantom (IBA Blue Phantom). While, in Persahabatan Hospital measurements were carried out in the water phantom with a calibrated ionization chamber (TM type 30013) connected to PTW Unidos electrometer T 10001 with the calibration factor of the detector is 53.25cGy/nC.

3. Results and discussions

3.1. Comparison with ISIS TPS

Analysis of linear regression was used to the comparison between in-house TPS and ISIS TPS calculation, summarize in table 1. A linear fit produced a slope that indicated a ratio of treatment time calculated by both TPS, and a correlation coefficient that demonstrated a match level between both datasets [10]. In general, good agreement was acquired between calculations performed by in-house TPS and ISIS TPS, with the level of data variability achieved 96% that was represented by the coefficient of linear regression (R^2) up to 0.96.

3.2. Measurement of absorbed dose

First, we evaluated output calibration of each cobalt-60 teletherapy unit for several date of treatment. It showed dose rate deviation between in-house calculation and measurement inside the acceptance tolerance of ±2% [2]. In other words, it indicated good performance about exponential extrapolation of Cobalt-60 decay in in-house TPS.

For open field, dose deviation of cobalt-60 teletherapy unit A and B were shown in left side of figure 2 and figure 3. In general, limit tolerance of ± 2% dose deviation for the open field was achieved. However, high dose deviations were found for small field size and deep depth, such field of 3×3 cm^2 and depth of 15 cm. This results were affected by limitation of RDF as a basic beam data of in-house TPS. Data of RDF for cobalt-60 teletherapy unit A were limited only from field size of 4×4 up to 25×25 cm^2. RDF data for field size of 3×3 cm^2 and 30×30 cm^2 was determined by extrapolation and gained uncertainties. Then, an effect of the penumbra in small field size and deep depth should be considered [11]. The same indication was found for results of Cobalt-60 teletherapy unit B.

For treatment with beam modifiers such tray and wedge field, results were displayed by right side of figure 2. For tray field, it showed same indication with open field. In addition, deviations of tray factor (TF) both of tray with hole and tray without hole provided additional errors of dose deviations. For wedge field, it was acquired that acceptable tolerance of ±3 % for wedge field was reached for several data. Dose deviations increased for field sizes larger than the physical size of wedge (10×15

cm^2) and high wedge angle. Deviations of wedge factor (WF) for each wedge angle provided additional errors. Cozzi *et al.* have reported that wedge factor depends on field size [12]. It confirmed

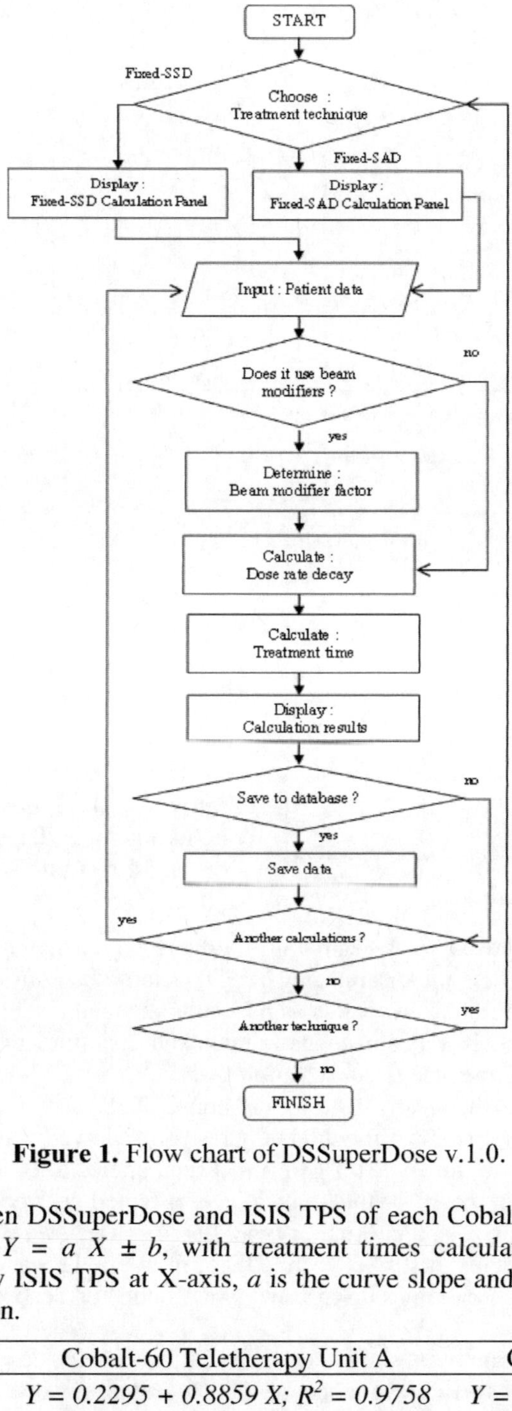

Figure 1. Flow chart of DSSuperDose v.1.0.

Table 1. Comparison between DSSuperDose and ISIS TPS of each Cobalt-60 teletherapy unit. A linear fit produced a linear equation $Y = a\ X \pm b$, with treatment times calculated by DSSuperDose at Y-axis, treatment times calculated by ISIS TPS at X-axis, a is the curve slope and b is the intecept, while R^2 is the coefficient of linear regression.

Beam conditions	Cobalt-60 Teletherapy Unit A	Cobalt-60 Teletherapy Unit B
Fixed SSD - Open field	$Y = 0.2295 + 0.8859\ X;\ R^2 = 0.9758$	$Y = 0.1074 + 0.9956X;\ R^2 = 0.9901$
Fixed SSD - Tray field	$Y = 0.1460 + 0.9357\ X;\ R^2 = 0.9607$	$Y = 0.1084 + 0.9719\ X;\ R^2 = 0.9894$
Fixed SAD - Open field	-	$Y = 0.1556 + 0.9545\ X;\ R^2 = 0.9721$
Fixed SAD - Tray field	-	$Y = 0.2363 + 0.8859\ X;\ R^2 = 0.9597$

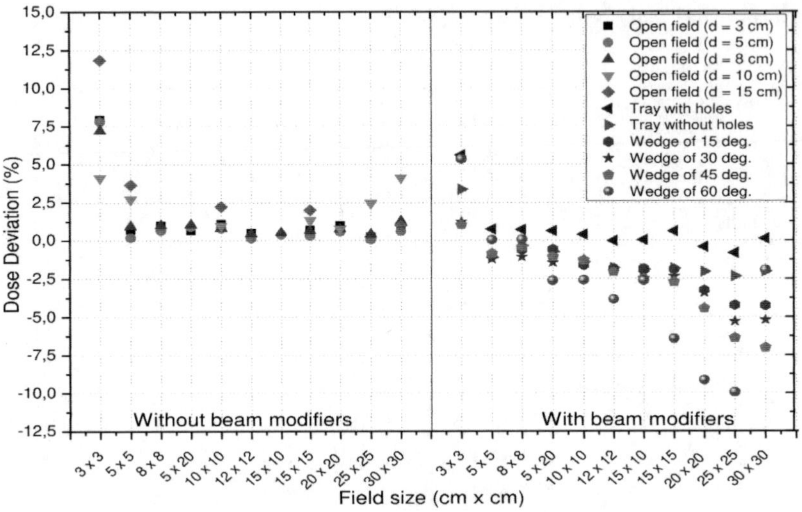

Figure 2. Dose deviation of absorbed dose measurement under fixed SSD technique for varied beam condition of Cobalt-60 teletherapy unit A. Measurements with beam modifiers were done using wedge and tray for varied beam field size at depth of 5 cm in water phantom.

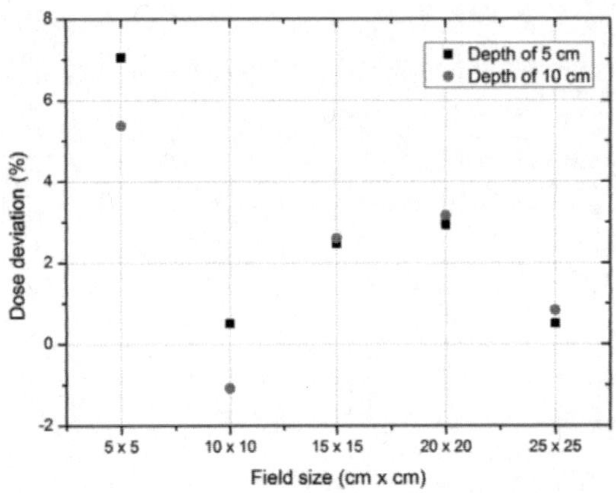

Figure 3. Dose deviation of absorbed dose measurement under fixed SSD technique for open field conditions of Cobalt-60 teletherapy unit B.

that dose deviation would increase for a beam with field size larger than the physical size of wedge as an effect of scattered dose. Then, the discrepancy of PDD should be considered. The use of open field PDD data to represent wedge field PDD data was not recommended because the beam hardening to be expected with wedges in a cobalt beam depends upon the spectrum of scattered beam, especially photon beam, as well as upon the wedge construction [8].

In general, discrepancies between PDD of in-house TPS and table data of each machine commissioning would provide errors. Although many research showed that discrepancy of PDD for others machine was only 1% up to 2% [8]. However, discrepancies of PDD for another machine were affected by characteristics of beam output which was affected by the construction of Cobalt-60 teletherapy unit, such source size and collimator type. Large source size would provide large effect of penumbra [13]. We found discrepancies between PDD of in-house TPS and table data of each machine commissioning were up to 3.61%, high discrepancy was found for deep depth, such depth of 15 cm, and field size of 25×25 cm^2.

Summary of dose deviation was shown in table 2. Confidence level (CL) 95% was taken to improve the data confidence. Results were acquired 80% of data for open field, 91% of data for tray field, and 65% of data for wedge field which achieved the limit tolerance for each beam condition. In addition, evaluation of equivalent square field showed discrepancies of dose deviation between a rectangular field with an equivalent square field that occurred fewer than 1%. It was confirmed that

Day and Sterling have reported that equivalent square demonstrated same dosimetry characteristics with own rectangular field [14].

Table 2. Summary of dose deviation with respect to varied beam conditions.

Beam conditions	|Dose deviations| (%)					
	number of fields	Min.	Max.	Mean	SD	CL
Open field	54	0.06	11.84	1.97	2.42	1.97 ± 3.63
Tray field	22	0.02	5.61	1.32	1.30	1.32 ± 1.94
Wedge field	43	0.05	9.95	2.91	2.36	2.91 ± 3.53
Overall	119	0.02	11.84	2.19	2.29	2.19 ± 3.44

4. Conclusion

In this study, we have done the validation of DSSuperDose v.1.0. In-house TPS gives optimal planning for open and tray beam conditions with depth fewer than 10 cm (≤ 10 cm) and field size up to 20×20 cm^2, while for wedge beam conditions with field size fewer than the physical size of the wedge. Comparison of in-house TPS and ISIS TPS demonstrated a good match of 96%. From the results, it is concluded that DSSuperDose v.1.0 is adequately accurate for treatment planning of radiotherapy. The accuracy of in-house TPS was affected by basic beam data.

5. References

[1] Erwin B P 2005 *Radiation oncology physics: A handbook for teacher and students* (Vienna: International Atomic Energy Agency) pp 161-272

[2] IAEA 2004 *Commissioning and quality assurance of computerized planning systems for radiation therapy of cancer* (Vienna: International Atomic Energy Agency) pp 47-54

[3] George S, Roy E S, Nathan H W, Laura O, Linda A M and Issac I R 2000 *J. Appl. Clin. Med. Phys.* **1** pp 86-94

[4] Jack V, Hans W and Ben M 2001 *Rad. and Onc.* **60** pp 191-201

[5] Jackson C, David R, Victor G P and Thomas J F 2002 *J. Appl .Clin. Med. Phys.* **3** pp 293-301

[6] Konrad W and Peter B D 2000 *J. Appl. Clin. Med. Phys.* **1** pp 120–5

[7] Sellakumar P, Arun C, Sanjay S S, and Ramesh S B 2011 *Phys.Med.* **27** pp 21-9

[8] McKenzie A L 1996 *Cobalt-60 gamma-ray beams Brith. J. Rad.* **25** pp 46-51

[9] IAEA 2000 *Absorbed dose determination in external beam radiotherapy: An international code of practice for dosimetry based on standars of absorbed dose to water* (Vienna: International Atomic Energy Agency) pp 71-77

[10] Zuo K, Tuncali K and Silverman S G 2003 *Radiology* **227** pp 617-28

[11] Indra J D, George X D and Anders A 2008 *Med. Phys.* **35** pp 206-15

[12] Cozzi A F, Cozzi L and Garavaglia G. 1996 *Rad.Onc.* **39** pp 31-4

[13] Dawson D J and Akinradewo A C 1985 *Med. Phys.***12** pp 389-92

[14] Day M J and Aird E G A 1996 *Brith. J. Rad.* **25** pp 138-51

Acknowledgements

This study was funded by Hibah PUPT 2015 with contract number 0540/UN2.R12/HKP.05.00/2015. We also gave thanks to the head and medical physicist of Department of Radiotherapy Cipto Mangunkusumo Hospital and Persahabatan Hospital who contributed in measurement procedures.

A fully automated calculation of size-specific dose estimates (SSDE) in thoracic and head CT examinations

C Anam[1,2], F Haryanto[2], R Widita[2], I Arif[2] and G Dougherty[3]

[1] Department of Physics, Faculty of Mathematics and Sciences, Diponegoro University, Indonesia
[2] Department of Physics, Faculty of Mathematics and Natural Science, Institut Teknologi Bandung, Indonesia
[3] California State University (CSU) Channel Islands, California, USA

E-mail: anam@fisika.undip.ac.id

Abstract. The purpose of this study is to automatically calculate and then investigate the size-specific dose estimate (SSDE) in thoracic and head CT examinations undertaken using standard imaging protocols. The effective diameter (D_{eff}), the water equivalent diameter (D_W), and the SSDE were calculated automatically from patient images. We investigated sixteen adult patients who underwent a CT head examination and thirty adult patients who underwent a CT thorax examination. Our results showed that the D_W value in the thoracic region was 4.5% lower than the value of D_{eff}, while the D_W value in the head region was 8.6% higher than the value of D_{eff}. The relationships between diameter (D_{eff} and D_W) and $CTDI_{vol}$ were distinctive. In the head region, decreasing the patient diameter resulted in a constant $CTDI_{vol}$ due to the tube current modulation (TCM) being off, while in the thoracic region decreasing the patient diameter resulted in a decrease in value of $CTDI_{vol}$ due to TCM being on. In the head region, decreasing the patient diameter resulted in an increase in the value of SSDE, while in the thoracic region decreasing the patient diameter resulted in a decrease in the value of SSDE.

1. Introduction

The radiation dose from CT examinations is greater than other radiological modalities [1, 2]. Therefore, it is very important to accurately estimate the patient dose and to reduce it to as low as possible. In order to estimate and compare CT radiation doses received by patients, the medical community uses volume CT dose index ($CTDI_{vol}$) as an indicator [3]. The $CTDI_{vol}$ is measured using a standard phantom made from polymethylmethaacrylate (PMMA) with a diameter of 32 cm to represent the patient's body and 16 cm to represent the patient's head and a 100 mm pencil ionization chamber [3]. The $CTDI_{vol}$ depends on exposure parameters, e.g., tube voltage, tube current, pitch, and so on. However, $CTDI_{vol}$ is considered to be an output dose indicator only and not a patient dose indicator, because the dose to the patient not only depends on output dose but also depends on the patient's characteristics [4, 5]. The patient dose indicator which takes into account both output dose and patient characteristics is the size-specific dose estimate (SSDE) [6].

The most obvious patient characteristic is the effective diameter (D_{eff}) [6]. But this alone is not sufficient to determine the patient characteristic because different parts of the patient are composed of different materials [7-9]. Air (lung) is the biggest contributor in the thoracic region, and soft tissue is the biggest contributor in the abdominal region. Therefore, although the thorax and abdomen may have the same effective diameter, the dose received by each of them will not be the same. The revised descriptor for patient characterization is the water equivalent diameter (D_W) [7, 10].

To reduce the patient dose, new methods and techniques have been introduced, one of which is the tube current modulation (TCM) technique. TCM changes the tube current continuously to deliver an approximately constant noise level for the complete scan range [11]. In TCM, the user selects a noise level instead of a tube current value. The scan is started with a pre-selected tube current value which ensures an adequately recorded signal for the most attenuating region, and then the tube current is reduced in the regions where the attenuation is lower [12]. It was reported that the patient dose was significantly decreased using TCM compared to non-TCM approaches [11, 12].

This study will automatically calculate effective diameter (D_{eff}), water equivalent diameter (D_W), and size-specific dose estimate (SSDE). We will investigate the relationship between D_{eff} and D_W in the thoracic and head examinations. We will also investigate the relationships between patient diameter (D_{eff} and D_W) and dose ($CTDI_{vol}$ and SSDE) in the thoracic and head CT examinations using standard imaging protocols.

2. Method

2.1. The patient images

In this study, we calculated the D_{eff}, D_W and SSDE of thirty adult patients who underwent thoracic examinations and sixteen adult patients who underwent head examinations. The patients were scanned using a 128-slice multi-detector CT (Aquilion 128, Toshiba, Japan) installed at Kensaras Hospital, Semarang, Central Java, Indonesia. We used standard protocols both for thoracic and head examinations. The standard protocol for the thoracic examinations used a voltage of 120 kVp, a pitch of 1.438, a collimation beamwidth 0.5 mm × 64, and 3D tube current modulation. The standard protocol for the head examinations used a voltage of 120 kVp, a tube current of 300 mA, an exposure of 225 mAs, a pitch of 0.688, and a collimation beamwidth 0.5 mm × 32. The tube current modulation (TCM) was switched off for the head examinations.

2.2. The automated D_{eff} and D_W calculation

The first step to automate the calculation of D_{eff} and D_W is to contour the patients automatically. The algorithm used a combination of basic segmentation techniques and specific information about the border of the patient body [13]. The first step was thresholding using a HU value of -200. However, thresholding alone was not be able to contour the patient accurately because of the presence of objects inside the patient with HU values lower than -200. To overcome this problem, we used edge detection to identify these objects and labeled them using their areas. The largest area identified was considered to be the border of the patient.

After automated contouring, D_{eff} and D_W could be calculated. The calculation of D_{eff} was completed in several steps. The first step was the determination of the center position of the patient. The second step was the determination of the diameters in the anterior-posterior (AP) and lateral (LAT) directions. And the final step was the calculation of the effective diameter from the root of the product of the AP and LAT dimensions.

$$D_{eff} = \sqrt{\mathrm{AP} \times \mathrm{LAT}} \qquad (1)$$

The calculation of D_W was also completed in several steps. The results of the automated contouring were used to crop the original images, and the area of the cropped image and the average HU value of the patient were calculated. D_W was then computed using equation (2):

$$D_W = 2\sqrt{\left[\frac{1}{1000}\overline{\mathrm{HU}} + 1\right]\frac{A}{\pi}} \qquad (2)$$

where *A* is the area of the patient after cropping.

2.3. The SSDE calculation

The values of $CTDI_{vol}$ were extracted from the DICOM header with the tag [0018, 9345]. The SSDE was calculated by multiplying the value of volume CT dose index ($CTDI_{vol}$) with the appropriate conversion factors (*f*).

$$\mathrm{SSDE} = \mathrm{CTDI}_{\mathrm{vol}} \times f \tag{3}$$

The conversion factor (*f*) depends on the diameter of the patient (D_{eff} and D_W) based on AAPM report 204 [6]. To calculate the SSDE value for the head, we used the conversion factor from a PMMA phantom 16 cm in diameter, and for the thoracic region, we used a conversion factor from a PMMA phantom 32 cm in diameter.

3. Results

3.1. The relationship between D_{eff} and D_W

The relationship between D_{eff} and D_W in the thoracic CT examinations is shown in figure 1a. D_W has a strong linear correlation ($R^2 = 0.93$) with D_{eff}. On average the D_W value in the thoracic region is 4.5% smaller than the value of D_{eff}. The relationship between D_{eff} and D_W in the head CT examinations is shown in figure 1b. D_W is also linearly correlated with D_{eff}, but not as strongly ($R^2 = 0.73$). The D_W value in the head region is on average 8.6% greater than the value of D_{eff}.

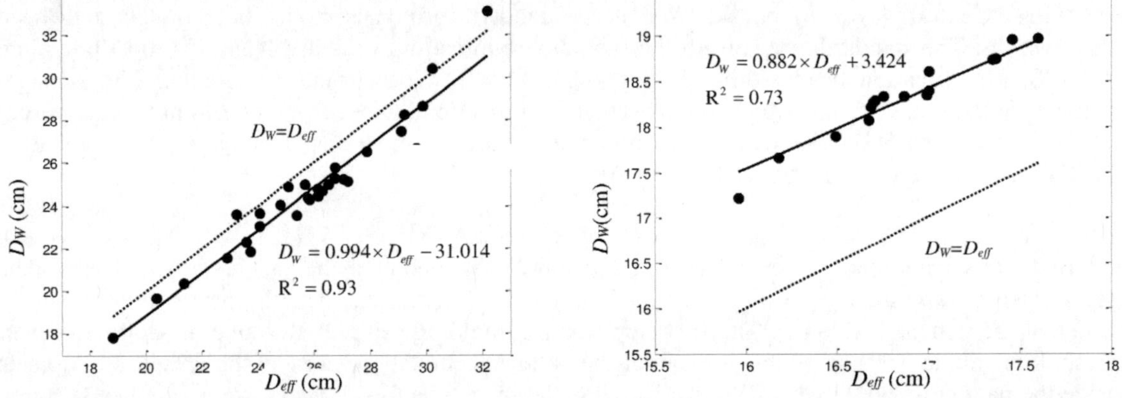

Figure 1. (a) The relationships between D_{eff} and D_W for the thoracic examinations, (b) for the head examinations.

3.2. Relationship between diameter (D_{eff} and D_W) and dose ($CTDI_{vol}$ and SSDE)

The relationship between D_{eff} and dose ($CTDI_{vol}$ and SSDE) for the thoracic examinations are shown in figure 2a, and the relationships between D_W and dose ($CTDI_{vol}$ and SSDE) are shown in figure 2b. Figure 2a shows that the value of $CTDI_{vol}$ increases with increasing diameters (D_{eff} and D_W). The increased $CTDI_{vol}$ leads to an increase in SSDE. From figures 2a and 2b, it seems that the correlation between diameters (D_{eff} and D_W) and $CTDI_{vol}$ is stronger (with R^2 about 0.8) than the correlation between diameters (D_{eff} and D_W) and SSDE (with R^2 about 0.5). This is due to the calculation of $CTDI_{vol}$ being directly based on the average mAs which fluctuates along the z-axis in the TCM mode. Thus, the correlation between diameters (D_{eff} and D_W) and $CTDI_{vol}$ is strong. $CTDI_{vol}$was converted into SSDE using a conversion factor (*f*), which reduced the correlation between diameters and SSDE in the TCM mode.

From figures 2a and 2b, SSDE seems to be highest at 24 cm then drops when the diameters (D_{eff} and D_W) increase. This is most likely due to the exponential nature of the conversion factor (f), whose value decreases with increasing diameters (D_{eff} and D_W).

The relationships between D_{eff} and dose ($CTDI_{vol}$ and SSDE) for the head examinations are shown in figure 3a, and the relationships between D_W and dose ($CTDI_{vol}$ and SSDE) are shown in figure 3b. The value of $CTDI_{vol}$ is constant with increasing diameters (D_{eff} and D_W). The constant value of $CTDI_{vol}$ and the increasing diameters (resulting in decreasing f values) make the value of SSDE decrease. Figure 3a shows a cross-over value of D_{eff} between 16 and 16.5 cm. When D_{eff} is smaller than the cross-over value the SSDE value is greater than $CTDI_{vol}$, and vice versa. On the other hand, all the SSDE values based on D_W are smaller than the $CTDI_{vol}$ values.

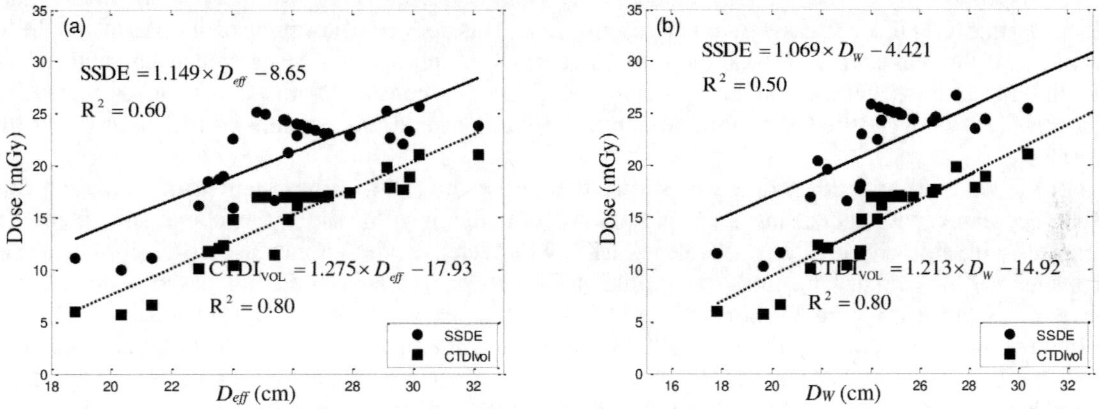

Figure 2. (a) Relationship between D_{eff} and SSDE; D_{eff} and $CTDI_{vol}$, (b) D_W and SSDE; and D_W and $CTDI_{vol}$ for the thoracic examinations.

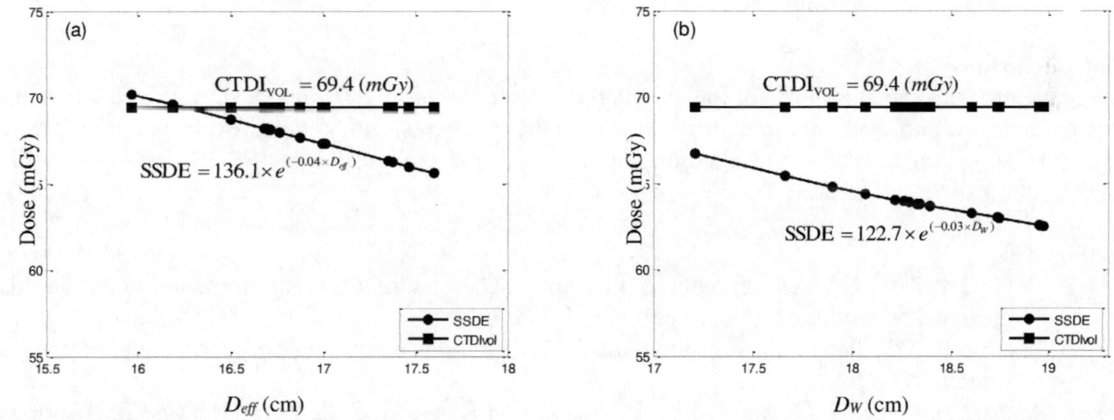

Figure 3. (a) Relationship between D_{eff} and SSDE; and D_{eff} and $CTDI_{vol}$, (b) D_W and SSDE; and D_W and $CTDI_{vol}$ for the head examinations.

4. Discussion

We have developed software for the automated estimation of D_{eff}, D_W, and SSDE. The software works without any user intervention.

Figure 1a shows that in thoracic examinations, the value of D_W is smaller than D_{eff}. This is due to the presence of the lung in the thorax. The HU value for air in the lung is approximately -1000. The presence of air in the lung makes the value of D_W smaller, whereas the value of D_{eff} is calculated only from the effective diameter and it is assumed that the patient composition is close to water with an HU value around 0. On the other hand, figure 1b shows that in head examinations, the value of D_W is greater than D_{eff}. This is because the biggest contributors to the head are soft tissue with HU value around 0 and bone with HU value around 1000+.

Since D_W considers not only the diameter of the patient, but also the composition of the patient, D_W is considered to be more appropriate than D_{eff}. However, the calculation of D_W is more difficult than D_{eff}. In practice, the value of D_W can be estimated from D_{eff} using equation 4 and 5 for thoracic and head examinations, respectively.

$$D_W = 0.994 \times D_{eff} - 31.014 \tag{4}$$

$$D_W = 0.882 \times D_{eff} + 3.424 \tag{5}$$

The relationships between patient diameters (D_{eff} and D_W) and $CTDI_{vol}$ are very interesting. Figure 2a shows that $CTDI_{vol}$ increases with increasing D_{eff}. This reflects the effect of TCM. In the TCM technique, if the diameter decreases, the tube current will proportionally decrease and this will reduce $CTDI_{vol}$. In contrast, with a non-TCM technique $CTDI_{vol}$ is constant with decreasing or increasing diameters (D_{eff} and D_W) because the tube current is constant and independent of the diameter of the patient.

For the head examinations, figure 3a shows that although $CTDI_{vol}$ is constant with increasing D_{eff}, SSDE decreases with increasing D_{eff}. We know from the AAPM table, the conversion factor (f) decreases with an increase of D_{eff}, therefore SSDE will decrease with an increase of D_{eff} in the case of a constant $CTDI_{vol}$ value. The cross-over value at D_{eff} between 16 and 16.5 cm is based on the AAPM conversion factor (f). Figure 3b shows that all SSDE values are lower than $CTDI_{vol}$ values.

Figure 3a and 3b show that the $CTDI_{vol}$ is constant with either decreasing or increasing diameter (D_{eff} and D_W). This is because TCM was not activated in the standard protocol for adult head examinations. Activating TCM in a head examination would not result in a high variation of dose due to the small variation in head diameter for our patients (D_{eff} between 15.96 and 17.60 cm, and D_W between 17.21 and 18.97 cm). However, activating TCM for the standard head protocol will decrease the dose (SSDE), and thus improve patient safety.

5. Conclusions

We have proposed an algorithm for the automated calculation of D_{eff}, D_W, and SSDE. We conclude that for thoracic examinations, the radiation dose (SSDE) decreases with a decrease in patient diameter when TCM is activated. For examinations of the head, the radiation dose (SSDE) increases with a decrease in patient diameter, if TCM is not activated.

References

[1] Slovis T L 2002 CT and computed radiography: The picture are great, but is the radiation dose greater than required? *Am. J. Roentgenol.* **179** 39-41

[2] Dawson P 2004 Patient dose in multislice CT: Why is it increasing and does it matters? *Br. J. Radiol.* **77** S10-S13

[3] Bauhs J A, Vrieze T J, Primak A N, Bruesewitz M R and McCollough C H2008 CT dosimetry: comparison of measurement techniques and devices *RadioGraphics* **28** 245-253

[4] McCollough C H, Leng S, Lifeng Y, Cody D D, Boone J M and McNitt-Gray M F 2011 CT dose index and patient dose: they are not the same thing *Radiology* **259** 311–316

[5] Brink J A and Morin R L 2012 Size-specific dose estimation for CT: how should it be used and what does it mean? *Radiology* **265** 666–668

[6] AAPM 2011 *Size-specific dose estimates (SSDE) in pediatric and adult body CT examinations* (College Park: AAPM)

[7] Wang J, Duan X, Christner J A, Leng S, Yu L and McCollough C H 2012 Attenuation-based estimation of patient size for the purpose of size-specific dose estimation in CT. I. Development and validation of methods using the CTimage *Med. Phys.* **39** 6764-6771

[8] Wang J, Christner J A, Duan Y, Leng S, Yu L and McCollough C H 2012 Attenuation-based estimation of patient size for the purpose of size-specific dose estimation in CT. II. Implementation on abdomen and thorax phantoms using cross sectional CT images and scanned projection radiograph images *Med. Phys.* **39** 6772-6778

[9] Bostani M, McMillan K, Lu P, Kim H J, Cagnon C H, Demarco J J, McNitt-Gray M F 2015 Attenuation-based size metric for estimating organ dose to patients undergoing tube current modulated CT exams *Med. Phys.* **42** 958-968

[10] AAPM 2014 *Use of Water Equivalent Diameter for Calculating Patient Size and Size-specific Dose Estimates (SSDE) in CT* (College Park: AAPM)

[11] Kalender W A, Wolf H and Suess C 1999 Dose reduction in CT by anatomically adapted tube current modulation II: phantom measurements *Med. Phys.* **26** 2248–2253

[12] McCollough C H 2005 Automatic exposure control in CT: are we done yet? *Radiology* **237** 755-756

[13] Anam C, Haryanto F, Widita R, Arif I 2015 Automated estimation of patient's size from 3D image of patient for size-specific dose estimates (SSDE) *Adv. Sci. Eng. Med.* **7** 892-896

Acknowledgements
This work was supported by the Research and Innovation Program, Bandung Institute of Technology (ITB) 2015, No. 237h/I1.C01/PL/2015. The authors would like to thank Mr. Sanggam Ramantisan from Kensaras Hospital, Semarang, Indonesia.

Assessment of knowledge and awareness among radiology personnel regarding current computed tomography technology and radiation dose

M K A Karim[1], S Hashim[1], D A Bradley[2], N A Bahruddin[1], W C Ang[1] and N Salehhon[1]

[1] Department of Physics, Universiti Teknologi Malaysia, 81310 Skudai, Johor, Malaysia
[2] Centre for Nuclear & Radiation Physics, Department of Physics, University of Surrey, Guildford, GU2 7XH, U.K

E-mail: khalis.karim@gmail.com

Abstract. In this paper, we evaluate the level of knowledge and awareness among 120 radiology personnel working in 7 public hospitals in Johor, Malaysia, concerning Computed Tomography (CT) technology and radiation doses based on a set of questionnaires. Subjects were divided into two groups (Medical profession (Med, n=32) and Allied health profession (AH, n=88). The questionnaires are addressed: (1) demographic data (2) relative radiation dose and (3) knowledge of current CT technology. One-third of respondents from both groups were able to estimate relative radiation dose for routine CT examinations. 68% of the allied health profession personnel knew of the Malaysia regulations entitled 'Basic Safety Standard (BSS) 2010', although notably 80% of them had previously attended a radiation protection course. No significant difference ($p<0.05$) in mean scores of CT technology knowledge detected between the two groups, with the medical professions producing a mean score of (26.7 ± 2.7) and the allied health professions a mean score of (25.2 ± 4.3). This study points to considerable variation among the respondents concerning their understanding of knowledge and awareness of risks of radiation and CT optimization techniques.

1. Introduction

Since its introduction in 1974, Computed Tomography (CT) has revolutionized the field of medicine, allowing physicians to diagnose the disease with enhanced confidence as a result of the greatly improved image resolution compared to conventional projection imaging. These advantages over other diagnostic imaging modality have made CT a preferred imaging tools among physicians and the number of CT examinations is increasing every year. CT technology has seen rapid developments in recent times, with the introduction of multi-slice CT (MSCT) in 2001 [1–3].

CT scans are recognized to be a high radiation dose modality with the concern that the doses could in some cases be clinically unjustified [4–6]. Many researchers have highlighted issues regarding choice of CT parameters and the influence on the dose to patients, the discussions narrowing down on how to optimize the dose [7–10], noting that there are many possible combinations of choice to optimize CT parameters. Radiology personnel and other users need to be knowledgeable about the various means in controlling radiation output of CT parameters via used tube potential, tube current, slice thickness, pitch, the detector used and etc.

Nowadays, most of modern CT equipped with automatic tube current modulation (ATCM) technology with a recommended protocols that have been setup by the manufacturer. However, the optimisation technique must also consider patient characteristics including patient's size, age and region of scanning[9, 11, 12]. Hence, this study aimed to describe the optimization of CT procedure and to evaluate the knowledge of radiology personnel and their awareness regarding radiation dose in Johor public hospitals.

2. Methodology

A total of 120 questionnaire forms were distributed to radiology personnel in the seven Departments of Radiology, of which 10, 22, 7, and 81 were distributed to Radiologists, Medical officers, Physicists and Radiographers, respectively. The questionnaire was adapted from Foley *et al.* study where the level of competency among radiology personnel, both in terms of knowledge and awareness in radiation safety in CT procedures and of current CT technology was assessed [11]. After given a day to complete the form, we collected the form from each participant [13].

The questionnaires consisted of 24 questions divided into three sections. The first section provided general demographic information such as level of experience in handling CT and in the second section, the questions were related to the radiation awareness. In the third section, the formats were multiple-choice questions (MCQs) which contained 40 sets of questions regarding current CT technology. Correct responses were given a score of 1, whereas questions with incorrect or incomplete answers were given a score of 0. Before distributing the questionnaires, a pilot test study was performed, involving five persons with different backgrounds; one radiologist, one physicist and three radiographers.

The data was analyzed by using the statistical package software SPSS, version 17.0, utilizing the Student t-test with a p-value of 0.05. This value was considered significant for data validation.

3. Results

3.1. Demography

A total of 120 questionnaires were completed by all participants. These give us a response rate of 100%. The respondents were divided into two groups; Medical professions (Radiologists and Medical officers) and Allied Health professions (Physicists and Radiographers), with a total of 32 and 88 respectively. The distribution of respondents represents the current ratio of radiology personnel in the local practices. The demographic features are shown in table 1, with 'Med' indicating the category for Radiologists and Medical officers and 'AH' representing the category for Allied Health professions.

Table 1. The demographic information obtained from the questionnaire.

General Information	Med[a] n = 32(w%)	AH[b] n = 88(w%)	Total, n = 120(w%)
Experience in handling CT (years)			
≤ 3	13(41)	54(61)	67(56)
3 to 6	3(9)	22(25)	25(21)
6 to 9	8(25)	7(8)	15(12)
> 9	8(25)	5(6)	13(11)
Confidence in altering CT parameters correctly			
Excellent		2(2)	2(2)
Good	5(16)	15(17)	20(17)
Moderate	21(66)	35(40)	56(46)
Poor	3(9)	25(28)	28(23)
Very poor	3(9)	11(13)	14(12)

[a] Med – indicating category for radiologist and medical officer in Department of Radiology
[b]AH – indicating category for radiographers in Department of Radiology

A total of 67(56%) out of the 120 respondents had a less than 3 years of experience, with an average experience of 4.5 years. A total of 98(82%) of the respondents believed that they do not know how to favourably alter the CT scan parameters in operating the scanners. As observed in table 2, two-thirds of allied health profession (68%) involved in radiology have heard about basic safety standard (BSS) 2010 regulations, although 80% of them have previously attended the associated course.

Table 2. Mean score value of radiation awareness among respondents.

Questions	Med[a]	AH[b]
Have you ever attended a Radiation Protection course?	13 (42)	70 (80)
Have you heard about the Basic Safety Standard (BSS 2010) Malaysia regulations under Act 304, 1985?	16 (50)	60 (68)

[a] Med – indicating category for radiologist and medical officer in Department of Radiology
[b]AH – indicating category for radiographers in Department of Radiology

3.2. Radiation dose

Respondents were asked to choose the correct dose in the equivalent of chest X-ray (CXR) examination of radiation for some commonly requested diagnostic imaging procedures. We set the concern based on this standard scale; from 0 to 300, 301 to 500 and 501 to 800 of CXR equivalent. Of the 7 multiple choice questions (MCQs) concerned, the analysis of responses is shown in table 3. The correct estimates of equivalent dose obtained were similar between two groups which ranged from 25% to 92% for Med while for the AH the ranged was from 17% to 92%.

Table 3. Mean score in percentage concerning knowledge of dose in CXR equivalent for particular procedures.

Type of examinations	Med (%)			AH (%)		
	Under estimated	Correct	Over estimated	Under estimated	Correct	Over estimated
CT Brain	34	58	8	31	40	29
CT Thorax	8	25	67	16	26	58
CT Abdomen	58	25	17	69	23	8
CT Lumbar spine	58	42	0	72	17	11
Skull X-Ray	0	92	8	8	92	0
Intravenous-urography (IVU)	25	42	33	31	69	0
Lumbar spine X-Ray	25	75	0	2	89	9

3.3. Knowledge on current CT technology

Comparative analysis of knowledge on current CT technology within the groups concerning over radiation dose and their experience is shown in table 4. Of 40 questions, the mean knowledge score of the medical professions was 26.7 ± 2.7 (ranging from 22 to 32), while the mean score for allied health professions was 25.2 ± 4.3 (ranging from 12 to 32). No significant difference was found between the scores for the two categories of the profession ($p<0.05$).

Table 4. Mean score for all correct responses on knowledge of CT technology.

Profession	n	Mean ± SD	Min - Max	*p*
Medical	32	26.7 ± 2.7	22 - 32	
Allied Health	88	25.2 ± 4.3	12 - 32	$p < 0.05$
Total	120	25.4 ± 4.1	12 - 32	

4. Discussion

4.1. Radiation dose considerations

The use of CT in medicine has been rapidly growing due to advances in technology in producing high-quality images. The employing of multiple array detectors in modern CT has resulted in briefer scan-acquisition times than that for conventional single array CT units. Radiation dose from MSCT has significantly increased compared to the use of SSCT, with additional and occasionally unjustified use of this capability in clinical practice [14], an observation also made by Brix *et al.* in a survey of German CT clinical practice [2]. Staff who operate CT should be independently capable of deciding the radiological protocol to be used during scanning procedure to avoid excessive CT doses to patients.

In this study, the respondents rank the relative radiation dose from CT for each procedure and other routine radiological examinations (such as skull X-ray, IVU and lumbar spine). The majority have estimated correctly for the skull x-ray, both groups scoring 92%. However, both groups have underestimated the dose for CT Brain, CT Abdomen and CT Lumbar with a respective response rate of 34%, 58% and 58% for the Med and 31%, 69% and 72% for the AH profession. Although, it has been observed that 67% and 58% of Med and AH, respectively overestimated the dose for CT Thorax but in regard to non-CT procedures, two-thirds of all respondents scored correctly. Therefore, it is recommended for continuous medical education (CME) regarding CT optimization techniques and knowledge of radiation doses in CT so that further benefit to all radiology personnel thus reducing dose to patients.

4.2. CT technology considerations

In the final part of the questionnaire, respondents were asked regarding their knowledge of CT technology, particularly on use of the ATCM and noise level as well as its influence on radiation dose. Typically, lowering the CT dose by adjusting acquisition parameter, the greater will be the commensurate image noise hence degraded image quality [15]. Concerning current ATCM technology, as in table 5, 92% of both Med and AH groups are aware of the purpose of the ATCM to control the image quality and dose to patients.

Table 5. Mean score for correct answer on current ATCM technology.

Questions	Med (w %)	AH(w %)
ATCM has been shown to decrease patient dose on average	92	92
ATCM is affected by centering of the patient within the gantry	67	62
ATCM can increase the patient dose in the pelvic region	58	42

In this regard, it is apparent that both groups of respondents scored within the restrictive range 40 – 50%, displaying a lack of knowledge of CT technology and ways to implement appropriate techniques in CT aimed at optimizing image quality and dose. To date, this study is considered as the first step towards enhancing our understanding of how radiology practice in CT optimization techniques is influenced by staff knowledge and awareness about CT scan.

5. Conclusion

In this study, knowledge and awareness of CT technology among radiology personnel were evaluated. The findings of this study demonstrate that most of the radiology personnel are aware of radiation dose to the patients but are not well versed in CT optimization techniques. This is due to their understanding in obtaining high-quality images and importantly commensurate control of radiation exposure. It is recommended that studies focusing on the understanding of CT image quality to be conducted in the future.

References

[1] Huda W, Ravenel J G and Scalzetti E M 2002 *Semin. Ultrasound, CT MRI* **23** 411-22
[2] Brix G, Nagel H D, Stamm G, Veit R, Lechel U, Griebel J and Galanski M 2003 *Eur. Radiol.* **13** 1979-91
[3] Jessen K A, Shrimpton P C, Geleijns J, Panzer W and Tosi G 1999 *Appl. Radiat. Isot.* **50** 165-72
[4] McCollough C H, Primak A N, Braun N, Kofler J, Yu L and Christner J 2009 *Radiol. Clin. North Am.* **47** 27-40
[5] Dougeni E, Faulkner K and Panayiotakis G 2012 *Eur. J. Radiol.* **81** 665-83
[6] Holmberg O, Malone J, Rehani M, McLean D and Czarwinski R 2010 *Eur. J. Radiol.* **76** 15-9
[7] Nagel H D 2007 *Radiation Dose from Adult and Paediatric Multidetector Computed Tomography* (Berlin) 1-26
[8] Chang Y L, Lee C M, Hsiao W T, Hsu F Y, Kuo M J and Chiu J H 2010 *Radiat. Meas.* **45** 701-3
[9] Ngaile J E and Msaki P 2006 *J. Appl. Clin. Med. Phys.* **7** 80-94
[10] Thornton F J, Paulson E K, Yoshizumi T T, Frush D P and Nelson R C 2003 *Acad. Radiol.* **10** 379-85
[11] Foley S J, Evanoff M G and Rainford L A 2013 *Insights Imaging* **4** 637–46
[12] Keijzers G B and Britton C J 2010 *Med. J. Aust.* **183** 450-3
[13] Goodman R 1997 *J. Child Psychol. Psychiatry* **38** 581–6
[14] Rehani M M 2011 *ICRP 2011 Proceedings* 154–60
[15] Lee T Y and Chhem R K 2010 *Eur. J. Radiol.* 76 28–35

Acknowledgements

The authors wish to thank the Ministry of Education (Malaysia) (MOE) for providing financial assistance through Research University Grant Scheme (RUGS), project number (Q.J130000.2526.10H60) and also to Dr. S.J Foley for his valuable suggestions.

Journal of Physics: Conference Series **694** (2016) 012032 doi:10.1088/1742-6596/694/1/012032

Occupational radiation doses during interventional procedures

N Nuraeni[1,2], E Hiswara[1], D Kartikasari[1], A Waris[2] and F Haryanto[2]

[1] Centre for Technology of Radiation Safety and Metrology, National Nuclear Energy Agency (BATAN), Jl. Lebak Bulus Raya Kav 49, 12440, Jakarta Selatan, Indonesia
[2] Department of Physics, Faculty of Mathematics and Natural Sciences, Institut Teknologi Bandung, Jl. Ganesa 10, Bandung 40132, Indonesia

E-mail: nunung@batan.go.id

Abstract. Digital subtraction angiography (DSA) is a type of fluoroscopy technique used in interventional radiology to clearly visualize blood vessels in a bony or dense soft tissue environment. The use of DSA procedures has been increased quite significantly in the Radiology departments in various cities in Indonesia. Various reports showed that both patients and medical staff received a noticeable radiation dose during the course of this procedure. A study had been carried out to measure these doses among interventionalist, nurse and radiographer. The results show that the interventionalist and the nurse, who stood quite close to the X-ray beams compared with the radiographer, received radiation higher than the others. The results also showed that the radiation dose received by medical staff were var depending upon the duration and their position against the X-ray beams. Compared tothe dose limits, however, the radiation dose received by all these three medical staff were still lower than the limits.

1. Introduction

Digital subtraction angiography (DSA) is a type of fluoroscopy technique used in interventional radiology to clearly visualize blood vessels in a bony or dense soft tissue environment. Images are produced using contrast medium by subtracting a 'pre-contrast image' or the mask from later images, once the contrast medium has been introduced into a structure. Hence the term 'digital subtraction angiography'. DSA procedures are increasingly used in medical practice. Both patients and medical staff receive a noticeable radiation dose. Each patient undergoing DSA procedures could receive radiation dose up to 685 times of dose received by patient undergoing PA chest radiograph, or around 13.7 mSv [1]. In case of medical staff, particularly interventionalist, eye lenses that caught a radiation dose can develop cataract. It was reported, for example, that typical eye dose to medical staff from cerebral angiography procedure is 0.014 mSv [2]. The threshold dose for cataract has recently been found lower than previously considered, so that the International Commission on Radiological Protection (ICRP) has lowered the annual dose limit for the eye lens from 150 mSv to 20 mSv for occupational exposures [3]. This paper present the study on occupational radiation dose received by medical staff during DSA interventional procedures performed in one hospital in Jakarta, Indonesia. This study is part of a more comprehensive one with the aim to get information on the level of radiation doses received medical staff during various interventional procedures performed in Indonesia.

2. Methods

The measurements were performed using TLD-100 (LiF:Mg,Ti) chip from Thermo Scientific Harshaw. Before use, TLD chip was annealed in 400°C for 1 h, and after cooled down the annealing was repeated in 80°C for 20 h. TLD was then grouped according to standard deviation of ±5% readout after anneal. Calibration of TLD was performed in Calibration Facility of PTKMR BATAN. All TLDs was irradiated with X-ray of 70 kVp in distance of 50 cm and field size of 10×10 cm^2. The irradiated TLD was read by Harshaw model 3500 TLD reader. The conversion of TLD reading for the dose was used by calibration factor obtained in calibration. The whole process is presented in figure 1.

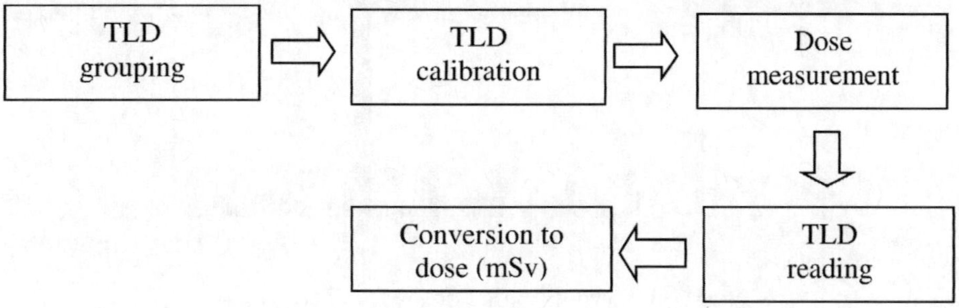

Figure 1. Process of measurements using TLD.

The position of TLD during measurements was seen in figure 2. An electronic pen dose was placed in the trunk behind the lead apron to measure Hp(10), meanwhile the measurement of thyroid and gonad dose was carried out by placing TLDs outside and inside the lead apron in the respective organ positions. The Hp(0.07) was measured by TLD placed in the finger of medical staff, and the measurement of lens dose (Hp(3)) was performed by TLD that was inserted in a special holder of EYE-D.

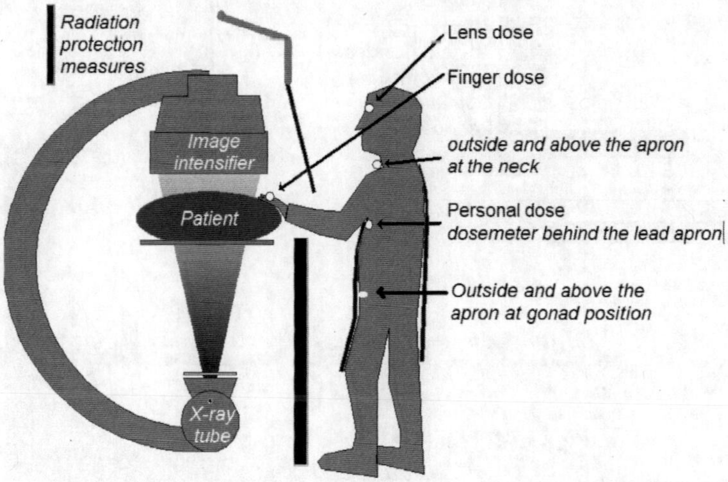

Figure 2. Occupational dose measurement.

The X-ray machine used for DSA interventional procedure is Philips Allura Xper FD20/20 biplane system. This system could be used both for diagnostic examinations and neurovascular interventions.

3. Results and discussion

The measurements of occupational doses in interventional procedures were carried out for medicalstaff that consists of one interventionalist, two nurses and one radiographer. There were eight patients undergoing these DSA procedures with varied exposure of time. The dose in every measurement position in 8 patients was received by each medical staff were shown in figures 3 to 6.

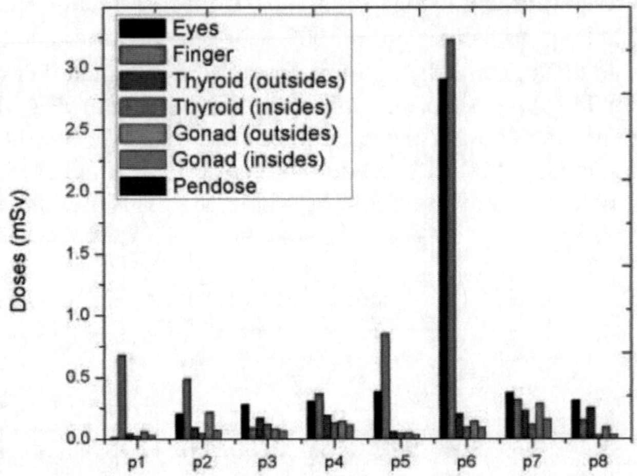

Figure 3. Interventionalist doses for each patient.

As can be seen in figure 3, interventionalist received around 3.2 mSv in his finger when he was performing DSA procedure for patient number 6. The fluoro time in this case was 17 min 49 s, which was the longest time compared to the others. Compared to much lower dose received in other parts of his body, it could be suggested that the interventionalist probably put his finger closer to the direct beam of the X-ray that was used.

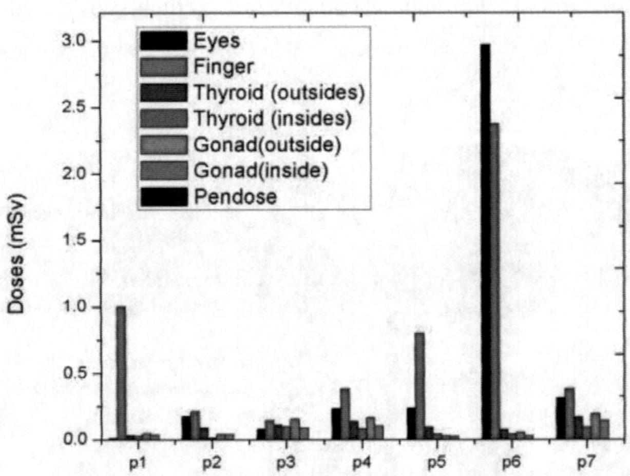

Figure 4. Nurse #1 dose for each patient

Comparing dose received by nurse #1, with dose that was received by nurse #2 in figure 4 and 5, respectively, it could be seen that nurse #1 received more dose than nurse #2. This was understandable since during conducting the procedure, nurse #1 always stood right next to the interventionalist, i.e. the nurse #1 stood closer to the X-ray beam than nurse #2.

This particular nurse #1, however, received radiation dose mostly at her eyes. This suggested that this particular nurse often bowed her head and came closer to the X-ray beam during administered the procedure.

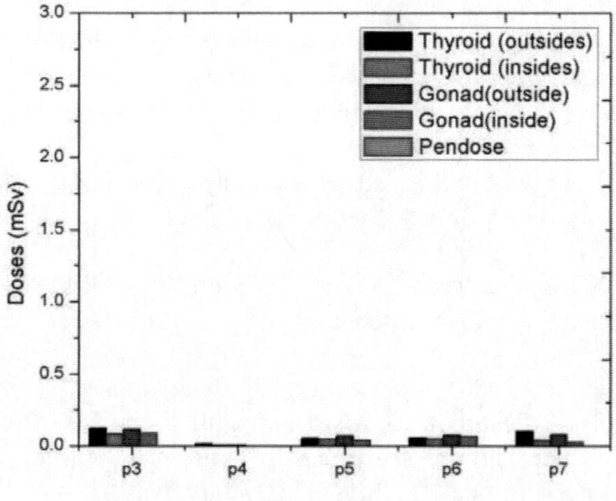

Figure 5. Nurse #2 dose for each patient.

In case of radiographer, figure 6 shows that the highest dose he received was also in his finger. This suggested that his finger sometimes got close to the X-ray beam, probably when the interventionalist asked the radiographer to help him to do certain actions.

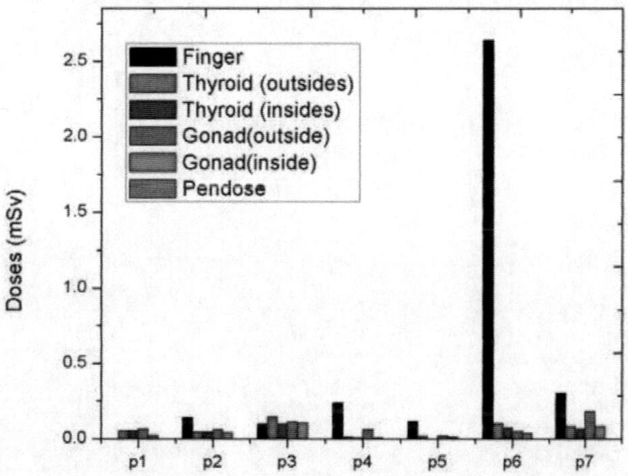

Figure 6. Radiographer dose for each patient.

Overall, the highest dose received by interventionalist, nurse and radiographer in terms of Hp(10), represented by the dose recorded by pen dose, respectively were 1, 2 and 1 μSv. These numbers wereless than the dose limit of 20 mSv per year recommended by the IAEA [4] and adopted in Indonesia [5], or around 10 μSv/h. With the dose limit for hand [Hp(0.07)] and eye lens [Hp(3)] of 500 mSv per year and 20 mSv per year [4,5], respectively, or around 0.25 mSv/h and 0.01 mSv/h, the highest dose received by interventionalist, nurse and radiographer obtained from study were also much lower than those limit.

4. Conclusion

The radiation doses received by medical staff varied depending upon the duration and their position against the X-ray beam. Interventionalist and nurse were close enough to the beam so that they received quite high doses. In order to reduce the doses they accumulated during the course of procedures, interventionalist and nurse should be equipped with personal protective equipment, and the X-ray machine should also be shielded properly.

References

[1] International Atomic Energy Agency https://rpop.iaea.org/RPOP/RPoP/Content/InformationFor/HealthProfessionals/5_InterventionalCardiology/index.htm.

[2] Marshall, N W, Noble J and Faulkner K 1995 Patient and staff dosimetry in neuroradiological procedures *Br. J. Radiol.* **68** 495-501

[3] International Atomic Energy Agency 2011 Radiation Protection and Safety of Radiation Sources: International Basic Safety Standards. Safety Standards Series No. GSR Part 3 (Vienna: IAEA Press)

[4] Chairman of BAPETEN Regulation No. 4 Year 2013 on Radiation Protection and Safety in the Application of Nuclear Energy (in Indonesian)

Acknowledgements

The authors thank to Ms. Dyah D. Kusumawati, Ms. Helfi Yuliati and Ms. Suyati for performing TLD measurements in the hospital. The authors also wish to express their gratitude for the hospital for permitting them to use its interventional facility. This study was financially supported by DIPA PTKMR BATAN.

Journal of Physics: Conference Series **694** (2016) 012033 doi:10.1088/1742-6596/694/1/012033

Establishment of multi-slice computed tomography (MSCT) reference level in Johor, Malaysia

M K A Karim[1], S Hashim[1], K A Bakar[2], H Muhammad[3], A Sabarudin[4], W C Ang[1] and N A Bahruddin[1]

[1] Department of Physics, Universiti Teknologi Malaysia, 81310 Skudai, Johor, Malaysia
[2] Department of Radiology, Hospital Sultanah Aminah 80100 Johor Bahru, Johor, Malaysia
[3] Department of Radiology, Hospital Segamat 80100 Segamat, Johor, Malaysia
[4] Faculty of Health Sciences, Universiti Kebangsaan Malaysia, 56000 Kuala Lumpur Malaysia

E-mail: khalis.karim@gmail.com

Abstract. Radiation doses from computed tomography (CT) are the highest and most hazardous compared to other imaging modalities. This study aimed to evaluate radiation dose in Johor, Malaysia to patients during computed tomography examinations of the brain, chest and abdomen and to establish the local diagnostic reference levels (DRLs) as are present with the current, state-of-art, multi–slice CT scanners. Survey forms were sent to five centres performing CT to obtain data regarding acquisition parameters as well as the dose information from CT consoles. CT-EXPO (Version 2.3.1, Germany) was used to validate the dose information. The proposed DRLs were indicated by rounding the third quartiles of whole dose distributions where mean values of $CTDI_w$ (mGy), $CTDI_{vol}$ (mGy) and DLP (mGy.cm) were comparable with other reference levels; 63, 63, and 1015 respectively for CT Brain; 15, 14, and 450 respectively for CT thorax and 16, 17, and 590 respectively for CT abdomen. The study revealed that the CT practice and dose output were revolutionised, and must keep up with the pace of introductory technology. We suggest that $CTDI_{vol}$ should be included in current national DRLs, as modern CTs are configured with a higher number of detectors and are independent of pitch factors.

1. Introduction

Computed tomography (CT) scan are well known for their contribution to high doses of radiation, with a resultant increased risk of cancer [1]. Although, CT examination has been acknowledged as the preferred method for diagnosis compared to other imaging modality. In 2008, according to the United Nations Scientific Committee's effect of Atomic Radiation (UNSCEAR) report that the total effective dose from CT examinations of whole populations have increased slightly, from 41% of the total annual population dose in the period 1991–1996, to 43% in the period 1997 – 2007 [2]. The contribution of CT examinations towards the population dose has steadily increased since its inception in 1974, where the organ absorbed dose from CT was conveyed approaching levels of cancer probability, with some reports on the deterministic effects on the skin [3–5].

To date, the number of CT scan installed showing the inclining trend as well as dose exposure [2]. These trends are due to the introduction of spiral CT technology which its capability to shorten the time of image acquisition and solved some of the limitation scopes in diagnosis. Although CT images are known to have a higher spatial resolution, misdiagnosis and unjustified requests from physicians still occurs [6]. As reported elsewhere, there are wide variations in CT doses between different facilities for similar procedures, although they used similar scanners [7, 8]. This information allows for the understanding of dose reduction activities and changes in the parameters and protocols that could be used without affecting the quality of the images. Consequently, there should be an encouragement in ensuring the principles of radiation protection, justification, and optimisation is continuously applied, in order to enhance the safety of the provided service. A simple principal of optimisation is the establishment of diagnostic reference levels (DRLs), which were first proposed by the International Commission on Radiation Protection (ICRP) in 1996, and which was initially implemented in Europe [9] and later in Malaysian regulation in 2013 [10].

Multi-slice CT (MSCT) scanner was only recently introduced to many developing countries and developed countries. Developed countries have well experienced in CT technology while most of the public hospitals in developing countries have started acquiring the latest technology MSCT scanners only later. Therefore, the purpose of this study was to evaluate patient dose and establishing local DRLs which focusing MSCT scanner in Johor, Malaysia.

2. Methodology

The survey was performed on five CT scanners from five different public hospitals in four regional cities in Johor, by means of adapted questionnaires during August 2014 to December 2014. The hospitals included were: Department of Radiology of Hospital Sultanah Aminah Johor Bahru (HSAJB); Permai Psychiatric Hospital (HPER); Hospital Enche' Hajah Besar Kalthom (HEHBK), Hospital Pakar Sultanah Fatimah Muar (HPSF), and Hospital Segamat (HSEG). Details of the scanners are summarized in table 1.

Table 1. Details of scanners in this study.

Hospital	CT scanner				
	Manufacturer	Brand	Detector configuration	Year installation	Cities
HSAJB	Siemens	Definition AS	64-slice	2010	Johor Bahru
HPER	Siemens	Somatom Emotion 16	16-slice	2010	Johor Bahru
HEBHK	Siemens	Somatom Emotion 16	16-slice	2013	Kluang
HPSF	Toshiba	Activion 16	16-slice	2010	Muar
HSEG	Siemens	Somatom Emotion 16	16-slice	2014	Segamat

In accordance with the approach used elsewhere, the questionnaires were completed by radiographers who had direct contact with the CT operations of each facility. The following information was obtained from CT consoles: tube potential (kV), tube current (mA), time (s), effective tube current (mAs), nominal collimation beam width (N*hcol), table feed, slice thickness, scan range and dose output. It is noted that all the CT scanners included in table 1 are subjected to Planned and Preventive Maintenance (PPM) and have all passed the annual quality assurance (QA) carried out by consultant physicists. It can, therefore, be assumed that all of the scanners were performing optimally.

Patient radiation doses from CT were calculated using the formalism implemented in the program CT-EXPO (Version 2.3.1, Germany), which is described in detail elsewhere [7, 8, 11]. This software provides the automatic calculation of organ doses, by specifying the scanner model, scanner manufacturer, and scanning parameters as input. All scanning parameters, including patient characteristics and calculated results, were collected and registered in a Microsoft Excel spreadsheet before being transferred to the statistical package SPSS (version 17) for performing the analysis.

3. Results

The region of scanning, as well important technical factors pertaining to the patient characteristics' data in this present study, are summarized in table 2. A total of 614 CT examinations' data were collected from the 262, 163, and 189 patients who underwent routine CT brain, thorax, and abdomen examinations, respectively. Patient characteristics, especially body mass index (BMI), will serve as the factors determining the CT doses in further analysis of this work. As can be observed from table 2, the weight of the patients in this study ranged from 40 kg to 87 kg, with the highest weight of a patient observed in CT abdomen. Parameter settings for different CT protocols obtained in the survey is presented in table 3. It is observed that the acquisition parameters used in this survey are widely diversified.

Table 2. Patient characteristics in this study.

Region of examination	Patient characteristics			
	n	Weight (kg)	Height (m)	BMI (kg.m^{-2})
Brain	262			
Thorax	163	40.0 - 80.0	1.55 - 1.75	16.6 - 31.2
Abdomen	189	48.0 - 87.0	1.52 - 1.78	18.3 - 28.4

Table 3. Range of CT scan parameters performs in this study.

Region of examination	Scan parameter				
	n	Tube output (kV)	Tube current (mAs)	Pitch	Scan range (cm)
Brain	262	110 - 130	270 - 420	0.6 - 1.0	8.3 - 20.1
Thorax	163	40.0 - 80.0	57 - 262	0.8 - 1.2	24.0 - 53.4
Abdomen	189	48.0 - 87.0	54 - 328	0.8 - 1.0	15.3 - 41.9

Table 4 provides the details of the descriptive statistics for patient dose for $CTDI_w$, $CTDI_{vol}$ and DLP. CT brain has the highest value in mean $CTDI_w$ value, ranging from 31.1 to 70.8 mGy, whereas the lowest values ranging from 4.3 to 31.1 mGy were noted for CT thorax examinations.

Table 4. Statistics value of radiation doses from CT examinations in this study.

Region of body	$CTDI_w$ (mGy) Mean ± SD (Min - Max)	$CTDI_{vol}$ (mGy) Mean ± SD (Min - Max)	DLP (mGy.cm) Mean ± SD (Min - Max)
Brain	58.2 ± 9.8 (31.1– 70.8)	59.1 ± 10.2 (31.1 – 70.8)	841.7 ± 179.8 (426.4 – 1167.5)
Thorax	12.0 ± 4.2 (4.3 – 31.1)	10.9 ± 5.2 (3.2 – 31.2)	401.3 ± 196.4 (167.9 – 902.6)
Abdomen	12.4 ± 6.1 (3.6 – 33.4)	12.2 ± 4.6 (4.0 – 34.6)	497.6 ± 186.0 (210.9 – 892.9)

The proposed local DRLs are compared with current national diagnostic reference levels (NDRLs) and other references in table 5. It is observed that the third-quartile (Q3) values of $CTDI_w$ for CT brain and CT abdomen exceed the NDRLs but Q3 values of $CTDI_{vol}$ were comparable to MSCT reference level; EC2004 and UK2005.

Table 5. Comparison of third-quartile values of this study with other reference levels.

Region	Dose descriptors	This study[a]	EU1999[9]	EU2004	UK2005[12]	Malaysia 2013[10]
Brain	$CTDI_w$	63	60	-	57	46.8
	$CTDI_{vol}$	63	-	60	56	-
	DLP	1015	1050	990	690	1050
Thorax	$CTDI_w$	15	30	-	14	19.9
	$CTDI_{vol}$	14	-	12	10	-

	DLP	450	650	430	400	600
Abdomen	$CTDI_w$	16	35	-	16	12.8
	$CTDI_{vol}$	17	-	16	12	-
	DLP	590	780	726	350	450

[a]By rounding up 3rd quartile value

4. Discussion

The radiation dose from CT mostly depends on the CT acquisition parameters, including tube potential (kVp), tube current (mAs), shift per rotation ratio (pitch), and the scan length (L). The modern CT scanners are usually equipped with ATCM technology that aids in reducing the mAs per rotation through angular and longitudinal modulation during scanning [7, 13]. This study revealed a significant variation in acquisition parameters among the radiology departments of each site and the scanners. Furthermore, the mean values of the dose quantities and CT acquisition parameters obtained are in agreement with other findings from previous researchers.

As shown in table 3 and 4, CT acquisition settings and radiation doses received for the same examination and patient type were varied. The highest value of cumulative CTDI was noted in the CT brain examinations, with a value of about 70.8 mGy, whereas the lowest value of 4.3 mGy was recorded for the CT thorax. These variations may be due to the use of non-optimised imaging protocols, even when using a scanner from the same manufacturer, and different practices, especially in the case of CT brain examinations, with or without the use of contrast enhancement. The exposure parameters that depend on ATCM thus need to be reconsidered for reference noise, or reference mAs for adjusting, and hence would reduce the doses significantly, rather than relying on the standard protocols only.

It has been noted that current Malaysian DRLs are based on a survey taken from 2007 to 2009; this being data obtained mostly from SSCT. Merely through monitoring and survey studies of CT doses, the influencing features that affect patient doses in CT can be identified and it is usually the first step in optimisation strategies. The health care level I countries are usually consistent in providing the needed information, such as numbers of CT examinations, dosage levels from scanners, and their optimisation strategies, particularly for MSCT scanners. However, such information, in the majority of developing countries, is uncommon. Therefore, this has allowed us to propose local DRLs, as well as national DRLs, for different types of scanners, whether they are multi-slice or single-slice CT (SSCT). As a result, radiation dose surveys have been suggested in order to constantly monitor CT examinations on a national or local basis, and that the practise undergoes regular reviewing.

There are two limitations in this study: the first is the small sample size, and the second the protocols practice not being standardised. The diverse of the data because of the different used of practices where for example, in CT thorax examination, some centres include part of the liver, whereas, in the usual diagnostic CT thorax, the posterior costophrenic of the lung is usually the lower limit of the scan volume. However, this study is considered as the first step in the optimisation techniques which focusing on current, state-of-art, MSCT scanners.

5. Conclusion

In this study, patient CT doses, from four regional cities, involving five CT scanners, have been presented in terms of $CTDI_w$, $CTDI_{vol}$ and DLP and effective dose values. Proposed local DRLs of routine examination were highlighted with the purpose of presenting current CT practice, which were assumed to be ensembles for multi-slice CT reference levels in accordance with EU 2004. Within this same report, $CTDI_{vol}$ was mentioned and found to be comparable with all other literature focusing on MSCT. The establishment of $CTDI_{vol}$ as part of DRLs is vital in order to cope with current practice and changes of CT protocols, especially with the involvement of modern CT scanners.

References

[1] Hall E J and Brenner D J 2008 *Br. J. Radiol.* **81** 362–78
[2] UNSCEAR 2010 Report of the United Nations Scientific Committee on the Effects of Atomic

Radiation *United Nations Assem. 2011 New York*

[3] National Academy of Sciences 2006 *BEIR VII Health Risks from Exposure to Low Levels of Ionizing Radiation*

[4] Einstein A J, Henzlova M J and Rajagopalan S 2007 *JAMA* **298** 317–23

[5] Rehani M M 2011 *ICRP 2011 Proceedings* 154–60

[6] Coles D R, Smail M A, Negus I S, Wilde P, Oberhoff M, Karsch K R, Baumbach A 2006 *J. Am. Coll. Cardiol.* **47** 1840–5

[7] Brix G, Nagel H D, Stamm G, Veit R, Lechel U, Griebel J and Galanski M 2003 *Eur. Radiol.* **13** 1979–91

[8] Muhogora W E, Nyanda A M, Ngoye W M and Shao D 2006 *Eur. J. Radiol.* **57** 461–7

[9] Commission European 1997 *European Guidelines on Quality Criteria For Computed Tomography EUR 16262 EN*

[10] Ministry of Health Malaysia 2013 *Guidelines in Malaysia Diagnostic Reference Levels in Medical Imaging (Radiology)* (Putrajaya)

[11] Stamm G and Nagel H D 2002 *Rofo* **174** 1570–6

[12] Shrimpton P C, Hillier M C, Lewis M A and Dunn M 2006 *Br. J. Radiol.* **79** 968–80

[13] Kalra M K, Maher M M, Toth T L, Schmidt B, Westerman B L, Morgan H T and Saini S 2004 *Rad* **233** 649–57

Acknowledgements
The authors wish to thank the Ministry of Higher Education (MOHE) for providing financial assistance through Research University Grant Scheme (RUGS), project number (Q.J130000.2526.10H60).

Profile of CT scan output dose in axial and helical modes using convolution

C Anam[1,2], F Haryanto[2], R Widita[2], I Arif[2] and G Dougherty[3]

[1] Department of Physics, Faculty of Mathematics and Sciences, Diponegoro University, Indonesia
[2] Department of Physics, Faculty of Mathematics and Natural Sciences, Institut Teknologi Bandung, Indonesia
[3] California State University (CSU) Channel Islands, California, USA

E-mail: anam@fisika.undip.ac.id

Abstract. The profile of the CT scan output dose is crucial for establishing the patient dose profile. The purpose of this study is to investigate the profile of the CT scan output dose in both axial and helical modes using convolution. A single scan output dose profile (SSDP) in the center of a head phantom was measured using a solid-state detector. The multiple scan output dose profile (MSDP) in the axial mode was calculated using convolution between SSDP and delta function, whereas for the helical mode MSDP was calculated using convolution between SSDP and the rectangular function. MSDPs were calculated for a number of scans (5, 10, 15, 20 and 25). The multiple scan average dose (MSAD) for differing numbers of scans was compared to the value of CT dose index (CTDI). Finally, the edge values of MSDP for every scan number were compared to the corresponding MSAD values. MSDPs were successfully generated by using convolution between a SSDP and the appropriate function. We found that CTDI only accurately estimates MSAD when the number of scans was more than 10. We also found that the edge values of the profiles were 42% to 93% lower than that the corresponding MSADs.

1. Introduction

The first descriptor for estimating CT dose was a multiple scan average dose (MSAD) [1]. MSAD was the average of multiple scan dose profiles (MSDP). The main drawback of measuring MSDP using thermoluminescent dosimeters (TLDs) is that it requires a very long time. The CTDI concept introduced by Shope *et al.* [2] was a clever way to estimate the MSAD at the center of a series of axial scans. The CTDI consists of measuring the integral of the single-scan dose profile (SSDP) for a single axial scan in the standard cylindrical phantom by using a 100 mm long cylindrical ion chamber [3]. This method has been used as the standard descriptor to characterize the output dose of CT. Because CTDI was measured only for the standard phantom, it was considered as the output dose of CT, not as dose to the patient [3]. The patient dose was calculated as the product of output dose (CTDI) and conversion factor (*f*) based on the size and composition of the patient [4].

As an output dose descriptor, CTDI has limitations. First, the value of CTDI approximates the MSAD value only when a considerable number of scans is taken [1, 2], otherwise the CTDI value is higher than the MSAD value. Second, CTDI estimates the output dose only for the center of the scanning area. CTDI neglects the profile output dose along the scanning axis. Obtaining the MSDP and MSAD in a more convenient way is challenging. Boone *et al.* [5] had proposed using convolution to get the MSDP from the SSDP using Monte Carlo simulation. Dixon [6] derived a theoretical expression to give a simpler and more complete treatment for axial and helical modes, and obtained the SSDP curve from TLD measurements.

Nowadays, a dosimeter for measuring SSDP is available [7]. The purpose of this study is to implement the Dixon formula to compute MSDP from SSDP measured by such a dosimeter, and to investigate MSDP and MSAD both for axial and helical modes.

2. Methods

2.1. The single-scan output dose profile (SSDP)

SSDP was measured using a solid-state detector (RTI Electronics, Sweden). The probe is shaped like a CT pencil ionization chamber and fits in the CTDI phantom in the same holes used for the CT pencil ionization chamber. The probe consists of a thin-walled aluminum tube filled with plastic and containing a sub-millimeter thick solid-state detector. The solid-state detector is then coupled to a high sampling-rate electrometer, which records the dose measured in very small time intervals. Instead of making measurements by using the axial mode in one tube rotation, the probe and phantom are passed through the x-ray beam in the spiral scan mode [7].

SSDP was measured in the multi-detector CT (MDCT) Siemens Sensation 64 with the following scanning parameters: tube voltage 120 kVp, tube current 370 mA, tube rotation time 1 s, collimation 1.0 cm, and pitch 1. The SSDP was measured in the head CTDI phantom (16 cm in diameter and 15 cm in length), and the solid-state detector was placed in the center hole of the phantom.

2.2. The multiple scans output dose profile

The multiple-scans output dose profile (MSDP) was calculated for both axial and helical modes. For the axial mode it was calculated using equation (1). *f(z)* is the SSDP and *δ(z-nd)* is the delta function (see figure 1). The distance between adjacent scans is 1 cm.

$$D(z)=\sum_{n=-J}^{J} f(z-nd)=f(z)\otimes\sum_{n=-J}^{J}\delta(z-nd) \qquad (1)$$

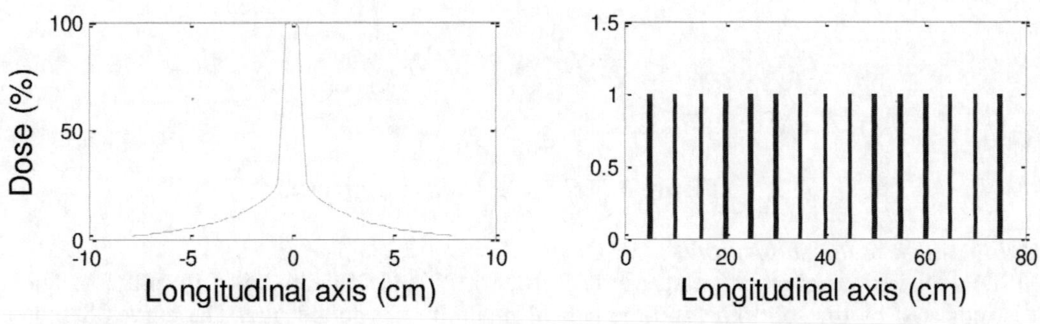

Figure 1. SSDP (left) and delta function shows the position of the scan (right).

The MSDP for the helical mode was calculated using equation (2). *f(z)* is SSDP and **Π***(z/L)* is the rectangular function (figure 2).

$$D(z)=\frac{1}{vt}\int_{-L/2}^{L/2} f(z-y)dy=\frac{1}{vt}f(z)\otimes\prod(z/L) \qquad (1)$$

where *v* is the velocity of the patient or phantom along the longitudinal axis and *t* is the time for one rotation. The value of pitch is 1.0.

The MSDPs for axial and helical modes were calculated for various numbers of scans: 5, 10, 15, 20 and 25. In this study, we calculated CTDI, MSAD and the edge dose of MSDP.

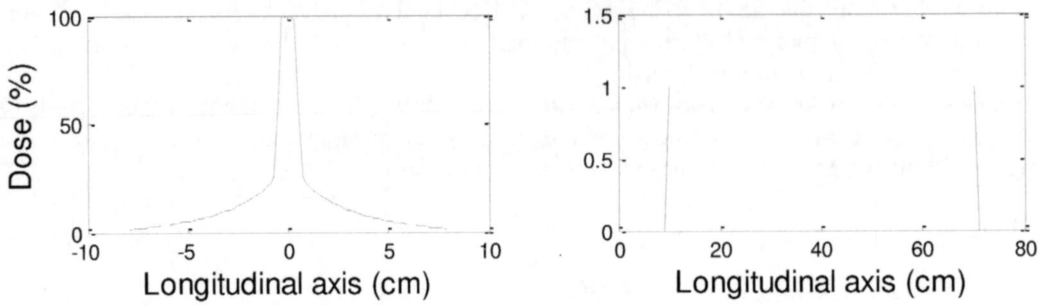

Figure 2. MSDP (left) and rectangular function (right).

3. Results

3.1. Single-scan output dose profile (SSDP)

The measured SSDP for a beam width of 1.0 cm is shown in figure 3. It shows that the tails in the output dose profile are very symmetrical. The curve demonstrates that there are areas in which the dose is lower than the surrounding areas (as indicated by arrows). This phenomenon is the result of the neck support that affected the dose profiles. If the radiation beam comes from the upper of phantom, the beam strikes the phantom directly and it leads to a higher dose in the center of phantom. On the other hand, if the beam comes from below the phantom, it strikes the neck support before the phantom and the dose in the center of phantom is lower.

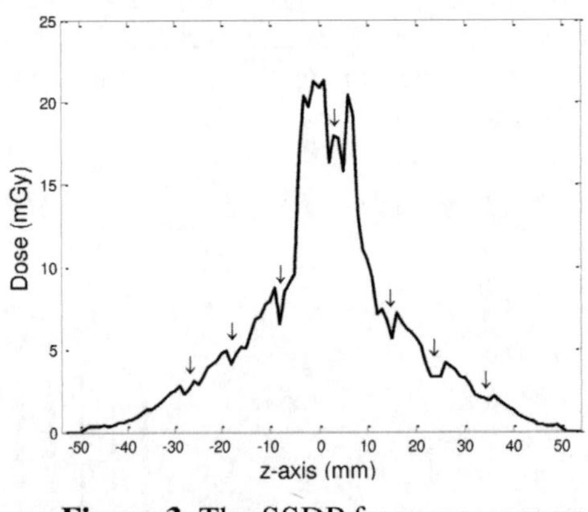

Figure 3. The SSDP from measurement.

3.2. Multiple-scans output dose profiles (MSDP)

The MSDPs for various numbers of scans (1, 5, 10, 15, 20 and 25) are shown in figure 4. The axial mode is indicated by the solid line and the helical mode by the dotted line. The curves demonstrate that increasing numbers of scans leads to the increases at the top of the MSDP. These increases are due to the contribution of the many scattered radiation tails. The MSDP curves in helical mode are flat, whereas in the axial mode they are wavy or non-flat. The flatness of the MSDP curves in helical modes is due to continuous scanning, whereas in the axial mode discontinuous scanning results in a non-flat MSDP.

The MSAD is calculated as the average of the top of MSDP curve. The MSAD values for different numbers of scans are indicated in figure 5. It shows that for scan numbers lower than 10, the value of MSAD for the helical mode is higher than for the axial mode. This is because to acquire a certain number of helical scans requires more length of tissue (For example for 5 helical scans, a 6 cm length

Journal of Physics: Conference Series **694** (2016) 012034 doi:10.1088/1742-6596/694/1/012034

of tissue is scanned). For scan numbers more than 10, the MSAD value for helical and axial modes is equal. For this number of scans, the MSAD value is equal to the CTDI value.

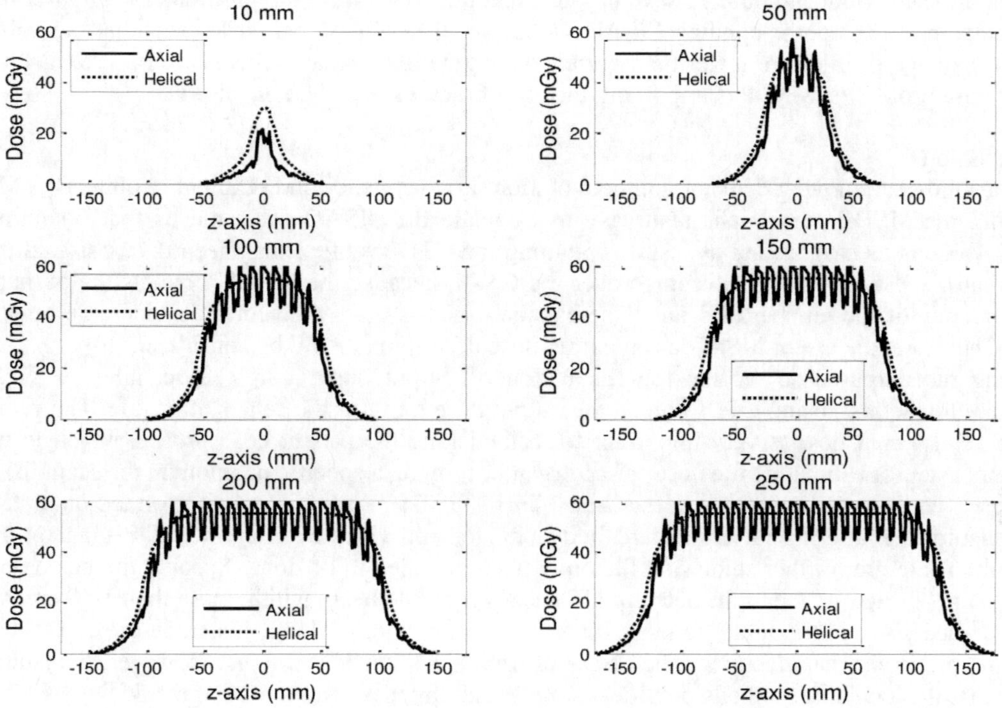

Figure 4. MSDP for helical (solid line) and axial (dotted line) modes.

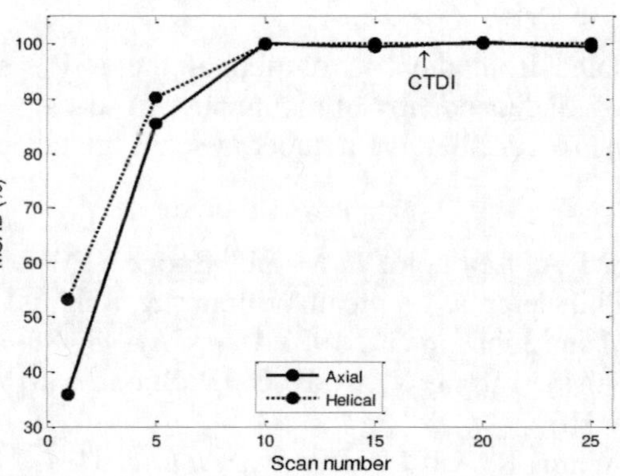

Figure 5. MSAD vs. number of scans for helical and axial modes.

Table1. Edge dose of MSDP vs. Scan number

Scan number	Edge dose of MSDP (%)	
	Axial	Helical
1	64.5	93.3
5	49.6	72.8
10	56.7	65.7
15	42.5	65.7
20	42.2	65.7
25	42.6	65.7

Figure 4 shows that the shape of the edge profiles is sigmoidal. Hence, the dose values at the edge are lower than the corresponding MSAD values. The dose values at the edge of the profiles (at the position of beam width and number scan product, e.g. for 10 scans, in the positions z = -5 cm and z = 5 cm) compared to the corresponding MSAD values are listed in table 1. It is clear that the values of dose at the edge profiles are lower than corresponding MSAD values. Therefore, for establishing the patient dose profile the MSDP is important, instead of only using CTDI or MSAD.

4. Discussion

Using a solid-state moving detector and convolution, it is very easy and efficient to obtain the MSDP. By using the MSDP curve, it is also easy to calculate the MSAD. This means that obtaining the MSAD value is as easy as and as fast as obtaining the CTDI value. It has been demonstrated by this study that the MSAD has an advantage over the CTDI, because the MSAD predicts the output dose independently of the number of scans, whereas the CTDI is only valid for a considerable number of scans. Therefore, the use of MSAD as an output dose descriptor should be considered.

Using the output dose profile, the fluctuation of output dose along the scanning axis can be observed. In contrast, using CTDI predict the output dose only for the center of profile. The profile of the CT scan output dose is very important for establishing the patient dose profile, which in turn is very important for estimating the dose for an organ occupying a specific position on the scanning axis.

There were several limitations in this study. First, the SSDP scan is only performed on a standard head phantom. Scattering on a standard body phantom provides a different SSDP than on a head phantom. Therefore, further studies on the body phantom need to be done. Second, the beam width is only 1.0 cm. Currently, the multi-detector CT has a very wide beam width, more than 10.0 cm, so that it will be necessary in the future to evaluate a wider beam width. Third, this study uses a CT scanner from one particular manufacturer. The SSDP of each CT scanner is unique. However, this study has shown that the available methods provide a simple and effective way to obtain the SSDP, MSDP, and MSAD.

5. Conclusions

MSDP is very easily calculated from the convolution of the SSDP and the appropriate function. MSAD predicts the output dose independently of the number of scans, whereas CTDI is only valid for predicting the output dose when a considerable number of scans are taken.

References

[1] Bushberg J T, Seibert J A, Leidholdt E M and Boone J M 2002 *The Essential Physics of Medical Imaging* (Philadelphia: Lippicott Williams & Wilkins)
[2] Shope T B, Gagne R M and Johnson G C 1981 *Med. Phys.* **8** 488–495
[3] McCollough C H, Leng S, Lifeng Y, Cody D D, Boone J M and McNitt-Gray M F 2011 *Radiology* **259** 311–316
[4] Anam C, Haryanto F, Widita R, Arif I 2015 *Adv. Sci. Eng. Med.* **7** 892-896
[5] Boone J M, Cooper III V N, Nemzek W R, McGahan J P and Seibert J A 2000 *Med. Phys.* **27** 2393–240
[6] Dixon R L 2003 *Med. Phys.* **30** 1272–1280
[7] RTI Electronics 2009 *CT dose profile analyzer manual version 4.0B* (Sweden: RTI Electronics)

Acknowledgement

This work was supported by the Research and Innovation Program, Bandung Institute of Technology (ITB) 2015, No. 237h/I1.C01/PL/2015.

Journal of Physics: Conference Series **694** (2016) 012035 doi:10.1088/1742-6596/694/1/012035

Radiation dose to physicians' eye lens during interventional radiology

N A Bahruddin[1], S Hashim[1], M K A Karim[1], A Sabarudin[2], W C Ang[1], N Salehhon[1] and K A Bakar[3]

[1] Department of Physics, Universiti Teknologi Malaysia, 81310 Johor, Malaysia
[2] Faculty of Health Sciences, Universiti Kebangsaan Malaysia, 50300 Kuala Lumpur, Malaysia
[3] Department of Radiology, Hospital Sultanah Aminah, 80100 Johor, Malaysia

E-mail: ashiqinbahruddin@gmail.com

Abstract. The demand of interventional radiology has increased, leading to significant risk of radiation where eye lens dose assessment becomes a major concern. In this study, we investigate physicians' eye lens doses during interventional procedures. Measurement were made using TLD-100 (LiF: Mg, Ti) dosimeters and was recorded in equivalent dose at a depth of 0.07 mm, Hp(0.07). Annual Hp(0.07) and annual effective dose were estimated using workload estimation for a year and Von Boetticher algorithm. Our results showed the mean Hp(0.07) dose of 0.33 mSv and 0.20 mSv for left and right eye lens respectively. The highest estimated annual eye lens dose was 29.33 mSv per year, recorded on left eye lens during fistulogram procedure. Five physicians had exceeded 20 mSv dose limit as recommended by international commission of radiological protection (ICRP). It is suggested that frequent training and education on occupational radiation exposure are necessary to increase knowledge and awareness of the physicians' thus reducing dose during the interventional procedure.

1. Introduction

The upper hand of interventional radiology in medical imaging has contribute to the growth of interventional procedures. It involves sophisticated techniques to visualize the inside of the human body with the guidance of a variety of catheters, needles, and guidewires. An important concept of interventional procedures which differ from other procedures is that the procedures are minimally invasive with shorter time of recovery, makes it more preferable for the physician. However, despite the benefits, some interventional procedure can deliver high dose of radiation to both patient and staff due to long fluoroscopy time and high-dose fluoroscopy modes.

Physicians are of particular concern as they are exposed to scattered radiation from the patient and in some cases, they are directly exposed to x-ray even though they are not supposed to be. Multiple previous studies showed high radiation exposure to the hands of physicians during fluoroscopy-guided procedures which might be due to direct exposure during the procedures [1–3]. Eye lens are also in particular interest as it is sensitive to radiation and overexposure of eyes can lead to cataract, vision loss, or even blindness. Many studies have been done to investigate the eye lens dose of physicians during interventional radiology and one study found that it was higher than the limits [4]. In 2011, ICRP has alerted the radiological community by lowering threshold dose of staff eye lens from 150 mSv per year to 20 mSv per year [5]. Even though many eye lens doses are well below the limit, it will be exceeded if the doses are compared to the latest ICRP recommendation of 20 mSv per year. Thus, continuous review of physicians eye lens doses during interventional procedures are recommended [4, 6].

Most data found were based in European Countries and inadequate study of physician's eye lens doses during interventional procedures in Malaysia has become motivation to initiate the current study. The objectives include to assess eye lens doses of physician during interventional procedures and to evaluate the compliance of current practice with dose limit proposed by ICRP.

2. Methodology

In Hospital Sultanah Aminah (Johor, Malaysia), the eye lens radiation exposure of 30 physicians during various interventional procedures were collected. All examinations include antegrade pyelography, barium swallow, cystogram, endoscopic retrograde cholangiopancreatography (ERCP), hysterosalpingogram (HSG), percutaneous nephrostomy (PCN), and urethrogram, were done in the same fluoroscopy room, equipped with MultiDiagnost Eleva's C-Arm. Double dosimetry method were used where thermoluminescence dosimeter, TLD-100 chips, worn on two position: under the lead apron at the waist level, and at both sides of the eyes. TLD-100 chips which are made of LiF: Mg, Ti, with the dimension of $3.2 \times 3.2 \times 3.2$ mm^3, were initially calibrated at Sondary Standard Laboratory (SSDL), Malaysian Nuclear Agency, for the purpose of coefficient factor determination. It was then annealed at 400°C for one hour and 100°C for two hours before being used.

The eye lens radiation exposure were recorded in terms of equivalent dose, Hp(0.07), and dose under the lead apron were recorded as Hp(10). The method was chosen based on the recommendation by the previous study where it was found to give better dose estimation compared to single dosimetry method [7]. Effective dose was calculated based on the following equation;

$$E = 0.84\,Hp(10) + 0.051\,Hp(0.07) \quad (1)$$

Equation (1) was known as Von Boetticher algorithm, derived based on ICRP 103 [8]. Annual eye lens dose and annual effective dose were estimated based on workload estimation for a year. Both annual doses were then evaluated to determine the compliance of the dose received with the dose limit as proposed by ICRP.

3. Results

In this study, eight different interventional procedures were selected, involving a total of 28 examinations and 30 physicians. 28 patients' variation in age and size is shown in table 1.

Table 1. Mean values (range) of patients' age and size.

Characteristics	Age (years)	Height (m)	Weight (kg)	BMI (kg/m^2)
Mean values (range)	48.75 (18-76)	1.55 (1.41-1.75)	53.18 (40.-75)	21.03 (15.63-26.89)

All patients are adult with average BMI of 21.03 which categorized as normal for adults. Protocol details were recorded for all investigated procedures as in table 2. Maximum and minimum values of DAP values and fluoroscopy time were in large variations. DAP ranged from 6 μGy.m^2 (HSG) to 639μGy.m^2 (antegrade pyelography). The shortest fluoroscopy time was recorded during cystogram (20 s) while the longest procedure was antegrade pyelography (577 s). Tube voltage setting for all procedures was 75 kV except for barium swallow (80 kV). For most of the procedures, tube current used was 220 mAs while the highest (400 mAs) was recorded during ERCP examinations.

Table 2. Mean values (range) of patients' radiation dose in terms of μGy.m^2 and other protocol details.

Protocol	Tube current (mAs)	Tube voltage (kV)	Fluoroscopy time (s)	DAP (μGy.m^2)
Mean values (range)	244.0 (220-400)	75.5 (75-80)	167.4 (20-577)	143.3 (6-639)

Overall, eye lens dose, Hp(0.07), and waist dose, Hp(10), were assessed for 30 physicians (table 3). In all procedures, all subjects use lead apron, thyroid collar, ceiling-suspended lead screen and table curtain. Hp(0.07) ranged from 0.02 mSv (urethrogram) to 0.96 mSv (fistulogram) for left eye and 0.02 mSv (urethrogram) to 0.43 mSv (ERCP) for right eye.

Table 3. Radiation dose exposure in terms of equivalent dose (mSv) to interventional radiology physician.

	Position	Mean	Median	Min.	Max.	Stand. Dev.
Hp(0.07)	Left	0.33	0.30	0.02	0.96	0.23
	Right	0.20	0.27	0.02	0.43	0.11
Hp(10)	Left	0.27	0.25	0.02	0.48	0.14
	Right	0.25	0.31	0.01	0.35	0.11

Even though the highest eye lens dose were recorded during fistulogram and ERCP, the highest average dose calculated by procedure was barium swallow (figure 1). Urethrogram has the lowest average eye lens doses compared to other procedure involved in this study. In addition, left eye lens dose were higher than right eye lens dose in most of the procedure.

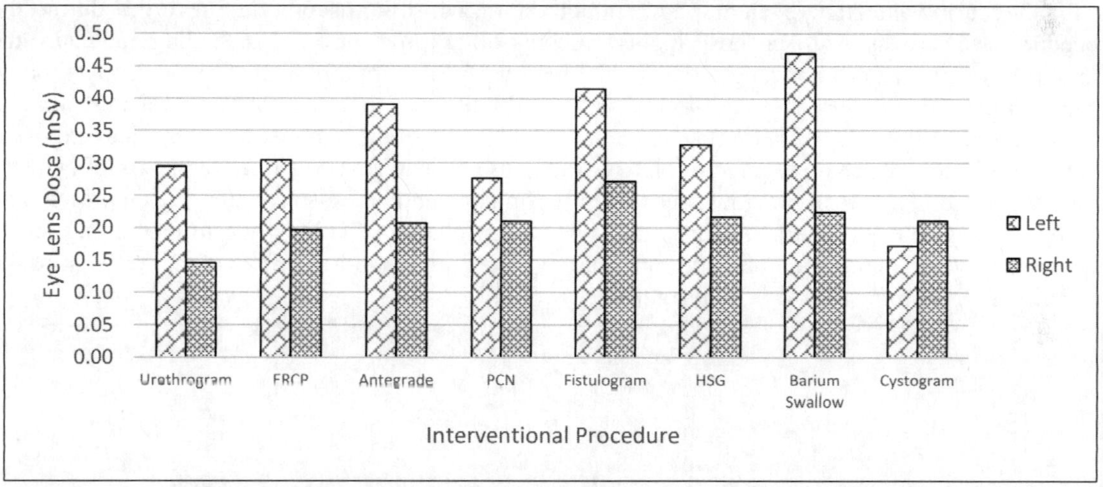

Figure 1. Average eye lens dose for each interventional procedures.

The data of eye lens dose were then used to assess annual eye lens dose of the physician. Annual eye lens dose represents the cumulative dose received by physician on eye lens for a year while annual effective dose reflects the risk of non-uniform dose distribution in terms of a uniform, whole body, and exposure for a year. Figure 2 shows the annual eye lens dose and annual effective dose estimated for all physicians. Mean values of annual eye lens dose were found to be 12.76 mSv with the highest value recorded was 29.33 mSv. For annual effective dose, mean values calculated was 10.99 mSv with the highest value of 17.32 mSv.

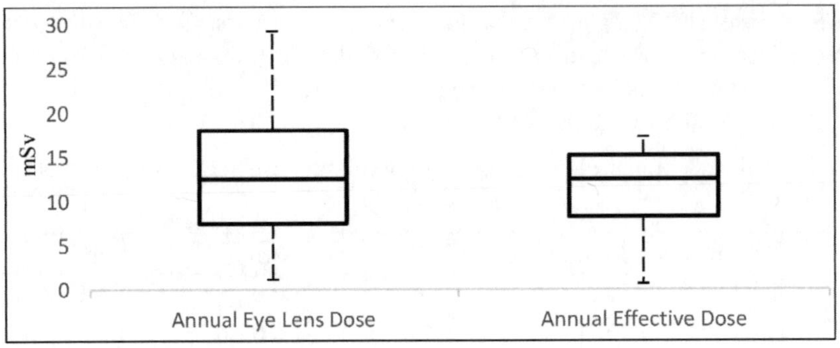

Figure 2. Estimated annual eye lens dose and annual effective dose.

4. Discussion

Overall view of the results gave a hint of radiation dose level received by physicians working in interventional suites. Wide variations of patients' dose may due to difference in BMI, duration of direct x-ray exposure (fluoroscopy time) and imaging protocol used by technologists. DAP were reported as they also attribute to occupational dose (scattered radiation) other than direct exposure of X-ray. High DAP value during antegrade pyelography is in correlation with fluoroscopy time recorded during the procedure, which was 577 seconds, highest among other procedure. This finding indicates the importance of physician knowledge, experience, and early preparation so that fluoroscopy time or exposure time can be reduced. In addition, it was found I all cases that none of the physicians wear an eyes protection or lead glasses during the procedures. Thus, contribute to unnecessary eye lens exposure where it could be reduced up to a factor of 10 with the use of lead glasses [9]. In fact, present eye lens dose were found higher than previous study [6, 10]. Similar trend was also found between present and previous dose where left eye lens were higher than right eye lens. This is due to the setting in the fluoroscopy room where x-ray tube and monitor are placed in a position that required left side of the physician received more exposure than right side.

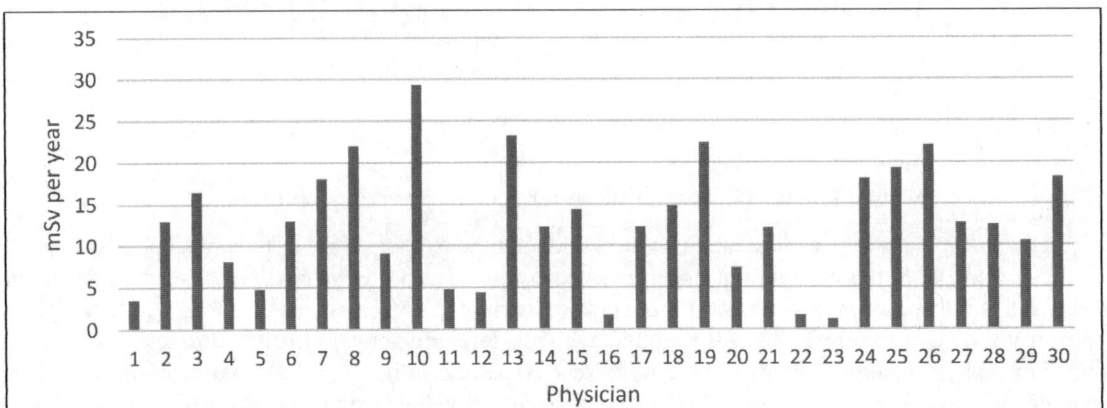

Figure 3. Annual eye lens dose received by each physician.

Five physicians had exceeded eye dose limit proposed by ICRP (20 mSv) with the highest reading of 29.33 mSv as in figure 3. These findings again emphasized the need of changing in physicians practice especially the use of lead glasses and appropriate training in particular procedure to reduce fluoroscopy time during interventional procedure are suggested. Training in radiation protection is also recommended for physicians to increase their awareness regarding radiation protection safety. For annual effective dose, all physicians dose are well below the dose limit recommendations of ICRP. However, precautions should be taken as biological effects of ionizing radiation (BEIR VII) proposed that the generally accepted estimated excess relative cancer risk from an acute single dose of 100 mSv is 10%, or in other words, no radiation exposure are safe and they might contribute to the future cancer risk [11]. Therefore, proper radiation protection, minimize fluoroscopy time and keep away from exposure whenever possible are very important for physicians.

5. Conclusion

Our study has established data on physicians' eye lens dose using double dosimetry method. Physicians' eye lens dose was found higher as compared to previous studies. In fact, five physicians had exceeded annual eye dose limit of 20 mSv per year. The findings indicate the needs to reduced physicians' dose during interventional procedure where unnecessary exposure could be avoided by training and wearing proper protective device (lead glasses/googles). The present study involved many interventional procedure that varies in many aspects. Therefore, future work should concentrate on particular complex procedure to obtain much accurate result.

References

[1] Vosbikian M M, Ilyas A M, Watson D D and Leinberry C F 2014 *J. Hand Surg. Am* **39** 1–5
[2] Hafez M A, Smith R M, Matthews S J, Kalap G and Sherman K P 2005 *Arch. Orthop. Trauma Surg.* **125** 330–5
[3] Singer G 2005 *J. Hand Surg. Am.* **30** 795–797
[4] Efstathopoulos E P, Pantos I, Andreou M, Gkatzis A, Carinou E, Koukorava C, Kelekis N L and Brountzos E 2011 *Br. J. Radiology* **84** 70–77
[5] ICRP Publication 103 2007 *The 2007 Recommendations of the International Commission on Radiological Protection*
[6] Vanhavere F, *et. al.* 2011 *Radiat. Meas.* **46** 1243–47
[7] Chida K, Kaga Y, Haga Y, Kataoka N, Kumasaka E, Meguro T and Zuguchi M 2013 *Am. J. Roentgenol.* **200** 138–141
[8] Boetticher H V, Lachmund J and Hoffmann W 2010 *Health Physics* **99** 800–805
[9] Ingwersen M, Drabik A, Kulka U, Oestreicher U, Fricke S, Krankenberg H, Schwencke C and Mathey D 2013 *JACC. Cardiovasc. Interv.* **6** 1095–102
[10] Domienik J, Rusicka D and Szubert W 2013 *Radiat. Measurements* **51–2** 62–66
[11] National Academy of Sciences 2006 *BEIR VII Health Risks from Exposure to Low Levels of Ionizing Radiation*

Acknowledgements
The authors wish to thank The Ministry of Education Malaysia (MOE) for providing financial assistance through Research University Grant Scheme (RUGS), project number (Q.J130000.2526.10H60), to Miss Diana and all physicians involved during the study.

Journal of Physics: Conference Series **694** (2016) 012036 doi:10.1088/1742-6596/694/1/012036

Computer-aided classification of Alzheimer's disease based on support vector machine with combination of cerebral image features in MRI

C Jongkreangkrai[1], Y Vichianin[1], C Tocharoenchai[1], H Arimura[2] and for the Alzheimer's Disease Neuroimaging Initiative[*]

[1] Department of Radiological Technology, Faculty of Medical Technology, Mahidol University, Thailand
[2] Faculty of Medical Science, Kyushu University, Fukuoka, Japan
[*] Data used in preparation of this article were obtained from the ADNI database (adni.loni.usc.edu). As such, the investigators within the ADNI contributed to the design and implementation of ADNI and/or provided data but did not participate in analysis or writing of this report

E-mail: chonnikan.jon@student.mahidol.ac.th

Abstract. Several studies have differentiated Alzheimer's disease (AD) using cerebral image features derived from MR brain images. In this study, we were interested in combining hippocampus and amygdala volumes and entorhinal cortex thickness to improve the performance of AD differentiation. Thus, our objective was to investigate the useful features obtained from MRI for classification of AD patients using support vector machine (SVM). T1-weighted MR brain images of 100 AD patients and 100 normal subjects were processed using FreeSurfer software to measure hippocampus and amygdala volumes and entorhinal cortex thicknesses in both brain hemispheres. Relative volumes of hippocampus and amygdala were calculated to correct variation in individual head size. SVM was employed with five combinations of features (H: hippocampus relative volumes, A: amygdala relative volumes, E: entorhinal cortex thicknesses, HA: hippocampus and amygdala relative volumes and ALL: all features). Receiver operating characteristic (ROC) analysis was used to evaluate the method. AUC values of five combinations were 0.8575 (H), 0.8374 (A), 0.8422 (E), 0.8631 (HA) and 0.8906 (ALL). Although "ALL" provided the highest AUC, there were no statistically significant differences among them except for "A" feature. Our results showed that all suggested features may be feasible for computer-aided classification of AD patients.

1. Introduction

Alzheimer's disease (AD) is a neurodegenerative disorder, which is the most common type of dementia, causing confusion, memory decline, and loss of cognitive function [1]. The incidence of AD is strongly related with the age; therefore, AD becomes one of the major public health problems in countries with the longer life expectancy [1, 2]. To diagnose AD, magnetic resonance imaging (MRI) is one of imaging techniques which allows quantitative estimation of brain features for assessing AD in a non-invasive way [3]. Nevertheless, in routine clinical practice, subjective estimating the degree of AD by analyzing the morphological changes on MR images is very difficult and time-consuming [2], then various computer-aided diagnosis methods for AD have been developed to assist the radiologists in evaluating MR images. Many cerebral image features have been used to classify AD patients such as cortical thickness, volume and shape of gray matter, white matter, cerebrospinal fluid and hippocampus [4-6].

In this study, the classification of AD patients and normal subjects were performed by supplying the cerebral image features obtained from MRI including hippocampus volume, amygdala volume and entorhinal cortex thickness as input to the support vector machine (SVM).

2. Subjects and MR data

Data used in the preparation of this article were obtained from the Alzheimer's Disease Neuroimaging Initiative (ADNI) database. The ADNI was launched in 2003 as a public-private partnership, led by Principal Investigator Michael W. Weiner, MD. For up-to-date information, see www.adni-info.org. According to ADNI, T1-weighted (MPRAGE) sequence with 1.2 mm thickness were acquired from study volunteers using standard ADNI protocols (http://adni.loni.usc.edu). In this paper, T1-weighted MR brain images of 100 AD patients (47 females and 53 males, age range: 56-89 years, mean age: 74 years) and those of 100 normal subjects (53 females and 47 males, age range: 56-89 years, mean age: 74 years) were studied.

3. Methodology

Figure 1 represents an overall scheme of proposed method. First, T1-weighted MR brain images were processed using FreeSurfer software to measure cerebral image features including hippocampus and amygdala volumes and entorhinal cortex thicknesses in both brain hemispheres. Next, the relative volumes of hippocampus and amygdala were calculated by using total intracranial volume (TIV) to normalize the effect of individual head size. Then, cerebral image features were used as input into the SVM for classification of AD patients and normal subjects. Finally, we evaluated the performance of proposed method using receiver operating characteristic (ROC) analysis.

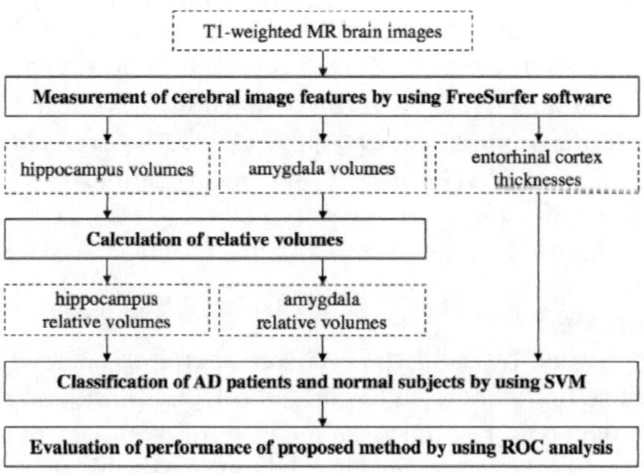

Figure 1. The overall scheme of proposed method.

3.1. Measurement of cerebral image features by using FreeSurfer software

FreeSurfer is a set of software tools for analysis structural brain imaging data developed by the Laboratory for Computational Neuroimaging at Martinos Center (http://surfer.nmr.mgh.harvard.edu). FreeSurfer provides a fully automatic structural imaging stream for processing data in cortical and subcortical brain structure. In subcortical-based stream, FreeSurfer automatically assigned a neuroanatomical label to each voxel in MRI volume based on probabilistic information estimated from a manually labeled training set [7]. Figure 2(a) represents an example of hippocampus and amygdala regions labeled by using FreeSurfer. To measure entorhinal cortex thickness, FreeSurfer constructed models of boundary between gray matter and white matter as well as between gray matter and pia matter overlaid on original brain volume [8] as shown in figure 2(b). The distance between two surfaces was computed and considered as the thickness at each location of cortex. Therefore, applying this tool to T1-

weighted MR brain images, hippocampus and amygdala volumes and entorhinal cortex thicknesses in left and right brain hemispheres were obtained.

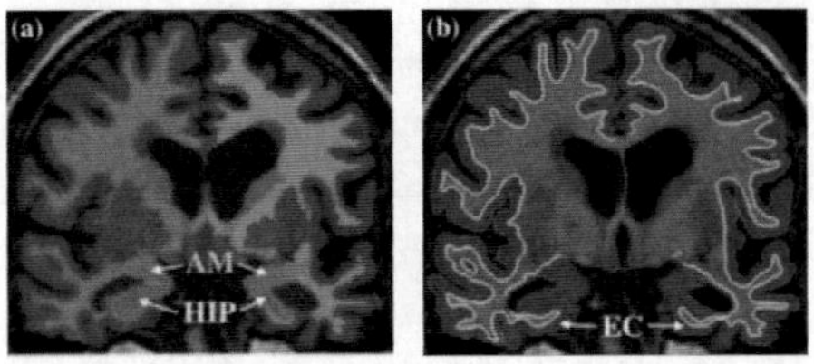

Figure 2. Hippocampus and amygdala regions (a) and entorhinal cortices (b) in left and right brain hemispheres obtained from an AD case by using FreeSurfer software.

3.2. Calculation of relative volumes of hippocampus and amygdala
To analyse the volume of brain structure, it should be realized that individual head sizes of each subject are markedly different. So some processes are required to account for head size variation. In this study, total intracranial volume (TIV) obtained by volume-based stream of FreeSurfer software was used to calculate the relative volumes of hippocampus and amygdala by using equation (1). Note that this correction is only useful for measurement of volume, not thickness because volume scales with head size which depends mostly on changes in surface area while thickness scales to much less degree.

$$\text{relative volume} = \frac{\text{hippocampus volume or amygdala volume}}{\text{TIV}} \tag{1}$$

3.3. Classification of AD patients and normal subjects by using SVM
In classification step, support vector machine (SVM), discriminative classifier developed by Cortes and Vapnik in 1995, was used. In this case, we implemented SVM algorithm by using SVM light, open source software package (http://svmlight.joachims.org). Six cerebral image features, i.e., left and right hippocampus relative volumes, left and right amygdala relative volumes and left and right entorhinal cortex thicknesses, obtained from previous steps were normalized from 0 to 1 and then used as input features for classification. In this study, SVM with radial basis function (RBF) was trained with five combinations of those cerebral image features: H (left and right hippocampus relative volumes), A (left and right amygdala relative volumes), E (left and right entorhinal cortex thicknesses), HA (left and right hippocampus and amygdala relative volumes) and ALL (all features) as shown in table 1 to determine an SVM model which distinguished AD patients and normal subjects efficiently. For RBF kernel, parameter C and γ were estimated using a grid-search method [9]. Finally, SVM model obtained from training session was used to classify new unknown subjects into one of two groups (i.e. AD or normal).

3.4. Evaluation of proposed method
We evaluated the proposed method by using the receiver operating characteristic (ROC) analysis. The resulting data from SVM classifier were used to create an ROC curve and then analyzed to determine area under the curve (AUC) which represented as a measure of overall performance of classification. In this study, we applied SVM to 100 AD patients and 100 normal subjects using leave-one-out method, and ROC curves were obtained from ROCKIT, the software package for ROC analysis developed at the University of Chicago (http://metz-roc.uchicago.edu/MetzROC). Finally, the area under ROC curve (AUC) of each combination was calculated. The AUC differences between five different combinations were compared using a paired t-test with 95% confidence interval ($p < 0.05$).

4. Results
Figures 3(a), (b) and (c) show relationships between relative volumes of left and right hippocampus, relative volumes of left and right amygdala, and left and right entorhinal cortex thicknesses obtained from AD patients and normal subjects, respectively. Figure 4 shows ROC curves and AUC values which represented the performance of proposed method using SVM with five combinations of cerebral image features for classification of AD patients and normal subjects. The AUC values for five combinations including H, A, E, HA and ALL were 0.8575, 0.8374, 0.8422, 0.8631 and 0.8906, respectively. Table 2 shows statistically significant differences among AUC values of five combinations.

Journal of Physics: Conference Series **694** (2016) 012036 doi:10.1088/1742-6596/694/1/012036

Table 1. Five combinations (Codes) of cerebral image features.

	Hippocampus relative volume		Amygdala relative volume		Entorhinal cortex thickness	
Code	Left	Right	Left	Right	Left	Right
H	✔	✔				
A			✔	✔		
E					✔	✔
HA	✔	✔	✔	✔		
ALL	✔	✔	✔	✔	✔	✔

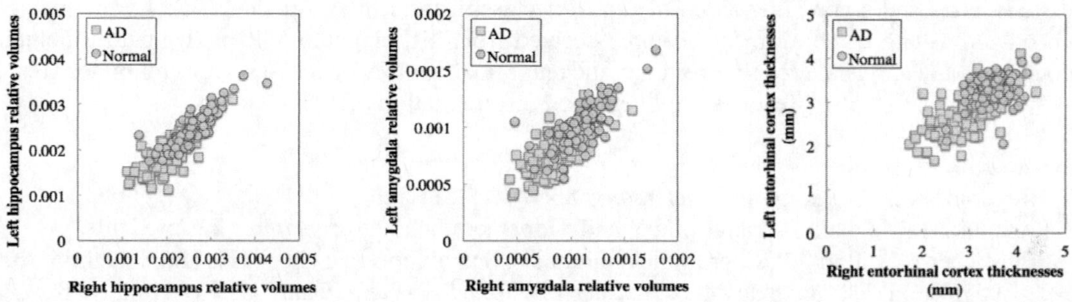

Figure 3. Relationships between cerebral image features of AD patients and normal subjects including left and right hippocampus relative volumes (a), left and right amygdala relative volumes (b) and left and right entorhinal cortex thicknesses (c).

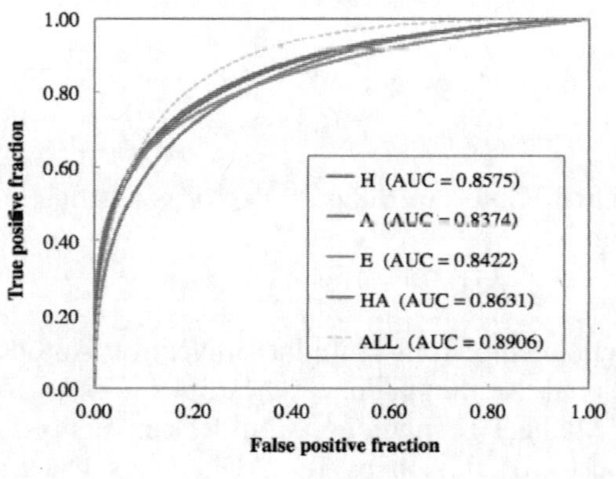

Figure 4. ROC curves and AUC values representing the performance of proposed method for classification of AD patients and normal subjects by using SVM with five combinations of cerebral image features.

Table 2. Comparison of AUC of five combinations.

	P-value				
Code	H	A	E	HA	ALL
H	-	0.2905	0.6151	0.4152	0.1441
A	-	-	0.7655	**0.0328**	**0.0199**
E	-	-	-	0.4305	0.0602
HA	-	-	-	-	0.2220
ALL	-	-	-	-	-

5. Discussion

The results of our study showed that all of five combinations (H, A, E, HA and ALL) provided high AUC values indicating good classification of AD. Although, using "ALL" feature yielded the highest classification with AUC of 0.8906, there were no statistically significant differences among the rest, except for "A" feature as shown in table 2.

The advantage of the proposed method is its achievably high discriminant ability. The overall process includes feature extraction and classification which can be performed automatically using open source software together with T1-weighted MR brain images. However, the limitation of this study is the relatively small number of subjects used in the classification training step. The larger number of subjects may be included for the training process in the future studies. Other machine learning techniques should be investigated in order to improve the performance of AD classification process.

6. Conclusion

This study proposed a computer-aided method for classification of AD patients and normal subjects based on SVM with cerebral image features derived from T1-weighted MR brain images including hippocampus and amygdala relative volumes and entorhinal cortex thicknesses. Our preliminary results suggested that using "ALL" features provided good differentiation of AD.

References

[1] Reynolds A 2013 *Radiologic Technology* **85** 187CT-211CT
[2] Arimura H, Magome T, Yamashita Y and Yamamoto D 2009 *Algorithms* **2** 925-52
[3] Luccichenti G, Ran P P, Cherubini A, Giugni E, Scarabino T, Hagberg G E and Sabatini U 2006 High-field 3 T imaging of Alzheimer disease *High-Field Brain MRI* (Germany: Springer Berlin Heidelberg) chapter 17 p 201
[4] Arimura H, Yoshiura T, Kumazawa S, Tanaka K, Koga H, Mihara F, Honda H, Sakai S, Toyofuku F and Higashida Y 2008 *Academic Radiology* **15** 274-84
[5] Magnin B, Mesrob L, Kinkingnehun S, Pellegrini-Issac M, Colliot O, Sarazin M, Dubois B, Lehericy S and Benali H 2009 *Neuroradiology* **51** 73-83
[6] Gerardin E *et al.* 2009 *Neuroimage* **47** 1476-86
[7] Fischl B, Salat D H, van der Kouwe A J, Makris N, Segonne F, Quinn B T and Dale A M 2004 *Neuroimage* **23** S69-84
[8] Desikan R S *et al.* 2006 *Neuroimage* **31** 968-80
[9] Hsu C W, Chang C C and Lin C J 2010 A Practical Guide to Support Vector Classification http://csie.ntu.edu.tw/~cjlin/papers/guide/guide.pdf

Acknowledgements

This study was supported by Postgraduate Exchange Scholarship from Mahidol University. Authors would like to acknowledge Laboratory for Computational Neuroimaging at Martinos Center for FreeSurfer software. Also, we are grateful to Faculty of Medical Technology, Mahidol University for all facilities. Data collection and sharing for this project was funded by the Alzheimer's Disease Neuroimaging Initiative (ADNI) (National Institutes of Health Grant U01 AG024904) and DOD ADNI (Department of Defense award number W81XWH-12-2-0012). ADNI is funded by the National Institute on Aging, the National Institute of Biomedical Imaging and Bioengineering, and through generous contributions: see www.adni.loni.usc.edu/about/funding. The Canadian Institutes of Health Research is providing funds to support ADNI clinical sites in Canada. Private sector contributions are facilitated by the Foundation for the National Institutes of Health (www.fnih.org). The grantee organization is the Northern California Institute for Research and Education, and the study is coordinated by the Alzheimer's Disease Cooperative Study at the University of California, San Diego. ADNI data are disseminated by the Laboratory for Neuro Imaging at the University of Southern California.

Parameter-based estimation of CT dose index and image quality using an in-house android™-based software

S Mubarok[1], L E Lubis[1] and S A Pawiro[1]

[1] Department of Physics, Faculty of Mathematics and Natural Sciences, University of Indonesia, Depok, 16424 West Java, Indonesia

E-mail: sayid.mubarok@sci.ui.ac.id

Abstract. Compromise between radiation dose and image quality is essential in the use of CT imaging. CT dose index (CTDI) is currently the primary dosimetric formalisms in CT scan, while the low and high contrast resolutions are aspects indicating the image quality. This study was aimed to estimate $CTDI_{vol}$ and image quality measures through a range of exposure parameters variation. CTDI measurements were performed using PMMA (polymethyl methacrylate) phantom of 16 cm diameter, while the image quality test was conducted by using catphan ® 600. CTDI measurements were carried out according to IAEA TRS 457 protocol using axial scan mode, under varied parameters of tube voltage, collimation or slice thickness, and tube current. Image quality test was conducted accordingly under the same exposure parameters with CTDI measurements. An Android™ based software was also result of this study. The software was designed to estimate the value of $CTDI_{vol}$ with maximum difference compared to actual $CTDI_{vol}$ measurement of 8.97%. Image quality can also be estimated through CNR parameter with maximum difference to actual CNR measurement of 21.65%.

1. Introduction

Tomographic imaging is commonly used to get 3D image for diagnostic purposes. CT-scan contributes on 10-50 times, even 500 times greater radiation dose than conventional X-ray [1]. Until recently, CT-dose index (CTDI) is a dosimetry parameter used in CT imaging. CTDI is measured by using homogeneous cylindrical Polymethyl Methacrylate (PMMA) phantom [2–4]. There were two diameter sizes of standard PMMA phantom; 16 cm to represent adult heads and paediatric bodies, and 32 cm to represent adult bodies.

Beside the dosimetry, image quality, being directly afflicted by exposure parameters (kV/HVL, mAs, and slice thickness), is also essential in determining the performance of a CT-scanner. As a matter of principle, clinically acceptable image and patient dose being as low as reasonably achievable (ALARA) should be achieved for optimization purposes. This work was aimed to estimate $CTDI_{vol}$ and image quality measures through a range of exposure parameters variation (kV, mAs and axial thickness). In addition, Android™ based software was developed to provide relatively quick assessment to perform preliminary optimization.

2. Materials and methods

The CT-scanner used in this work was GE Brightspeed Series 317010HM7 in Cipto Mangunkusumo National Central Hospital, Jakarta. The maximum tube rate of this CT was 140 kV and 345 mA. Image quality was evaluated using Catphan® 600 as an object scan.

2.1. Dose measurement

Dose was assessed by measuring $CTDI_{vol}$ on 16 cm PMMA phantom using Unfors Xi-CT. The exposures were done using 2 second rotation time and axial scan mode. The parameter variation used can be observed on table 1.

CT dose index (CTDI) or CT air kerma index was calculated through equations (1), (2), and (3)

$$C_{a,100} = \frac{1}{nT}\int_{-50}^{+50} K(z)dz \quad (1)$$

$$C_W = \frac{1}{3}\left(C_{PMMA,100,c} + 2C_{PMMA,100,p}\right) \quad (2)$$

$$C_{VOL} = C_W\frac{nT}{l} = \frac{C_W}{pitch} \quad (3)$$

with $C_{a,100}$ refers to CT air kerma in free air; $C_{PMMA,100,c}$ refers to CT air kerma in centerof PMMA phantom meanwhile $C_{PMMA,100,p}$ refers to CT air kerma in periphery; C_w refers to weight CT air kerma; C_{vol} refers to volume CT air kerma that includes helical pitch or axial scan spacing in calculation; nT is nominal beam width; $K(z)$ is air kerma value along z-axis; and l being the distance for a single rotation gantry [5].

2.2. Image quality measure

The analysed image quality aspects were low contrast resolution and spatial resolution. Low contrast resolution was evaluated using Catphan® 600 single scans on CTP 515 marked position. CNR (Contrast to Noise Ratio) was chosen as the quantity to represent low contrast resolution. The CNR was evaluated using Region of Interest (ROI) analysis on supra-slice section with 15 mm in diameter on 1% contrast level (Figure 1a) and was subsequently calculated using equation (4)

$$\text{CNR} = \frac{(\bar{x}_s - \bar{x}_{bg})}{\sigma_{bg}} \quad (4)$$

with $\bar{x}_s$ being mean of ROI gray-scale value on each image sample (supra-slice 15 mm in diameter), $\bar{x}_{bg}$ being mean value of ROI gray scale value on background and σ_{bg} being the background noise [3].

On the other hand, high spatial resolution was evaluated using Catphan® 600 scans on point CTP 528 and was evaluated using Modulation Transfer Function (MTF) curve. The MTF curves were evaluated by using ROI on test pattern beginning with 1.0 *linepair/cm* frequency to the last visually-separable objects (as seen on figure 1b).

Table 1. Exposure parameter variations.

kV	mAs	Slice thickness, no. of image/acquisition
120	400	10 mm, 1 image
100	300	5 mm, 1 image
80	200	0.625 mm, 2 images

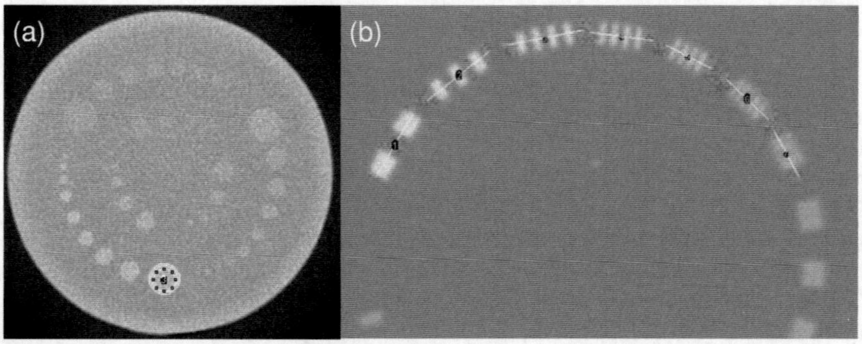

Figure 1. ROI placement on image analysis performed to evaluate (a) low contrast resolution and (b) spatial resolution.

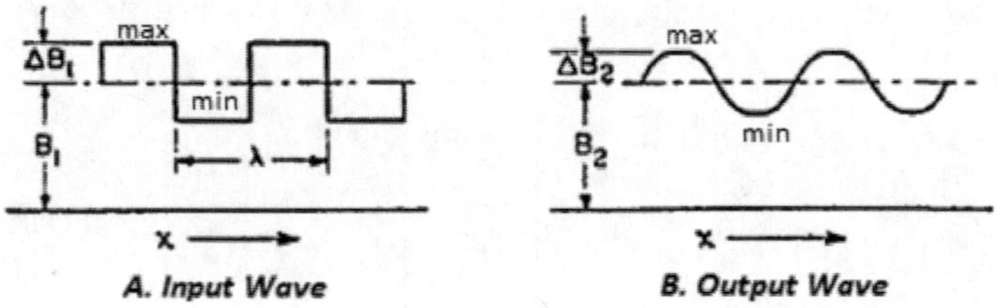

a. b.

Figure 2. Graph of optical density (gray value) (a) in object or test pattern and (b) in image system [6]

From each ROI, Graph of optical density (gray value) was observed. Square Wave Response Function (SWRF) would be observed in accordance with the graph. SWRF *r(u)* (*u* was test pattern frequency) was evaluated using equation 5. Figure 2 was given to explaining this paragraph.

$$r(u) = \frac{\Delta B_2/\Delta B_1}{B_2/B_1} = \frac{\left(\frac{max-min}{max+min}\right)_2}{\left(\frac{max-min}{max+min}\right)_1} \quad (5)$$

Each SWRF values were normalized on values of smallest test pattern frequency (1.0 *linepair/cm*). Afterwards, curves would be made of frequency (*linepair/cm*) versus normalized SWRF using MATLAB curve fitting through equation 6.

$$f(u) = \frac{c.\exp\left(-\frac{\pi^2 u^2}{d}\right) + a.\left(1+4\pi^2 u^2/b^2\right)^{-1}}{c+a} \quad (6)$$

a, b and c were the values to be solved for [7]. By performing Fourier Transform on the curve fitting $f(u)$, MTF curve was estimated. Term-12 Fourier Transform was applied in the calculation (equation 7).

$$R(u) = \frac{\pi}{4}\left[r(u) + \frac{r(3u)}{3} - \frac{r(5u)}{5} + \frac{r(7u)}{7} + \frac{r(11u)}{11} - \frac{r(13u)}{13} - \frac{r(15u)}{15} - \frac{r(17u)}{17} + \frac{r(19u)}{19} + \frac{r(21u)}{21} + \frac{r(23u)}{23} - \frac{r(29u)}{29}\right] \quad (7)$$

3. Results and discussion

3.1. Parameter-based CTDI estimation

Correlation between each exposure parameters (kV, mAs, and beam collimation) to $CTDI_{vol}$ or *nCTDI* ($CTDI_{vol}$/mAs) was first investigated, particularly the correlation between mAs and $CTDI_{vol}$, kV and *n*CTDI, as well as collimation with *n*CTDI. Each parameter correlations is given on figure 3, allowing interpretation of linear correlation between mAs and $CTDI_{vol}$ as well as $kV^{2.5}$ and *n*CTDI with R^2 exceeding 0.9 (figure 3). An equations relating between collimation and *n*CTDI was also formulated. The linearity between kV and *n*CTDI as well as collimation and *n*CTDI allowed equation 8 to be applied to estimate *n*CTDI *($nCTDI_{etimation}$)* or $CTDI_{estimation}$.

$$nCTDI_{estimation} = \frac{f_{kV}(k\prime)}{f_{kV}(k)} \times \frac{kV\prime^{2.5}}{kV^{2.5}} \times nCTDI_{kV,k} \quad (8)$$

The deviation between $CTDI_{estimation}$ and actual CTDI measurement was found to be below 5 %. However, two data points have demonstrated deviations of above 5 %. The deviation ranges between $CTDI_{estimation}$ and actual CTDI measurement were in the range of 0.36% to 8.97%.

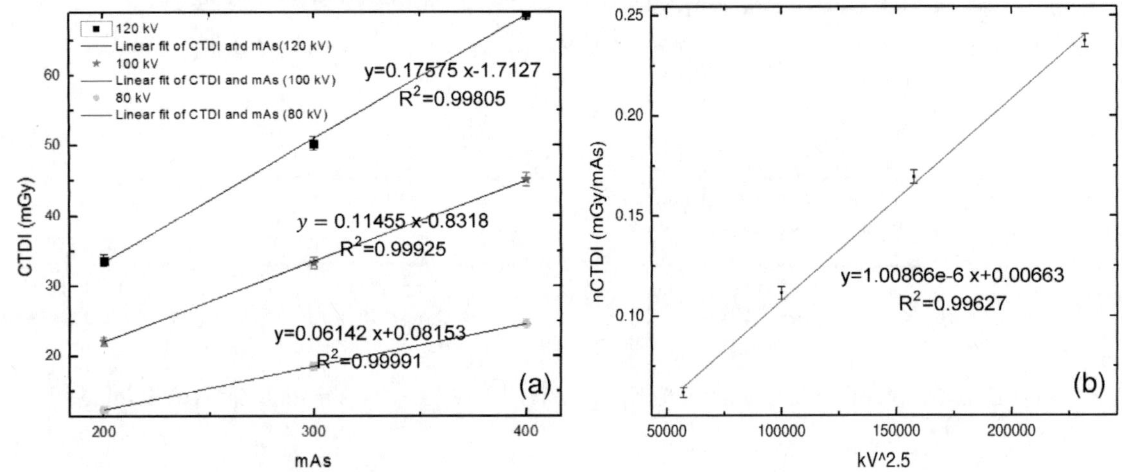

Figure 3. Correlations between (a) mAs and CTDI and (b) $kV^{2.5}$ and nCTDI

3.2. Predicting image quality from exposure parameter

Image quality can be represented either by SNR (signal to noise ratio) or CNR (Contrast to Noise Ratio). As discussed by (Bushberg, *et al.*, 2012), dose is proportional with squared SNR divided by slice thickness [3]. In this work, correlation between $CTDI_{estimation}$ and CNR squares divided by slice thickness was chosen to predict image quality. Linear curve fitting between $CTDI_{est}$ and $\frac{CNR^2}{T}$ demonstrated the regression coefficient value of exceeding 0.9. Therefore, the equation was chosen to be applied in estimating CNR with the formula being given on equation 9.

$$CNR_{est} = \sqrt{(0.01966\ CTDI_{calc} + 0.11167) \times slice\ thickness} \quad (9)$$

The deviation between $CNR_{estimation}$ and actual CNR measurement was found to be below 5.0% in thirteen data points, while other data points have deviations of above 5.0%. The deviation ranges between $CNR_{estimation}$ and actual CNR measurement were in the range of 0.04% to 21.65%. In this work, correlation between exposure parameters and high contrast resolution was not identified using MTF curve due to the inability of the varied exposure parameters to provide a variation of image quality, therefore requiring further study. However, the idea of Goldman (2007) that the primary factors affecting spatial resolution are focal spot size, patient motion, matrix size and reconstruction filter [8] supports our result.

3.3. Android™ based software to estimate CTDI and CNR

The Android™-based software allowing user to estimate dose information ($CTDI_{vol}$ and DLP) and image quality (CNR) through parameter exposure input (kV, mAs, collimation or slice thickness) was developed using `MIT app Inventor`. Whenever selected parameter combination yields on dose index exceeding European Commission EUR 1626 (2000) a certain warning message will appear. Image quality in terms of CNR was grouped (labelled) into categories A, B, and C, each for images yielding CNR ≥ 2.7, $1.67 < \text{CNR} < 2.7$, and CNR ≤ 1.67, respectively, to provide quick decision making. The software layouts are given on figure 4.

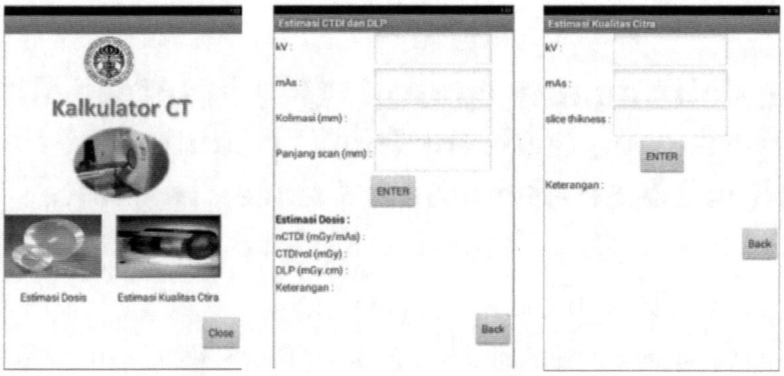

Figure 4. The android software layouts

4. Conclusion

In-house Android™-based software was developed and can be used to estimate $CTDI_{vol}$ or CNR through exposure parameters combination input. The software have deviation in the range of 0.36% to 8.97% and 0.04% to 21.65% to estimate $CTDI_{vol}$ and CNR on adult head examination, respectively. The algorithm was tailored to the unit used to commission the software.

References

[1] Cederquist B, Båth M and Hansson J 2008 *Evaluation of two thin CT dose profile detectors and a new way to perform QA in a CTDI head phantom*

[2] Bauhs J A, Vrieze T J, Primak A N, Bruesewitz M R and McCollough C H CT dosimetry: Comparison of measurement techniques and devices *Radiographics* **28** 245–53

[3] Bushberg J T, Seibert J A, Leidholdt E M and Boone J M 2011 *The essential physics of medical imaging* (Lippincott Williams & Wilkins)

[4] AAPM 2008 *The measurement, reporting, and management of radiation dose in CT (AAPM Report no.69)* (College Park: American Association of Physicists in Medicine)

[5] IAEA 2007 *Dosimetry in diagnostic radiology: An international code of practice (Report Technical Reports Series No.457)* (Vienna: IAEA)

[6] Coltman J W 1954 The specification of imaging properties by response to a sine wave input *J. Opt. Soc. Am.* **44** 468

[7] Hassan W and Saridan W M 2002 Calculation of the modulation transfer function from the square wave response function data with an interactive curve fitting software *Phys. Medica* **18** 25–31

[8] Goldman L W 2007 Principles of CT: Radiation dose and image quality *J. Nucl. Med. Technol.* **35** 213–25

Establishing daily quality control (QC) in screen-film mammography using leeds tor (max) phantom at the breast imaging unit of USTH-Benavides Cancer Institute

K J C Acaba[1,2], L D Cinco[2] and J N Melchor[3]

[1] The Graduate School, University of Santo Tomas, España Blvd., Manila, Philippines
[2] Breast Imaging Unit – Benavides Cancer Institute, University of Santo Tomas Hospital, Manila, Philippines
[3] Center for Device Regulation, Radiation Health, and Research – Food and Drug Administration, Department of Health, Manila, Philippines

E-mail: akb.karl@gmail.com

Abstract. Daily QC tests performed on screen film mammography (SFM) equipment are essential to ensure that both SFM unit and film processor are working in a consistent manner. The Breast Imaging Unit of USTH-Benavides Cancer Institute has been conducting QC following the test protocols in the IAEA Human Health Series No.2 manual. However, the availability of Leeds breast phantom (CRP E13039) in the facility made the task easier. Instead of carrying out separate tests on AEC constancy and light sensitometry, only one exposure of the phantom is done to accomplish the two tests. It was observed that measurements made on mAs output and optical densities (ODs) using the Leeds TOR (MAX) phantom are comparable with that obtained from the usual conduct of tests, taking into account the attenuation characteristic of the phantom. Image quality parameters such as low contrast and high contrast details were also evaluated from the phantom image. The authors recognize the usefulness of the phantom in determining technical factors that will help improve detection of smallest pathological details on breast images. The phantom is also convenient for daily QC monitoring and economical since less number of films is expended.

1. Introduction

The Breast Imaging Unit of USTH- Benavides Cancer Institute has been conducting QC program to constantly evaluate the mammography system performance with the goal of improving the detection of punctate calcifications, tumour masses, fibrous extensions and other pathological details to even smaller dimensions.

Daily QC on the SFM equipment in the facility is performed by the assigned QC radiologic technologist following the test protocols in the IAEA Human Health Series No.2 manual. Two films, at least, are used for light sensitometry and for AEC constancy. Though presently, the facility has acquired the Leeds TOR (MAX) breast phantom through the IAEA Coordinated Research Project (CRP) E13039. Only one exposure of the phantom is done to accomplish the two tests, hence, the economy of films used.

The objective of this paper is to verify that measurements made using Leeds phantom are comparable with that obtained from the usual conduct of tests and can be assessed with the performance criteria specified in the IAEA manual. Also, other quantifiable details in the phantom are separately evaluated against the limiting criteria provided in the user manual.

2. Methodology

Daily tests were performed on the same time every morning before any patient films were processed. The films used for the AEC constancy and light sensitometry tests were processed first. The film for the Leeds phantom image is followed. These films were processed at 31 ± 1°C developer temperature. All films used were taken from the same box and were exposed using the same cassette ID.

Film exposures were carried out with Mo/Mo target/filter combination at 28 kVp, "0" density control setting, and with AEC chamber at chest wall position. The attenuator stacks used in the AEC constancy test and Leeds breast phantom exposure are the semi-circular acrylic plates of 22 cm in diameter. In AEC constancy test, 45 mm- thick plates (four pieces 10 mm plus 5 mm) are mounted on the breast support and sufficient amount of compression is applied. The same compression force is applied to the 10 mm- thick Leeds test plate on top the stack of 35 mm- thick acrylic plates. The first 10 mm- thick of acrylic plate placed at the bottom of the stack has a lead disc for measurement of the true base + fog level of the film and a 1 cm- hole to serve as high density test point.

The leeds phantom set up simulates the "standard breast" recommended by the British Institute of Physical Sciences in Medicine (IPSM). It simulates the average breast of 45 mm thickness with 50% glandular tissue and 50% adipose tissue-composition under compression [1].

The displayed mAs on the console were recorded and the film densities were measured with a densitometer. The mean background optical density (OD) on the film image was consistently measured at a point 40 mm from the chest wall margin and laterally centered. Test forms available in the IAEA manual and in the Leeds user manual were utilized for data recording. The baseline values were determined over a six-day period. Data gathered in the months of June and July 2015 (a total of 40 days) were used in the assessment.

3. Results

Figure 1. This is a sample image of the Leeds TOR (MAX) breast phantom. The phantom image is drawn from this study. Several test objects are embedded on the Leeds test plate to produce this image. It has two sets of resolution gratings (each with 1 to 20 lp/mm), low-contrast bar patterns (1.8 to 5 lp/mm), micro-particle stepwedge in 3 grades, 10-point greyscale stepwedge, base + fog test point, 1 cm- hole for the high density test point, twelve circular details of 5.6 mm- diameter in diminishing contrasts, two groups of eleven 0.5 mm and 0.25 mm- sized details in diminishing contrasts. The high contrast details (0.5 mm and 0.25 mm sizes) are too small to be seen in this image. But in actual layout, they are lined up in series, between rows of low-contrast circular details.

A 9x magnifying lens available with the phantom set is used in aid of the visualization of these test details.

Tables 1 and 2 show the results of the mAs output, the film density measurements and the evaluation of film images over a six-day period for baseline values.

Table 1. Baseline values for the mAs output, mean background OD and essential film ODs in sensitometry.

Tests	mAs output	Density Measurement			
		Mean background OD	B+F	Speed index	Contrast index
AEC constancy & light sensitometry	66.9	1.91	0.24	1.40	0.87
leeds phantom image assessment	54.3	1.88	0.24	1.42	2.12

Table 2. Baseline scores for other quantifiable details from the Leeds phantom image.

Density measurement (OD)			Visual contrast [1- (speed index/mean bkgd. OD)]	Unsharpness measurement (lp/mm)			Low-con sensitivity	Small detail visibility		Micro-particles
B+F test pt.	HD test pt.	Scatter		∥Rhs Grating	⊥ Rhs Grating	Low-conbars	5.6 mm diam.	0.50 mm	0.25 mm	Grade Number
0.24	3.54	0.0	0.25	20	20	10	7	9	7	3
							Numbers detected			

Tables 3 and 4 present the observed range of measured and calculated data as well as the results from the evaluation of the image quality parameters within the 40-day period.

Table 3. Observed range in mAs and in measured film ODs.

Tests	Observed mAs range	Observed range in the measured ODs			
		Mean background OD	B+F	Speed index	Contrast index
AEC constancy & light sensitometry	63.7 – 67.9	1.78 – 2.04	0.23 – 0.25	1.30 – 1.51	0.83 – 1.00
Leeds phantom image assessment	51.0 – 55.7	1.75 – 2.07	0.24 – 0.25	1.32 – 1.55	1.88 – 2.18

Table 4. Observed range in OD measurements and number of frequently resolved circular details.

B+F test pt.	0.24 – 0.26	Circular details		Numbers detected				
HD test pt.	3.34 – 3.64			6	7	8	9	10
Scatter	0.00 – 0.01	Low-contrast	5.6 mm		52.5%	47.5%		
Visual contrast	0.19 – 0.28	High contrast	0.5 mm			37.5%	45.0%	17.5%
			0.25 mm	37.5%	57.5%	5.00%		

Data from unsharpness measurement and grade number of micro-particles were observed to not vary from the baseline scores over this period.

4. Discussion

The baseline output from the exposure of the 45 mm Leeds phantom was observed to be lower by 12.6 mAs relative to the baseline output from exposure of the routine 45 mm acrylic plates (refer to table 1). This could be due to the presence of test objects embedded in the test plate. Hence, the degree of attenuation is different between the two set-ups.

The contrast index also differ since 21-step sensitometer is used following the IAEA protocol, while Leeds phantom has 10-point greyscale step wedge. The contrast index from Leeds phantom image is determined by the difference between the high-density (HD) point and the speed index. Thus, a film gamma calculation can be drawn from the data. But in the IAEA protocol, it suggests average gradient to indicate film contrast [2]. Nevertheless, other results for the mean background OD, base plus fog, and speed index did not differ considerably between two set-ups.

Other quantifiable details from the Leeds phantom image were presented in table 2. The base plus fog (B+F) test point OD is compared to the OD from step 1 in the grey scale wedge to determine the presence of scattered radiation. The baseline scatter measurement was zero. On the other hand, the measurement for visual contrast estimate changes with applied tube kV and there was no performance criterion available, yet. Nonetheless, baseline visual contrast in the phantom image was observed at 0.25.

Resolution limit of the screen-film combination was observed to consistently demonstrate maximum number of lines per millimeter (lp/mm) under a 9x magnifying lens. However, the number of low contrast circular details that was frequently resolved is 7 out of 12. The numbers of resolved 0.5 mm and 0.25 mm small details were 9 and 7, respectively. There is a need for improvement in the detection of these circular details.

Observed range in data from each evaluated parameter were presented in tables 3 and 4. Deviations of measurements against the baseline values can be observed from these tables. Over the 40- day assessment, the mAs output data were within the ±10% of baseline mAs and the mean background ODs were within ±0.2 tolerance as specified in the IAEA protocol for AEC constancy. Also, the sensitometry data met the minimum B+F, speed index and contrast index tolerances indicated in the IAEA protocol. Results from the evaluation of the image quality parameters were consistent to the baseline scores.

5. Conclusion

It is verified that Leeds TOR (MAX) phantom is convenient to use for daily QC monitoring. Measurements from using the phantom were agreeable with that of the measurements obtained from the separate tests on AEC constancy and light sensitometry. Results from the evaluation of the image quality parameters were observed to demonstrate consistency with the baseline scores. However, this could still be improved by determining technical factors that can resolve the smallest details that the phantom has provided.

References

[1] Leeds Test Objects TOR MAX user manual.

[2] IAEA 2009 Human Health Series No.2 Quality Assurance Programme for Screen Film Mammography (Vienna, Austria: IAEA)

Acknowledgements

The authors would like to acknowledge the cooperation and support of the USTH Department of Radiological Sciences.

Effect of filter on average glandular dose and image quality in digital mammography

C Songsaeng[1]**, A Krisanachinda**[1] **and K Theerakul**[2]

[1] Department of Radiology, Faculty of Medicine, Chulalongkorn University, Bangkok, Thailand
[2] King Chulalongkorn Memorial Hospital, Bangkok Thailand

E-mail: chatsuda.song@gmail.com

Abstract. To determine the average glandular dose and entrance surface air kerma in both phantoms and patients to assess image quality for different target-filters (W/Rh and W/Ag) in digital mammography system. The compressed breast thickness, compression force, average glandular dose, entrance surface air kerma, peak kilovoltage and tube current time were recorded and compared between W/Rh and W/Ag target filter. The CNR and the figure of merit were used to determine the effect of target filter on image quality. The mean AGD of the W/Rh target filter was 1.75 mGy, the mean ESAK was 6.67 mGy, the mean CBT was 54.1 mm, the mean CF was 14 1bs. The mean AGD of W/Ag target filter was 2.7 mGy, the mean ESAK was 12.6 mGy, the mean CBT was 75.5 mm, the mean CF was 15 1bs. In phantom study, the AGD was 1.2 mGy at 4 cm, 3.3 mGy at 6 cm and 3.83 mGy at 7 cm thickness. The FOM was 24.6, CNR was 9.02 at thickness 6 cm. The FOM was 18.4, CNR was 8.6 at thickness 7 cm. The AGD from Digital Mammogram system with W/Rh of thinner CBT was lower than the AGD from W/Ag target filter.

1. Introduction

Mammography is the process of using low-dose X-ray to examine the human breast. It is used as a diagnostic as well as a screening tool. The goal of mammography is early detection of breast cancer, typically through detection of characteristic of the masses and micro-calcifications. In this respect, mammography is believed to reduce mortality of breast cancer. The radiation exposure associated with mammography is a potential risk of screening which is greater in younger women. Research has shown the risk of breast cancer for asymptomatic women under 35 years is not high enough to warrant the risk of radiation exposure [4]. For this reason, and because the radiation sensitivity of breast tissue in women under 35 years is possibly greater than that of older women, screening mammography is not recommended in women under 40 years old. However, if there is high risk of cancer in patients, mammography may still be important. Digital Mammography system, in 2003, molybdenum target and molybdenum or rhodium filters were introduced. Currently, tungsten-rhodium and tungsten-silver target-filters are used in the latest development, resulting in reduction in the average glandular dose and the entrance surface air kerma [5, 7]. The spectrum was suit for various breast thicknesses. The optimal energy band and breast thickness was 14-18 keV for 2 cm, 17-21 keV for 4 cm, 19-23 keV for 6 cm and 20.5-23.5 keV for 8 cm. The K-edge Rhodium was 20.2 - 22.8 keV. The characteristics of the new digital detectors use anode filter combinations W/Rh, W/Ag which have some advantages for imaging dense or thick breasts. Breast doses associated with these combinations (Rh/Rh, W/Rh, W/Ag, W/Al) are lower than those delivered with Mo/Mo or Mo/Rh. The threshold value for breast thickness where the spectrum changes depends on the AEC calibration. The correct selection of the X-ray beam will strongly influence the dose and image quality. Williams and colleagues [5] studied the impact of selection of exposure factors on image quality and dose for range of breast thickness and tissue type. They use the figure of merit (FOM) (contrast to noise ratio^2/ mean glandular dose) to compare the technique factors of all available target filter combination.

2. Material and methods

2.1. Patient group

Our retrospective study involved recruitment from 28 January to 1 September 2015 and compilation of patient data collected 400 mammograms from 200 women and included contact, automatic exposure control (AEC) mammograms of cranio-caudal and medio-lateral projections from preventive examinations, screening and after care.

2.2. Mammographic system

The study was performed using the Digital Mammography system (Hologic Selenia Dimension). This system has a dual track X-ray tube with a filter combination Tungsten (W) anode with Rhodium (Rh) filter and Tungsten (W) anode with Silver (Ag) filter. Focal spot sizes are 0.3 mm for contact examinations and 0.1 mm for magnification examinations. The 18 cm × 14 cm digital detector was used with a pixel size is 100 μm. The use of this digital detector offers the opportunity to directly acquire digital mammograms with a 14-bit resolution resulting in more than 16,000 gray levels. After exposure the images are displayed on a 2 k × 2.5 k monitor. All images were acquired with a grid. The following information was recorded for each mammogram: tube voltage (kVp); post exposure tube loading (mAs), target and filter materials, compressed breast thickness (mm), compression force (1bs.), the patient dose: AGD (mGy), ESAK (mGy) for CC and MLO projections for both breasts. The records were analyzed and AGD and ESAK between tungsten-rhodium target-filter and tungsten-silver target-filter were compared.

2.3. Determination of the image quality in phantom

The contrast to noise ratio (CNR) and the figure of merit (FOM) are estimated for the image quality using BR12 phantom and ACR phantom [2]. At 7 cm of compressed breast thickness the W/Rh filter was automatically changed to W/Ag filter. The study was performed by using ACR phantom (4 cm) and BR12 phantom was added 1,2 cm under ACR phantom. The system was operated by the automatic exposure control. The CNR for each filter were calculated by drawing the square of ROI (3×3 mm^2) in object and background. The CNR and FOM were calculated according to the work of Y. Ruiz-Gonzalez *et al.* [2] (1), (2) and based on the definition of the European guidelines for quality assurance in mammography screening [4].

$$\text{CNR}=\frac{I_{obj} - I_{back}}{I_{back}}\times 100 \tag{1}$$

I_{obj} = the mean intensity values of a given ROI
I_{back} = the intensity values of background selected

$$\text{FOM}=\frac{\text{CNR}^2}{\text{AGD}} \tag{2}$$

CNR = Contrast to noise ratio and AGD = Average glandular dose.

3. Results

The mean and standard deviation of important parameters in the ACR and BR12 phantoms are summarized in table 1.The phantom average glandular dose (AGD) in different thickness are provided in figure 1. The figures of merit (FOM) of phantom in different breast are provided in figure 2. The mean important parameters of patient are summarized in Table2. The patient average glandular dose (AGD) of both filters in different position (Cranio-caudal and Medio-lateral view) are provided in figures 3 and 4.

Table 1. Summary of parameters in ACR phantom and BR12 phantom

Compressed breast thickness (cm)	Target/Filter	AGD (mGy)	ESAK (mGy)	kVp	CNR (%)	FOM (mGy^{-1})
4	W/Rh	1.2	3.7	28	9.97 ± 0.13	83 ± 2.5
6	W/Rh	3.3	10.1 ± 0.3	28	9.02 ± 0.41	24.6 ± 1.9
7	W/Ag	3.83 ± 0.3	10.6 ± 2.1	33	8.6 ± 0.42	18.4 ± 2.4

AGD,average glandular dose; ESAK,entrance surface air kerma; SNR, signal to noise ratio; CNR, contrast to noise ratio.

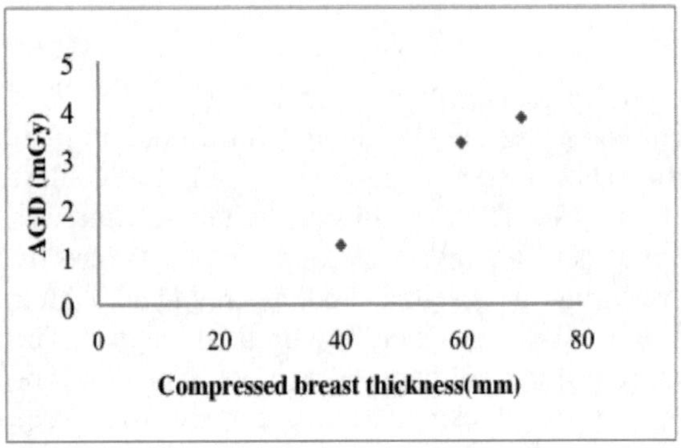

Figure 1. AGD per exposure as a function of CBT in phantom study.

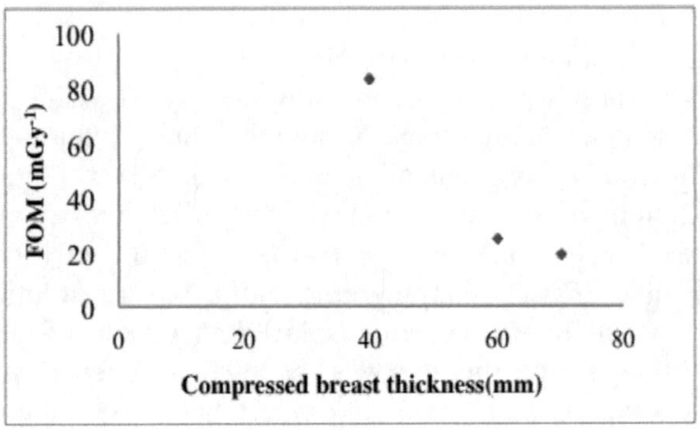

Figure 2. FOM as a function of CBT in phantom study.

Table 2. Summary of parameters of W/Rh and W/Ag in 200 patients

	AGD (mGy)	ESAK (mGy)	CBT (mm)	CF (1bs.)	kVp	mAs
W/Rh target-filter (CC view)						
Ave	1.7	6.67	54	14.05	29.5	145.06
Min	0.7	1.8	32	4.4	26	61
Max	3.79	16.45	69	22	32	289
W/Ag target-filter (CC view)						
Ave	2.6	12	75.5	14.9	31	196.4
Min	1.76	7.86	64	4.3	30	115
Max	5.73	25.3	86	26.5	33	398

AGD average glandular dose; ESAK,entrance surface air kerma; CBT, compressed breast thickness; CF, Compression force

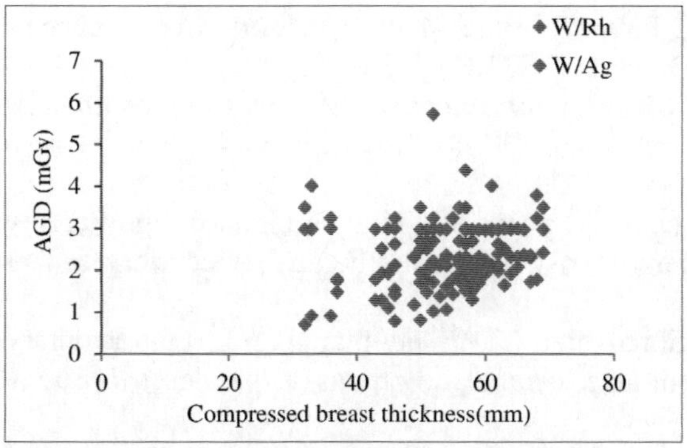

Figure 3. AGD per exposure as a function of breast thickness in CC view in patient study.

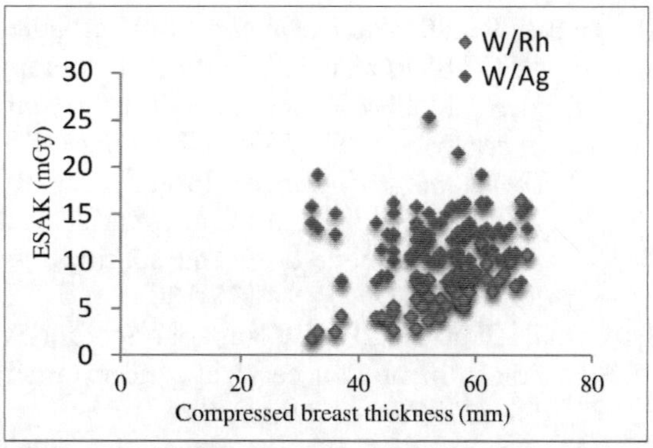

Figure 4. AGD per exposure as a function of breast thickness in MLO view in patient study.

3.1. In phantom study

The FOM of W/Rh target filter was 24.6 ± 1.9 kVp was 28, CNR was 9.02 ± 0.41, the AGD was 3.3±0.09 mGy, and the ESAK was 10.1±0.3 mGy at phantom thickness 6 cm. The FOM of W/Ag target filter was 18.4 ± 2.4 at kVp 33, CNR was 8.6± 0.42, the AGD was 4±0.3 mGy, and the ESAK was 10.6±2.1 mGy at phantom thickness 7 cm. The relationship between CBT and patient AGD in both CC and MLO views are given in figures 3 and 4, respectively. The lines of best fit demonstrate that the AGD increases with thick breast. The AGD of W/Ag target filter is higher than W/Rh target filter as increasing the CBT, however, as a function of breast thickness. The relationship between CBT and AGD in phantom study are given in Figures 1, the AGD increases with the phantom thickness. The relationship between CBT and FOM in phantom study is given in figures 2, the FOM is affected by kVp, target filter and the phantom thickness.

4. Discussion

Our results are similar to S. Emanuelli1 *et al.* [6] who compared the AGD from digital mammography systems using W/Rh and W/Ag target filters. The mean AGD of W/Rh was 1.37 mGy, the mean CBT was 48.3 mm. The mean AGD of W/Ag was 2.84 mGy, the mean CBT was 75mm. In our study, the mean AGD of W/Rh was 1.7 mGy, the mean CBT was 54 mm. The mean AGD of W/Ag was 2.6 mGy, the mean CBT was 75.5 mm. Flynn's *et al.* [3] studied the relationship between FOM and tube voltage of W/Rh and W/Ag. FOM is highest at kVp 28 for W/Rh target filter at 33 and 22 mGy^{-1} at kVp 26 for W/Ag target filter. Our study, the FOM decreased from 24 to 18 at thickness 6 to 7 cm for W/Rh and W/Ag target filter. The kVp was 28 for phantom thickness 4 and 6 cm of W/Rh target filter and 32 for phantom thickness 7 cm of W/Ag target filter. The AGD was 1.7 mGy for W/Rh, 2.6 mGy for W/Ag at higher CBT. All are less than the IAEA DRL of 3.0 mGy with grid for CC projection [7].

5. Conclusion

The AGD from Digital Mammogram system with W/Rh of thinner compressed breast thickness was lower than the AGD from W/Ag, although this difference may not hold for patients with similar breast thickness. The FOM and CNR decrease with the compressed breast thickness.

References

[1] Haie-Meder C *et al.* 2005 Recommendations from Gynaecological (GYN) GEC-ESTRO Working Group (I): Concepts and terms in 3D image based 3D treatment planning in cervix cancer brachytherapy with emphasis on MRI assessment of GTV and CTV *Radiother Oncol* **74** 235-245

[2] Cancer Care Ontario 2014 *Imaging Strategies for Definitive Intracavitary Brachytherapy of Cervical Cancer: Recommendation Report* (Ontario, Canada: Cancer Care Ontario)

[3] Pötter R 2002 *Vienna method* In: Gerbaulet A, Potter R, Haie-Meder C,editors. Cervix cancer. GEC ESTRO handbook of brachytherapy (Brussels: ESTRO) p. 331–8.

[4] Tanderup K 2014 *Physics aspects of treatment planning intracavitary +/- interstitial techniques in cervix cancer. ESTRO Teaching Course Image Guided Radiotherapy & Chemotherapy in Gyneacological Cancer* (Firenze: ESTRO)

[5] Tanderup K *et al.* 2010 From point A to the sculpted pear: MR image guidance significantly improves tumour dose and sparing of organs at risk in brachytherapy of cervical cancer *Radiother. Oncol.* **94** 173-180

[6] Onal C *et al.* 2009 Comparison of conventional and CT-based planning for intracavitary brachytherapy for cervical cancer: target volume coverage and organs at risk doses *J. Exp & Clinical Cancer Research* **28** 95

[7] Madan R *et al.* 2014 Comparative evaluation of two-dimensional radiography and three dimensional computed tomography based dose-volume parameters for high-dose-rate intravitary brachytherapy of cervical cancer: A prospective study *Asian Pacific J. Cancer Prevention* **15** 4717-21

Journal of Physics: Conference Series **694** (2016) 012040 doi:10.1088/1742-6596/694/1/012040

Calculation of size specific dose estimates (SSDE) value at cylindrical phantom from CBCT Varian OBI v1.4 X-ray tube EGSnrc Monte Carlo simulation based

M Nasir[1], D Pratama[1], C Anam[1,2] and F Haryanto[1]

[1] Department of Physics, Faculty of Mathematics and Natural Sciences, Institut Teknologi Bandung, Jl. Ganesa 10, Bandung 40132 Indonesia
[2] Department of Physics, Faculty of Mathematics and Natural Sciences, Diponegoro University, Semarang, Indonesia

E-mail: agamnashir@gmail.com

Abstract. The aim of this research was to calculate Size Specific Dose Estimates (SSDE) generated by the varian OBI CBCT v1.4 X-ray tube working at 100 kV using EGSnrc Monte Carlo simulations. The EGSnrc Monte Carlo code used in this simulation was divided into two parts. Phase space file data resulted by the first part simulation became an input to the second part. This research was performed with varying phantom diameters of 5 to 35 cm and varying phantom lengths of 10 to 25 cm. Dose distribution data were used to calculate SSDE values using trapezoidal rule (trapz) function in a Matlab program. SSDE obtained from this calculation was compared to that in AAPM report and experimental data. It was obtained that the normalization of SSDE value for each phantom diameter was between 1.00 and 3.19. The normalization of SSDE value for each phantom length was between 0.96 and 1.07. The statistical error in this simulation was 4.98% for varying phantom diameters and 5.20% for varying phantom lengths. This study demonstrated the accuracy of the Monte Carlo technique in simulating the dose calculation. In the future, the influence of cylindrical phantom material to SSDE would be studied.

1. Introduction

The accuracy in radiotherapy is influenced by many factors, one of which is the technology used. In the case of tumor radiotherapy, tumor deformation (movement) is one factor that must be considered. Cone Beam Computed Tomography (CBCT) can be used as an imaging modality to record a tumor movement. CBCT consists of a tube and a detector mounted on a rotatable tool. CBCT system enables a sequence of 2D radiographic projection images to be acquired from a kV source and flat panel detector imaging system [1]. Used daily, CBCT allows a more precise positioning of the treatment isocenter by shifting images performed before each session on the initial dosimetric imaging [2]. The dosimetry for the CBCT has become more important due to its widespread applications for IGRT and multislice CT scanners. CBCT has some kind of modalities, one of which is Varian On Board Imager (OBI) v1.4 released in 2008.

The radiation dose received by a patient during a scanning process is highly variable and depends on many technical factors. The technical factors that affect the radiation dose generally comprise a function of the X-ray beam quality, geometry, limitation devices, and acquisition imaging dose rate settings. The amount of the radiation dose received from the scanning also depends on the phantom geometry, shape, and material [2]. The radiation dose received by the patient will cause changes in his or her biological systems and can increase the risk of cancer in the sensitive organs. Therefore, it is very important to estimate accurately the patient's allowed dose generated by CBCT. CTDI is a standardized measurement of the radiation dose output from the scanning process. In fact, the dose received by a patient from a scanning process depends on the patient's characteristics. SSDE is better for estimating the dose given to the patient [3].

Monte Carlo (MC) method is a numerical method for statistical simulation which utilizes sequences of random numbers to model a process of experimentation through a process simulation system. MC simulation has been widely applied in various fields, such as medical physics, particularly radiation dosimetry, and also for analyzing nuclear reactors. In medical physics, this method has been widely used in nuclear medicine detector and dose calculation in radiotherapy. MC is the most accurate method for simulating dose distributions in homogeneous and inhomogeneous tissues when the electron transport effects cannot be accurately calculated by conventional methods [4]. Therefore, the aim of this research was to calculate Size Specific Dose Estimates (SSDE) generated by variant OBI CBCT v1.4 X-ray tube working at 100 kV and using EGSnrc Monte Carlo simulations. Results of this study will provide information about the SSDE generated by the X-ray tube Varian OBI CBCT v.1.4 deposited by the cylindrical phantom.

2. Method

2.1. Set up of X-ray tube and cylindrical phantom

The EGSnrc Monte Carlo code is a well-known code used to simulate the biography of photon and electron through a medium [5]. The particles simulated have the energy range from a few keV up to hundreds of GeV. EGSnrc has enormous flexibility allowing the user to run the simulation even with only little understanding about the program. In EGSnrc, transports of photons and electrons were simulated in multiple steps where each step has a random length [4]. To get the final result of the simulation, a lot of history is recommended to be simulated.

MC simulation for determining dose distribution in the cylindrical phantom generated by Varian OBI CBCT v1.4 X-ray tube performed using the EGSnrc program package consists of BEAMnrc and DOSRZnrc. BEAMnrc was used to design and simulate X-ray tube; DOSRZnrc was used to design and perform simulations in the cylindrical phantom. The dose distribution obtained from the simulation was used to calculate SSDE using Matlab. The schema of this research is presented in figure 1.

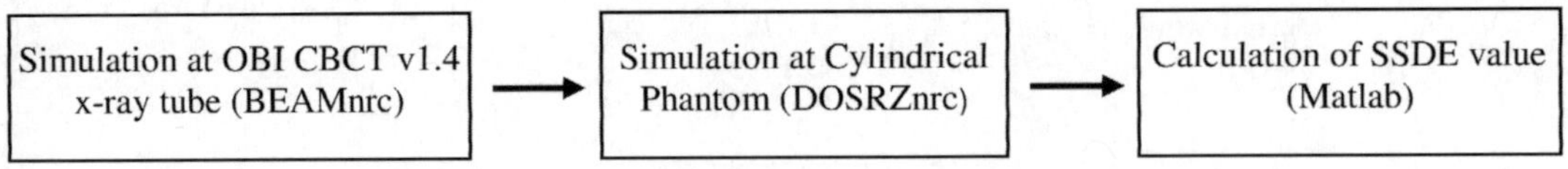

Figure 1. Schema of research methodology.

Monte Carlo code used in this simulation was divided into two parts. The first part was simulation of varian OBI CBCT v1.4 X-ray tube using BEAMnrc. The tube consists of several components namely target, exit windows, pre-filter, pre collimator, upper blades, glass, bowtie filter and collimator.

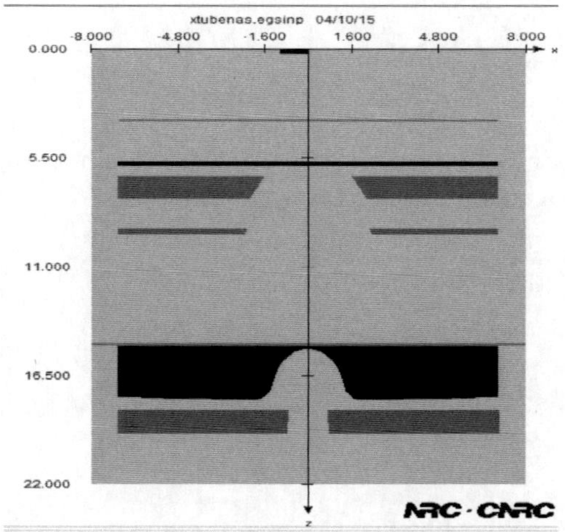

Figure 2. Setup of simulated geometry for Varian OBI CBCT v1.4 X-ray tube.

Figure 2 shows the setup of the simulated geometry for the Varian OBI CBCT v1.4 X-ray tube used in this simulation. The particles that have been transported in the X-ray tube will be scored on the scoring plane named phase space (phsp) file. Beam parameters for the CBCT X-ray tube simulation were as follows: ISOURC = 10 (parallel circular beam incident from side) and slice thickness = 1 cm. The number of histories was set to 250 million. Simulation process was run on Linux Ubuntu with the CPU specifications of Intel core i5 3.4 GHz processor and 2 GB RAM. Simulations were performed using a 100 kV voltage.

The second part was the simulation in the cylindrical water phantom using DOSRZnrc. After the BEAMnrc simulation was finished, the phase space file obtained was used as an input source in DOSRZnrc simulations to determine the dose distribution in the cylindrical phantom. This research was performed with varying phantom diameters of 5, 10, 15, 20, 25, 30 and 35 cm. This research was also performed with varying phantom lengths of 10, 15, 20 and 25 cm. The number of histories was set to 500 million. The design of the cylindrical phantom used in this simulation is presented in figure 3.

(AAPM Report No. 204, 2011)

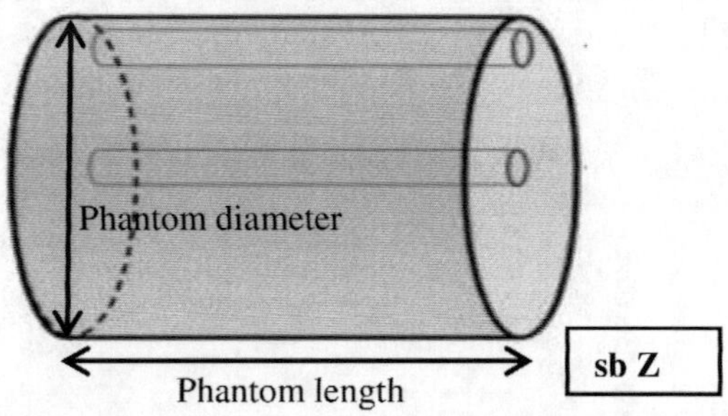

Figure 3. Design of cylindrical phantom.

Dose calculations were performed on two areas, namely the edge phantom (peripheral) region and the middle of the phantom (center) region. Dose distribution data obtained were used to calculate SSDE using Matlab.

2.2. CTDI and SSDE

The computed tomography dose index (CTDI) is a standardized measure of radiation dose output of a CT scanner. The CTDI can be used in conjunction with the patient size to estimate the absorbed dose [3]. The equation of CTDI is

$$CTDI = \frac{1}{T}\int_{-\infty}^{\infty} D(z)dz \qquad (1)$$

where T is the slice thickness and $D(z)$ is the radiation dose measured at position z along the scanner's main axis. The dose distribution imparted by a CT scan is still somewhat larger near the skin than in the center of the body [3]. The weighted CTDI was introduced to account for this

$$CTDI_W = \frac{1}{3}CTDI^{center} + \frac{2}{3}CTDI^{peripheral} \qquad (2)$$

In this case, $CTDI_W$ is equal to SSDE

$$SSDE = CTDI_W \qquad (3)$$

3. Results

SSDE values obtained were compared with the AAPM 2011 data and the results of the experiment contained in figure 4.

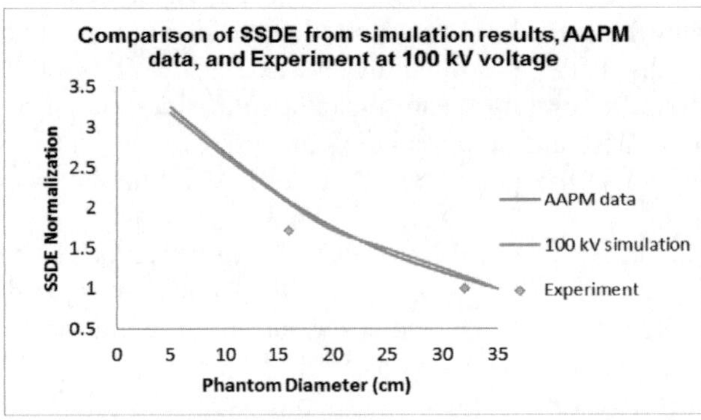

Figure 4. SSDE curve comparison with varying phantom diameter.

As can be seen in figure 4 that the SSDE value decreases with increasing diameter of the cylindrical phantom. Furthermore, SSDE value were calculated for simulation with varying phantom length as shown in figure 5 below.

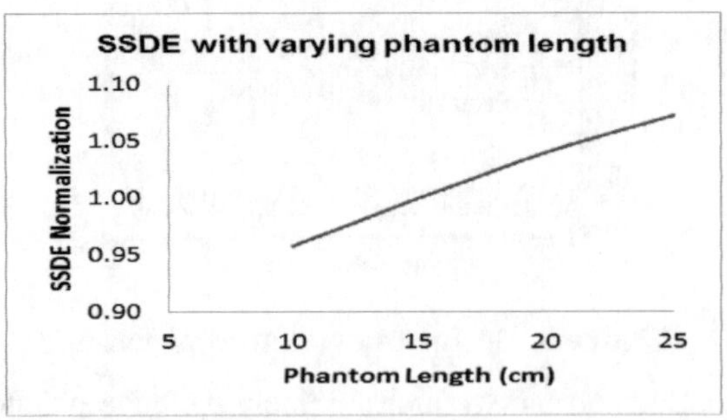

Figure 5. SSDE curve with varying phantom length.

As can be seen in figure 5 that the length of cylindrical phantom affected the dose deposited by phantom. SSDE for varying phantom length increased with phantom length.

4. Discussion

Journal of Physics: Conference Series **694** (2016) 012040 doi:10.1088/1742-6596/694/1/012040

The radiation dose deposited by a phantom during a scanning process depends on some factors. It can be affected by the phantom geometry, size, shape, and material [2].
Phantom diameter has a significant effect on the dose received by the phantom. This is because the dose deposited by a radiation beam into water phantom decreases with phantom depth [6]. The deeper radiation beam entering phantom, the lower dose deposited by phantom. By using a cylindrical phantom with a larger size, the radiation beam will travel a farther distance when entering the phantom and thus the dose will be lower [7]. It was also obtained that the dose at the peripheral region was higher than that at the center region.

It can be seen in figure 4 that the good agreement between the simulated and experimental data was obtained. The SSDE values obtained through the simulation are hardly very different from the SSDE values of AAPM 2011 data and the experimental results. This is indicated by the values of a relatively small deviation error; the average deviation is 2.18%.

As can be seen in Fig. 5, the varying phantom lengths of SSDE for shorter phantoms were much greater than those for longer sizes. The area of the penumbra region has an essential effect in this phenomenon. When using a cylindrical phantom with a longer size, the penumbra region would increase in such a way that the contribution to the total dose calculation due to the dose in the penumbra region also increased [6].

5. Conclusions

SSDE values decreased with the increasing diameter of cylindrical phantom, but contrastively increased with rose of phantom length. Monte Carlo simulation is good enough to use in performing the calculation of the dose at the cylindrical phantom. For the future work, it is suggested to develop this research using heterogeneous phantom.

References

[1] American Cancer Society 2014 The science behind radiation therapy

[2] IAEA Technical Reports Series No 457 2007 *Dosimetry in Radiodiagnostic Radiology: An International Code of Practice* (Vienna: IAEA)

[3] AAPM 2011 *Size Specific Dose Estimates (SSDE) in Pediatric and Adult Body CT Examinations* (College Park: AAPM)

[4] Nakagawa K *et al.* 2008 *Int. J. Radiat. Oncol. Biol. Phys.* **72** *S623*

[5] Hing K M, Rogers D, Tessier F and Walters B R B 2011 *The EGSnrc Code System: Monte Carlo Simulation of Electron and Photon Transport* (Ottawa: National Research Council Canada)

[6] Boutoleau B, Boissonnade O and Al Garcia 2013 Integration of geometric data recorded during irradiation sessions for post-treatment dosimetry calculation in external radiotherapy Unite de Physique Medicale Poitiers

[7] Lewis and Maria 2005 Radiation dose issues in multi-slice CT sacnning" ImPACT Technology Update No 3

Acknowledgement

This study has been financially supported by RIK (Riset & Inovasi KK), LPPM, Institut Teknologi Bandung, No. 237h/I1.C01/PL/2015.

Patient radiation dose from computed tomography angiography and digital subtraction angiography of the brain

Y Netwong[1] and A Krisanachinda[1]

[1] Department of Radiology, Faculty of Medicine, Chulalongkorn University, Bangkok Thailand

E-mail: yutthana_net@hotmail.com

Abstract. The 64-row multidetector computed tomography angiography (64-MDCTA) provides vascular image quality of the brain similar to digital subtraction angiography (DSA), but the effective dose of CTA is lower than DSA studied in phantom. The purpose of this study is to evaluate the effective dose from 64-MDCTA and DSA. Effective dose (according to ICRP 103) from 64-MDCTA and DSA flat panel detector for cerebral vessels examination of the brain using standard protocols as recommended by the manufacturer was calculated for 30 cases of MDCTA (15 male and 15 female).The mean patient age was 49.5 (23–89) yrs. 30 cases of DSA (14 male and 16 female), the mean patient age was 46.8 (21–81) yrs. For CTA, the mean effective dose was 3.7 (2.82- 5.19) mSv. For DSA, the mean effective dose was 5.78 (3.3-10.06) mSv. The effective dose of CTA depends on the scanning protocol and scan length. Low tube current can reduce patient dose whereas the number of exposures and number of series in 3D rotational angiography (3D RA) resulted in increasing effective dose in DSA patients.

1. Introduction

Cerebrovascular disorders can create the clinical problem to the brain. The most common types are Stroke, Aneurysm and Arteriovenous malformations (AVMs). Computed Tomographic Angiography (CTA), Magnetic Resonance Angiography (MRA) and Digital Subtraction Angiography (DSA) are procedures used to diagnose the diseases.

MDCT provides similar diagnostic information of cervicocranial vasculature as compared to DSA. The DSA should be considered primarily when peripheral vessels or ICA segments close to the skull base [1]. However, the previous study showed that effective dose in phantom with CTA examinations for the cerebral vessels yields 5 times lower than the same examination performed by DSA [2].

Biological effects of radiation can be grouped as stochastic and deterministic effects. The stochastic effect depends on the probability rather than its severity and it increases with dose. Radiation-induced cancer and genetic effects are stochastic [3].

The effective dose, absorbed and organ doses are important for the procedures that either involve high doses or include sensitive tissues in the primary radiation beam. Effective dose provides an approximate indicator of potential detriment from ionizing radiation and should be used as one parameter in evaluating the appropriateness of examinations involving ionizing radiation [4].

The purpose of this study is to evaluate the patient effective dose from CTA and DSA performed in the cerebrovascular region of the brain.

2. Materials and methods

2.1. Imaging equipment and protocols
CTA was performed by using a 64-row MDCT (Somatom Definition 64; Siemens, Erlangen, Germany). For DSA, a biplane angiography equipment (Axiom Artis Flat panel biplane; Siemens, Erlangen, Germany) was used. The optimized protocols of the CTA and DSA for cerebral vessels had been employed in our department as a routine procedure. Table 1 and 2 provide the parameters of the protocol for CTA and DSA examinations, respectively.

Table 1. Parameters for CTA examination.

Scanning parameters	kVp	mAs	pitch
Topogram	120	35	
Control scan	120	381/450	
Non-Contrast	120	332/450	0.55
Pre-monitoring	120	20	
Monitoring	120	20	
Head Angiogram	100	125/175	1.4
Post contrast	100	125/175	1.4
Rotation time (sec)	1.0 s		
Slice Acquisition (mm)	64×0.6		
Slice collimation (mm)	0.6		

Note: CTA of the brain scan from Common carotid artery bifurcation (C-4) to the vertex.

Table 2. Parameters for DSA examination.

Parameters	2D mode	3D RA mode
kVp	77	70
Pulse width (ms)	40	12.5
kVp filter	81	OFF
kVp mask	96	
Dose (μGy/pulse)	3.6	0.36
Phase 1	4F/s	4s
Phase2	1F/s	16s
Mask	30F/s	5s
Fill	30F/s	5s
Washout	0.5F/s	2s

2.2. Patient population
Thirty cases (15 male, 15 female) at the mean age of 49.5 (23–89) yrs. were studied by the 64-row MDCTA and thirty cases (14 male, 16 female) at the mean age of 46.8 (21–81) yrs. were studied by the DSA flat panel detector for cerebral vessels of the brain from October 2014 to June 2015 at Prasat Neurological Institute, Bangkok, Thailand.

2.3. Radiation dose determination
The effective dose of CTA was calculated by using DLP (mGy.cm) × k-factor (mSv/mGy.cm) [5]. For CTA brain, k-factor was 0.0019 mSv/mGycm [6, 7] and DSA used DAP (mGy.cm^2) × Dose conversion coefficient (mSv/mGy.cm^2) [5]. The dose conversion coefficient for cerebral angiogram was 0.087 mSv/mGy.cm^2 [8]. The results were shown in section 3.1

3. Results

3.1. Comparison of the effective dose from CTA and DSA

The mean effective dose related to patient characteristic is 3.7 (2.82 - 5.19) mSv for CTA and 5.78 (3.3 – 10.06) mSv for DSA as shown in table 3.

Table 3. Effective dose and patient characteristics for CTA and DSA.

	CTA	DSA	p value
Effective dose (Mean±SD)	3.70 ±0.66	5.78 ±1.53	<.001
Gender (% male)	15 (50%)	14 (46.67%)	.800
Age (Mean±SD)	49.53 ±17.3	46.8 ±16.12	.529
BMI (Mean±SD)	24.02 ±4.01	22.82± 4.27	.265

After adjusting for gender, age and BMI differences in our sample, we demonstrate the effective dose, on average was significant (Mean difference adjust = 2.193, $p<0.001$).The effective dose comparing between procedures by box plots is shown in figure 1.

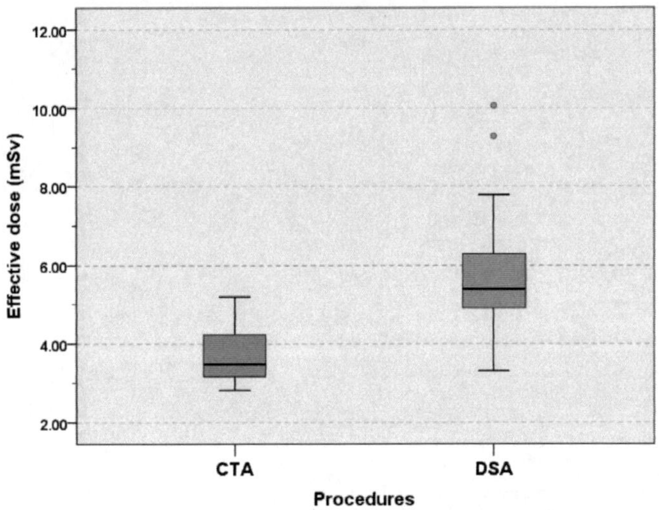

Figure 1. Box plots show the distribution of effective dose for the CTA and DSA procedures.
The effective dose for CTA, Mean = 3.7, Min = 2.82, Max = 5.19, Mod = 2.82, Median = 3.48, Range = 2.37, SD = 0.66, Percentiles; 25th = 3.16, 75th = 4.23 mSv, respectively.
The effective dose for DSA, Mean = 5.78, Min = 3.3, Max = 10.06, Mod = 3.3, Median = 5.39, Range = 6.76, SD = 1.53, Percentiles; 25 = 4.86, 75 = 6.34 mSv, respectively.

*Outliers are the maximum effective dose over the third quartiles.

3.2. Correlations between effective dose and patient characteristics of CTA

When we investigated the correlation between effective dose and patient characteristics for CTA procedure, the results show a moderately strong linear relationship between the effective dose and body weight ($r=0.642$, $p<0.001$), BMI ($r=0.552$, $p=0.002$) and fair linear relationship between the effective dose and height ($r=0.445$, $p=0.014$), respectively. Table 4 shows Spearman correlation.

Table 4. Correlations between effective dose and patient characteristics of CTA.

	r	p value
Age	-.353	.056
Weight	.642**	<.001
Height	.445*	.014
BMI	.552**	.002

Spearman's Correlation is significant at the 0.01 level (2-tailed) **, or 0.05 level (2-tailed).*

Journal of Physics: Conference Series **694** (2016) 012041 doi:10.1088/1742-6596/694/1/012041

3.3. Correlations between effective dose and patient characteristics of DSA
For DSA, we found poor linear relationship no significant between the effective dose and patient characteristics as Table 5 for Spearman correlation coefficient.

Table 5. Correlations between effective dose and patient characteristics of DSA.

	r	p value
Age	.243	.196
Weight	.203	.283
Height	.109	.556
BMI	.172	.362

Spearman's Correlation is significant at the 0.01 level (2-tailed) **, or 0.05 level (2-tailed).*

3.4. Correlations between effective dose and factors of DSA
We also investigated other factors that might be associated with effective dose for DSA. Table 6 gives Spearman correlation coefficient for several potential predictors of effective dose in the DSA procedure.

Table 6. Correlations between effective dose and factors of DSA.

	r	p value
Number of exposures	.755**	<.001
Fluoroscopic time	.330	.075
Number of Vessels selection	.282	.130
Number of 3D RA	.406*	.026
Experience of Radiologist	.337	.362

Spearman's Correlation is significant at the 0.01 level (2-tailed) **, or 0.05 level (2-tailed).*

We found a moderately strong correlation between the effective dose and number of exposures ($r=0.775$, $p<0.01$) and a fair correlation between the effective dose and number of 3D rotational angiography ($r=0.406$, $p<0.05$), respectively.

4. Discussion
The purpose of this study was to evaluate the effective dose from 64-row MDCTA and diagnostic DSA of the brain. Cohnen M *et al.* [9] measured the average effective dose in comprehensive CT of the brain in acute stroke at 3.6 mSv. Our study, the mean effective dose was 3.7 mSv.The key factors that affecting to high dose in CT were scan length, tube voltage, and effective mAs.

The reduction of scan length can significantly result in patient dose reduction in CT. However, in case of scan length reduction is impossible, using of low tube voltage should be considered [10, 11]. CTA of the brain obtained by using 80 kVp not only reduces patient radiation exposure by approximately 40% lower than 120 kVp but also give superior in higher contrast and contrast-to-noise ratio [12] [13]. Optimization in the acquisition parameters such as gantry rotation time, tube current, pitch and dose modulation can also reduce effective dose.

Manninen A L *et al.* [2] compared radiation exposure between diagnostic CTA and DSA examinations of brain and neck vessels by exposed anthropomorphic phantom. The effective dose for DSA of cerebral vessels was approximately 5 times of the CTA (0.67 mSv for CTA and 2.71 mSv for DSA). Our study, the mean effective dose of DSA was 1.5 times of the CTA. The number of exposures and number of 3D RA were the major factors affecting effective dose in DSA. Patients received a higher dose from 2D mode rather than 3D mode due to the differences in imaging parameters, higher frame rates in 2D mode result in high radiation dose during procedure.

For DSA, using lower pulse rates during fluoroscopy and frame rates are techniques to reduce the radiation dose to the patient from cerebral angiography procedures.

According to our study, CTA procedure should be primarily considered in order to reduce radiation dose to the patient.

5. Conclusion

Our study shows, the effective doses for CTA ranged from 2.82 - 5.19 mSv and the average of 3.7 mSv. Whereas, for DSA, the average effective dose was 5.78 mSv and doses ranged from 3.3 – 10.06 mSv, respectively. The key factors that affecting to high dose in CTA were effective mAs, kVp and scan length. Increasing patient height, weight and BMI result in increasing effective dose from CTA examination.

For DSA, the number of exposures and a number of 3D RA were the factors affecting the effective dose. Most of the dose is high as a result of a number of radiographic DSA exposures. The patient dose increases with a number of 3D RA of the procedure.

References

[1] Klingebiel R, Kentenich M, Bauknecht H C, Masuhr F, Siebert E, Busch M and Bohner G 2008 Comparative evaluation of 64-slice CT angiography and digital Subtraction angiography in assessing the cervicocranial vasculature *Vasc. Health Risk Manag.* **4** 901-07

[2] Manninen A L, Isokangas J M, Karttunen A, Siniluoto T and Nieminen M T 2012 A comparison of radiation exposure between diagnostic CTA and DSA examinations of cerebral and cervicocerebral vessels *AJNR.* **33** 2038-42

[3] Mahadevappa M 2001 Fluoroscopy: Patient radiation exposure issues *Radio. Graphics.* **21** 1033-45

[4] Mettler F A, Huda W, Yoshizumi T T and Mahesh M 2008 Effective Doses in Radiology and Diagnostic Nuclear Medicine: A catalog *Radiol.* **248** 254-63

[5] *American Association of Physicists in Medicine* 2008 The Measurement Reporting and Management of Radiation Dose in CT AAPM Report: 96 (New York) p 13

[6] Deak P D, Smal Y and Kalender W A 2010 Sex- and age specific conversion factors for effective dose at CT *Radiol.* **257** 158-66

[7] Huda W, Magill D and He W 2011 CT effective dose per dose length product using ICRP 103 weighting factors *Med Phys.* **38** 1261

[8] *National Council on Radiation Protection and Measurements 2009* Ionising Radiation Exposure of The Population of The United States: NCRP Report No. 160 (Bethesda)

[9] Cohnen M, Wittsack h J, Assadi S, Muskalla K, Ringelstein A, Poll L W, Saleh and Modder U 2006 Exposure of patients incomprehensive computed tomography of the head in acute stroke *AJNR.* **27** 1741-45

[10] Waaijer A, Prokop M, Velthuis B K, Bakker C J G, de Kort G A P and van Leeuwen M S 2007 Circle of Willis at CT angiography: dose reduction and image quality reducing tube voltage and increasing tube current settings *Radiol.* **242** 832-9

[11] Smith A B, Dillon W P, Gould R and Wintermark M 2007 Radiation dose-reduction strategies for neuroradiology CT protocols *AJNR.* **28** 1628-32

[12] Imai K, Ikeda M, Kawaura C, Aoyama T, Enchi Y and Yamauchi M 2012 Dose reduction and image quality in CT angiography for cerebral aneurysm with various tube potentials and current settings *BJNR.* **85** 673-81

[13] Bahner M L, Bengel A, Brix G, Zuna I, Kauczor H U and Delorme S 2005 Improved vascular opacification in cerebral computed tomography angiography with 80 kVp *Invest. Radiol.* **40** 229–34

Acknowledgements

We would like to extremely grateful Associate Professor Cameron Paul Hurst, Ph.D. (Clinical Epidemiology Unit, Faculty of Medicine, Chulalongkorn University, Thailand) for his helpful recommendations in statistical analysis.

Three-dimensional image reconstruction of macula from stratus optical coherence tomography (OCT) for diagnosis of macular degeneration

Arinilhaq[1] **and R Widita**[1]

[1] Department of Physics, Faculty of Mathematics and Natural Sciences, Institut Teknologi Bandung, Jalan Ganesa 10, Bandung 40132, Jawa Barat, Indonesia

E-mail: arinilhaqueen@gmail.com

Abstract. Diagnosis of macular degeneration using a Stratus OCT with a fast macular thickness map (FMTM) method produced six B-scan images of macula from different angles. The images were converted into a retinal thickness chart to be evaluated by normal distribution percentile of data so that it can be classified as normal thickness of macula or as experiencing abnormality (e.g. thickening and thinning). Unfortunately, the diagnostic images only represent the retinal thickness in several areas of the macular region. Thus, this study is aims to obtain the entire retinal thickness in the macula area from Status OCT's output images. Basically, the volumetric image is obtained by combining each of the six images. Reconstruction consists of a series of processes such as pre-processing, segmentation, and interpolation. Linear interpolation techniques are used to fill the empty pixels in reconstruction matrix. Based on the results, this method is able to provide retinal thickness maps on the macula surface and the macula 3D image. Retinal thickness map can display the macula area which experienced abnormalities. The macula 3D image can show the layers of tissue in the macula that is abnormal. The system built cannot replace ophthalmologist in decision making in term of diagnosis.

1. Introduction

Macular degeneration is a deterioration or breakdown of the eye's macula. The macula is a small area in the retina, the light sensitive tissue lining at the back of the eye. The macula is the part of the retina that is responsible for your central vision, allowing you to see fine details clearly. There are two types of macular degeneration: (1) dry macular degeneration with drusen; this condition is caused by aging and thinning of the tissues of the macula, and (2) wet macular degeneration, which occurs when abnormal blood vessels begin to grow underneath the retina, characterized by thickening of the retina.

Diagnosis of macular degeneration using a Stratus OCT with a fast macular thickness map (FMTM) method produced 6 macular cross-section images from different angles [1]. The images were converted into a retinal thickness chart to be evaluated using normal distribution percentile of data so that it can be classified as normal thickness of the macula or as experiencing abnormality (e.g. thickening and thinning). Unfortunately, the diagnostic images only represent the retinal thickness in several areas of the macular region, namely the area of data acquisition at a certain angle. Due to the limitated the data obtained were not able to provide data on the overall macular thickness deviation, and therefore, the proposed methods to reconstruct the 6 OCT output images into a 3D image. The aim of this study is to obtain the entire retinal thickness in the macula area from Status OCT's output images.

2. Method

This study uses a three-dimensional reconstruction method to obtain the image of the entire area of the macula (volumetric image). Input image that is used in the form of 6 macular cross-section images and 6 retinal thickness charts is obtained from 3 random samples. Basically the volumetric image is obtained by combining each of the six images to obtain three-dimensional images by using software MATLAB 2013a.

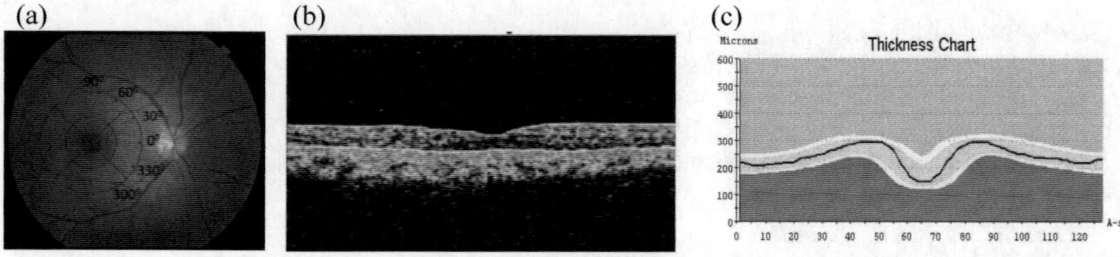

Figure 1. Left to right: (a) Data collection schemes of macular cross section images; (b) Macular cross section image; (c) Retinal thickness chart.

The reconstruction consists of a series of processes such as preprocessing, segmentation, and interpolation as described in flowchart below.

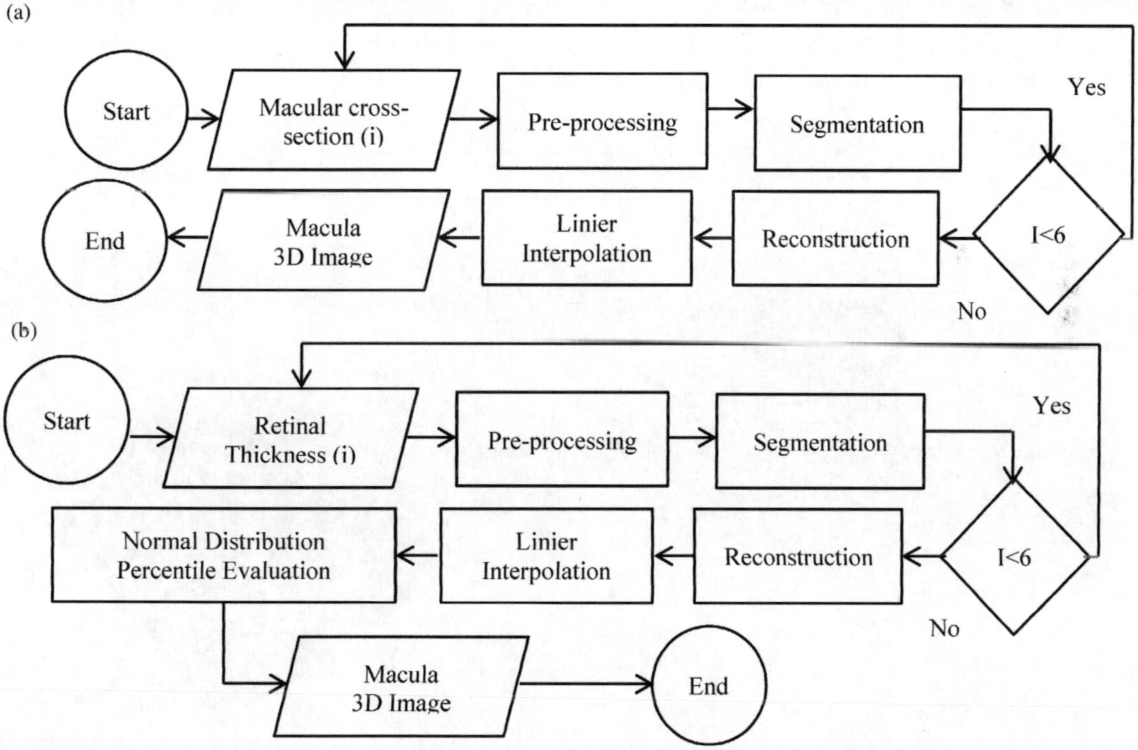

Figure 2. (a) Flowchart of macular cross-section processing; (b) Flowchart of retinal thickness processing.

2.1. Macular cross-section processing

Some The FMTM scan protocol was used for data acquisition. The protocol consists of 6 linear scans in a spoke pattern separated by 30° intervals centered at the fovea [1]. The line scans were 6 mm in the transverse direction, had a 2-mm axial depth, and was composed of 512 axial scans each [2-5]. In this case, linear interpolation techniques are used to fill the empty pixels (no value) in the three-dimensional matrix through MATLAB 2013a software using the following equation:

$$Z(r,\theta_1) = \frac{\sin\theta_1}{\sin 30} Z(r,0) + \frac{\sin(30-\theta_1)}{\sin 30} Z(r,30) \quad (1)$$

where Z (r, θ) is the pixels value, 30 is interval of current data in degree, (r, θ_1) is pixel coordinate.

2.2. Retinal thickness processing
Stratus OCT can change the macular cross-section image into retinal thickness chart. Then, this data is extracted to obtain the thickness value of retina at each angle. Finally, the data from previous process are then reconstructed by linear interpolation in equation (1), resulting surface topography image of the macula. Reconstructed images were evaluated by a healthy macula normative data percentile to provide visualization of macular abnormalities.

3. Result and discussion
This study has two forms of results: (1) retinal thickness map on the surface of the macula, (2) macula 3D image. Retinal thickness map that were evaluated by the normal data distributions can display the macula area which has abnormalities such as thickening or thinning of the macula. The macula 3D image can show the layers of tissue in the macula based on the increase in thickness. The high refraction index tissue is represented by bright colors.

Table 1. Macula 3D images and retinal thickness map on the surface of the macula as result of macular cross-section image and retinal thickness processing.

No	Three-dimension Image	
	Macular cross-section image processing	**Retinal Thickness Processing**
1.	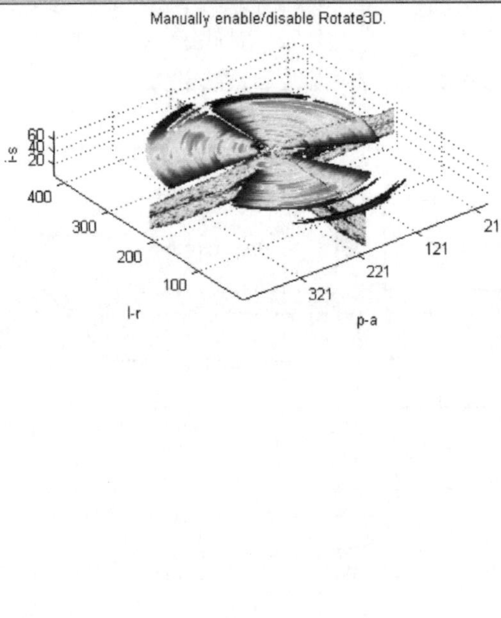	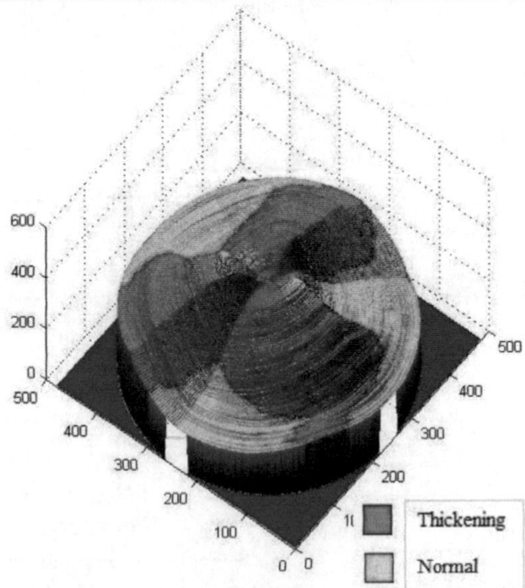 Thickening = 19.61 × 10^{-2} Thinning = 32.902 × 10^{-7}

2	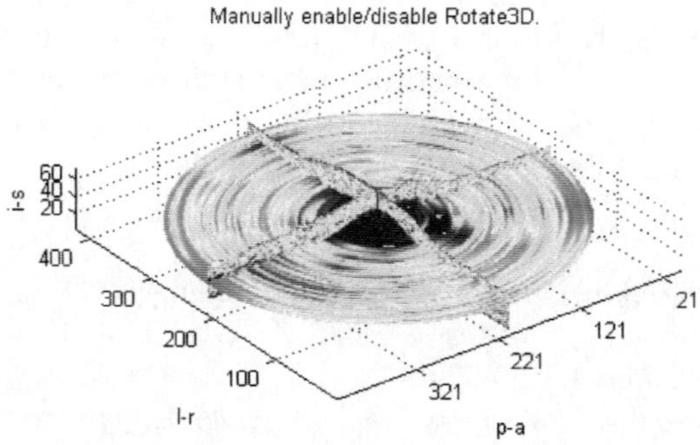	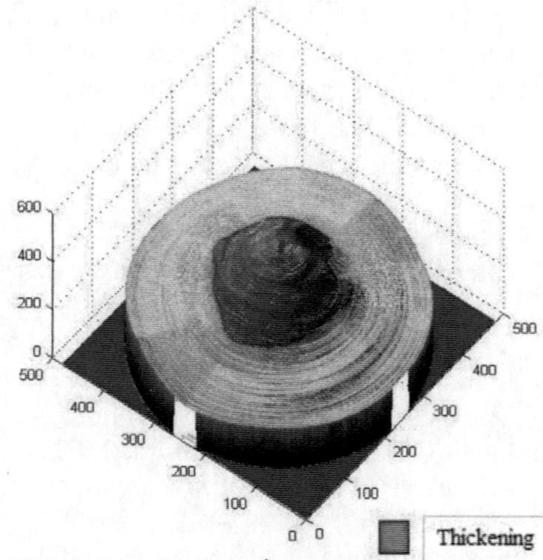 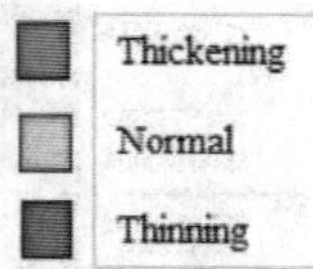 Thickening = 8.84×10^{-2} Thinning = 22.25×10^{-6}
3	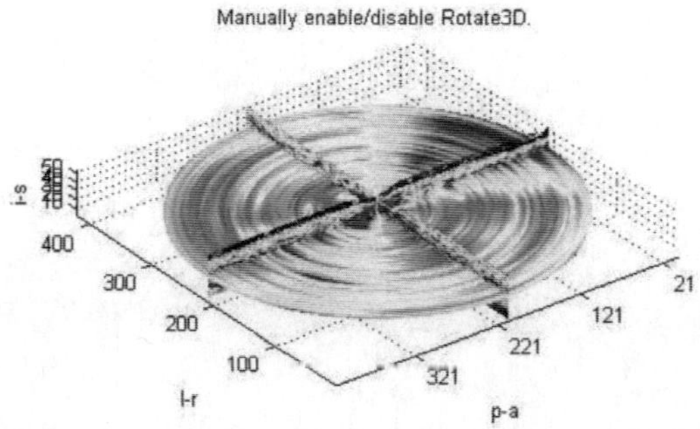	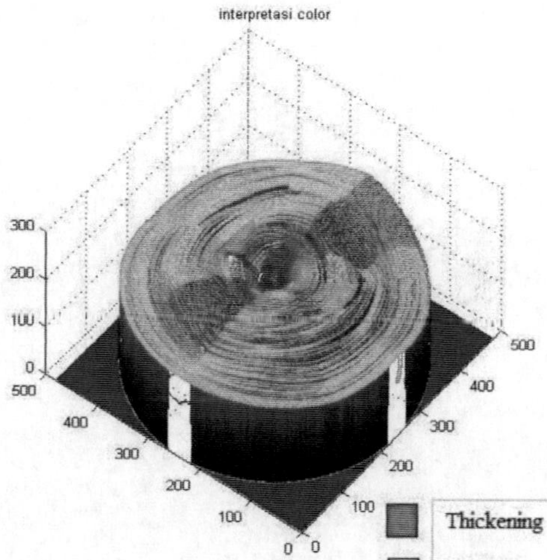 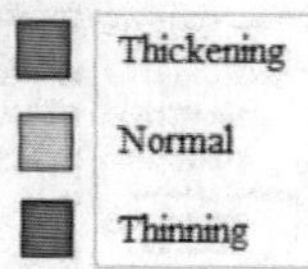 Thickening = 0.00 Thinning = 0.40×10^{-2}

Based on table 1, retinal thickness processing may provide an illustration of retinal thickness map of the macula surface. Then, macula normal distribution data contained in the Stratus OCT system was used to evaluate the retinal thickness map. Thus, the researchers can calculate the percentage of the thickening and thinning of the macula that have occurred in the sample per unit volume. The result of macular cross-section image processing can provide illustration layers of the macula. A high reflective tissue, such as the RNFL, appears green and yellow, whereas a less reflective tissue has darker colors such as blue [2].

4. Conclusion

The reconstruction method with linier interpolation is able to provide retinal thickness map on the surface of the macula and the macula 3D image. Retinal thickness map that were evaluated by the normal data distributions can display the macula area which is have abnormalities such as thickening or thinning of the macula. The macula 3D image can show the layers of tissue in an abnormal macula. However, the system built can't replace ophthalmologist and doctors in the decision making diagnosis.

References

[1] Peter M B *et al.* 2009 *The official journal of the international society for imaging in the eye* **40(1)** 25-31
[2] Gianmarco V *et al.* 2011 *Indian Journal of Ophthalmology* **59** 59-68
[3] Mark E B 2006 *Optical Coherence Tomography: Principles and Applications* (Boston, MA: Massachusetts Institute of Technology)
[4] James G F *et al.* 2000 *Neoplasia* **2** 9-25
[5] Michelle L G *et al.* 2011 *Investigative Ophthalmology & Visual Science* **52(5)**

Acknowledgments

The authors want to thank the other members of the biomedical imaging group Choirul Anam who indirectly supported the project as well as Erwin Iskandar from Department of Retina, Cicendo Eye Hospital, Bandung, Indonesia for supporting data rendering.

Journal of Physics: Conference Series **694** (2016) 012043 doi:10.1088/1742-6596/694/1/012043

Segmentation of solid subregion of high grade gliomas in MRI images based on active contour model (ACM)

P Seow[1, 2], Win M T[1, 2], J H D Wong[1,2], N A Abdullah[3] and N Ramli[1, 2]

[1] Department of Biomedical Imaging, Faculty of Medicine, University of Malaya, Kuala Lumpur, Malaysia
[2] University of Malaya Research Imaging Centre, Faculty of Medicine, University of Malaya, Kuala Lumpur, Malaysia
[3] Department of System and Computer Technology, Faculty of Computer Science and Information Technology, University of Malaya, Kuala Lumpur, Malaysia

E-mail: jeannie.wong@ummc.edu.my

Abstract. Gliomas are tumours arising from the interstitial tissue of the brain which are heterogeneous, infiltrative and possess ill-defined borders. Tumour subregions (e.g. solid enhancing part, edema and necrosis) are often used for tumour characterisation. Tumour demarcation into substructures facilitates glioma staging and provides essential information. Manual segmentation had several drawbacks that include laborious, time consuming, subjected to intra and inter-rater variability and hindered by diversity in the appearance of tumour tissues. In this work, active contour model (ACM) was used to segment the solid enhancing subregion of the tumour. 2D brain image acquisition data using 3T MRI fast spoiled gradient echo sequence in post gadolinium of four histologically proven high-grade glioma patients were obtained. Pre-processing of the images which includes subtraction and skull stripping were performed and then followed by ACM segmentation. The results of the automatic segmentation method were compared against the manual delineation of the tumour by a trainee radiologist. Both results were further validated by an experienced neuroradiologist and a brief quantitative evaluations (pixel area and difference ratio) were performed. Preliminary results of the clinical data showed the potential of ACM model in the application of fast and large scale tumour segmentation in medical imaging.

1. Introduction

Glioma is a type of tumour that initiates from the interstitial tissue of the brain. Astrocytic tumour can be grouped to diffuse astrocytoma, anaplastic astrocytomas or malignant glioblastoma multiforme (GBM) respectively labelled by grade II, III or IV according to the World Health Organization (WHO) guidelines [1], where high grade gliomas are grade III and IV. Glioma grading facilitates the management of the disease and establishes the prognosis of the patient. Glioma is characterised by heterogeneous neoplasms consisted of several subregions that include the solid part, cystic (necrotic) part, varying degrees of tumour infiltration in normal tissue as well as microvascular cellular proliferation and edema [2]. The enhancing solid subregion is the wanted part identified by the clinicians because it is the part where biopsy sites were taken as it is considered to resemble the histopathology of the brain tumour most closely. The growth of the solid subregion primarily comes from cellular proliferation [3]. Magnetic Resonance Imaging (MRI) is commonly used for biopsy management and

Journal of Physics: Conference Series **694** (2016) 012043 doi:10.1088/1742-6596/694/1/012043

characterisation of gliomas. A combination of selected MRI sequences (pre- and post-contrast T1-weighted (T1W), T2-weighted (T2W) and T1 Fluid-Attenuated Inversion Recovery (FLAIR)) is often employed for tumour subregions delineation [4]. Manual delineation of gliomas is not only time-consuming but also subjected to intra- and inter-observer variability which potentially leads to substantial inconsistency in the delineation of the region of interest (ROI). Furthermore, similar results are not reproducible even though for the same observer. These problems can be addressed by engaging automatic segmentation methods. Accurate segmentation is challenging due to high diversities of tumour properties and characteristics even within the same tumour grade, occurrence of heterogeneity within the tumour, unpredictable growth of the tumourous cells and vague boundaries. To date, there are many segmentation methods available for brain tumour and active contour model (ACM) is one of the most robust techniques [5].

2. Methodology

2.1. MRI data retrieval and pre-processing

The brain MRI data for pre- and post-gadolinium T1W, T2W and FLAIR sequences of four histologically proven high-grade gliomas (n=1 for grade III & n=3 for grade IV) were retrieved from the database after ethical approval was obtained. These glioma patients were scanned on a 3T MRI (Signa HDx, General Electric, USA). The pre- and post-contrast T1W MRI images were used for ACM segmentation where all the slices were gone through to obtain a single slice image with the largest region of the solid subregion while the T2W and FLAIR MRI images were used to assist the manual delineation. The solid subregion was determined as the portion of the tumour that exhibits isointense signal on T1W images and iso- to hyperintense signal on T2W images, with presence of tumour enhancement. The MRI images were pre-processed prior to segmentation by employing image subtraction followed by skull-stripping to remove the non-cerebral tissue region such as skull, scalp, and meninges.

2.2. Active contour model segmentation

Local region based ACM is preferable for medical image segmentation due to robustness to initialization and noise [6]. ACM for objects detection in an image is based on techniques of curve evolution and the level set method. The basic idea of ACM or snakes is to evolve a curve in a given image until it is guided to stop at the boundary of the desired object. Region based models exploit global region information by employing features such as intensity, colour, and texture. ACM is advantageous because it is able to detect objects whose boundaries are not necessarily defined by the gradient. The general formalism of the ACM model is based on the snake model [7] and is expressed as equation (1) [6]:

$$F(c_1, c_2, C) = \int_{\Omega_1=\omega} (u_0(x,y) - c_1)^2 dxdy + \int_{\Omega_2=\Omega-\omega} (u_0(x,y) - c_2)^2 dxdy + v|C| \quad (1)$$

where u_0 is the given image, $C = \delta\omega$ is the parameterized curve with c_1 and c_2 are the unknown constants. The first two terms are described as two forces. The goal is to minimize the energy forces with respect to the constants and curve so that contour C reaches its equilibrium. The first term is the force to shrink the contour while the second term is the force to expand the contour. These two forces get balanced when the contour arrives at the boundary of the desired object. The MRI images were retrieved into MATLAB R2013a (MathWorks, Massachusetts, USA) for ACM segmentation implementation which took around less than 20 minutes for solid subregion segmentation in one sequence of a patient.

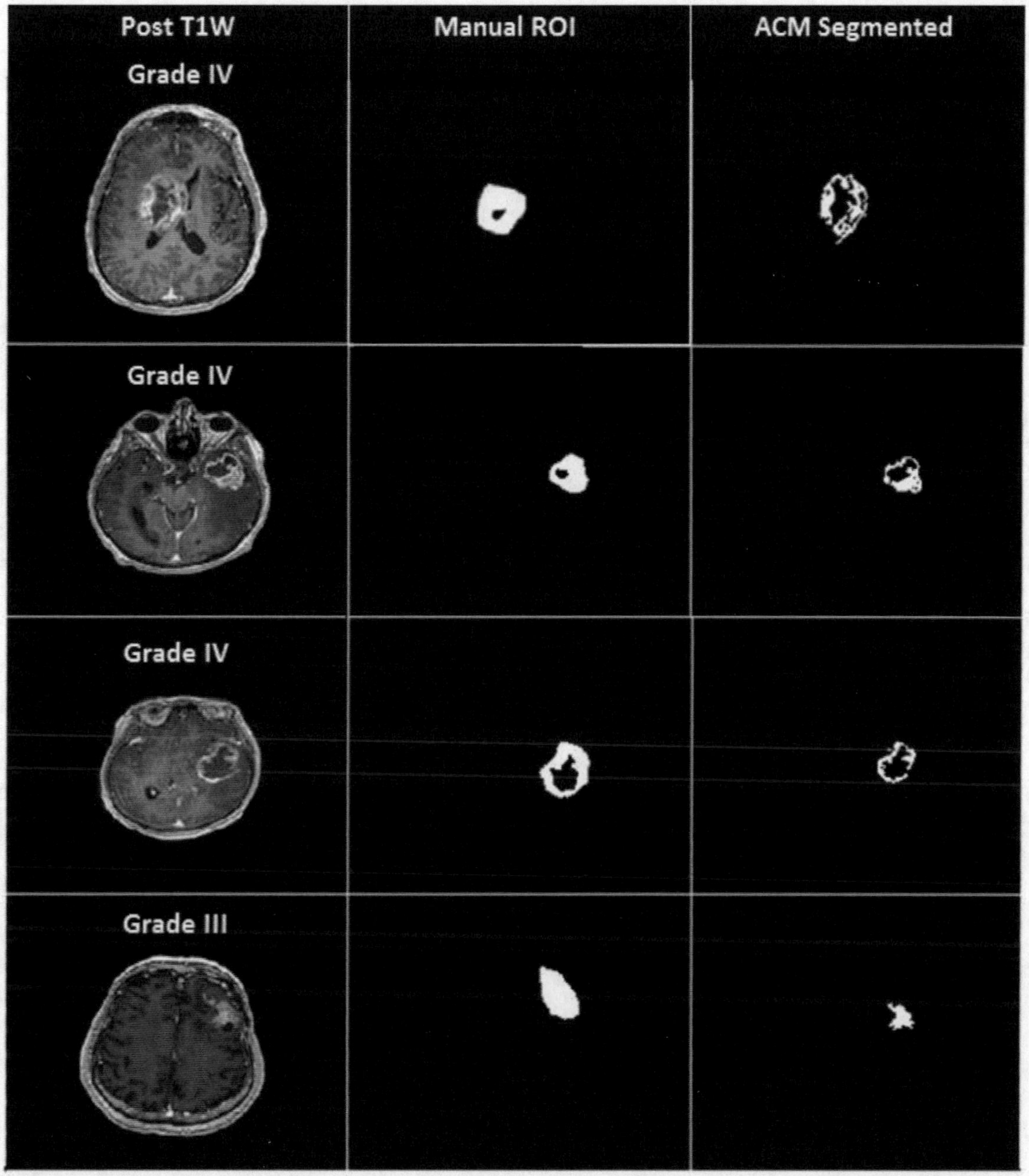

Figure 1. Post-gadolinium T1W MRI images of four high grade glioma patients (left), the manually delineated ROIs of the solid subregions by the radiologist (center) and the ACM segmented solid subregions of the tumour (right). The gliomas grade IV are glioblastoma multiforme (GBM) and the glioma grade III is anaplastic oligodendroglioma. The cystic portion of the grade III patient is too small to be delineated manually.

2.3. Comparison with manual segmentation

Manual delineations of the solid subregions were performed by a trainee radiologist (WMT) for comparison with the ACM segmented results. The average time spent for manual volumetric delineation of the tumour into its subregions (solid, cystic and edema part) was roughly 1 hour per sequence for one single patient.

2.4. Validation of segmented solid subregions

The manual delineated ROIs and ACM segmented solid subregions were randomized and presented to an experienced neuroradiologist (NR) who was blinded to the segmented results and the histological grades of the images. The neuroradiologist then decided on which segmented results that better represent the enhancing solid subregions. The preferred segmented results serve as the validation of the ACM model.

3. Results and discussion

Figure 1 shows better morphological structures on the solid subregions depicted using ACM segmentation when compared to the manually delineated ROI of the solid portions of all the four images. The segmented solid subregions showed clear boundaries between the solid subregions and the cystic subregions. The region based ACM model make use of the intensity characteristics emphasized from the subtraction process of the pre-contrast from the post-contrast T1W MRI image. Large differences for the pixel areas between the manual and automatic segmented results were obtained (table 1). The pixel area of the manually delineated ROIs were always larger than the ACM segmented solid subregions because the ROIs of the manual delineations are merely estimations of the subregions. Validation by the clinician showed that ACM segmentation was preferred compared to manual segmentation as the ACM segmented results were chosen to best represent the solid subregions in all the four images. Difference ratios describe about the dissimilarities between the manual and the automatic segmented subregions where high difference ratio indicates high dissimilarity between them. The segmentation results for grade III showed the highest difference ratio implying a large difference between the manual and ACM segmented subregions. The current ACM model can be improved for segmentation of cystic and edema subregions.

The subtraction process of the pre-contrast from the post-contrast MRI image facilitates the segmentation technique by accentuating the intensity characteristic of the solid subregion, a feature manipulated by the ACM model to carry out the segmentation of the solid portion of the tumour. The limitation of using the ACM model is that the initialization depends on the placement of the curve and the size of the initial contour. The model works better when the initial contour was placed nearer to the region of the solid portion of the tumour. Other than that, the ACM model does not work so well for the glioma cases which possess high degree of heterogeneity and weak borders between the tumour subregions. ACM segmentation is efficient for lesion segmentation since manual delineation could never achieve such detailed segmentation results that showed clear structures and border of the solid portion as ACM. Furthermore, ACM segmentation is much faster than manual segmentation. Precise manual delineation following the outlines of the enhancing solid subregion is time consuming, laborious and not practical in clinical practice.

Table 1. The validation measures of the four glioma patients for the segmentation of the solid subregions as compared to the manually delineated ROIs. Data 1 till 3 correspond to grade IV gliomas while data 4 refers to grade III glioma. Validation described the decision to choose which segmentation results that best represent the solid enhancing subregions based on the MRI images where value 1 means represent while value 0 means did not represent.

	Segmented Area (pixel number)		Validation by Neuroradiologist		Difference Ratio
	Manual	ACM	Manual	ACM	$\left(\frac{Manual-ACM}{ACM}\right)$
1	973	712	0	1	0.367
2	408	281	0	1	0.452
3	554	227	0	1	1.441
4	722	181	0	1	2.989

4. Conclusion

Preliminary results showed that the ACM model is feasible for time efficient and large scale segmentation of solid enhancing subregion from the tumour core of brain gliomas. The ACM method is a better method to represent the solid enhancing portion of the tumour when compared to manual delineation. Brain tumour segmentation has great impact on diagnosis, monitoring, treatment planning for patients and clinical trials. Improved accuracies and performance in the ACM model will favour the realization of computer aided detection (CAD) for detection and diagnosis of gliomas.

References

[1] Kleihues P, Louis D N, Scheithauer B W, Rorke L B, Reifenberger G, Burger P C and Cavene W K 2002 The WHO classification of tumors of the nervous system *J Neuropathol Exp Neurol* **61(3)** 215-25

[2] Swanson K R, Bridge C, Murray J D and Alvord Jr E C 2003 Virtual and real brain tumors: using mathematical modeling to quantify glioma growth and invasion *Journal of the Neurological Sciences* **216(1)** 1-10

[3] Guzman-De-Villoria J A, Mateos-Perez J M, Fernandez G P, Castro E and Desco M 2014 Added value of advanced over conventional magnetic resonance imaging in grading gliomas and other primary brain tumors *Cancer Imaging* **14(1)** 35

[4] Liu J, Li M, Wang J, Wu F, Liu T and Pan Y 2014 A Survey of MRI-Based Brain Tumor Segmentation Methods *Tsinghua Science and Technology* **19(6)** 578-95

[5] Gordillo N, Montseny E and Sobrevilla P 2013 State of the art survey on MRI brain tumor segmentation *Magnetic Resonance Imaging* **31(8)** 1426-38

[6] Tony F C and L A Vese 2001 Active Contour Without Edges *IEEE Transactions on Image Processing* **10(2)** 1-12

[7] Kass M, Witkin A and Terzopoulos D 1988 Snakes: Active contour model *International Journal of Computer Vision* **1(4)** 321-31

Acknowledgments

This work was supported by University of Malaya Research Grant (UMRG) (RP008D-13HTM). The authors gratefully acknowledge the essential contributions of the research staff of University Malaya Research Imaging Centre (UMRIC).

Journal of Physics: Conference Series **694** (2016) 012044 doi:10.1088/1742-6596/694/1/012044

Comparison of segmentation using fast marching and geodesic active contours methods for bone

A Bilqis[1] and R Widita[1]

[1] Department of Physics, Faculty of Mathematics and Natural Sciences, Institut Teknologi Bandung, Jalan Ganesha 10, Bandung 40132, Indonesia

E-mail: ayeshabilqis@gmail.com, rena@fi.itb.ac.id

Abstract. Image processing is important in diagnosing diseases or damages of human organs. One of the important stages of image processing is segmentation process. Segmentation is a separation process of the image into regions of certain similar characteristics. It is used to simplify the image to make an analysis easier. The case raised in this study is image segmentation of bones. Bone's image segmentation is a way to get bone dimensions, which is needed in order to make prosthesis that is used to treat broken or cracked bones. Segmentation methods chosen in this study are fast marching and geodesic active contours. This study uses ITK (Insight Segmentation and Registration Toolkit) software. The success of the segmentation was then determined by calculating its accuracy, sensitivity, and specificity. Based on the results, the Active Contours method has slightly higher accuracy and sensitivity values than the fast marching method. As for the value of specificity, fast marching has produced three image results that have higher specificity values compared to those of geodesic active contour's. The result also indicates that both methods have succeeded in performing bone's image segmentation. Overall, geodesic active contours method is quite better than fast marching in segmenting bone images.

1. Introduction

Image processing plays an important role in diagnosing diseases or damages of the human organs. One of the important stages of image processing is segmentation process. Medical image segmentation is used to simplify image by separating an image into regions of certain similar characteristics so that it is easier to analyze [1]. The case raised in this study is the image segmentation of bones. Bone's image segmentation is a way to get bone dimension, which is needed for making prosthesis to treat broken or cracked bones. Bone images are used in this study as the input for the segmentation process.

Image segmentation is generally based on two types of approaches, discontinuity and similarity [2]. The similarity approach separates image into regions that are similar in accordance with a set of pre-defined criteria. The discontinuity approach separates an image based on sudden changes intensity, such as edges in an image. The method that is used in this study is based on edges detection technique: Level Set method. Edge detection is an operation of detecting edges that confine two homogenous image regions which have different levels of brightness [3]. The aim is to improve the appearance of the boundary line of a particular object in the image. Level Set method can be used to track the evolution of contours and surfaces [4]. This method is chosen because it can do segmentation by forming deformable curves so that this method can be suitable for all geometric shapes. Two of several methods from level set segmentation used in this study is fast marching and geodesic active contours. Therefore, the aim is to compare segmentation results from both of the methods to analyze how successful these methods are in segmenting bone images. And after that, a better method will be known.

2. Methods

This study uses ITK (Insight Segmentation and Registration Toolkit) software with DICOM file as the input. DICOM data is segmented, using Fast Marching and Geodesic Active Contours method that is available in ITK's library, by inserting parameters required by the ITK. After the image segmentation result is obtained, the accuracy, sensitivity, and specificity values are calculated. The result will be compared with the original image using ROC (Receiver Operating Characteristics) measurement [5].

2.1. Fast marching method

Fast marching (FM) is a method that performs a propagation spread from a starting point to all possible directions [6]. Every time the propagation happens, this algorithm always calculates and stores the values of the distance of a point to the starting point. The illustration is shown on figure 1 below.

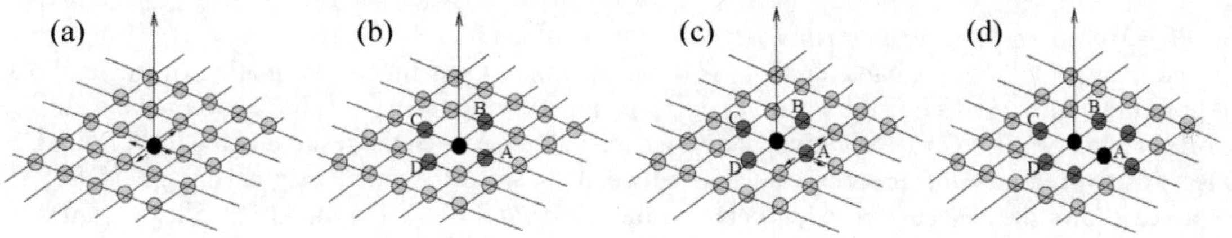

Figure 1. The illustration for overall process of fast marching method [7].

In figure 1(a), there is a black sphere, which marks the point of the initial value where u's value is known, and the gray spheres show the still-unknown u's value. When the FM algorithm starts the propagation from the origin point, it will calculate a possible new value in each of the four nearest neighbors (dark gray spheres A, B, C, and D), as seen on figure 1(b). The same rule also applies for the next steps. As seen on figure 1(c), consider sphere A has the smallest value. After that, three new possible values will be obtained in each of the three closest neighbors, as seen on Figure 1(d).

There are some parameters needed as the input for ITK. For the FM method, there is a *seed index* which must be specified by a user, stating the position of the starting point where the contour propagation process will begin. Then, there are σ as the smoothing factor value for the segmentation result's curve, α as an elasticity constant value of the segmentation result's curve, β as a flexibility constant value of the segmentation result's curve, *threshold* as scale to determine the grey level of the segmented sections, and the *stopping value* as a value to determine a time limit for the propagation spreads.

2.2. Geodesic active contours method

Active contours (AC) method produces a flexible curve that can adapt dynamically to the edges of the desired object [8]. Users should determine the initial contour estimation that almost matches the shape of the object feature, as seen in figure 2 below. Furthermore, the initial contour will be extracted toward the features in the image due to the influence of the internal energy that produces the image. The energy function used in the AC method consists of two components, which are the internal energy and the external energy. The energy minimization function can generally be expressed as in the following equation:

$$E^*_{snake} = \int_0^1 E_{snake}(v(s))ds = \int_0^1 \{E_{\text{int}}(v(s)) + E_{image}(v(s)) + E_{con}(v(s))\}ds \quad (1)$$

where E_{int} is the internal energy of the curve, E_{image} is the energy of the image, and the E_{con} is external energy. Internal energy makes compact curve (elastic force). While external energy tends to make the curve moves toward the object boundary.

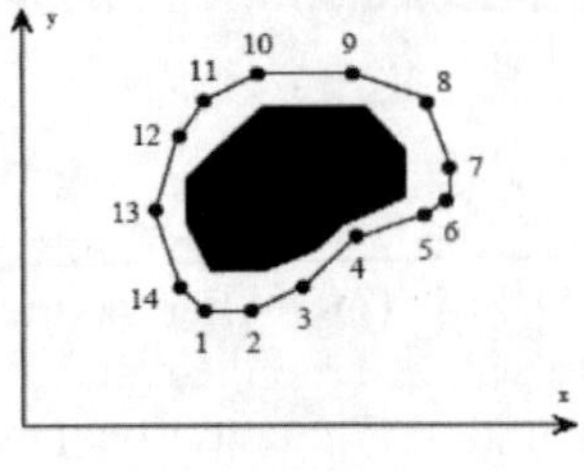

Figure 2. Illustration of geodesic active contours segmentation method, curve formed by point 1 to 14 is the initial contour [8].

There are some parameters needed as the input for ITK. For AC method, seed index, σ, α, and β parameters which show the same functions as those in the FM method. The different parameters are distance which declares the spread's distance and *propagation scaling* that states the scale of the segmentation's curve.

2.3. Receiver operating characteristics (ROC)

ROC (Receiver Operating Characteristics) is a measurement used for the evaluation of medical tests [9]. In figure 3, there is a TP (True Positive) component, showing the truth value between the result and the original images. TN (True Negative) shows a truth value between the result and the background. FP (False Positive) shows the inaccuracy value between the result and the original image. FN (False Negative) shows the inaccuracy value between the result and the background [5]. Those values are calculated based on the number of pixels covering the components.

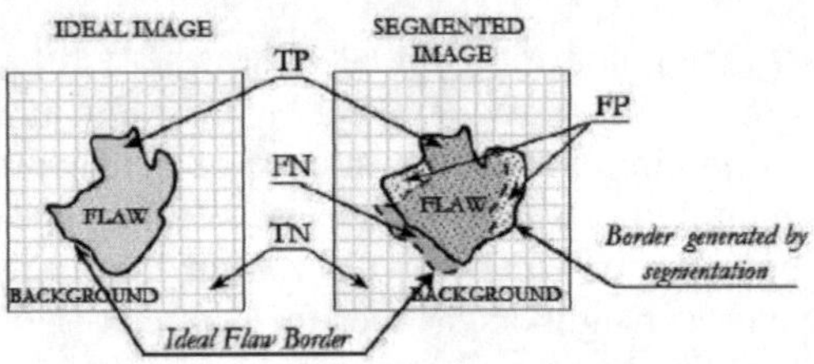

Figure 3. Illustration of regional division of TP, TN, FN, and FP on the original image and segmentation's image [9].

The calculation of accuracy, sensitivity, and specificity shown on these equations:

$$Accuracy = \frac{(TP+TN)}{(TP+TN+FP+FN)} \tag{2}$$

$$Sensitivity = \frac{TP}{P} = \frac{TP}{(TP+FN)} \tag{3}$$

$$Specificity = \frac{TN}{N} = \frac{TN}{(FP+TN)} \tag{4}$$

3. Result and discussion

Comparison of image segmentation based on the image results of both methods can be seen in table 1 below. Visually, it appears that the image segmentation results using the fast marching still have some hollows in them. For the edges, it does not look too much different from the image segmentation results using the Active Contours method. But it appears that there are some parts that are not filled on the results of the image.

The values of the accuracy, sensitivity, and specificity are shown in table 2 below. In this study, the accuracy values for the results with the AC method (from image 1 to 5) are 0.9932, 0.9984, 0.9993, 0.9903, and 0.9865. The accuracy values for the results using the FM method are 0.9878, 0.9968, 0.9984, 0.9714, and 0.9802. It shows that both of methods have succeeded in segmenting an image. Each accuracy value almost reaches a value of 1 (meaning 100%). However, the accuracy values using AC is slightly higher than FM's. In term of sensitivity values, the result of sensitivity shows how well the bone part being portrayed as bone on the segmentation image result is. The sensitivity values for the results using the AC method are 0.9416, 0.9016, 0.9617, 0.9490, and 0.8351. Meanwhile, the sensitivity values obtained using the FM method are 0.8864, 0.7866, 0.6885, 0.8486, and 0.7348. It appears that sensitivity values using the AC method are still slightly higher than those of the FM's.

Table 1. Comparison of the segmentation results by fast marching and geodesic active contours.

	Original image	Segmentation's image			
		Fast Marching		Geodesic Active Contours	
1	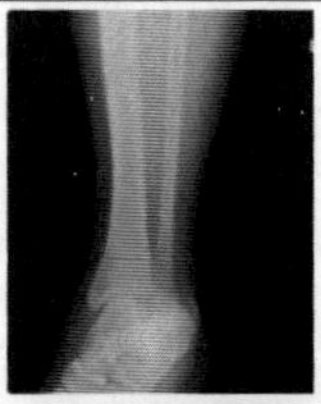	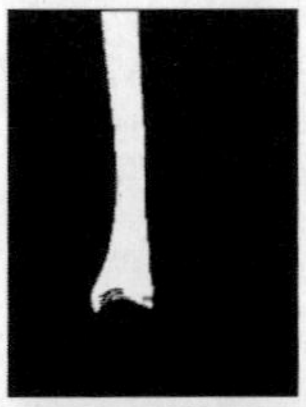	*Seed index* : (121, 171) σ : 0.5 α : -0.5 β : 3.0 *Threshold* : 200 *Stopping value* : 180	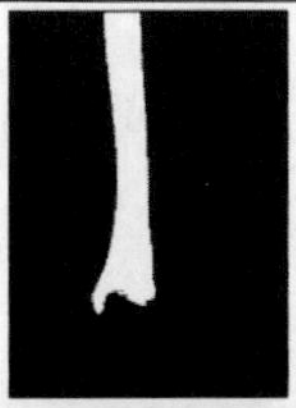	*Seed index* : (121, 171) *Distance* : 5.0 σ : 0.5 α : -0.5 β : 3.0 *Propagation scaling* : 2.0
2	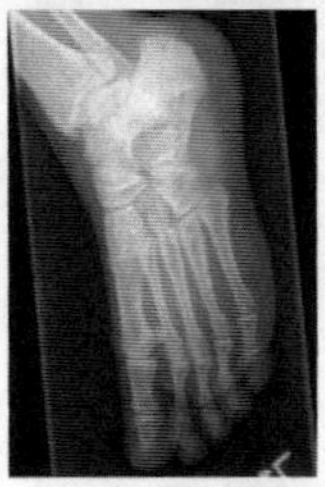		*Seed index* : (164, 171) σ : 0.5 α : -0.5 β : 3.0 *Threshold* : 100 *Stopping value* : 100		*Seed index* : (164, 171) *Distance* : 10.0 σ : 0.5 α : -0.5 β : 4.0 *Propagation scaling* : 1.0
3	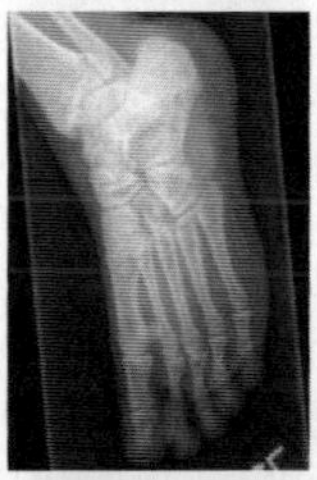	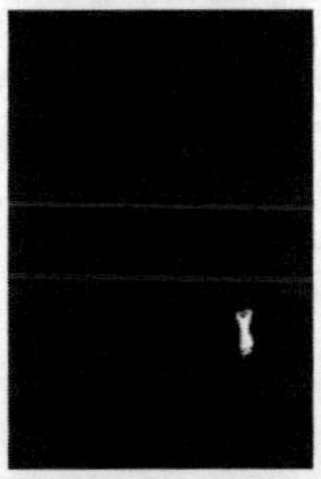	*Seed index* : (231, 309) σ : 0.5 α : -0.5 β : 3.0 *Threshold* : 250 *Stopping value* : 100	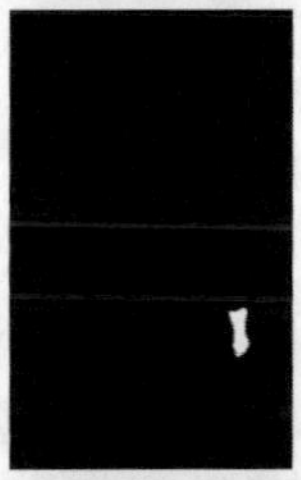	*Seed index* : (231, 309) *Distance* : 5.0 σ : 0.5 α : -0.5 β : 5.0 *Propagation scaling* : 1.0
4	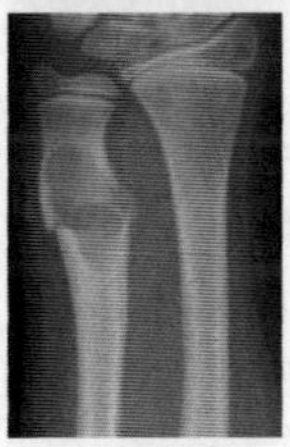		*Seed index* : (150, 263) σ : 0.5 α : -0.5 β : 3.0 *Threshold* : 5000 *Stopping value* : 300		*Seed index* : (150, 263) *Distance* : 15.0 σ : 0.5 α : -0.5 β : 8.0 *Propagation scaling* : 10.0
5	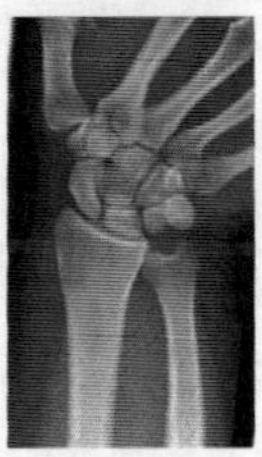	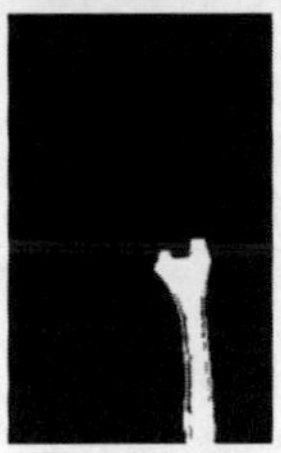	*Seed index* : (167, 341) σ : 0.5 α : -0.5 β : 3.0 *Threshold* : 5000 *Stopping value* : 800		*Seed index* : (167, 341) *Distance* : 10.0 σ : 0.5 α : -0.5 β : 6.0 *Propagation scaling* : 2.0

Specificity shows how well the non-bone part being portrayed as non-bone on the segmentation result is. For the specificity value, there is quite a difference from the two previous parameter values. The specificity values obtained by using the AC are 0.9984, 0.9998, 0.9994, 0.9999, and 0.9971. And the specificity values using the FM are 0.9981, 0.9999, 0.9997, 0.9998, and 0.9974. It can be seen that for the images of 1 and 4, the specificity values by using AC are still slightly higher than those using the

FM method. However, for the images of 2, 3, and 5, the specificity values using FM have slightly higher values than AC's.

Table 2. Data of accuracy, sensitivity and specificity.

				Fast Marching			
Picture	TP	TN	FP	FN	Accuracy	Sensitivity	Specificity
1	11057	123043	228	1416	0.9878	0.8864	0.9981
2	1519	130676	1	412	0.9968	0.7866	0.9999
3	378	132027	32	171	0.9984	0.6885	0.9997
4	15076	76752	11	2689	0.9714	0.8486	0.9998
5	5140	99822	255	1855	0.9802	0.7348	0.9974
				Geodesic Active Contours			
Picture	TP	TN	FP	FN	Accuracy	Sensitivity	Specificity
1	11745	123081	190	728	0.9932	0.9416	0.9984
2	1741	130660	17	190	0.9984	0.9016	0.9998
3	528	131992	67	21	0.9993	0.9617	0.9994
4	16860	76756	7	905	0.9903	0.9490	0.9999
5	5842	99788	289	1153	0.9865	0.8351	0.9971

4. Conclusion

The determination of the parameter values in the ITK needs a certain strategy. Users must insert these parameter values based on the experiments until the desired result is obtained. Based on the comparison from the accuracy, sensitivity, and specificity values, both methods Active Contours and Fast Marching have successfully segmented bone images. It can be seen from the values of accuracy, sensitivity, and specificity that the results almost reach the value of 1 (which means 100%). In this study, the segmentation results using active contours method have slightly higher accuracy and sensitivity values than those using fast marching's segmentation results. For the active contours based on the five sample images, the accuracy values are 0.9932, 0.9984, 0.9993, 0.9903, and 0.9865. The sensitivity values are 0.9416, 0.9016, 0.9617, 0.9490, and 0.8351. As for the value of specificity, fast marching produces three image results that have higher specificity values when compared with those of geodesic active contour's. Overall, the geodesic active contours is quite better than the fast marching in segmenting bone images.

References

[1] Pratt W K 1978 *Digital image processing: PIKS Inside* (New York: John Wiley & Sons)
[2] Gonzalez R C and Woods R E 2002 *Digital image processing* (New Jersey: Prentice Hall)
[3] Yunus M 2012 Perbandingan metode-metode edge detection untuk proses segmentasi citra digital *Jurnal Teknologi Informasi: Teori, Konsep, dan Implementasi* **3** 146–160
[4] Johnson H J, McCormick M M and Ibanez L 2014 *The ITK software guide book 2: Design and functionality fourth edition updated for ITK version 4.7*
[5] Metz C E 2006 Receiver operating characteristics analysis: A tool for the quantitative evaluation of observer performance and imaging systems *J. American College of Radiology* **3** 413–422
[6] Sethian J A 1999 *Level set methods and fast marching methods: Evolving interfaces in computational geometry, fluid mechanics, computer vision, and materials science* (Cambridge University Press)
[7] Sethian J A 1999 Fast marching method *SIAM Review* **41** 208–209
[8] Bakos M 2007 Active contours and their utilization at image segmentation *5th Slovakian-Hungarian Joint Symp. on Appl. Machine Intelligence and Informatics* (Poprad, Slovakia) pp 313–314
[9] Mardhiyah A and Harjoko A 2013 Metode segmentasi paru-paru dan jantung pada citra X-ray thorax *IJEIS – Indonesian J. of Electronics and Instrumentation Systems* **1** 41

Journal of Physics: Conference Series **694** (2016) 012045 doi:10.1088/1742-6596/694/1/012045

Contrast medium usage reduction in abdominal computed tomography by using high-iodinated concentration contrast medium

A Suwannasri[1,2], R Kaewlai[2] and S Asavaphatiboon[2]

[1] School of Medical Physics, Faculty of Medicine, Ramathibodi Hospital, Mahidol University, Bangkok, Thailand
[2] Department of Diagnostic and Therapeutic Radiology, Faculty of Medicine, Ramathibodi Hospital, Mahidol University, Bangkok, Thailand

E-mail: apisit_de_arsenal@hotmail.com

Abstract: This study was to determine if administration of a low volume high-concentration iodinated contrast medium can preserve image quality in comparison with regular-concentration intravenous contrast medium in patient undergoing contrast-enhancement abdominal computed tomography (CT). Eighty-four patients were randomly divided into 3 groups of similar iodine delivery rate; A: 1.2 cc/kg of iomeprol-400, B: 1.0 cc/kg of iomeprol-400 and C: 1.5 cc/kg of ioversol-350. Contrast enhancement of the liver parenchyma, pancreas and aorta was quantitatively measured in Hounsfield units and qualitative assessed by a radiologist. T-test was used to evaluate contrast enhancement, and Chi-square test was used to evaluate qualitative image assessment, at significance level of 0.05 with 95% confidence intervals. There were no statistically significant differences in contrast enhancement of liver parenchyma and pancreas between group A and group C in both quantitative and qualitative analyses. Group C showed superior vascular enhancement to group A and B on quantitative analysis.

1. Introduction

Clinical use of multidetector computed tomography (MDCT) has increased in the past decades because of its accuracy in diagnosis of many diseases, fast acquisition and wide availability. In many CT studies, particularly abdominal CT, intravenous iodinated contrast medium administration is essential to improve image quality and to enhance image interpretation. However, contrast medium poses additional risk of contrast-induced nephropathy (CIN) because the development of CIN is dependent on the volume of administered iodinated contrast medium [4], reduction of the contrast medium volume could potentially help alleviating this phenomenon. Reduction of iodinated contrast volume needs to be balanced with the degree of parenchymal organ and vascular enhancement, given that they are proven to be directly dependent on each other [5]. Fenchel S *et. al.* [7] reported that the use of a high-concentration iodinated contrast medium, such as iomeprol-400, could help reduce the contrast volume while preserving the degree of contrast enhancement. Because contrast medium volume reduction is preferable to reduce the risk of CIN, the objective of this investigation was to compare image quality, in terms of parenchymal and vascular enhancement, between administration of low-volume iomeprol-400 and regular-volume ioversol-350 using weight-based injection protocols [2, 3].

2. Materials and Methods

2.1. Patients

This prospective investigation was approved by our hospital's Ethics Committee. All consecutive patients undergoing contrast-enhanced abdominal CT at the accident and emergency center, Ramathibodi Hospital, who gave informed consents, were prospectively enrolled in the investigation. Patients with a history of diabetes, pregnancy or nursing women, asthma, hypersensitivity to iodinated contrast medium, hyperthyroidism, eGFR lower than 60 ml/min/1.73 m^2, were excluded. A total of eighty-four patients (40 men; mean age 57 years) was randomly assigned to three groups, consisting of 28 in each group of different volume and concentration of intravenous contrast administration.

- Group A: 1.2 ml/kg of iomeprol-400 (equal to 480mgI/kg).
- Group B: 1.0 ml/kg of iomeprol-400 (equal to 400 mgI/kg).
- Group C: 1.5 ml/kg of ioversol-350 (equal to 525 mgI/kg).

At our institution, the standard technique for administering intravenous contrast medium in abdominal CT is to use weight-based dosage [1]. Because the total amount of iodine that is administered to a patient divided by patient's weight (mgI/kg) for optimal parenchymal enhancement is between 500 and 700 mgI/kg [6], our standard dose delivers a weight-base dose of 1.5 ml/kg of a low-osmolar contrast medium with concentration between 350 and 370 mgI/ml (equal to at least 525 mgI/kg). Ten-percent reduction of iodine administration below the standard dose was equal to 1.2 cc/kg iomeprol-400 at 480 mgI/kg (Group A). Group B represented a 23% reduction of iodine administration of the standard dose.

2.2. Imaging protocol

All scans were performed on a 64-row multidetector CT scanner (Toshiba Aquilion CX). Routine abdominal CT protocol included venous and delayed acquisitions. Scanning parameters for the venous phase were 120 kVp automatic tube current modulation (SD 15, Sure Exposure 3D) with minimum and maximum mA of 40 mA and 440 mA, rotation time 0.6s, pitch factor 0.906, collimation 0.5×64. Automatic contrast medium-injector was used to deliver intravenous contrast medium IV at a flow-rate of 2.6 ml/s, followed by 20 ml of normal saline solution at a flow rate 2.0 ml/s. The venous-phase images were acquired at 70s after contrast administration.

2.3. Image evaluation

Quantitative assessment of enhancement was performed by measuring CT attenuation in Hounsfield Unit with a 1-cm-diameters circular regions of interest (ROI) on the images viewed on Picture Archiving and Communication System (PACS). The ROIs were placed at the right and left liver lobes excluding hepatic vessels, abdominal aorta and pancreas on venous phase images. Qualitative assessment was performed by an experienced radiologist as a blind reader for overall image enhancement and evaluation of vessels. Image quality grading was performed using a 5-point scale (5=excellent, 4=good, 3=sufficient, 2=insufficient, 1=poor) Statistical analysis was performed using Predictive Analytic SoftWare (PASW) Statistics 18 at a significance level of 0.05 with 95% confidence intervals, by comparing group by group (group A vs group C, group B vs group C, group A vs group B). Independent T-test was to compare contrast enhancement and Chi-squared test for qualitative analysis.

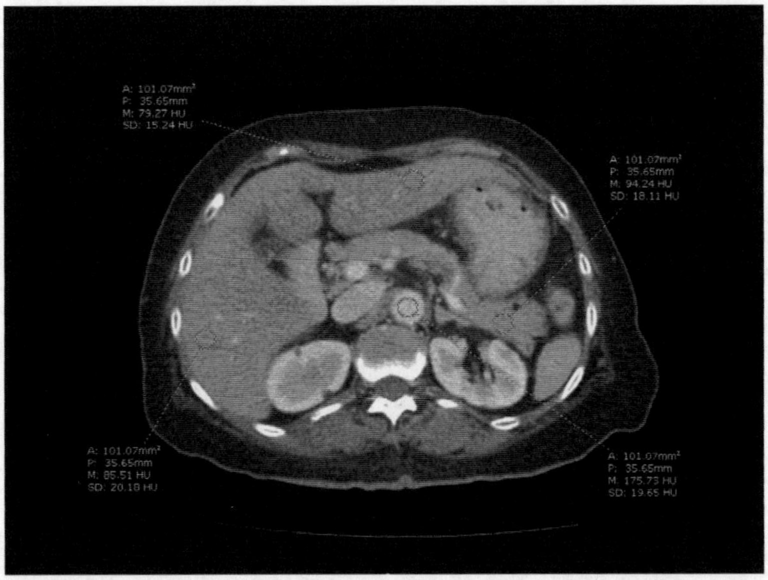

Figure 1. The ROIs were placed at the right and left liver lobes excluding hepatic vessels, abdominal aorta and pancreas on venous phase images.

3. Results

3.1. Quantitative analysis

The average HUs (±standard deviation) of the liver parenchyma, pancreas, and aorta were shown in table 1. Between group A and group C, there was no statistical difference of enhancement of the right-lobe liver parenchyma (97.9±13.4HU vs 101.5±13.2, p = 0.321), left-lobe liver parenchyma (101±15.5HU vs 107.7±12.8HU, p = 0.084) and pancreas (92.5±9.8HU vs 96.8±13.7HU, p = 0.193). However, there was a significantly difference in the attenuation of the aorta (144.9±21.6HU vs 164.5±28.7HU, p = 0.006). Between group B and group C, there was a significantly difference in all parts ($p < 0.05$). Between group A and group B, there was no statistical difference of the attenuation of the aorta (144.9±21.6HU vs 135.8±15.7HU, p = 0.078) and pancreas (92.5±9.8HU vs 86.4±14HU, p = 0.069). There was a significant difference between the right-lobe liver parenchyma (97.9±13.4HU vs 90.8±9.3HU, p = 0.001) and left-lobe liver parenchyma (101±15.5HU vs 93.3±9.3HU, p = 0.0001).

Table 1. Comparison of contrast enhancement in venous phase: group A (Iomeprol400 1.2ml/kg) v.s. group C (Ioversol350 1.5 ml/kg), group B (Iomeprol400 1.0 ml/kg) vs group C and group A vs group C, using independent t-test.

Quantitative Analysis	Average HU			*p*-value		
	Group A	Group B	Group C	A vs C	B vs C	A vs B
Right-Lobe Liver Parenchyma	97.9±13.4	90.8±9.3	101.5±13.2	0.321	0.001	0.026
Left-Lobe Liver Parenchyma	101±15.5	93.3±9.3	107.7±12.8	0.084	0.000	0.027
Aorta	144.9±21.6	135.8±15.7	164.5±28.7	0.006	0.000	0.078
Pancreas	92.5±9.8	86.4±14.4	96.8±13.7	0.193	0.008	0.069

3.2. Qualitative analysis

None of all groups contained 5-point scales of 5 and 1(table 2). Statistical calculation of all group had p-value > 0.05(table 3) that denoted no statistical difference in overall image enhancement and vessels.

Table 2. Total 5-point scale of all groups.

Qualitative Analysis	Group A			Group B			Group C		
	4	3	2	4	3	2	4	3	2
Overall image enhancement	18	10	0	18	8	2	17	10	1
Evaluation of vessels	21	7	0	19	9	0	21	7	0

Table 3. Group by group of statistical comparison using Chi-squared test.

Qualitative Analysis	p-value		
	A vs C	B vs C	A vs B
Overall image enhancement	0.598	0.747	0.329
Evaluation of vessels	1.000	0.554	0.554

4. Conclusion

Low volume of iomeprol-400 could provide parenchymal image quality similar to regular-volume ioversol-350 on both qualitative and quantitative assessment. For vascular enhancement, regular-volume ioversol-350 still demonstrated superior enhancement on quantitative analysis but not on qualitative assessment.

References

[1] Alec J M, Ginette J, Jay P H, Eric K P, Kenneth D H, Gregory S, Sanjay S, Bernard A B, Richard R and Eliot K F 2001 Quantitative and qualitative evaluation of volume of low osmolality contrast medium needed for routine helical abdominal CT *AJR.* **176** 583

[2] Liu Y, Xu X Q, Lin X Z, Song Q and Chen K M 2010 Prospective study comparing two iodine concentrations for multidetector computed tomography of the pancreas *Radiol. Med.* **115** 898

[3] Lomano L, Grazioli L, Bonomo L, Xu J R, Chen K M, Dore R, Vanzulli A and Catalano C 2009 Enhancement and safety of iomeprol-400 and iodixanol-320 in patients undergoing abdominal multidetector CT *BJR.* **82** 204

[4] Rajan K G and Tami J B 2010 Prevention of contrast-induced nephropathy (CIN) in interventional radiology practice *Semin. Intervent. Radiol.* **27** 348

[5] Francesco P, Isabella G, Lorenzo B, Andrea M, Athena G P, Ludovica R, Cristina P, Francesca M, Matteo P and Gian A R 2014 CT of the abdomen: Degree and quality of enhancement obtained with two concentrations of the same iodinated contrast medium with fixed iodine delivery rate and total iodine load *European Journal of Radiology* **83** 1995

[6] Marco R *et al.* 2012 High concentration (400 mgI/mL) versus low concentration (320 mgI/mL) iodinated contrast media in multi detector computed tomography of the liver: A randomized, single centre, non-inferiority study *European Journal of Radiology* **81** 3096

[7] Fenchel S, Fleiter T R, Aschoff A J, Gessell R V, Brambs H J and Merkle E M 2004 Effect of iodine concentration of contrast media on contrast enhancement in multislice CT of the pancreas *BJR* **77** 821

Acknowledgment

I would like to thank all radiological technologists and other staff in emergency radiological imaging center, department of diagnostic and therapeutic radiology, faculty of medicine, Ramathibodi Hospital, for their kind support of this work.

Measurement of computed tomography dose profile with pitch variation using Gafchromic XR-QA2 and thermoluminescence dosimeter (TLD)

S Purwaningsih[1], **L E Lubis**[1], **S A Pawiro**[1] **and D S Soejoko**[1]

[1] Department of Physics, Faculty of Mathematics and Natural Sciences, Universitas Indonesia, Depok, Indonesia

E-mail: septi.physics06@gmail.com

Abstract. This research was aimed to check the patterns of dose profile on adult and pediatric head scan. We compared measurement result on dose profile along the z- axis rotation at peripheries and center phantom with a variety of pitch, i.e. 0.75, 1, 1.5 for adult and pediatric head protocol, keeping the rest of the scan parameters constant. Measurements were performed on homogeneous, cylindrical PMMA phantom with diameters of 16 and 10 cm using XR-QA2 Gafchromic film and TLD as dosimeters. The measurement result indicated a decrease in the dose about 50% and 47% for adult and pediatric head scan with the increase of pitch. For 0.75 value of pitch adult head scan, dose range for each position were (2.4 – 5.0) cGy, (3.1 – 5.3) cGy, (2.2 – 4.5) cGy, (2.8 – 5.3) cGy, and (3.3 – 5.6) cGy for position of center, 3, 6, 9 and 12 o'clock peripheral phantom position respectively. Dose profile for adult and pediatric head scan protocols has pattern curve with the maximum dose in the middle and tendency of symmetry near the edges, with different the plateau length along z- axis direction in accordance to the measurement position in the phantom.

1. Introduction

CT-scan is an important part in diagnostic radiology. CT-scan is used in a variety of clinical applications, for example, intra-cranial examination in neurology, diagnostics in oncology, staging and treatment planning in radiotherapy. In addition, CT-scan is also used for cardiology, angiography, virtual endoscopy and as image-guidance in interventional procedures [1].

CT-scan can perform three dimension image, but it gives the patient dose 10-50 times more than the radiation of conventional X-ray [2]. Optimization of radiation protection and safety is important in order to get the low dose for patient but diagnostic purpose can be still achieved. The importance of information dose received by patients, measurements were taken directly along z-axis rotation of the scan, so we can know how the actual dose profile. Dose profile along z-axis rotation of the scan can be measured using TLD and film. Related research dose profile has not been done. Measurement of dose profile required dosimeter which can measure radiation dose along z-axis rotation of the scan. In 2011 Mourão, Gonçalves and Alonso have done dose profile measurements on the CT-scan. They used Toshiba Aquilion CT-scan and protocol PMMA head phantom with 16 cm in diameter and 15 cm in length. Measurements using Strip Gafchromic-XR film. Research results for all variations of pitch 0.8, 1 and 1.5 higher doses obtained on all edges than dose in center phantom position. For a tube voltage of 120 kV dose values obtained at the edge of 31% greater than the dose in the phantom center [3].

The difference with previous research, to get the profile dose more accurate in this study performed by variation dosimeter using Gafchromic XR-QA2 film and rod TLD, than using polymethyl methacrylate (PMMA) cylinder 16 cm in diameter which represents the adult head and 10 cm represents the pediatric head. Profile doses of both phantom and dosimeters will be analyzed. Dose profile obtained results are expected to complete the dose received information useful for patient and clinic.

2. Methodology

Dose profile measurement using film gafchromic performed on two variations of PMMA phantom, both 16 cm in diameter for adult head and 10 cm in diameter for pediatric head, variations of pitch 0.75; 1 and 1.5, as well as variations in the position of the phantom center, with four position clockwise: 3, 6, 9 and 12 o'clock position were shown in figure 1. The Gafchromic film placed crosswise in the middle of the shell phantom was shown in figure 2. After exposure to radiation is read using a scanner Gafchromic. The film was scanned using EPSON V700 scanner and then was evaluated using ImageJ software.

For the calibration process, annealing TLD advance in Medical Physics Laboratorium Universitas Indonesia with a temperature 400° C for one hour followed by oven 100° C for three hours. Than, TLD exposed to dose of 6 cGy, 120 kV x-ray BATAN. Radiation processes carried out by the SSD 100 cm in the air. After exposure, the TLD was saved approximately 1x24 hours to avoid instability in TLD readings. TLD was read using a Harshaw TLD reader. The readings obtained value in μC units, then TLD sorted from smallest to largest value reading. TLD grouped into three groups and each group has one calibration factor value. TLD dose profile measurement only performed at the center position for measurement adult head phantom with 0.75 and 1 pitch value was given in figure 3.

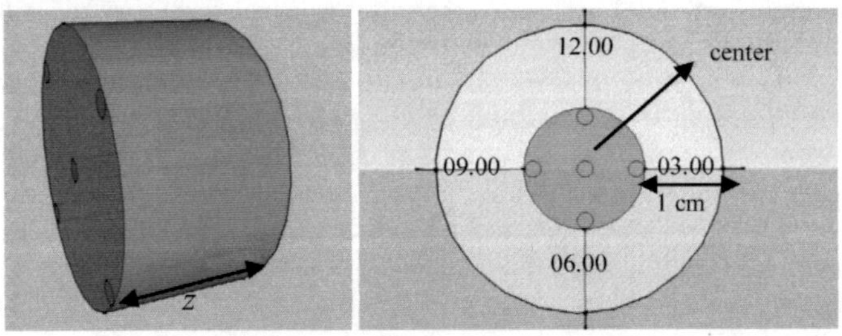

Figure 1. PMMA phantom.

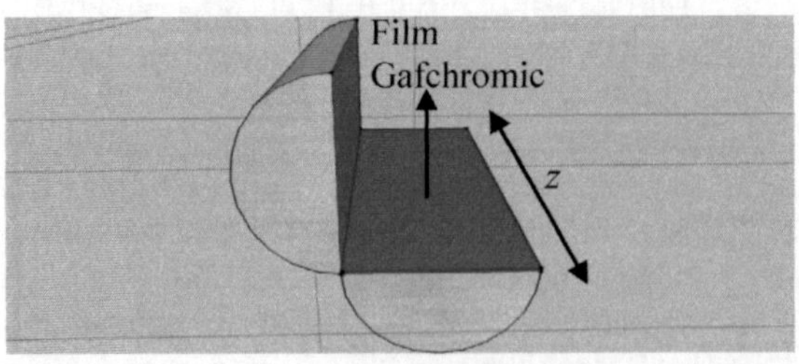

Figure 2. Position of Gafchromic film in holder at PMMA phantom.

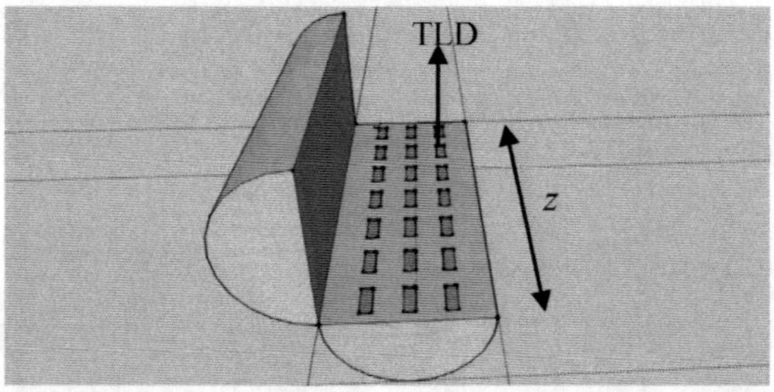

Figure 3. Position of TLD in holder at PMMA phantom.

3. Results and discussion

Dose obtained from TLD measurement values tended to be discrete, having a distribution and dose values being almost the same with measurement results in XR-QA2 Gafchromic film. A similar pattern was observed between pediatric and adult head scan protocols, with the maximum dose in the middle and tendency of symmetry near the edges was shown in figure 4.

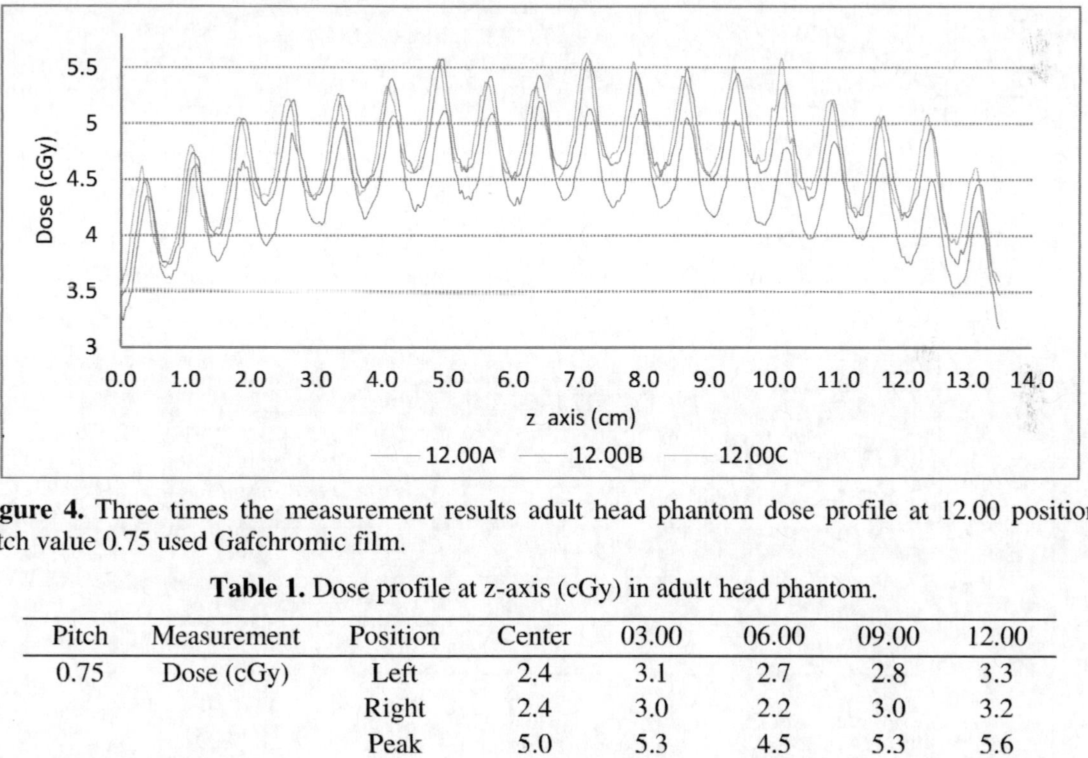

Figure 4. Three times the measurement results adult head phantom dose profile at 12.00 position, pitch value 0.75 used Gafchromic film.

Table 1. Dose profile at z-axis (cGy) in adult head phantom.

Pitch	Measurement	Position	Center	03.00	06.00	09.00	12.00
0.75	Dose (cGy)	Left	2.4	3.1	2.7	2.8	3.3
		Right	2.4	3.0	2.2	3.0	3.2
		Peak	5.0	5.3	4.5	5.3	5.6
1	Dose (cGy)	Left	1.3	2.2	1.4	2.1	2.1
		Right	1.5	2.2	1.5	2.2	2.2
		Peak	3.8	3.9	3.3	3.7	3.7
1.5	Dose (cGy)	Left	1.1	1.1	1.1	1.2	1.2
		Right	1.2	1.1	1.0	1.0	1.2
		Peak	2.6	2.9	2.4	2.8	3.0

Table 2. Dose profile at z-axis (cGy) in pediatric head phantom.

Pitch	Measurement	Position	Center	03.00	06.00	09.00	12.00
0.75	Dose (cGy)	Left	1.6	2.0	2.0	1.9	2.3
		Right	1.6	2.0	1.8	2.0	2.2
		Peak	3.3	3.6	3.2	3.5	3.8
1	Dose (cGy)	Left	1.1	1.2	1.2	1.4	1.3
		Right	1.2	1.5	1.2	1.4	1.6
		Peak	2.8	2.8	2.6	2.8	3.0
1.5	Dose (cGy)	Left	0.9	0.9	0.8	0.8	0.9
		Right	0.9	0.9	0.9	0.9	0.9
		Peak	1.7	1.7	1.5	1.7	1.8

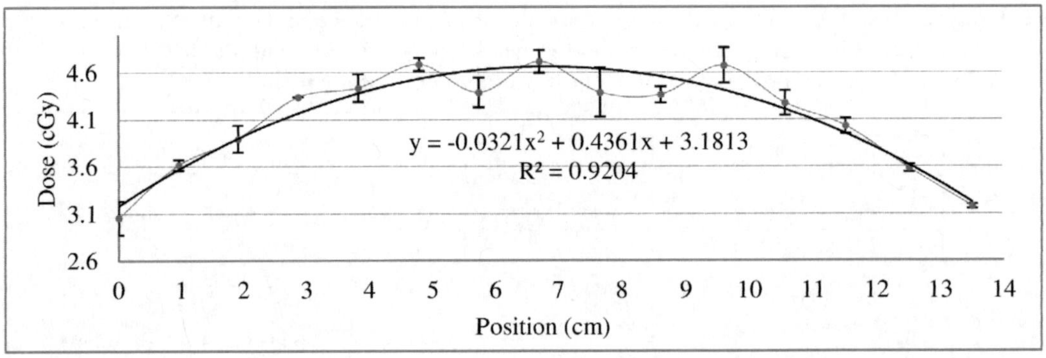

Figure 5. Dose profile of TLD at center adult head phantom with 0.75 pitch using TLD.

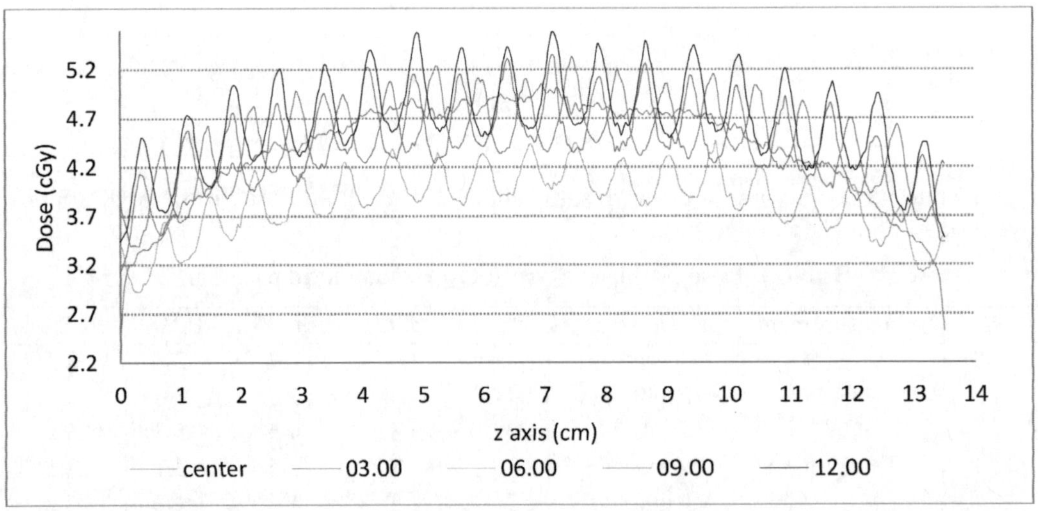

Figure 6. Dose profile of adult head phantom with 0.75 pitch using Gafchromic film.

For the adult head scan, dose range for each pitch values were (2.4-5.0) cGy, (1.3-3.8) cGy and (1.1-2.6) cGy for pitch values of 0.75, 1 and 1.5 respectively. For the pediatric head scan, dose were in the ranges of (1.6-3.3) cGy, (1.1-2.8) cGy and (0.9-1.7) cGy for the pitch values of 0.75, 1 and 1.5, respectively. A radiation dose of the scan was found proportional to the pitch value. For 0.75 value of pitch adult head scan, dose range for each position were (2.4 – 5.0) cGy, (3.1 – 5.3) cGy, (2.2 – 4.5) cGy, (2.8 – 5.3) cGy, and (3.3 – 5.6) cGy for position of center, 3, 6, 9 and 12 o'clock peripheral phantom position respectively. While. for 0.75 value of pitch pediatric head scan. The dose was in the ranges of (1.6 – 3.3) cGy, (2.0 – 3.6) cGy, (1.8 – 3.2) cGy, (1.9 – 3.5) cGy, and (2.2 – 3.8) cGy for position of center, 3, 6, 9 and 12 o'clock peripheral phantom position respectively. Radiation dose of the scan was found to be maximum at the 12 o'clock position and minimum at 6 o'clock position.

4. Conclusion

The measurement results indicated a decrease in the dose about 50% and 47% for adult and pediatric head scan with the increase of pitch. A radiation dose of the scan was found to be maximum at the 12 o'clock position and minimum at 6 o'clock position. For each position different length of the plateau was found due to the different amount of scattered radiation received. Dose profile for adult and pediatric head scan protocols has pattern curve with the maximum dose in the middle and tendency of symmetry near the edges with different plateau length along z- axis direction in accordance to the measurement position in the phantom.

References

[1] The imPACT Group 2009 *Buyer's Guided: Multi-Slice CT Scanners.* (*March* 2009). St. George's Healthcare Trust. The ImPACT Group

[2] Cederquist Björn 2008 *Evaluation Of Two Thin CT Dose Profile Detectors and A New Way To Perform Qa In A CTDI Head Phantom.* Department of Radiation Physics Göteborg University

[3] Mourão A P, Gonçalves Jr R G and Alonso T C 2011 Dose Profile Variation with Pitch in Head CT Scans Using Gafchromic Films *Recent Advances in Biomedical & Chemical Engineering and Materials Science*

Acknowledgements

The authors expressed special thanks to Mrs. Djarwani S.S, Mr. Supriyanto A. Pawiro, Mr. L. Evan L, and the staff at Department of Radiotherapy, Cipto Mangunkusumo Hospital.

Journal of Physics: Conference Series **694** (2016) 012047 doi:10.1088/1742-6596/694/1/012047

Validation of a paediatric thyroid phantom using different multidetector computed tomography models

M Alsabbagh[1], L Y Ng[1], A A Tajuddin[1, 2], M A Manap[1] and R Zainon[1]

[1] Advanced Medical and Dental Institute, Universiti Sains Malaysia, Bertam 13200 Kepala Batas, Pulau Pinang, Malaysia
[2] School of Physics, Universiti Sains Malaysia, 11800 Pulau Pinang, Malaysia

E-mail: moayyad@alsabagh.net

Abstract. The aim of this study was to compare the attenuation values of a fabricated paediatric thyroid phantom material using different MDCT models. A paediatric thyroid phantom was designed to mimic the shape and size of a paediatric patient with an age of 9 years using high-density Polyethylene as the phantom material. The fabricated phantom was scanned using two different multidetector CT scanners (16- and 128-row detectors). The CT numbers were evaluated and the mass attenuation coefficients (μ/ρ) of the phantom material were obtained at each applied energy from each scanner. The results were compared with the tables of the National Institute of Standards and Technology (NIST). The CTs of 16- and 128-row detectors showed that the obtained attenuation values are very similar to the NIST's values. However, the CT of the 128-row detectors showed a slightly much closer match to the NIST's values. This refers to the type and quality of the electronic connections between the detectors. Furthermore, the type and number of detectors (16- and 128-detectors) could affect the details and quality of the output images. The results show that different multidetector CTs can be used to validate the phantom and determine the mass attenuation coefficients of its material.

1. Introduction

In the early 1990s, the spiral CT was introduced and the usage of a wider array of detectors had improved, thus speeding up the scanning procedure. Sequentially, the manufacturing of wider detectors increased and rose with up to 320 rows. This increase of row detectors is known as Multidetector Computed Tomography (MDCT). It has a shorter scanning time as well as lesser artefacts in the output images and higher contrast resolution [1–3].

The improvement of the CT scanners focused on two major things, namely the number and size of detectors along the patient table or the z-axis directions, as well as the connection between these detectors that produce a set of multi-projection data in each gantry rotation around the desired organ. The design of detectors and the electronic connection of these detectors strongly rely on the scanner model and manufacturer [4]. As a result, the improvement of MDCT enhanced the CT number values. The CT numbers depend mainly on the linear attenuation coefficients of the scanned material. Few studies have been conducted to evaluate the difference in CT numbers between different scanner models.

The ultimate goal of this study was to validate a fabricated paediatric thyroid phantom using different MDCT and to investigate the effect of a number of row detectors on the obtaining mass attenuation coefficients of the phantom material.

2. Theoretical background

The material of the CT detectors is so sensitive to radiation and has a high detection efficiency to match the high speed of the gantry rotation [5]. The linear attenuation coefficient of the phantom material that can be used to obtain the CT numbers (measured by Hounsfield Unit, HU), is affected strongly by the energy of penetrated photons and the detection efficiency of the detectors. By using the known linear attenuation coefficient of water (μ_w) at any applied energy and the calculated linear attenuation coefficient of the material (μ_m), the CT number is computed [6,7].

$$CT\ number = \left(\frac{\mu_m - \mu_w}{\mu_w}\right) \times 1000 \quad (1)$$

Essentially, two types of multidetector scanners were introduced to the market, namely the Uniform and Non-uniform detectors. The Uniform design, also called Matrix or Fixed detectors, was invented and first manufactured by GE Healthcare, which the detectors are divided into equal elements. Whereas, the elements are divided by the Non-uniform design (also known as Variable or Adaptive detectors) into different sizes. Such designs were used by several manufacturers such as Siemens (Germany) and Philips (Netherlands) in their CT scanners [4,5]. Furthermore, another type invented by Toshiba (Japan) called Hybrid detectors in which a combination of two different dimensions of detector elements are used in the same row.

Recent studies found considerable variations of measured CT numbers from different CT models and different detector numbers, but still have useful clinical information for physicians [7,8]. In one of the interesting studies done by Grosjean *et al.* 2013 which is related to the scanning of human renal stones using different CT manufacturers, the evaluation of CT numbers showed substantial variation from four different models. It was found that at least one scanner model gave poor details to recognize the attenuation values of the human renal stones, whilst the others show very similar match of CT numbers [8].

In principle, data is provided by 16 detectors for 16 slices, with a thickness as small as 1.25 mm for each slice from the 16 rows. It is assumed that each row contains 1000 detectors and a typical scanner can achieve 1000 views per single rotation whilst 16 million measurements from the 16-row detectors in the z-direction could be made per single rotation [9]. The 128 rows of detectors will give 128 million measurements for the same number of views and number of detectors per one rotation.

3. Methods

In this study, the 128 detectors CT (Somatom Definition; Siemens AG, Wittelsbacherplatz Muenchen, Germany) and the 16 detectors CT (GE Healthcare, Waukesha, WI) were used to scan a fabricated paediatric thyroid phantom. The phantom consists of three parts that can be combined together as one unit as demonstrated in figure 1. The Neck and Trachea are made of Polymethylmethacrylate (PMMA) while the thyroid is made of high-density Polyethylene. The average size of each part of the phantom were taken for the same age (9 years) of female and male children. A diameter of 10 cm for the neck part and 1 cm for the trachea part were selected. Meanwhile, the thyroid part had a width of 2.4 cm and height of 2.5 cm [10–14].

Four voltages (80, 100, 120 and 140 kV) were applied to scan the phantom using both CT machines (the 16- and 128-detectors) to produce single images for each applied kV.

The current (mAs) was set automatically using CARE Dose 4D in the 128 Siemens CT and SMART mA in the 16 GE CT. The slice width was also set automatically to 1.5 mm and 5 mm for the 128 Siemens CT and 16 GE CT respectively.

Weasis Medical Viewer Free Software v.1.2.7 was used to read and analyse the output images from both machines in DICOM format. The CT numbers of each image were evaluated for each applied voltage and the average was taken from both CT scanners, where the mass attenuation coefficients (μ_p/ρ) of the thyroid phantom material were then calculated. The obtained results from both machines were compared with the tables of the National Institute of Standards and Technology (NIST).

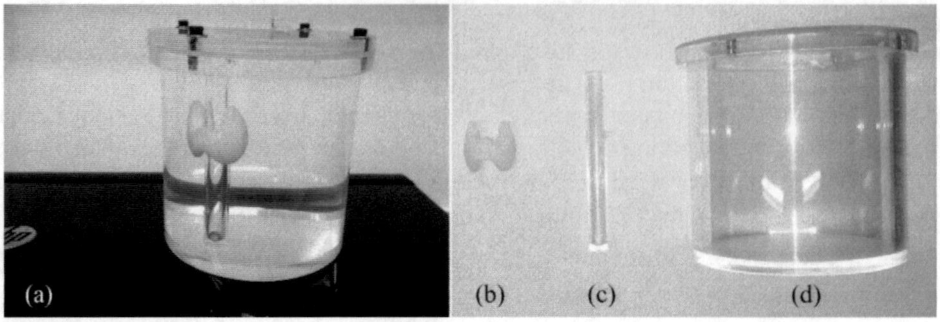

Figure 1. (a) The fabricated paediatric thyroid phantom consists of (b) thyroid phantom, (c) trachea phantom and (d) neck phantom.

4. Results and discussion

The area of the thyroid phantom was selected in each output image and the CT numbers were evaluated by Weasis software. Table 1 summarizes the results of the evaluated CT numbers of the phantom at each applied energy from both scanner models. The average difference in readings of CT numbers from both machines was -1.87 (2.42%), which showed insignificant difference between the CT numbers from the 16- and 128-row detectors CT scanners.

Table 1. The evaluated CT numbers from 16- and 128-detectors CTs.

kV	128 Detectors CT		16 Detectors CT		Difference in reading
	CT No. (HU)	STD (±)	CT No. (HU)	STD (±)	
80	-90.73	11.57	-93.38	7.87	-2.65
100	-76.44	9.52	-78.14	6.56	-1.70
120	-66.33	8.95	-68.84	6.16	-2.51
140	-60.22	7.36	-60.85	5.13	-0.63

Figure 2 shows a close match between the readings in CT numbers of the fabricated thyroid phantom from both scanner models. The SMART mA is used by GE, instead of the CARE Dose 4D in Siemens. However, it is noticeable that current has no effect on the measured CT numbers, where the current in this study varied from 84 mAs to 39.2 mAs in the 16 detectors GE and from 106 mAs to 133 mAs in the 128 detectors Siemens at different applied voltages.

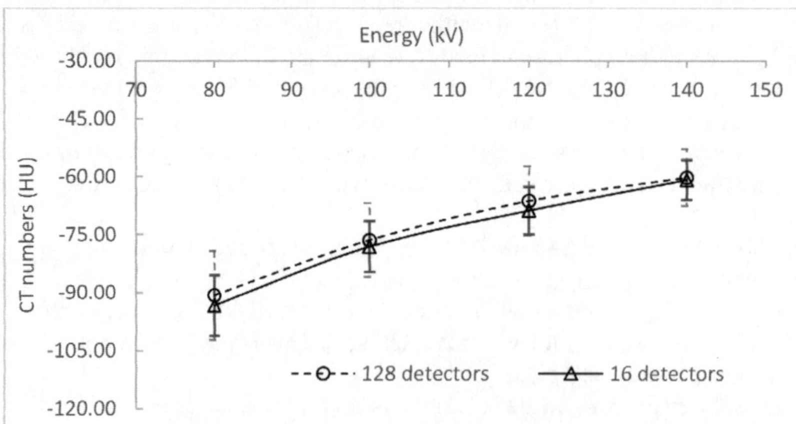

Figure 2. The CT numbers from 16- and 128-row detectors CTs scanners at different applied energies.

The calculated linear attenuation coefficients were used and divided by the density (ρ) of the thyroid phantom material (Polyethylene ≅ 0.92 g/cm^2) [15] to obtain the mass attenuation coefficients of Polyethylene (μ_p/ρ) for each applied energy in both machines. Table 2 shows the obtained (μ_p/ρ) and the percentage difference between each attenuation value at certain voltage with the values listed in the NIST's tables. The average percentage difference between the (μ_p/ρ) and the values of NIST were -0.02 % for the 128 Siemens and 0.18 % for the 16 GE, which showed no significant different between the two machines and can be ignored. The average of the mean percentage difference was also found to be very small (0.08 %).

Table 2. The mass attenuation coefficients from 16- and 128-row detectors CTs compared with the values of NIST.

kV	128 Detectors		16 Detectors			NIST
	μ_p/ρ	% Difference	μ_p/ρ	% Difference	Mean % Difference	μ/ρ
80	0.1816	0.41	0.1810	0.70	0.55	0.1823
100	0.1714	0.31	0.1710	0.50	0.41	0.1719
120	0.1650	-0.33	0.1646	-0.06	-0.19	0.1645
140	0.1579	-0.49	0.1578	-0.42	-0.45	0.1571

The results of the obtained mass attenuation coefficients of the phantom material from the 128 multidetector CT was expected to have a closer match to the values of NIST with the increase in the number of multidetector rows. On the contrary, the results from both CT scanners showed a very good match with the reference tables of NIST as illustrated in figure 3. However, the 128-row detectors CT showed a slightly much closer match to the values in NIST's tables. This refers to the type and quality of the electronic connections between the detectors. Furthermore, the type and number of detectors (16- and 128-detectors) could affect the quality of the output images since the number of output measurements from the 128-row detectors, which could be made per single rotation in the z-direction, is higher than the 16-row detectors.

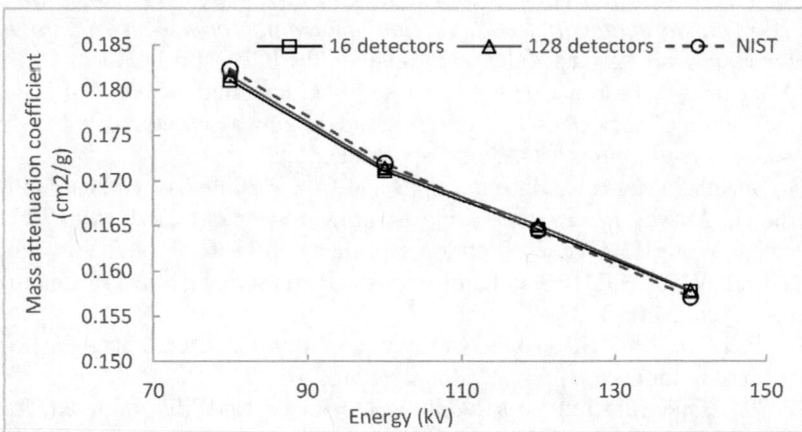

Figure 3. The obtained mass attenuation coefficient from 16 GE and 128 Siemens compared with the values of NIST at different applied energies.

For further comparison between the different multidetector CT models, the t-test for two-tails was used. All p-values showed insignificant difference (P-value > 0.05) between the results from the 16- and 128-row detectors, where the p-values between the mass attenuation coefficients from the 16- and 128-row detectors with the NIST's values were equal to 0.96 and 0.99 respectively, and 0.85 between the CT numbers from both scanner models.

5. Conclusion

As this study was performed using only two different CT models (Siemens and GE) with different numbers of multidetector rows, the results can vary when using other scanner models. Siemens CT and GE show very similar CT number readings and can be used to determine the mass attenuation coefficients of the phantom material, as they show a perfect match with the values of NIST. Therefore, the multidetector CT can be used to validate the fabricated paediatric thyroid phantom material.

A slightly better reading is obtained with the increase in the number of detectors. However, readings are varied from one manufacturer model to another due to the image reconstruction method, the number of electronic connections, the number of row detectors, the type and size of the detector elements as well as the scanner model.

References

[1] Neubauer J *et al.* 2014 Comparing the Image Quality of a Mobile Flat-Panel Computed Tomography and a Multidetector Computed Tomography: A Phantom Study *Invest Radiol.* **49(7)**

[2] Boone J M 2006 Multidetector CT: Opportunities, Challenges, and Concerns Associated with Scanners with 64 or More Detector Rows *Radiology* **241(2)** p 334–7

[3] Flohr T G, Schaller S, Stierstorfer K, Bruder H, Ohnesorge B M and Schoepf U J 2005 Multi–Detector Row CT Systems and Image-Reconstruction Techniques *Radiology* **235(3)** 756–73

[4] Cody D D and Mahesh M 2007 AAPM/RSNA physics tutorial for residents: Technologic advances in multidetector CT with a focus on cardiac imaging *Radiographics* **27**(6) 1829–37

[5] Ulzheimer S and Flohr T 2009 Multislice CT : Current Technology and Future Developments *Multislice CT* Springer Berlin Heidelberg; p. 3–23

[6] Bryant J, Drage N and Richmond S 2012 CT number definition *Radiat Phys Chem* **81(4)** 358–61

[7] Pelberg R and Mazur W 2011 CT Basics *Vascular CT Angiography Manual* (Springer-Verlag London Limited) p. 377

[8] Grosjean R *et al.* 2013 Pitfalls in urinary stone identification using CT attenuation values: are we getting the same information on different scanner models? *Eur J Radiol.* **82(8)** pp1201–06

[9] Goldman L W 2008 Principles of CT: Multislice CT. *J Nucl Med Technol.* **36(2)** 57–68

[10] ICRP-89 2002 *Basic anatomical and physiological data for use in radiological protection: reference values* vol 32, ed J Valentin (Annals of the ICRP) pp 1–277

[11] Katz SL, Vaccani J-P, Clarke J, Hoey L, Colley R C and Barrowman N J 2014 Creation of a reference dataset of neck sizes in children: standardizing a potential new tool for prediction of obesity-associated diseases? *BMC Pediatr.* **14(1)** 159

[12] Wheeler D, Spaeth J, Mehta R, Hariprakash S and Cox P 2009 Assessment and Management of the Pediatric Airway *Resuscitation and Stabilization of the Critically Ill Child SE*, ed DS Wheeler HR Wong TP Shanley (London: Springer) pp 1–30

[13] Griscom N T and Wohl M E B 1986. Dimensions of the growing trachea related to age and gender *Am J Roentgenol.* **146(2)** 233–7

[14] Marwaha R K *et al.* 2008 Ultrasound evaluation of thyroid size: A large nationwide study of choolchildren in India *Natl Med J India.* **21***(2)* 69–74

[15] CRU-44 1989 Tissue Substitutes in Radiation Dosimetry and Measurement. *ICRU Report 44. International Commission on Radiation Units and Measurements.* (MD, USA) 1989

Acknowledgments

We would like to thank and show our gratitude to the staff in the imaging department, Advanced Medical and Dental Institute (AMDI), Universiti Sains Malaysia whom provided insight and expertise that greatly assisted this research.

Journal of Physics: Conference Series **694** (2016) 012048 doi:10.1088/1742-6596/694/1/012048

Comparison of image segmentation of lungs using methods: connected threshold, neighborhood connected, and threshold level set segmentation

A R Amanda[1] and R Widita[1]

[1] Department of Physics, Faculty of Mathematic and Natural Science, Institut Teknologi Bandung

E-mail: amalia.amanda20@gmail.com, rena@fi.itb.ac.id

Abstract. The aim of this research is to compare some image segmentation methods for lungs based on performance evaluation parameter (Mean Square Error (MSE) and Peak Signal Noise to Ratio (PSNR)). In this study, the methods compared were connected threshold, neighborhood connected, and the threshold level set segmentation on the image of the lungs. These three methods require one important parameter, i.e the threshold. The threshold interval was obtained from the histogram of the original image. The software used to segment the image here was InsightToolkit-4.7.0 (ITK). This research used 5 lung images to be analyzed. Then, the results were compared using the performance evaluation parameter determined by using MATLAB. The segmentation method is said to have a good quality if it has the smallest MSE value and the highest PSNR. The results show that four sample images match the criteria of connected threshold, while one sample refers to the threshold level set segmentation. Therefore, it can be concluded that connected threshold method is better than the other two methods for these cases.

1. Introduction

Along the time, medical imaging systems have become more frequently used by health experts to diagnose a disease suffered by a patient. To be able to produce a medical image, modality of an image is needed, for example, CT-scan. The image generated by imaging modalities needs to be processed first in order to get the image quality as expected. One of the image processing is segmentation.

Segmentation is a technique to partition an image into areas that are more meaningful for a particular purpose; it is one of the first steps to analyze and interpret the image [1]. In the medical world, segmentation itself is often used to make it easier to analyze the shape, size, and borders of the target or region of interest (ROI) of the patient. But, segmentation is not easy; there are some issues that need to be considered when doing this process, such as: highlights, shadows, transparency, and object occlusion [2]. Then, to facilitate it, segmentation methods are divided into two basics, i.e. segmentation based on similarity and based on discontinuity. One method that is based on similarity is region growing, while the method based on the discontinuity is level set.

This study used the image of the lungs as the object of segmentation. The lung is an organ that is quite difficult to define in term of its limits. Therefore, it is necessary to facilitate the segmentation technique to analyze the results of the lung image. The accuracy of the identification of lung disorders depends on the efficiency of the lung segmentation techniques [3].

The image that has been successful in the segmentation needs to be evaluated in order to determine a better method that can segment the image appropriately. In the medical world there are some parameters that can be used to evaluate the image, such as: signal noise to ratio (SNR), peak signal noise to ratio (PSNR), timing run, histograms, mean square error (MSE), accuracy, sensitivity, specificity, etc. This research will use the evaluation parameters of MSE and PSNR. MSE [4] and PSNR [5] equation is as follows.

$$MSE = \frac{1}{NM} \sum_{i=1}^{N} \sum_{j=1}^{M} (f(i,j) - \hat{f}(i,j))^2 \tag{1.1}$$

$$PSNR = 10 \log_{10} \frac{L^2}{MSE} \tag{1.2}$$

Therefore, the purpose of this study is to compare the results of the image segmentation of lungs using several methods of segmentation with performance evaluation parameters: MSE and PSNR. Analysis on the segmentation results is used to determine the best method for this case.

2. Methods and material

In this research, segmentation process will be carried out using Insight Toolkit software or commonly referred to ITK. The methods used are connected threshold, neighborhood connected, and threshold level set segmentation.

2.1. Connected threshold

Connected threshold includes the part of region growing method. This technique is done by evaluating the intensity values at predetermined intervals by two threshold values (threshold value) or can be written as follows [6]:

$$I(x) \in [lower, upper] \tag{2.1}$$

That is, only the pixels nearby that have an intensity which is in the range of lower limit and upper limit that can be segmented. Those beyond that range will be considered as background.
The initial step of this process is to determine seed x and seed y manually; after the seed is determined and put in, the next step is to read the histogram to determine the upper and lower threshold. Then, the segmentation process begins. This process continues to spread to the nearest neighboring pixels until no pixels have similarities. The segmentation process in this method is very quick.

2.2. Neighbourhood connected

The next method is neighborhood connected, one of the region growing methods. This method is similar to connected threshold, only this method will add in the nearest neighboring pixels when pixels are included in the measurement defined at the beginning of the process. The main reason to evaluate the neighborhood of each pixel is to make even smaller structures [7]. Because there is a requirement of pixel size and intensity range that must be qualified, the neighborhood connected method is not suitable for segmentation of small-sized organ.

The working steps of neighborhood connected method are the same as those of connected threshold. It's just that there is a specified radius. Because this study uses ITK software, the large radius has been pre-determined and cannot be changed.

2.3. Threshold level set segmentation

The basic techniques of this method is discontinuity, but with an additional threshold parameter. The limit of the intensity used is the lower threshold and upper threshold. For the pixels belonging to the intensity range, the equation level set is applied where the connected components bounded surface evolution, which is commonly called the propagation [7]. The following equations were used.

$$P(x) = \begin{cases} g(x) - L, if \quad g(x) < \dfrac{(U-L)}{2} - L \\ U - g(x), \ others \end{cases} \tag{2.2}$$

Where P is the propagation, g (x) the function of the image, U is the upper threshold, and L is those below the threshold limit.

3. Results and discussion

In this study, the segmentation process begins by inserting the image of the lungs with a format of DICOM files. This research uses five samples of lung images with different pixel dimensions. Five sample images and histograms of each original image can be seen in table 1 below.

Table 1. Input image of the lungs of sample 1, sample 2, sample 3, sample 4, sample 5 and histograms of each sample.

Lungs	The histogram of the original image
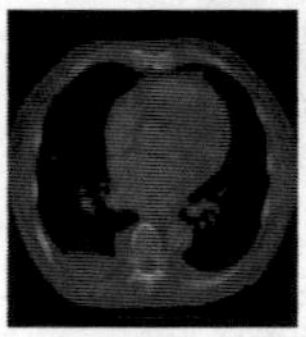	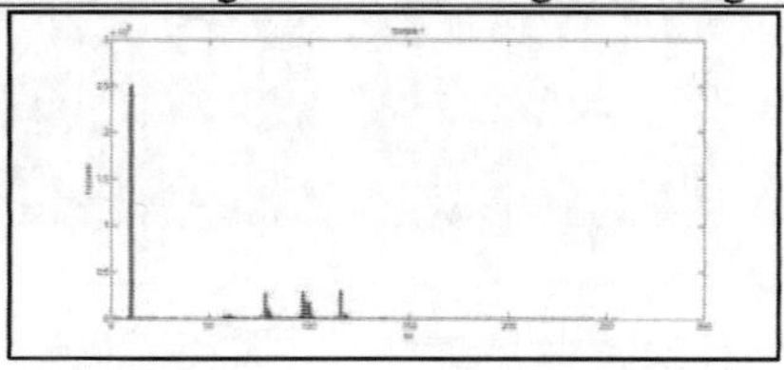
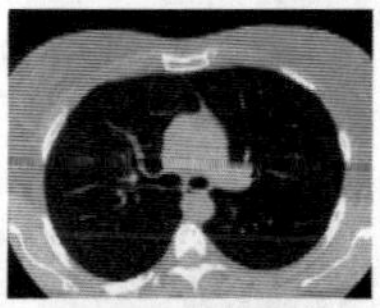	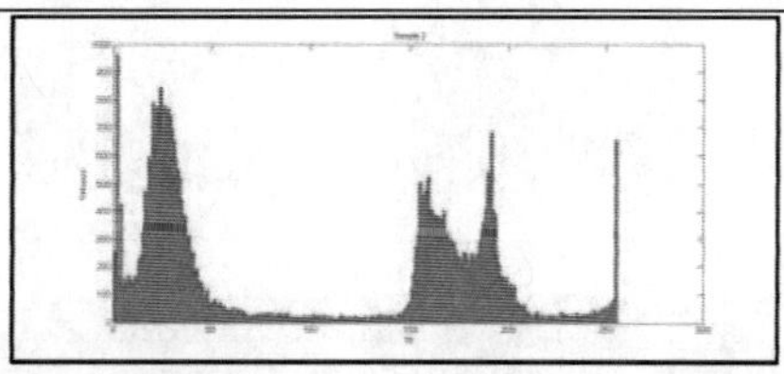
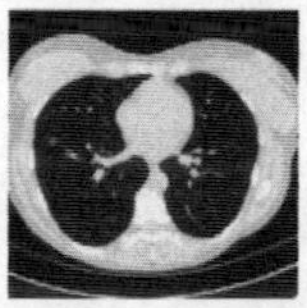	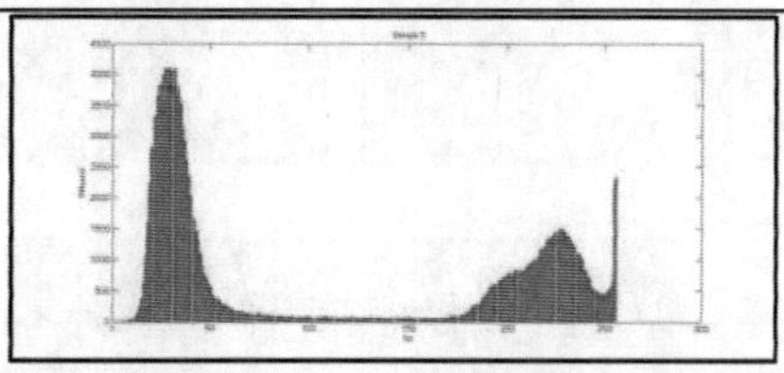
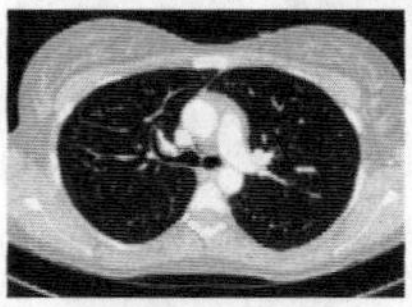	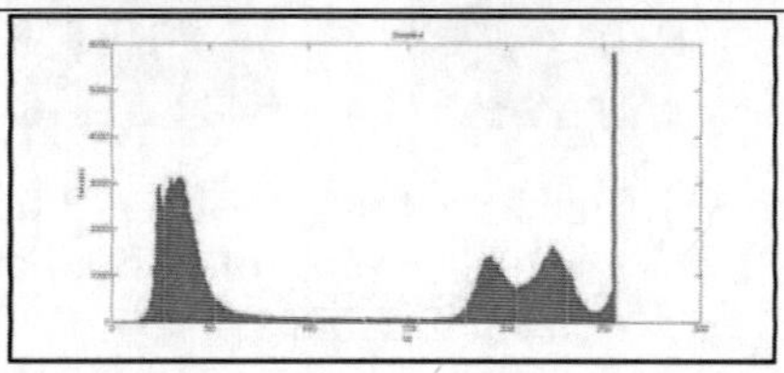
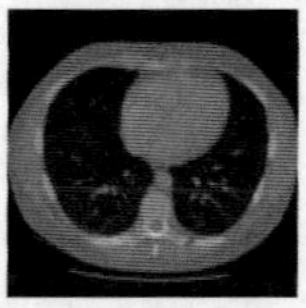	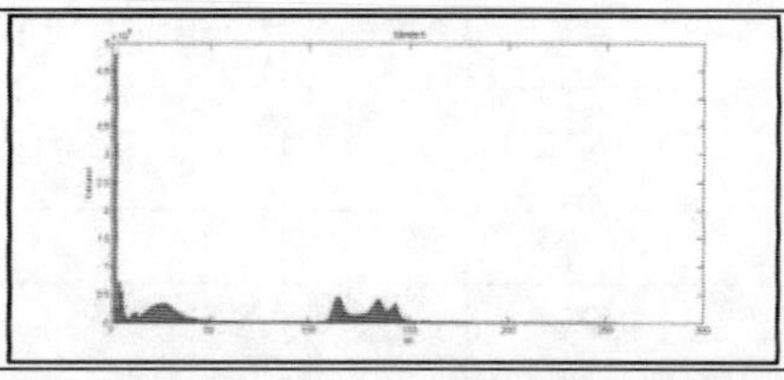

The histogram above shows the upper and lower thresholds of each image, where the threshold is indicated by the valleys of the histogram. Errors in determining the threshold range would affect the results of image segmentation in which the closer the distance of the threshold interval, the smaller the segmented region. However, when conducted excessively, the segmented region would be superfluous. After determining the location of the seed, threshold, and initial distance, the importance of the new image that has been segmented is as in table 2 below.

Table 2. Segmentation results obtained from each method on five samples.

Sample	Connected Threshold	Neighborhood Connected	Threshold Level set Segmentation	Parameter
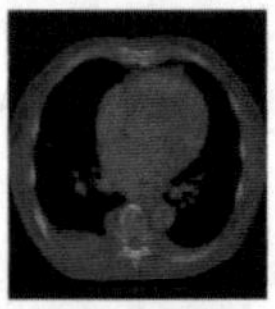				Dimension : 686 × 753 Seed : 170 388 Threshold : 0 13 Initial Distance: 1.0
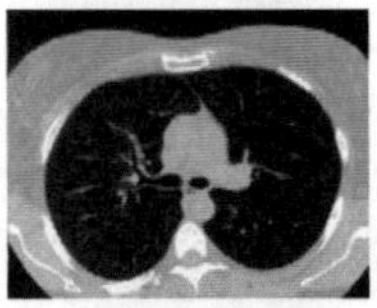				Dimension : 228 × 183 Seed : 45 86 Threshold : 5 46 Initial Distance: 5.0
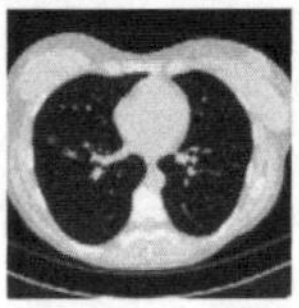			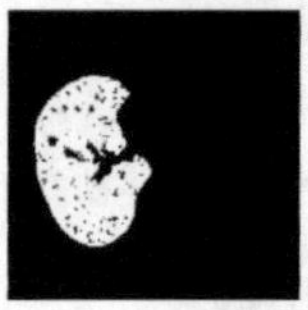	Dimension : 388 × 388 Seed : 82 201 Threshold : 13 69 Initial Distance: 5.0
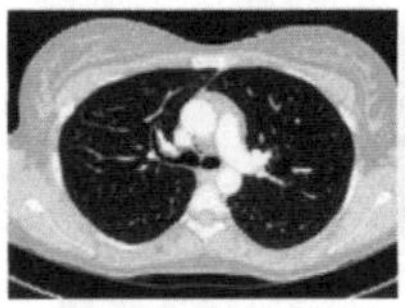				Dimension : 448 × 329 Seed : 127 147 Threshold : 15 81 Initial Distance: 5.0
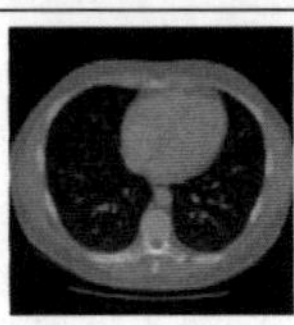			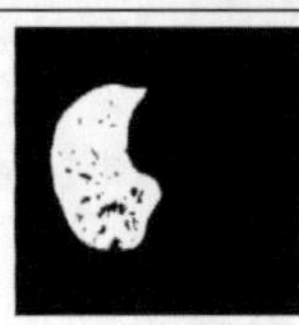	Dimension : 496 × 494 Seed : 150 200 Threshold : 0 51 Initial Distance: 1.0

Table 3 below the MSE and PSNR values obtained from each method in the fifth sample, where the smaller value of MSE and the higher value of PSNR have a good level of similarity with the original image.

Table 3. Result of MSE and PSNR

M / S	MSE			PSNR (MATLAB)			PSNR (PhotoDefiner)		
	C	N	T	C	N	T	C	N	T
P1	39602,71	40206,65	39599,86	2,154	2,088	2,154	2,154	2,088	2,154
P2	21294,41	24639,94	21748,66	4,848	4,214	4,756	4,848	4,214	4,756
P3	19060,19	20827,44	19217,22	5,330	4,944	5,294	5,330	4,944	5,294
P4	17422,36	19107,11	17446,22	5,720	5,319	5,714	5,720	5,319	5,714

P5	33869,14	34889,08	33942,95	2,833	2,704	2,823	2,833	2,704	2,823

In table 3 above, it can be seen that for sample two to sample five, the smallest MSE and PSNR value is highest in the connected threshold method; whereas, in sample one, there are the best values in the threshold level set segmentation method. In this case, by looking at the values above, the connected threshold method is deemed better than the other two methods.

4. Conclusion

After doing research on segmentation of five sample images that have different dimensions, it can be concluded as follows:

- After doing some research on this case, it can be concluded that the connected threshold method is superior to the other two methods. It is based on the faster time required for the segmentation process; MSE and PSNR show this method is superior, and it takes only the parameters of seed location and threshold range.

References

[1] Sonka, Milan and J Michael Fitzpatrick 2000. *Handbook of medical imaging (volume 2, Medical image processing and analysis)* (SPIE PRESS)

[2] Deserno T M 2011 *Biomedical image processing* (Heidelberg, Germany: Springer Berlin Heidelberg)

[3] Memon Ahmed N, Mirza, Majid A and Gilani S A M 2008 Segmentation of lungs from CT scan images for early diagnosis of lung cancer *International Scholarly and Scientific Research & Innovation*

[4] Moreno *et al.* 2013 Towards no-reference of peak signal to noise ratio estimation based on chromatic induction model *International Journal of Advanced Computer Science and Applications*

[5] Wang, Zhou, Bovik and Alan C 2006 *Modern image quality assessment* (USA: Morgan & Claypool)

[6] Johnson H J *et al.* 2014 *The ITK Software Guide Book 2: Design and Functionality Fourth Edition Updated for ITK version 4.7.* ITK

[7] Todd *et al.* 2009 *An analysis of medical image processing methods for segmentation of the inner ear* Researches online (Dubay: University of Wollongong)

Measuring radiation dose to patients undergoing fluoroscopically-guided interventions

L E Lubis[1] and M K Badawy[2,3]

[1] Department of Physics, Faculty of Mathematics and Natural Sciences, Universitas Indonesia, Kampus UI Depok, 16424 West Java, Indonesia
[2] Department of Medical Radiations, School of Medical Sciences, RMIT, Bundoora Victoria 3083, Australia
[3] Medical Physics Department, Austin Health, Heidelberg, Victoria 3084, Australia

E-mail: lukmanda.evan@sci.ui.ac.id

Abstract. The increasing prevalence and complexity of fluoroscopically guided interventions (FGI) raises concern regarding radiation dose to patients subjected to the procedure. Despite current evidence showing the risk to patients from the deterministic effects of radiation (e.g. skin burns), radiation induced injuries remain commonplace. This review aims to increase the awareness surrounding radiation dose measurement for patients undergoing FGI. A review of the literature was conducted alongside previous researches from the authors' department. Studies pertaining to patient dose measurement, its formalism along with current advances and present challenges were reviewed. Current patient monitoring techniques (using available radiation dosimeters), as well as the inadequacy of accepting displayed dose as patient radiation dose is discussed. Furthermore, advances in real-time patient radiation dose estimation during FGI are considered. Patient dosimetry in FGI, particularly in real time, remains an ongoing challenge. The increasing occurrence and sophistication of these procedures calls for further advances in the field of patient radiation dose monitoring. Improved measuring techniques will aid clinicians in better predicting and managing radiation induced injury following FGI, thus improving patient care.

1. Introduction

Global advances in healthcare and technology have significantly increased the usage of fluoroscopically guided interventions (FGI) for diagnosis and treatment. FGI has also increased in complexity over the years due to advancing technology and increased sophistication. Although it is well-known that there is a risk of radiation induced skin injuries in adult patients' [1–3], the highest chance of which occurs during cardiac catheterization procedures [4], injuries remain commonplace [5].

Worldwide literature has historically focused on adult cardiac interventional procedures, with some reporting radiation dose to the skin surpassing 1 Gy [6–11], and occasionally more than 9 Gy [12] with complex procedures. Despite the volume literature reporting on radiation dose to adults [13–21], a large variation in results means a general consensus is hard to reach [22]. In pediatric literature, reports show lower radiation exposure (< 1 Gy) than adults, but with the added increasing risk of stochastic effects [23, 24], which is believed to double or triple the risk of radiation-induced malignancy [25]. This increase in risk may be attributed to the need for long term follow up catheterization procedures in congenital cases [26] or to the prolonged procedure duration in high complexity cases working on small vessel sizes, which may take an average of 3-5 hours [22].

The increase in both FGI procedures and their complexity raises a need to review the radiation dose to patients. This paper aims to briefly review on radiation dose to patients undergoing FGI in catheterization laboratories, with a focus on measurement methods, advances in radiation monitoring and estimation tools, as well as to the current limitations and challenges in the field.

2. Patient dosimetry

As dosimetry in diagnostic radiology involves several quantities, there is a set of definitions that are utilized specifically for FGI. The specific geometry and configuration used in FGI procedures is used to estimate and measure the radiation dose emitted by fluoroscopic devices. Table 1 presents the quantities and measurements of radiation dosimetry utilized in both general diagnostic radiology and specifically for patient dosimetry in FGI.

The kinetic energy released per unit mass of medium irradiated by x-ray is denoted by the abbreviation, "kerma". Hence the entrance surface air kerma (ESAK; $K_{a,e}$), air kerma at reference point ($K_{a,r}$), and air kerma-area product ($P_{k,a}$) are all derived from kerma, each of which have different definitions and use (refer to table 1). The term 'dose' is used to denote the amount of energy absorbed in a unit of mass of a medium, from which the point absorbed dose (D), entrance surface absorbed dose ($D_{skin,e}$), and peak skin dose ($D_{skin,max}$) are derived. Correction factors may be needed to arrive at specific organ/tissue dose estimate (the mean organ/tissue absorbed dose, D_T). This long list of dose measures and quantities raises a need for a common standard used for communicating regulations. An example of this selection is the air kerma at reference point ($K_{a,r}$) and FDA (Food and Drug Administration, USA) compliance air kerma.

The peak skin dose (Dskin,max) is a measure of the highest absorbed dose in any part of a patient's skin accumulated during a procedure [22]. In addition to the dose from the primary x-ray beam, the quantity also includes the contributions of any existing tube leakage and back-scattered radiation. As skin injury may take place at a certain threshold, the peak skin dose information becomes useful for prediction or estimating any potential onset of radiation injury to the patient's skin. The concept of Ka,r was first described in the technical standard for interventional x-ray equipment published by the International Electrotechnical Commission (IEC) in 2000, being described as the air kerma measured on the beam's central axis at 15 cm away from the isocentre towards the x-ray tube [27]. This is meant to represent the entrance air kerma point on the patient's body. Subsequently the standard was adopted by the FDA in regulating the performance standards for any medical fluoroscopy system operating in the U.S., with the reference point described as 30 cm away from the image receptor with minimum distance between the x-ray focal spot and image receptor surface [28]. All medical fluoroscopy devices conforming with IEC standards or produced after 2006 in the U.S. have the feature of calculating and displaying $K_{a,r}$ on the interface [22].

Table 1. Radiation quantities involved in FGI radiation dosimetry[1].

Quantity	Description	Symbol	SI Unit	Reference
Incident air kerma	Air kerma from the incident x-ray beam at beam axis at a certain source-to-surface distance (SSD).	$K_{a,i}$	Gray (Gy)	[29]
Entrance-surface air kerma	Air kerma from the incident x-ray beam at beam axis at patient's entrance point, including contribution from back-scattered radiation.	$K_{a,e}$	Gray (Gy)	[29]

[1] Table reformulated from NCRP Report No. 168 (2010) [22]

Point absorbed dose	Energy absorbed per unit mass at a certain point in a medium including contribution from back-scattered radiation.	D	Gray (Gy)	[29]
Entrance-surface absorbed dose	Energy absorbed on the central x-ray beam axis at patient's entrance point including contribution from back-scattered radiation.	$D_{\text{skin,e}}$	Gray (Gy)	[29]
Mean organ/tissue absorbed dose	Integral or average absorbed energy over a whole tissue/organ mass.	D_{T}	Gray (Gy)	[30]
Peak skin dose	Energy absorbed per unit mass at the most irradiated local area of patient's skin including contribution from back-scattered radiation.	$D_{\text{skin,max}}$	Gray (Gy)	[29]
Air kerma at the reference point	The air kerma (K_a) without any contribution of back-scattered radiation on certain, regulated/standardized positions. In IEC standard, the position is defined at the beam central axis 15 cm away from the isocenter towards the x-ray tube.	$K_{\text{a,r}}$	Gray (Gy)	[27,28,31, 32]
Air kerma-area product or dose-area product	The integral of air kerma without any contribution of back-scattered radiation over the area of the x-ray beam in a plane perpendicular to the beam axis.	P_{KA}	Gy cm^2	[29]
FDA compliance air kerma rate	Air kerma per unit of time measured under FDA compliance conditions.	$\dot{K}_{FDA}$	mGy / min	[28,33]

The clinical setting introduces further variables that increase the complexity of measuring or estimating a patient's radiation dose. Factors such as patient size and positioning beam energy spectrum, gantry tilting and angulation, attenuation-governed radiological parameters, and variable distances are key in determining dose to patients. The factors may relate to patient, the size of the patients, or both.

A typical catheterization laboratory uses ceiling- or floor-mounted fluoroscopy device, where a high degree of freedom in setting patient position exists. This is due to the freely movable ("floating"), tiltable, and rotatable gantry, as well as the ability to adjust the patient table in x-, y-, and z-axis directions. Additionally the SID (Source to Image Distance), which indicates the distance between the x-ray tube's focal spot to the image receptor, is also variable with the image receptor being able to move toward or away from the x-ray tube. Furthermore, the use of Automated Exposure Rate Control (AERC) enables the device to set the radiological parameters (the tube voltage, kVp, and tube current, mA) according to the amount of radiation received by the AERC sensor, which is mounted adjacent to the image receptor. Thus a larger patient will attenuate more radiation, which in turn reduces the signal (radiation) to the AERC sensor, and reduces the image contrast. In such circumstances, the AERC system will automatically increase radiological parameters to improve image contrast, which increases the patient dose. Figure 1 demonstrates how the patient size has implications to patient positioning and, therefore, patient dose.

For a larger patient, operators commonly increase the SID as well as lower the patient table to facilitate the size of the patient. Such changes, in addition to high patient attenuation, contributes to an increase of patient dose. This occurs for several reasons. Firstly, the skin of the patient lying supine on the table will obviously get closer to the radiation source, increasing the radiation dose due to the inverse-square law (ISL). Secondly, by moving the image receptor away from the x-ray tube causes less radiation to be received by the receptor also due to the ISL. Both the high attenuation and SID will contribute to inducing the AEC to adjust the the tube voltage and/or current, which will have an effect on patient dose. The influence of patient size to patient radiation dose has been discussed by Kotre *et al.* (2005), involving an attempt to formulate a correction factor between Dose Area Product (DAP, or P_{KA}) and varying patient size, to incorporate them to standard ICRP reference patient [34].

Journal of Physics: Conference Series **694** (2016) 012049 doi:10.1088/1742-6596/694/1/012049

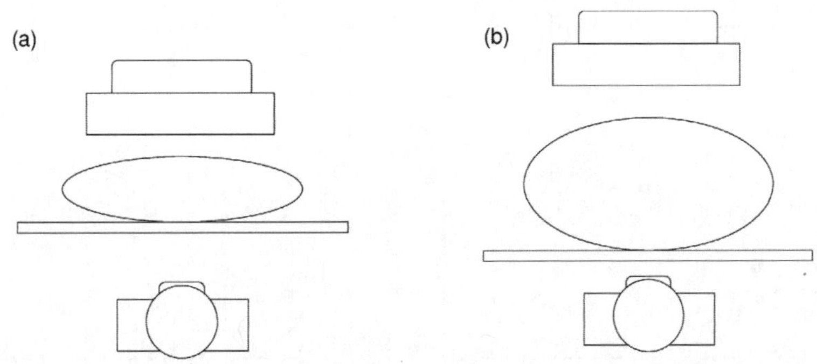

Figure 1. Distance setting of a (a) thin patient and (b) thick patient.

3. Current equipment advances

Real time patient dose monitoring has enabled operators to optimize dose during procedures, as well as provide accurate documentation in patient records for surveillance of deterministic effects. A number of tools are available to estimate patient skin dose in real time, including the Toshiba Dose Tracking System (DTS), Siemens CareGraph system, GE DoseMap and PEMNET. Of these, the PEMNET was the first to be developed. The PEMNET was an accurate and convenient method of monitoring patient dose, and was useful as a teaching tool in predicating deterministic effects, as well as practically evaluating the trade-off between image quality and patient dose [35]. However, the downsides include not taking into account overlapping fields from several sources [36] and the complexity involved in its installation.

The Siemens CareGraph used mathematical models to monitor real time skin dose, with reports of estimation of peak skin dose within 15% accuracy in the anterior posterior plane [37]. However, an underestimation of 67% in the lateral planes proved a downfall and a lack of demand eventually lead to discontinuation of the system. A similar system developed by Toshiba has been tested in studies of its efficacy in interventional procedures; with dose distribution was found to correlate well with directly measured values. The software was found to estimate dose rate at patient skin within 10% [38,39]. The GE DoseMap system was introduced in 2013 to monitor real time dose in fluoroscopic units, however at time of writing no studies validating this system currently exist.

4. Present limitation and challenges

Currently two machine dose metrics are used as a surrogate for patient dose: Reference Air Kerma and DAP (or Kerma-Area Product). However, these values, whilst a good indication for the radiation output of the system, are not accurate predictors for peak skin dose. Balter (2006) illustrates that based on patient size and reference point locations, the dose measurement point on units may not coincide with the patient's skin [40]. Chida *et al.* (2011) concluded that understanding the different regulations regarding interventional reference points may lead to better indirect dose estimations [41]. It was found that different manufacturers might use different reference points of measurements. This would lead an error in dose calculations if the reference point was assumed as the IEC standard.

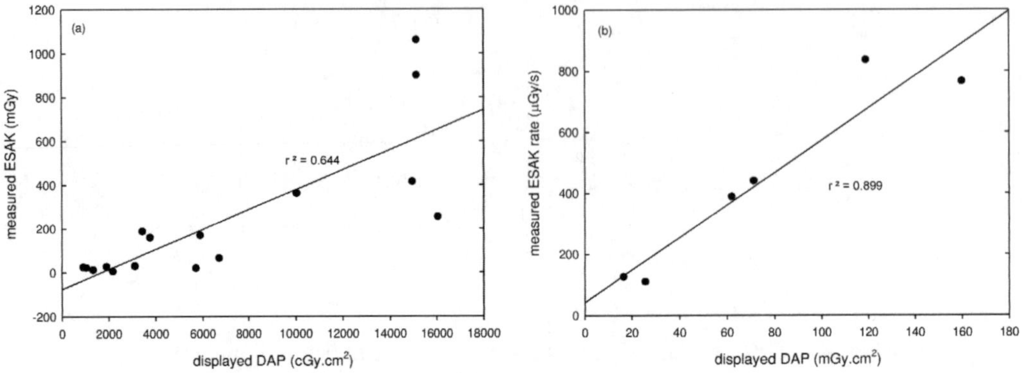

Figure 2. Correlation between displayed DAP and (a) accumulated ESAK in patient and (b) ESAK rate in phantom; both showing weak correlation ($r^2 < 0.95$). Graphs was reproduced from Yohannes *et al.* (2003) (a) and a recent work of the author (b).

DAP is currently the most commonly recorded and used to assess radiation safety in FGI. This value is used to compare the difference of radiation exposure between cases and to set threshold dose limits for deterministic effects. However, it is limited in predicting deterministic effects since the same DAP value could be used to describe a high intensity beam in a small area or a low intensity beam in a large area. Chu *et al.* (2006) used Gafchromic Film® to correlate peak skin entrance dose to the more readily available parameters: DAP and fluoroscopy time. Yohannes *et al.* (2003) had carried out a slightly similar work with TLD measurement, finding entrance surface dose correlation with DAP [42]. The results of both studies show weak correlations and that current estimation methods do not provide values which correspond closely to the direct measured dose at the patient's skin [43]. We also compared the work of Yohannes *et al.* (2003) with our recent measurement in phantom, as seen on figure 2 (b), imposing similar idea to that of Yohannes *et al.* Results from van de Putte *et al.* (2000) agree with this finding however they suggest that using DAP averaged over a large group was a reasonable indicator of skin dose rather than being a reliable method for per patient specific skin dose calculation [44].

5. Conclusion

With the increased prevalence on fluoroscopically-guided interventions, there lies a notable increase of concern on the delivered radiation dose to patients. Aside from the available direct, independent measurement tools all of which have their limitations, some modern x-ray units are equipped with sophisticated tools with ability to present real-time patient dosimetry. These tools, however, have demonstrated a range of shortcomings (i.e being manufacturer bound, poor accuracy) that warrant further investigation. As a result, radiation dose measurement to patients undergoing FGI remains a challenge. Improvements in this field will lead to optimising patient dose monitoring through improved measurement and estimation methods. Reliable accuracy will enable clinicians to ensure better patient care through ongoing radiation dose optimisation.

References

[1] Balter S, Shope Jr. T B, Fletcher D W, Kuan H M and Seissl H 2002 Techniques to estimate radiation dose to skin during fluoroscopically guided procedure *Medical Physics Monograph No. 28: Intravascular Brachytherapy, Fluoroscopically Guided Interventions* ed S Balter, R C Chan and T B Shope Jr. (Madison: Medical Physics Publishing)

[2] Vano E, Gonzalez L, Guibelalde E, Aviles P, Fernandez J M, Prieto C and Galvan C 2005 Evaluation of risk of deterministic effects in fluoroscopically guided procedures *Radiat. Prot. Dosim.* **117** 190–4

[3] Vano E, Goicolea J, Galvan C, Gonzalez L, Meiggs L, Ten J I and Macaya C 2001 Skin radiation

injuries in patients following repeated coronary angioplasty procedures. *Br. J. Radiol.* **74** 1023–31

[4] Miller S W and Castronovo F P 1985 Radiation exposure and protection in cardiac catheterization laboratories. *Am. J. Cardiol.* **55** 171–6

[5] Balter S and Miller D L 2014 Patient skin reactions from interventional fluoroscopy procedures *AJR. Am. J. Roentgenol.* **202** W335–42

[6] Padovani R and Quai E 2005 Patient dosimetry approaches in interventional cardiology and literature dose data review. *Radiat. Prot. Dosimetry* **117** 217–21

[7] Balter S, Miller D L, Vano E, Ortiz Lopez P, Bernardi G, Cotelo E, Faulkner K, Nowotny R, Padovani R and Ramirez A 2008 A pilot study exploring the possibility of establishing guidance levels in X-ray directed interventional procedures *Med. Phys.* **35** 673

[8] Trianni A, Chizzola G, Toh H, Quai E, Cragnolini E, Bernardi G, Proclemer A and Padovani R 2005 Patient skin dosimetry in haemodynamic and electrophysiology interventional cardiology. *Radiat. Prot. Dosimetry* **117** 241–6

[9] Rosenthal L S *et al*. 1998 Predictors of fluoroscopy time and estimated radiation exposure during radiofrequency catheter ablation procedures *Am. J. Cardiol.* **82** 451–8

[10] McFadden S L, Mooney R B and Shepherd P H 2002 X-ray dose and associated risks from radiofrequency catheter ablation procedures. *Br. J. Radiol.* **75** 253–65

[11] International Atomic Energy Agency 2009 *Safety Report Series No. 59 - Establishing guidance levels in X-ray guided medical interventional procedures : A pilot study* (Vienna: International Atomic Energy Agency)

[12] Suzuki S, Furui S, Kohtake H, Yokoyama N, Kozuma K and Yamamoto Y 2006 Radiation exposure to patient's skin during percutaneous coronary intervention for various lesions, including chronic total occlusion *Circ. J.* **70** 44–8

[13] Blaszak M A, Majewska N, Juszkat R and Majewski W 2009 Dose-area product to patients during stent-graft treatment of thoracic and abdominal aortic aneurysms. *Health Phys.* **97** 206–11

[14] Bleeser F, Hoornaert M-T, Smans K, Struelens L, Buls N, Berus D, Clerinx P, Hambach L, Malchair F and Bosmans H 2008 Diagnostic reference levels in angiography and interventional radiology: a Belgian multi-centre study. *Radiat. Prot. Dosimetry* **129** 50–5

[15] Brambilla M, Marano G, Dominietto M, Cotroneo A R and Carriero A 2004 Patient radiation doses and references levels in interventional radiology. *Radiol. Med.* **107** 408–18

[16] Dauer L T, Thornton R, Erdi Y, Ching H, Hamacher K, Boylan D C, Williamson M J, Balter S and St Germain J 2009 Estimating radiation doses to the skin from interventional radiology procedures for a patient population with cancer. *J. Vasc. Interv. Radiol.* **20** 782–8; quiz 789

[17] Miller D L *et al*. 2003 Radiation doses in interventional radiology procedures: The RAD-IR study: part I: overall measures of dose. *J. Vasc. Interv. Radiol.* **14** 711–27

[18] Miller D L *et al*. 2003 Radiation doses in interventional radiology procedures: The RAD-IR study: part II: skin dose. *J. Vasc. Interv. Radiol.* **14** 977–90

[19] Tsalafoutas I A, Goni H, Maniatis P N, Pappas P, Bouzas N and Tzortzis G 2006 Patient doses from noncardiac diagnostic and therapeutic interventional procedures. *J. Vasc. Interv. Radiol.* **17** 1489–98

[20] Vano E *et al*. 2008 Patient dose in interventional radiology: An European survey. *Radiat. Prot. Dosimetry* **129** 39–45

[21] Vano E *et al*. 2009 Patient dose reference levels for interventional radiology: a national approach. *Cardiovasc. Intervent. Radiol.* **32** 19–24

[22] National Council on Radiation Protection and Measurement 2010 *Report No. 168 - Radiation Dose Management for Fluoroscopically-Guided Interventional Medical Procedures* (Bethesda: National Council on Radiation Protection and Measurement)

[23] Papadopoulou D, Yakoumakis E, Sandilos P, Thanopoulos V, Makri T, Gialousis G, Houndas D, Yakoumakis N and Georgiou E 2005 Entrance radiation doses during paediatric cardiac catheterisations performed for diagnosis or the treatment of congenital heart disease *Radiat.*

Prot. Dosim. **117** 236–40

[24] Yakoumakis E N, Gialousis G I, Papadopoulou D, Makri T, Pappouli Z, Yakoumakis N, Papagiannis P and Georgiou E 2009 Estimation of children's radiation dose from cardiac catheterisations, performed for the diagnosis or the treatment of a congenital heart disease using TLD dosimetry and Monte Carlo simulation *J. Radiol. Prot.* **29** 251–61

[25] Committee to Assess Health Risks from Exposure to Low Levels of Ionizing Radiation 2006 *Health effects of exposure to low levels of ionizing radiation (BEIR VII Phase 2)* (Washington DC: National Academy Press)

[26] Chung T 1998 Techniques *Practical pediatric imaging, diagnostic radiology of infants and children* ed D R Kirks and N T Griscom (Lippincott-Raven) pp 1–63

[27] International Electrotechnical Commission 2000 *IEC 60601-2-43 ed 1.0 Medical electrical equipment – Part 2-43: Particular requirements for the safety of X-ray equipment for interventional procedures* (Geneva: International Electrotechnical Commission)

[28] U.S. Food and Drug Administration 2014 *Performance standards for ionizing radiation emitting products. Fluoroscopic equipment. 21 CFR Part 1020.32* (U.S. Food and Drug Administration)

[29] The International Commission on Radiation Unit and Measurement 2005 *ICRU Report 74: Patient dosimetry for X rays used in medical imaging* vol 5 (Oxford University Press)

[30] The International Commission on Radiological Protecition 1991 *ICRP Publication 60: 1990 Recommendations of the International Commission on Radiological Protection* vol 21 (Pergamon Press)

[31] International Electrotechnical Commission 2004 *IEC TR 60788 ed 2.0 Medical electrical equipment–Glossary of defined terms* (Geneva: International Electrotechnical Commission)

[32] International Electrotechnical Commission 2010 *IEC 60601-2-43 ed 2.0 Medical electrical equipment – Part 2-43: Particular requirements for the basic safety and essential performance of X-ray equipment for interventional procedures* (Geneve: International Electrotechnical Commission)

[33] U.S. Food and Drug Administration 1999 *Resource manual for compliance test parameters of diagnostic X-ray systems* (Rockville, Maryland)

[34] Kotre C J, Reay J and Chapple C-L 2005 The influence of patient size on patient doses in cardiology *Radiat. Prot. Dosim.* **117** 222–4

[35] Gkanatsios N A, Huda W, Peters K R and Freeman J A 1997 Evaluation of an on-line patient exposure meter in neuroradiology. *Radiology* **203** 837–42

[36] Mahesh M 2001 Fluoroscopy: patient radiation exposure issues. *Radiographics* **21** 1033–45

[37] Ozeroglu M A 2005 *Verification of caregraph(R) peak skin dose data using Radiochromic film*

[38] Bednarek D R, Barbarits J, Rana V K, Nagaraja S P, Josan M S and Rudin S 2011 Verification of the performance accuracy of a real-time skin-dose tracking system for interventional fluoroscopic procedures. *Proc. SPIE--the Int. Soc. Opt. Eng.* **7961**

[39] Rana V, Bednarek D, Josan M and Rudin S 2011 SU-F-BRA-09: Comparison of skin-dose distributions calculated by a real-time dose-tracking system with that measured by Gafchromic film for a fluoroscopic C-Arm Unit *Med. Phys.* **38** 3702

[40] Balter S 2006 Methods for measuring fluoroscopic skin dose. *Pediatr. Radiol.* **36 Suppl 2** 136–40

[41] Chida K, Inaba Y, Morishima Y, Taura M, Ebata A, Yanagawa I, Takeda K and Zuguchi M 2011 Comparison of dose at an interventional reference point between the displayed estimated value and measured value. *Radiol. Phys. Technol.* **4** 189–93

[42] Yohannes I, Adi R W and Soejoko D S 2003 *Entrance surface dose in cardiac region on cardiac catheterization patients* (Universitas Indonesia)

[43] Chu R Y L, Thomas G and Maqbool F 2006 Skin entrance radiation dose in an interventional radiology procedure. *Health Phys.* **91** 41–6

[44] van de Putte S, Verhaegen F, Taeymans Y and Thierens H 2000 Correlation of patient skin doses

in cardiac interventional radiology with dose-area product. *Br. J. Radiol.* **73** 504–13

The relationships between breast volume, breast dense volume and volumetric breast density with body mass index, body fat mass and ethnicity

N Zakariyah[1, 2], N B Pathy[3], N A M Taib[4], K Rahmat[1, 2], C W Judy[1, 2], F Fadzil[1, 2], S Lau[1, 2] and K H Ng[1, 2]

[1] Department of Biomedical Imaging, University of Malaya, Kuala Lumpur Malaysia
[2] University of Malaya Research Imaging Centre, Kuala Lumpur Malaysia
[3] Department of Social and Preventive Medicine
[4] Department of Surgery, University of Malaya, 50603, Kuala Lumpur, Malaysia

E-mail: ngkh@ummc.edu.my

Abstract. It has been shown that breast density and obesity are related to breast cancer risk. The aim of this study is to investigate the relationships of breast volume, breast dense volume and volumetric breast density (VBD) with body mass index (BMI) and body fat mass (BFM) for the three ethnic groups (Chinese, Malay and Indian) in Malaysia. We collected raw digital mammograms from 2450 women acquired on three digital mammography systems. The mammograms were analysed using Volpara software to obtain breast volume, breast dense volume and VBD. Body weight, BMI and BFM of the women were measured using a body composition analyser. Multivariable logistic regression was used to determine the independent predictors of increased overall breast volume, breast dense volume and VBD. Indians have highest breast volume and breast dense volume followed by Malays and Chinese. While Chinese are highest in VBD, followed by Malay and Indian. Multivariable analysis showed that increasing BMI and BFM were independent predictors of increased overall breast volume and dense volume. Moreover, BMI and BFM were independently and inversely related to VBD.

1. Introduction

Previous studies have shown that breast cancer has been linked to obesity [1, 2] and also to breast density [3, 4]. Breast tissue is often regarded as consisting of two components, adipose and fibroglandular (dense) tissues. The ratio of the fibroglandular tissue volume to the total breast volume, expressed in percentage, is termed volumetric breast density (VBD). Overweight and obese women tend to have larger breast volumes compared to slimmer women, and subsequently lower overall VBD. In addition, previous studies have shown that breast volume may be a useful metric for plastic surgery and breast conservation surgery, as it can help the surgeon to predict the aesthetic effect of various breast surgeries on women and offer the most appropriate treatment decision. Measurement of breast volume also allows for better surgical planning and implant selection in breast reconstruction.
The aim of this study was to investigate the association of breast volume, breast dense volume and VBD with anthropomorphic parameters such as body mass index (BMI), body fat mass (BFM) and ethnicity in an Asian women setting.

2. Materials and methods

2.1. Study population
We collected raw digital mammograms from 2450 women (1275 Chinese, 679 Malays and 496 Indians) who underwent mammography examination from June 2012 to August 2014 at the University of Malaya Medical Centre, Kuala Lumpur. Only women with no breast abnormality (BI-RADS 1 to 3) and no breast cancer cases were included in this study.

2.2. Measurement of body parameters
Anthropomorphic parameters were collected for all the subjects involved. The data included body weight, BMI and BFM, which were measured using a body fat analyser (Tanita, Japan).

2.3 Assessment of mammographic density
The mammograms for all the subjects were acquired on three digital mammography systems (General Electronic Essential, Siemens Novation and Hologic Selenia). The images were analyzed using a VBD assessment system (Volpara 1.5.1, Matakina Technology, New Zealand) to obtain breast volume, breast dense volume and VBD values.

2.4 Data analysis
Multivariable logistic regression (SPSS version 16.0) was used to determine the independent predictors of increased overall breast volume, breast dense volume and VBD.

3. Results and discussion
Table 1 shows the demographic information for the three ethnic groups (Chinese, Malay and Indian). The median age of participants was 57 years, with Chinese women (median = 58 years) being significantly older than the Malay (median = 53 years), and Indian women (median = 57 years). Chinese women had significantly lower BMI (median = 23.2 kgm^{-2}) compared to their Malay (26.9 kgm^{-2}) and Indian (26.1 kgm^{-2}) counterparts.

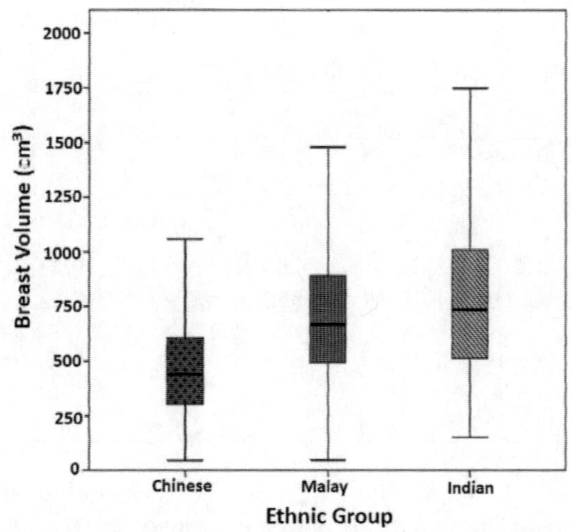

Figure 1. Box plot of breast volume for the three ethnic groups.

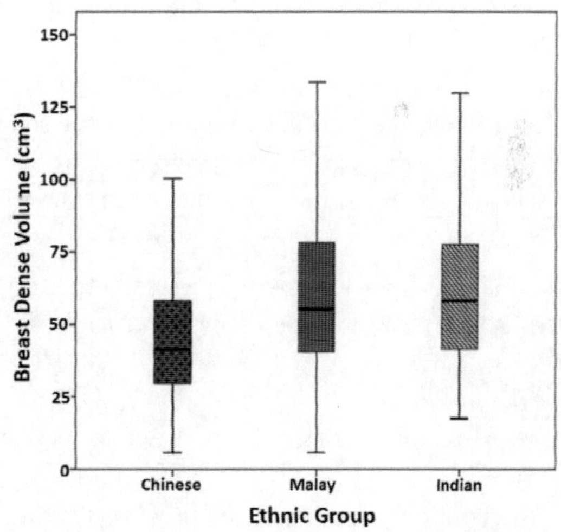

Figure 2. Box plot of breast dense volume for the three ethnic groups.

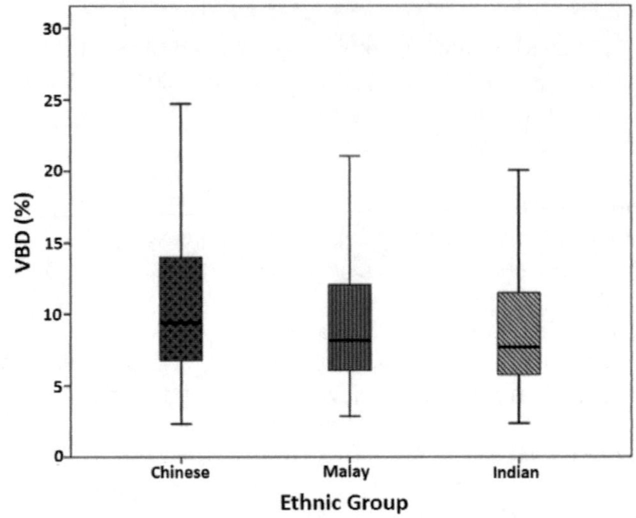

Figure 3. Box plot of VBD for the three ethnic groups.

The median overall breast volume was highest in Indian women (736 cm^3) followed by Malay (666 cm^3), and Chinese women (440 cm^3); $p<0.001$ (figure 1). Moreover, breast dense volume was highest in Indian women (58 cm^3) followed by Malay (55 cm^3), and Chinese women (48 cm^3); $p<0.001$ (figure 2). VBD was highest in Chinese women (median = 9.4%) followed by Malay (8.15%), and Indian women (7.69%); $p<0.001$ (Fig. 3). This finding is in agreement with the studies reported by Jamal *et. al.* [5], Zulfiqar *et. al.* [6] and Mariapun *et. al.* [7]. Multivariable analysis showed that increasing BMI and BFM were independent predictors of increased overall breast volume and breast dense volume. Moreover, BMI and BFM were independently and inversely related to VBD, whereas ethnicity was independently associated with VBD.

Table 1. Demographic, anthropomorphic and body composition data for the three ethnic groups. The values are presented as median (inter-quartile range).

	Overall (N= 2450)	**Chinese (N= 1275)**	**Malay (N= 679)**	**Indian (N= 496)**	**p-value**
Age (years)	57 (50, 63)	58 (52, 65)	53 (47, 59)	57 (50, 63)	<0.05
BMI (kgm^{-2})	24.9 (22.0, 28.1)	23.2 (21.0,26.05)	26.9 (24.1, 30.15)	26.1 (23.6, 26.1)	<0.001
Breast vol (cm^3)	549 (371, 786)	440 (302,606)	666 (492, 891)	736 (513,1010)	<0.001
Breast dense vol (cm^3)	48 (32.8, 72.3)	41 (26.4, 54.4)	55 (43.9, 82.0)	58 (46.5, 94.0)	<0.001
VBD (%)	8.7 (6.3,13.1)	9.4 (6.8,14.0)	8.2 (6.1,12.1)	7.7 (5.8,11.5)	<0.001

4. Conclusion

Higher BMI and BFM are predictive of larger breast volumes and larger breast dense (glandular) volumes in Indian women followed by Malay and Chinese women. We also observed that BMI and BFM are inversely associated with VBD across the three ethnic groups in our study.

References

[1] Stephenson G D and Rose D P 2003 Breast cancer and obesity: An update *Nutrition and cancer* **45** 1-16
[2] La Guardia M and Giammanco M 2001 Breast cancer and obesity *Panminerva medica* **43** 123-33
[3] Boyd N F *et al.* 2007 Mammographic density and the risk and detection of breast cancer *New England Journal of Medicine* **356** 227-36
[4] Boyd N F *et al.* 1998 Mammographic densities and breast cancer risk *Breast disease* **10** 113-26

[5] Jamal N, Ng K-H, McLean D, Looi L-M and Moosa F 2004 Mammographic breast glandularity in Malaysian women: data derived from radiography *American Journal of Roentgenology* **182** 713-7

[6] Zulfiqar M, Rohazly I and Rahmah M 2011 Do the majority of Malaysian women have dense breasts on mammogram? *Biomedical imaging and intervention journal* **7** e14

[7] Mariapun S, Li J, Yip C H, Taib N A and Teo S H 2015 Ethnic differences in mammographic densities: An Asian cross-sectional study. *PloS one* **10** e0117568

Acknowledgments

This study was supported by the HIR Grant UM.C/625/1/HIR/MOME/06, Ministry of Education, Malaysia. We also thank the radiologists and radiographers at the University of Malaya Medical Centre, Kuala Lumpur for their technical help and support for this project.

Journal of Physics: Conference Series **694** (2016) 012051 doi:10.1088/1742-6596/694/1/012051

Analysis of urinary stone based on a spectrum absorption FTIR-ATR

V Asyana[1], F Haryanto[1], L A Fitri[1], T Ridwan[2], F Anwary[2] and H Soekersi[2]

[1] Department of Physics, Faculty of Mathematics and Natural Sciences, Institut Teknologi Bandung, Jalan Ganesa 10 Bandung 40132, Indonesia
[2] Hasan Sadikin Hospital, Bandung, Indonesia

E-mail: vepyasyana@gmail.com

Abstract. This research analysed the urinary stone by measuring samples using Fourier transform infrared-attenuated total reflection spectroscopy and black box analysis. The main objective of this study is to find kinds of urinary stone and determine a total spectrum, which is a simple model of the chemical and mineral composition urinary stone through black box analysis using convolution method. The measurements result showed that kinds of urinary stone were pure calcium oxalate monohydrate, ion amino acid calcium oxalate monohydrate, a mixture of calcium oxalate monohydrate with calcium phosphate, a mixture of ion amino acid calcium oxalate monohydrate and calcium phosphate,pure uric acid, ion amino acid uric acid, and a mixture of calcium oxalate monohydrate with ion amino acid uric acid. The results of analysis of black box showed characteristics as the most accurate and precise to confirm the type of urinary stones based on theregion absorption peak on a graph, the results of the convolution, and the shape of the total spectrum on each urinary stones.

1. Introduction

Urinary stone is is the most painful diseases that is found in developing countries. In Indonesia, it becomes the highest visiting number of patients in urology clinic. According to health research in 2013, the highest prevalence of urinary stone disease occurs in the age group of 55-64 years (1.3%), which slightly decreased in the age group of 65-74 years (1.2%) and lesser on $\geq$ 75 years (1.1%). Urinary stone occurs more frequently in male (0.8%) better than in female (0.4%). Urinary stone which was formed caused disturbances in the urinary system. Therefore, it is very important to remove it in order to treat the disease, to prevent, and to hamper the recurrence of the stone formation. The surgery methods extracorporeal shock wave lithotripsy (ESWL) and percutaneous nephrolithotomy (PCNL), have been established for stone removal [1]. The preference and indication of success using these methods are affected by several factors. One of them is based on the chemical and mineral composition of urinary stones. In Indonesia, most of the hospital such as Hasan Sadikin Hospital Bandung, determined the urinary stones by using CT-Scan. CT-Scan is one of techniques to determine the chemical and mineral composition urinary stones. However, the measurements result showed that CT-Scan had the least accuracy and precision. This was due to the overlaping in Hounsfield Unit value for some types of urinary stones. Therefore, to get the most accurate and precise types of urinary stones by using CT-Scan. It is necessary to manufacture a standard operating procedure (SOP) to determine the types of urinary stones. Making the SOP was conducted with a CT-scan method and combined with other methods such as Fourier transform infrared spectroscopy (FTIR) method. The first stage to build the SOP was the determination of chemical and mineral composition model on urinary stone. The present study was conducted earlier by determining a total spectrum of a simple model of chemical and mineral composition in urinary stone through a black box analysis using convolution method.

2. Method and experiment

2.1. FTIR-ATR characterization

This research was performed using FTIR-ATR Alpha-P by Bruker. FTIR-ATR spectra were recorded in the range 4000-500 cm^{-1}. The evaluation menu that was provided in the OPUS software performed the entire process automatically.

A total of 25 urinary stones for the present study had been collected from Hasan Sadikin Hospital Bandung and it had dimension range from 0.7 cm – 3.2 cm. All together, the external layer of different parts of urinary stones were recorded and analyzed.

2.2. Black box analysis

Black box analysis was conducted by convolution method. Convolution is a mathematical way of combining two signals to form a third signal. Convolution is important because it relates the three signal of interest that is the input signal, the output signal, and the impulse response. On this research the impulse response signal was spectrum as a results of FTIR characterization and the input signal was infrared spectrum. The convolution was conducted using Matlab R2009a.

3. Results and discussion

3.1. Results of FTIR characterization

The information that was obtained from the infrared spectral pattern of samples is shown on figure 1-6, where BU is the number of samples urinary stone.

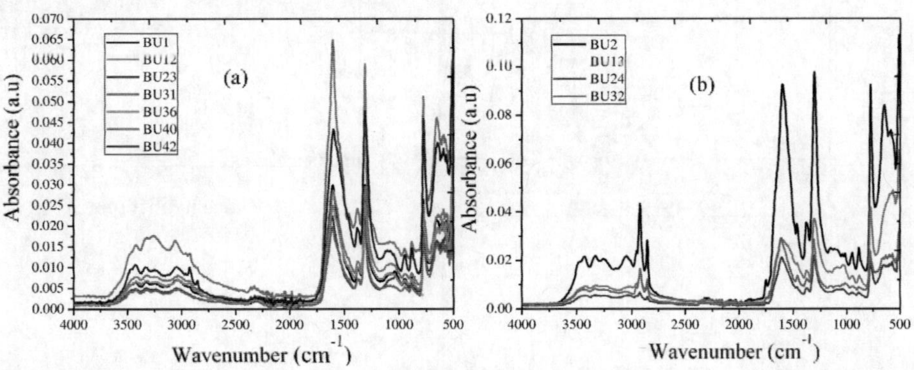

Figure 1. FTIR spectrum of urinary stone containing (a) pure calcium oxalate monohydrate, (b) ion amino acid calcium oxalate monohydrate.

The spectrum of a pure calcium oxalate monohydrate showed a high absorbance at 1616 - 1600 cm^{-1} and 1314 - 1302 cm^{-1} belonged to C=O and C-O stretching vibration, respectively. The frequency region was 779 - 775 cm^{-1} corresponding to C-H bending. The absorption band observed at 3446 - 3021 cm^{-1} which happened due to symmetric and asymmetric O-H bending. The absorption band at 1387– 1364 cm^{-1} was happened due to C-C dan C-O stretching, 891 - 874 cm^{-1} was due to C-C *stretching*, and 693 - 687 cm^{-1} was due to O-H bending. Visually spectrum of ion amino acid calcium oxalate monohydrate had shown similarity with figure 1a. However, in figure 1b, it showed absorbance at 2924 - 2850 cm^{-1} corresponding to N-H stretching vibration, which was the area that indicated ion amino acid. Spectrum in figure 1b, showed the chemical composition of urinary stone that consisted of ion amino acid and calcium oxalate monohydrate [1-5].

The spectrum mixture of calcium oxalate monohydrate and calcium phosphate showed a high absorbance at 1028 - 1001 cm^{-1} related to P-O stretching, the area indicated ion PO_4^{3-} [4, 5, 7]. The absorption at 3430 - 3043 cm^{-1} indicated symmetric and asymmetric O-H *stretching*, 1618 cm^{-1} - 1604 cm^{-1} indicated C=O stretching, 1316 - 1311 cm^{-1} indicated C-O stretching, 781 - 776 cm^{-1} indicated C-H bending, 886 - 872 cm^{-1} indicated C-C stretching [1,3].

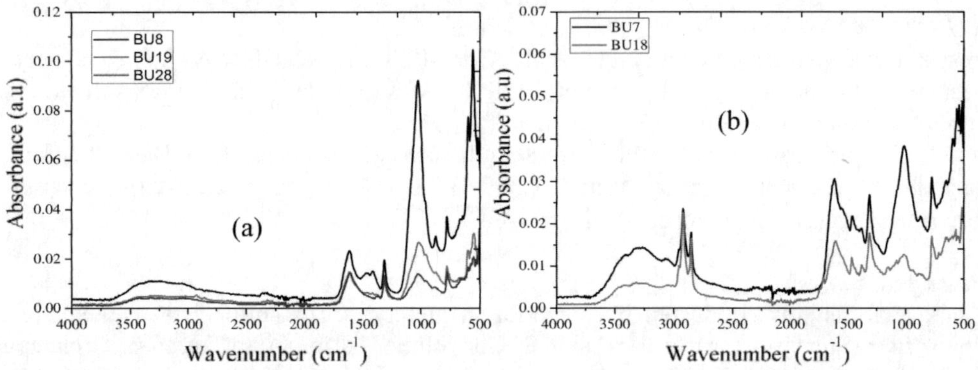

Figure 2. FTIR spectrum of urinary stone containing (a) mixture of calcium oxalate monohydrate with calcium phosphate, (b) mixture of ion amino acid calcium oxalate monohydrate and calcium phosphate.

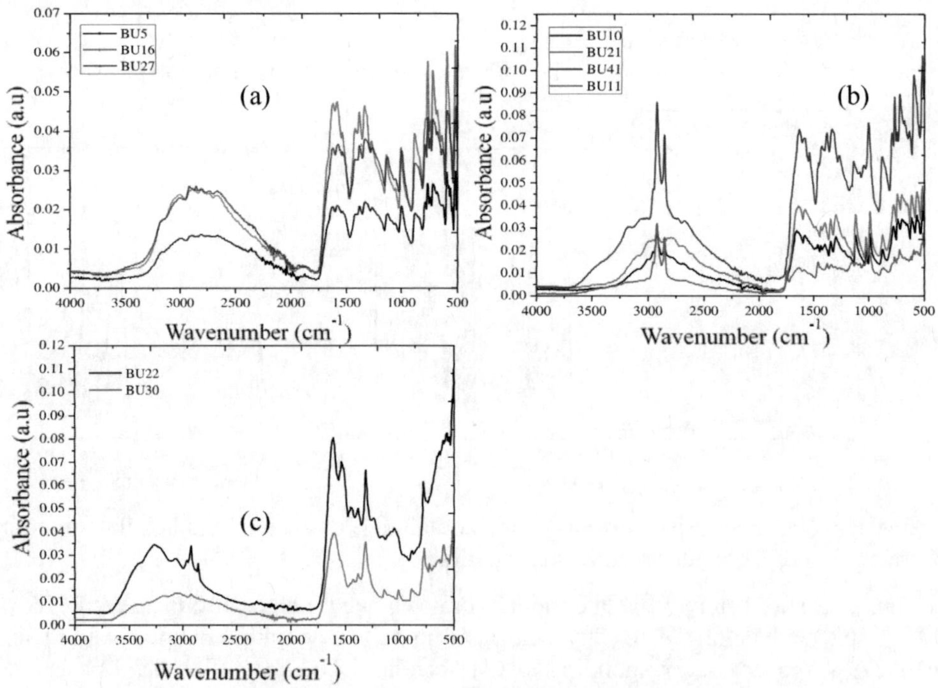

Figure 3. FTIR spectrum of urinary stone containing (a) pure uric acid, (b) ion amino acid uric acid, (c) mixture of calcium oxalate monohydrate with ion amino acid uric acid.

Visually, the spectrum of the mixture of amino acid calcium oxalate monohydrate ion with calcium phosphate showed similarity with figure 2a. However, figure 2b showed absorbance at 2922 - 2849

cm^{-1} related to N-H stretching, the area indicated amino acid ion. Spectrum in figure 2b showed that the chemical compositions of urinary stone consisted of amino acid ion, calcium oxalate monohydrate, and calcium phosphate.

The spectrum of thepure uric acid showed N-H bending vibration in the frequency region 1635 - 1622 cm^{-1}. In frequency region 1589 - 1574 cm^{-1} related to C-N stretching vibration. The absorption band that was observed at 1542 - 1538 cm^{-1} was due to N-O stretching. The absorption band at 1424 - 1417 cm^{-1} was due to O-H bending. The absorption band at 1385 - 1380 cm^{-1} was due to C-N stretching. The absorption band at 1337 - 1322 cm^{-1} was due to N-H stretching, 1137 - 1120 cm^{-1} was due to C-O stretching. C-N stretching and C-N bending vibration in frequency region 1057 - 993 cm^{-1} and 882 - 871 cm^{-1} respectively [3, 6]. Visually, the spectrum of amino acid and uric acid ion were similar with figure 3a. However, figure 3b showed absorbance at 2923 - 2850 cm^{-1} related to N-H stretching vibration indicated amino acid ion. Spectrum in figure 3b showed that the chemical composition of urinary stone consisted of amino acid and uric acid ion. And the figure 3c showed a mixture of amino acid and uric acid ion with calcium oxalate monohydrate. The spectrum of calcium oxalate monohydrate shows O-H stretching vibration in the frequency region was 3276 - 3252 cm^{-1}. In frequency region 1315 - 1311 cm^{-1} related C-O stretching, 778 - 776 cm^{-1} related to C-H bending. The spectrum of amino acid and uric acid ion showed N-H stretching vibration in the frequency region 2920 - 2849 cm^{-1}, the region indicated ion NH_3^+. The absorption band observed at 1627 - 1622 cm^{-1} was due to N-H bending, 1538 - 1537 cm^{-1} was due to N-O stretching, 1455 - 1436 cm^{-1} was due to O-H bending, 1388 - 1385 cm^{-1} was due to C-N stretching, and 1045 - 1041 cm^{-1} was due to C-N stretching [1, 3, 6].

3.2. Results of convolution method

The results of convolution from the infrared spectral pattern of samples were given in figure 4-5, where n is the number of convolution urinary stone.

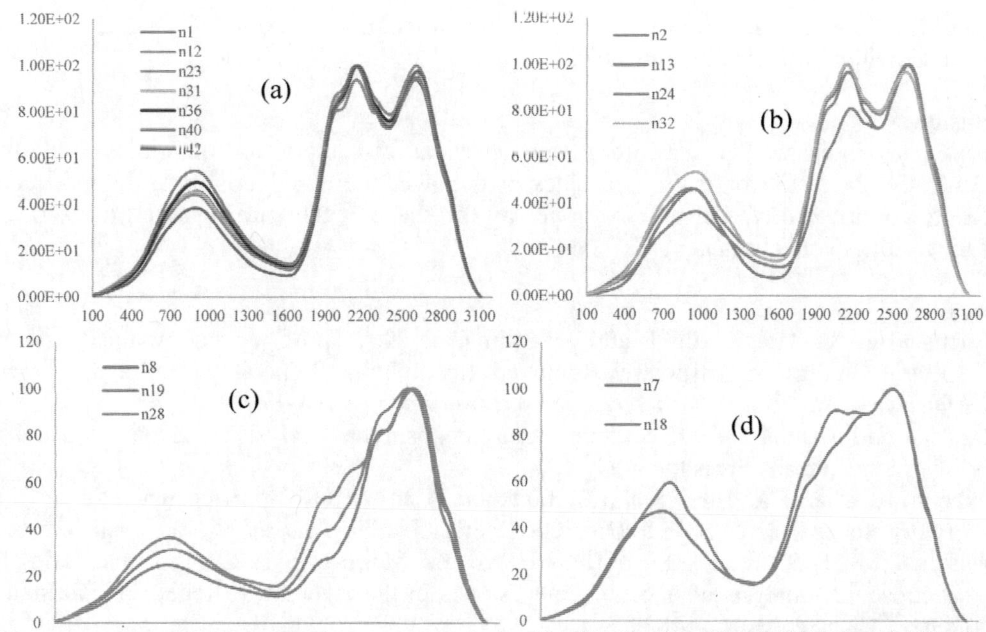

Figure 4. Result of convolution method (a) pure calcium oxalate monohydrate, (b) ion amino acid calcium oxalate monohydrate. (c) mixture of calcium oxalate monohydrate with calcium phosphate, (d) mixture of ion amino acid calcium oxalate monohydrate and calcium phosphate.

Figure 4 and 5 are the normalization of convolution results for urinary stone. The convolution pattern of the samples under investigation were analyzed, for each samples of urinary stone had a different intensity, this indicated a functional group density had different amounts. The normalization of the results for kinds of urinary stones had a different shape and maximum intensity.

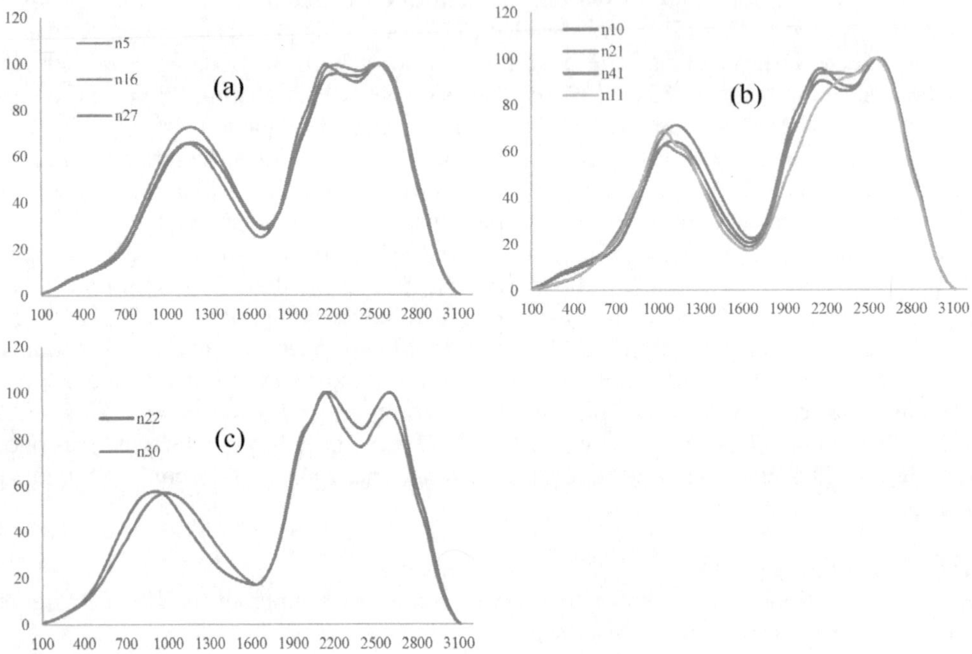

Figure 5. (a) Pure uric acid, (b) ion amino acid uric acid, (c) mixture of calcium oxalate monohydrate with ion amino acid uric acid.

4. Conclusions

The present study indicated that the information on chemical composition of urinary stone mostly varied. Generally, the results of analysis of black box showed the characteristics at the most accurate and precise to confirm the type of urinary stones by the shape of the total spectrum of each urinary stones as the result of convolution.

References

[1] Kanchana G, Sundaramoorthi P and Jeyanthi G P 2009 Bio-Chemical Analysis and FTIR-Spectral Studies of Artificially Removed Renal Stone Mineral Constituents *Journal of Minerals & Materials Characterization & Engineering* **8** 161-170

[2] Mandel I and Mandel N 2007 Structure and compositional analysis of kidney stones Totowa New Jersy Humana Press Inc

[3] Sekkoum K, Cheriti A, Taleb S and Belboukhari N 2011 FTIR Spectroscopic study of human *urinary* stones from El Bayadh District (Algeria) *Arabian Journal of Chemistry*

[4] Khaskheli M H, Sherazi S T H, Ujan H M and Mahesar S A 2012 Transmission FTIR spectroscopic analysis of human kidney stones in the Hyderabad Region of Pakistan *Turk Journal Chemical* **36** 477-483

[5] Shamema A A, Arul K T, Kumar R S and Kalkura S N 2014 Physicochemical analysis of *urinary* stones from Dharmapuri District *Spectrochimica Acta Part A: Molecular and Biomolecular Spectroscopy* **134** 442-448

[6] Selvaraju R, Thiruppathi G and Raja A 2011 FT-IR Spectral studies on certain human urinary stones in the patients of rural area *Spectrochimica Acta Part A: Molecular and Biomolecular* **93** 260-265

Using computer aided system to determine the maximum depth of visualization of B-Mode diagnostic ultrasound image

G Maslebu[1], K Adi[2] and Suryono[2]

[1] Physics Department, Satya Wacana Christian University, 52-60 Diponegoro Street, Salatiga 50711, Indonesia
[2] Instrumentation and Electronics Laboratory, Faculty of Science and Mathematics, Diponegoro University, Professor Haji Soedharto Street, Semarang 50275, Indonesia

E-mail: ginmaslebu@gmail.com

Abstract. In the service unit of radiology, ultrasound modality is widely used because it has advantages over other modalities, such as relatively inexpensive, non-invasive, does not use ionizing radiation, and portable. Until now, the method for determining the depth visualization on quality control program is through the visual observation of ultrasound image on the monitor. The purpose of this study is to develop a computer-aided system to determine maximum depth of visualization. Data acquisition was done by using B-Mode Diagnostic Ultrasound machine and Multi-purpose Multi-tissue Ultrasound Phantom model 040GSE. Phantom was scanned at fixed frequency of 1,8 MHz, 2,2 MHz, 3,6 MHz and 5,0 MHz with a gain variation of 30 dB, 45 dB, and 60 dB. Global thresholding and Euclidean distance method were used to determine maximum visualization depth. From this study, it is proved that the visualization depth using computer aided provide deeper visualization than visual interpretation. The differences between expert verification and the result of image processing are < 6%. Thus, computer aided system can be used for the purpose of quality control in determining maximum visualization depth of B-Mode diagnostic ultrasound image.

1. Introduction

In the service unit of radiology, ultrasound modality is widely used because it has advantages over other modalities, such as relatively inexpensive, non-invasive, non-ionizing radiation, and portable [1]. It has been reported that in 2000 there were 5 million ultrasound scanning around the world per week [2].

Quality Control (QC) is part of the Quality Assurance (QA) activities that emphasize its program on the techniques needed on supervision, maintenance, keeping the technical elements of radiography and imaging equipment system that affects image quality. Quality control is needed in radiology to convince a radiology specialist doctor that the used modality is in the best condition for reading (interpretation) the image [3]. In order to provide maximum service quality, then the quality assurance of ultrasound modality should be done within a certain period of time continuously [4]. In an effort to maintain the quality of ultrasound modality, the medical physicist is in charge of designing and implementing QA. Basic QA for the phantom tests are the visual evaluation of image uniformity, accuracy of quantitative evaluation of vertical distance, horizontal distance accuracy and maximum penetration depth [5].

Recent studies for the quality control of ultrasound has shown that the use of a computer for ultrasound image analysis ultrasound has advantages or the potentially increase its usefulness. Computer aided system can be used for distance measurement [6] and determining the image uniformity on the ultrasound quality control [7].

One of the QC parameters for ultrasound is the depth of visualization. Until now, the method for determining the depth visualization is through the visual observation of ultrasound image on the monitor [8].

Based on the problem above, a computer aided system was developed to analyse the visualization depth of ultrasound image for better quantification using a parameter of frequency and gain. It is expected to provide contributions for the QC program that impact on improving the quality of health services to society.

2. Methods

This study used B Mode medical ultrasound, multifrequency transducer, Curved Array Transducer CH5-2, Multi-purpose Multi-tissue Ultrasound Phantom to test the functionality and quality using ultrasound phantom CIRS brand 040GSE models, and computers equipped with software for image processing. During scanning of the phantom, coupling gel was applied to the surface of phantom or transducer as a medium that allows the energy coupling of ultrasonic waves to travel freely between the surface of the phantom and transducer. In this study, the frequency is set at a fixed value of 1.8 MHz, 2.2 MHz, 3.6 MHz and 5 MHz, while the gain variation in the value of 30 dB, 45 dB, and 60 dB. For the data analysis process, there are three main procedures, namely:

1. Making of Graphical User Interface (GUI) for image processing in the following order:
   1.1. Reading image from data acquisition on a computer using software developed for image processing.
   1.2. The region of interest (ROI) selection with ultrasound image cropping techniques for the determination of ROI as an area to be analyzed further.
   1.3. Image segmentation to separate the object from the background image based on intensity. Segmentation of ROI in this study is conducted by global thresholding method.
   1.4. Determination of the maximum visualization depth is determined according to Euclidean distance method, i.e. the distance between the surfaces to the deepest phantom object from segmentation results.
2. Maximum depth of visualization was determined by the expert.
   Maximum depth of visualization was compared between computational result and expert's visual interpretation.

3. Results and discussion

Figure 1 shows the results of a scanned phantom image with anechoic test target object using a transducer with a frequency of 3.6 MHz and gain of 45 dB and 60 dB. From those images, visually, increase in gain causes the image brighter so that objects that are far from the surface of the phantom can be observed [9]. It also occurs in the image of the scanned phantom with another frequency and gain variation.

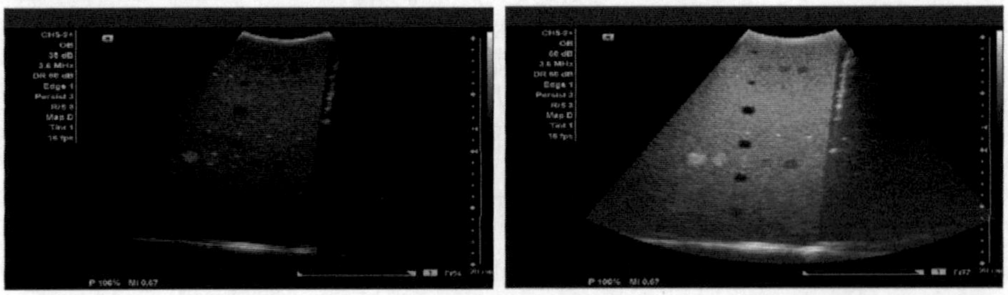

(a) (b)

Figure 1. Scanned phantom image using 3,6 MHz Frequency with gain variation of (a) 30 dB and (b) 60 dB.

An image is defined in 2-dimensional function *f(x,y)*, where x and y are the spatial coordinate and amplitude of *f* in certain *(x,y)* is called *gray level.* In image processing, this image is represented by a pixel (*picture element*) matrix M x N with the coordinate system is shown in figure 2 [10].

$$f(x,y) = \begin{bmatrix} f(0,0) & f(0,1) & \dots & f(0,N-1) \\ f(1,0) & f(1,1) & \dots & f(1,N-1) \\ \vdots & \vdots & & \vdots \\ f(M-1,0) & f(M-1,1) & \dots & f(M-1,N-1) \end{bmatrix}$$

Figure 2. Pixel coordinates system.

The spatial resolution is an indicator of the number of pixels per unit length [11]. Spatial resolution measurement is done by counting the number of pixels that exist between the center (centroid) of the two nearest objects. For the 8-bit image, the gray value of each pixel ranges from 0-255. A value of 0 means black and 255 means white. Determination of the centroid of the two nearest object using global thresholding method from the Region of Interest (ROI) and resulted in the object of 8 bits are converted into binary (black and white). Anechoic objects that are further analyzed in the form of a white object. The next two nearest object will be selected to analyze the value of spatial resolution, which is the number of pixels divided by the distance between the center (centroid) of the second object. Illustration of determining the spatial resolution is shown in figure 3.

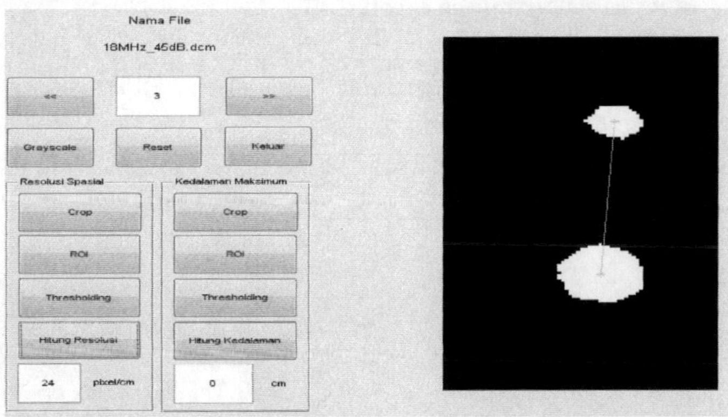

Figure 3. Determination of spatial resolution using computer aided system.

The spatial resolution has a constant value of 24 pixel/cm. Determination of maximum depth of visualization based on the Euclidean distance vertically position of the furthest object from the surface of the phantom image. Pixel coordinate of the phantom's surface and the deepest object are (x_1,y_1) and (x_2, y_2), respectively. Thus, the number of pixels determined by using Euclidean distance formula [12]

$$Number\ of\ pixel = \sqrt{(x_1 - x_2)^2 + (y_1 - y_2)^2)} \tag{1}$$

The distance in cm unit is determined using equation (2) and (3):

$$Spatial\ Resolution = \frac{Number\ of\ pixels\ (pixel)}{Distance\ (cm)} \tag{2}$$

$$Distance(cm) = \frac{Number\ of\ pixels\ (pixel)}{Spatial\ Resolution(pixel/cm)} \tag{3}$$

Table 1 shows the percentage error of visualization depth result from computer aided system and expert. The visualization depth of verification by the expert and computer aided system were represented by *a* and *b*. By using gain and frequency variation, percentage error for the determination of the visualizations depth with image processing techniques is <6%. Smaller percentage error means the better value of the visualization depth determination.

Table 1. Percentage error of visualization depth result from computer aided system and expert.

		Depth of Visualization (cm)		The difference (cm)	Percentage error
Freq. (MHz)	Gain (dB)	a^a	b^b	*b-a*	$\frac{b-a}{a} \times 100\ \%$
1,8	30	7,4000	7,6655	0,2755	3,588 %
1,8	45	10,2500	10,5047	0,2547	2,485 %
1,8	60	13,2500	13,3404	0,0904	0,682 %
2,2	30	7,0000	7,3883	0,3883	5,547 %
2,2	45	10,2500	10,5617	0,3117	3,041 %
2,2	55	10,2500	10,5486	0,2986	2,913 %
3,6	30	7,2500	7,44517	0,3252	2,692 %
3,6	45	10,2500	10,5354	0,2854	2,784 %
3,6	60	13,2500	13,3823	0,1323	0,998 %
5,0	30	7,2500	7,3992	0,1492	2,058%
5,0	45	10,2500	10,3830	0,1330	1,298 %
5,0	60	13,2500	13,4526	0,2026	1,529 %

[a]Maximum depth of visualization from an expert.
[b]Maximum depth of visualization from computer aided system adjusting with the expert.

Table 2. Comparison between computer aided system and expert observing result of maximum depth of visualization.

Freq. (MHz)	Gain (dB)	Maximum depth of visualization (cm)	Verification result of visualization depth (cm)	
			Expert I	Expert II
1,8	30	10,7300	7,40	7,40
1,8	45	13,5462	10,25	10,25
1,8	60	16,5073	13,25	13,25
2,2	30	10,2632	7,00	7,00
2,2	45	13,5197	10,25	10,25
2,2	55	13,6734	10,25	13,25
3,6	30	10,4597	7,25	7,25
3,6	45	13,4938	10,25	13,25
3,6	60	16,7157	13,25	13,25
5,0	30	10,3993	7,25	7,25
5,0	45	13,4663	10,25	10,25
5,0	60	16,2026	13,25	13,25

From table 2, it can be explained that the computer system can determine the deeper maximum depth of visualization than the human eye's ability to see objects on ultrasound phantom image. Two different experts determined the same value of visualization depth for almost all data, but still there are differences in using 2,2 MHz of the frequency with 55 dB of gain and using 3,6 MHz of the frequency with 45 dB of gain. It means that visual interpretation is possibly different among observers. Thus, it can be said that the computer as a tool to study images for maximum depth of visualization purposes has advantages in determining the farthest objects that are difficult to distinguish by the normal eyes so that it can be used for quality control purposes.

4. Conclusion

Based on a study that has been done, it can be concluded that computer aided system can be used for quality control purposes in determining maximum depth of visualization with thresholding method for segmentation of objects from the background image and Euclidean distance for determining the distance. In this case, maximum depth of visualization using computer aided system is deeper than the

Journal of Physics: Conference Series **694** (2016) 012052 doi:10.1088/1742-6596/694/1/012052

visual observations. The difference between the results of the expert verification and developed computer aided system is <6%.

References

[1] Tyovenda A A, Aiyohuyin E O and Akaagerger N B 2013 Diagnostic ultrasound quality assurance by means of machine output parameters *Int. J. Sci. Technol.* **3** 287-90
[2] Szabo T L 2004 Diagnostic ultrasound imaging: Inside out (UK: Elsevier Academic Press)
[3] Bushong S C 2001 Radiologic science for technologists physics, biology and protection, Seventh Edition (St Louis: Mosby Inc)
[4] Dudley N J, Griffith K, Houldsworth G, Holloway M, and Dunn M A 2001 A Review of two alternative ultrasound quality assurance programmes *European. J. of Ultrasound* **12** 233-45
[5] Tradup D J, Hangiandreou N J and Taubel J P 2003 Comparison of ultrasound quality assurance phantom measurements from matched and mixed scanner-transducer combination *J. App.Clin. Med. Phys* **4** 239-47
[6] Fallo F, Adi K and Suryono 2014 Perancangan Model Pengukuran Jarak Menggunakan Computer-Aided pada Kontrol Kualitas Ultrasonografi *Berkala Fisika* **17** 13-20
[7] Sudarsih K, Budi W S and Suryono 2014 Analisis keseragaman citra pada pesawat ultrasonografi (USG) *Berkala Fisika* **17** 33-8
[8] Choi J I, Jung S E, Kim P N, Cha S H, Jun J K, Lee H Y and Park E C 2014 quality assurance in ultrasound screening for Hepatocellular Carcinoma using a standardized phantom and standard clinical images: A 3-year national investigation in Korea, *J. Ultrasound Med.* **33** 985-95
[9] Thijssen J M, Weijers G and De Korte L C 2006 Objective performance testing and quality assurance of medical ultrasound equipment *Ultrasound and Med. & Biol.,* **33** 460-71
[10] Prasetyo E 2011 Pengolahan citra digital dan aplikasinya menggunakan MATLAB (Yogyakarta: Penerbit ANDI)
[11] Aksoy P and DeNardis L 2008 *Information technology in theory* (Vancouver: Thomson Learning.Inc)
[12] Maslebu G, Adi K and Suryono 2015 Effect of gain changing to maximum visualization depth on diagnostic ultrasound B-mode image *Int. J. App. Eng. Research ISSN 0973-4562* **10** 33449-52

Dual energy micro CT SkyScan 1173 for the characterization of urinary stone

L A Fitri[1], V Asyana[1], T Ridwan[2], F Anwary[2], H Soekersi[2], F D E Latief[1] and F Haryanto[1]

[1] Department of Physics, Faculty of Mathematics and Natural Sciences, Institut Teknologi Bandung, Jalan Ganesa 10, Bandung 40132 Indonesia
[2] Hasan Sadikin Hospital, Jl. Pasteur 38, Bandung 40161

E-mail: aeefleni@gmail.com

Abstract. Knowledge of the composition of urinary stones is an essential part to determine suitable treatments for patients. The aim of this research is to characterize the urinary stones by using dual energy micro CT SkyScan 11173. This technique combines high-energy and low-energy scanning during a single acquisition. Six human urinary stones were scanned in vitro using 80 kV and 120 kV micro CT SkyScan 1173. Projected images were produced by micro CT SkyScan 1173 and then reconstructed using NRecon (in-house software from SkyScan) to obtain a complete 3D image. The urinary stone images were analysed using CT analyser to obtain information of internal structure and Hounsfield Unit (HU) values to determine the information regarding the composition of the urinary stones, respectively. HU values obtained from some regions of interest in the same slice are compared to a reference HU. The analysis shows information of the composition of the six scanned stones obtained. The six stones consist of stone number 1 (calcium+cystine), number 2 (calcium+struvite), number 3 (calcium+cystine+struvite), number 4 (calcium), number 5 (calcium+cystine+struvite), and number 6 (calcium+uric acid). This shows that dual energy micro CT SkyScan 1173 was able to characterize the composition of the urinary stone.

1. Introduction

Urinary stones are composed of of waste products or residual substances that the body does not need in urine. When there is excessive waste, or not enough fluid to flush it out, it comes to form stones [1, 2]. The stones have a crystalline structure with the size of more than 1 mm; about 95% of the stones are crystalline components and 5% are organic components [3]. Urinary stones come in different varieties: calcium containing stones, uric acid stones, struvite stones, and cystine stones [1, 2]. Not only can urinary stones be removed by invasive surgery but also by Extra-Corporeal Shockwave Lithotripsy (ESWL) [4], percutaneous nephrolithotripsy, and ureteroscopy. Uric Acid (UA) stones can better be treated with medical dissolution therapy rather than intervention [5]. Therefore, the characterization of urinary stones is necessary to provide suitable treatment parameters during ESWL, which helps correctly diagnose the causal disease, and implement the correct treatment to prevent the recurrence of the disease [4].

Dual energy (DE) computed tomography is becoming a more available imaging method. DE is increasingly being utilized for evaluation of urolithiasis due to its unique ability to determine a urinary stone composition as either UA or non-UA [5]. DE utilizes data from two energy spectra to discriminate tissues and characterize tissue composition based on differences in the attenuation of photons with different energies by materials [6, 7].

Based on NIST graph, high energy scanning was unable to show the difference of cystine and struvite Hounsfield Unit (HU) that overlapped each other. dr. Ridwan has discovered that cystine was not detected on the urinary stones by using the high energy CT scanning 120 kV at Hasan Sadikin Hospital. Region of Interest (ROI) from the images appeared homogeneous. Thus, a combination of high energy and low energy for scanning the urinary stones by micro CT SkyScan 1173 is deemed necessary.

Micro CT SkyScan 1173 has made it possible to reach a spatial resolution of 8 μm corresponding to near 5×10^{-7} cubic mm voxel size [8]. DE micro CT acquisition is achieved by using different voltages. The object should not be moved between scans and image magnification and pixel size should be the same. We performed a study test of the ability of dual energy micro CT SkyScan 1173 to characterize the urinary stones.

2. Material and methods

Any given material will have different attenuation values when imaged at low and high kV, resulting in a predictable change in the attenuation when imaged by two known X-ray energies. Thus, two materials with similar attenuation values on a single energy but different chemical compositions can be distinguished from one another by analyzing their energy dependent changes in attenuation. The degree of material separation depends on the spectral separation between the high and low energy X-rays, accurate temporal registration and spatial correlation. This combination of high and low energy X-rays is known as dual energy scanning [5].

The method for dual energy micro CT acquisition is done by performing two scans of urinary stones at higher and lower X-ray energy. This can be achieved by using different X-ray filters and applied voltage. The object should not be moved between scans and image magnification and pixel size should be the same [8].

2.1. Preparation of samples

The experimental setup consisted of six urinary stones consecutively obtained from patients who underwent surgery at Hasan Sadikin Hospital. The six urinary stones were scanned using high energy CT of 120 kV. The six urinary stones were arranged as shown in figure 1 below.

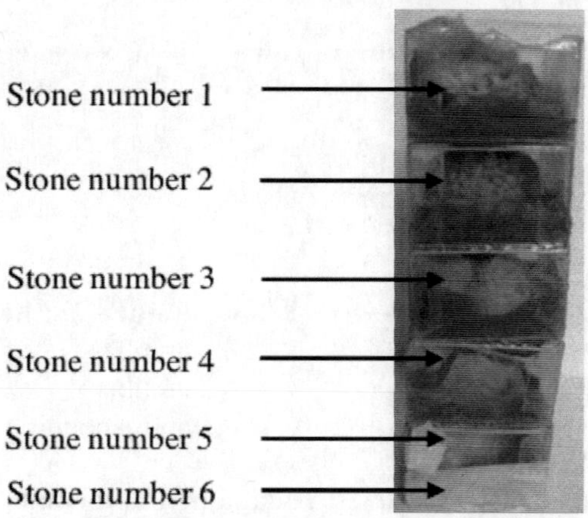

Figure 1. The six urinary stones would be scanned by dual energy micro CT SkyScan 1173.

The six urinary stones were arranged tier up in the box like figure 1. The material of box was made from mica. For keeping the urinary stones unmoved during the scanning process, there were waxed micas between them. The sample was ready to scan and put into the scanning stage inside the sample chamber to be scanned by the machine.

2.2. Acquisition of data

Dual energy micro CT SkyScan 1173 performed alternately. First, a scanning was performed with high energy of 120 kV, 66 μA, 1000 ms, using a filter, 14.96 μm/pixel, with a rotation step of 0.2°, and a rotation of 180°. The second, a scanning with low energy of 80 kV, 100 μA, 1000 ms, using a filter, 14.96 μm/pixel, with a rotation step of 0.2°, and a rotation of 180°. Post scanning involved reconstruction of the production raw (projection) images obtained, which were different from the scanning direction by NRecon software. The basic of reconstruction was performed based on the principle of Feldkamp back projection algorithm [8]. The reconstruction process produced a set of trans-axial greyscale 2D slice of the sample. Each of the reconstruction image of the urinary stones consisted of hundreds of trans-axial slices with 0.015 mm slice thickness.
A slice was taken from the center of trans-axial urinary stones. A number of ROI were selected inside the inner contours of the urinary stones margins. The ROI geometry was circular with a 15 pixels diameter [9] and the number of pixels involved was 83. The ROI position was selected in the homogeneous regions of the slice. The mean HU from several ROIs on the slice is shown in the table 1 and 2 below.

2.3. Determination of the chemical composition
Several ROI of mean HU were compared with HU from the graph of NIST (National Institute of Standards and Technology). The determination of chemical composition was performed by using percentage. The equation of percentage is represented below [10]:

$$Percentage = \frac{P_{ca}}{P_{ca} + P_{str} + P_{cyst} + P_{ua}} \times 100\% \qquad (1)$$

where P_{ca} is calcium probability in the urinary stones. To obtain P_{ca} the following formula is used:

$$P_{ca} = \frac{1}{\sqrt{\left(HU_{stone80kV} - HU_{refcal80kV}\right)^2 - \left(HU_{stone120kV} - HU_{refcal120kV}\right)^2}} \qquad (2)$$

Equation (2) is the probability of calcium on a certain ROI. Similar equation also used to calculate probability of cystine, struvite and uric acid.

3. Results and discussion
Figure 2 is a reconstruction of slice image scanned by using a voltage of 80 kV and a 0.25 mm Cu filter..

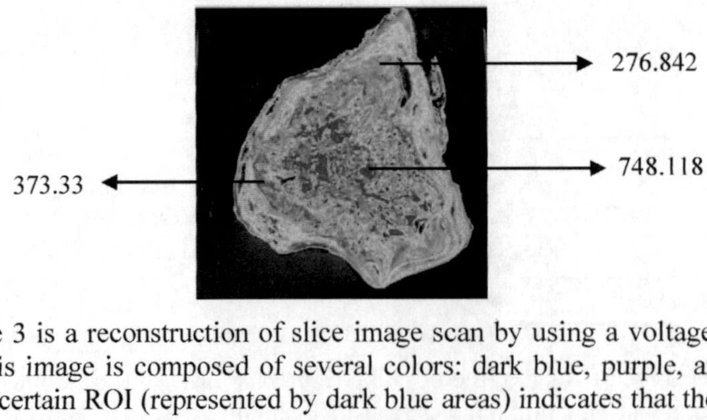

Figure 2. Slice on the stone number 1 by 80 kV. The numbers are the corresponding HU values.

Figure 3 is a reconstruction of slice image scan by using a voltage of 120 kV and a 0.25 mm Cu filter. This image is composed of several colors: dark blue, purple, and yellow. The higher mean of HU on a certain ROI (represented by dark blue areas) indicates that the X-ray was highly absorbed by the object. In the image, the mean HU of the dark blue areas was 630.359. The means HU of the shades purple are 335.137; 193.357; 163.624.

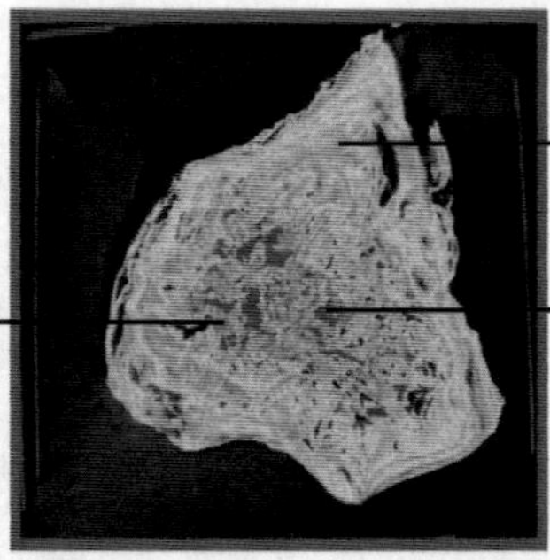

Figure 3. Slice on the stone number 1 by 120 kV. The numbers are the corresponding HU values.

The mean HUs from several ROIs from each stone are showed in the table 1 and table 2 below.

Table 1. Mean HU from several ROI for stones number 1-3.

kV	Stone number 1				Stone number 2					Stone number 3		
	ROI 1	ROI 2	ROI 3	ROI 4	ROI 1	ROI 2	ROI 3	ROI 4	ROI 5	ROI 1	ROI 2	ROI 3
80 kV	748	277	249	372	190	639	1162	309	618	401	264	153
120 kV	630	194	164	335	15	389	615	202	172	252	72	92

Table 2. Mean HU from several ROI for stones number 4-6.

kV	Stone number 4			Stone number 5			Stone number 6			
	ROI 1	ROI 2	ROI 3	ROI 1	ROI 2	ROI 3	ROI 1	ROI 2	ROI 3	ROI 4
80 kV	684	608	598	580	302	971	537	676	0	0
120 kV	305	172	457	297	78	513	161	69	-75	-33

The determination of the chemical composition was performed by using percentage. The percentage describes the probability of calcium, cystine, struvite, or uric acid on the urinary stones. To calculate the percentage, several ROIs of mean HU on the table 1 and table 2 were substituted in the equation (2). The calculation of chemical composition percentage is showed on the table 3 and table 4 below.

Table 3. The chemical composition percentage for the stones number 1 until 3.

Composition	Stone number 1				Stone number 2					Stone number 3		
	ROI 1	ROI 2	ROI 3	ROI 4	ROI 1	ROI 2	ROI 3	ROI 4	ROI 5	ROI 1	ROI 2	ROI 3
Calcium	47.67	13.27	7.30	37.52	8.35	74.78	39.11	22.33	50.63	38.79	13.10	4.26
Cystine	19.78	55.97	70.17	30.31	32.68	9.83	22.15	35.75	19.12	26.12	37.63	46.95
Struvite	18.41	21.24	16.12	20.18	44.92	9.02	21.25	28.27	17.98	22.31	35.41	41.95
Uric acid	14.14	9.53	6.40	11.99	14.05	6.37	17.49	13.64	12.28	12.79	13.86	6.84

Based on the percentage of the chemical composition percentage from several ROIs, stone number 1 consists of calcium (47.67% and 37.52%) and cystine (55.97% and 70.17%). Stone number 2 consists of calcium (74.78%; 39.11%; 50.63%) and struvite (44.92%). Stone number 3 is composed of struvite (41.95% and 35.41%), cystine (46.95% and 37.63%), and calcium (38.79%). The number 4 is of calcium (65.14%; 48.23%; 66.04%). The number 5 consists of calcium (77.63% and 44.29%), struvite (33.09%), cystine (35.67%). The last stone is composed of calcium (45.42% and 40.70%) and uric acid (48.47% and 40.93%).

Table 4. The chemical composition percentage for the stones number 4 until 6.

Composition	Stone number 4			Stone number 5			Stone number 6			
	ROI	ROI	ROI	ROI	ROI	ROI	ROI	ROI	ROI	ROI

Journal of Physics: Conference Series **694** (2016) 012053 doi:10.1088/1742-6596/694/1/012053

	1	2	3	1	2	3	1	2	3	4
Calcium	65.14	48.23	66.04	77.63	16.49	44.29	45.42	40.70	6.52	6.60
Cystine	13.44	22.86	14.46	8.86	35.67	20.61	21.61	22.26	18.99	21.64
Struvite	12.51	17.22	11.42	8.08	33.09	19.58	20.07	21.54	26.01	30.84
Uric acid	8.91	11.69	8.08	5.44	14.76	15.51	12.90	15.49	48.47	40.93

4. Conclusion

Dual energy micro CT SkyScan 1173 can be used to characterize the urinary stones based on the mean HU values. The six stones are stone number 1 (calcium+cystine), stone number 2 (calcium+struvite), stone number 3 (calcium+cystine+struvite), stone number 4 (calcium), stone number 5 (calcium+cystine+struvite), and stone number 6 (calcium+uric acid).

References

[1] Rodman J S, Sosa R E, Seidman C and Jones R 2007 *No more kidney stones* (New Jersey: John Wiley & Sons Inc) p 5 39

[2] Bonn A H, Stockholm H T, Bonn R S and Cologne B H 2009 *Urinary stones* (Switzerland: Reinhard Druck) p 4

[3] Schubert G 2006 Stone analysis *J. Urology* **34** 146-150

[4] Dalun L 2010 *Characterization of kidney stones* (Singapore: National University of Singapore) p 7

[5] Jepperson M A, Cernigliaro J G, Sella D, Ibrahim E, Thiel D D, Leng S and Haley W E 2013 Dual energy CT for the evaluation of urinary calculi Image interpretation pitfalls and stone mimics *J. Clinical Radiology* **68** e707-e714

[6] Purysko A S, Primak A N, Baker M E, Obuchowski N A, Remer E M, John B and Herts B R 2014 Comparison of radiation dose and image quality from single energy and dual energy CT examinations in the same patients screened for hepatocellular carcinoma *J. Clinical Radiology* **69** e538-e544

[7] Matlaga B R, Kawamoto S and Fishman E 2008 Dual source computed tomography a novel technique to determine stone composition *J. Urology* **72** 1164 –68

[8] Deveci S, Coskun M, Tekin M I, Peskircioglu L, Tarhan N C and Ozkardes H 2004 Spiral computed tomography role in determination of chemical compositions of pure and mixed urinary stones an in vitro study *J. Urology* **64** 237–240

[9] Claessence *Selecting filter and voltage SkyScan 1173* (Belgium: Bruker micro CT)

[10] Dougherty G 2013 *Pattern recognition and classification* (USA: Springer)

Acknowledgments

The authors would like to thanks Hasan Sadikin Hospital Bandung for the urinary stones samples. This study was fully supported by RIK (Riset & Inovasi KK), LPPM, Institut Teknologi Bandung (237h/I1.C01/PL/2015). The authors would also like to thank the Basic Science Advanced Laboratory of the Faculty of Mathematics and Natural Science Institut Teknologi Bandung for providing the access to the SkyScan 1173 micro CT scanning device.

Eye dose to staff involved in interventional and procedural fluoroscopy

D McLean[1], D Hadaya[1] and J Tse[1]

[1] Medical Physics and Radiation Engineering, The Canberra Hospital, Garran ACT, Australia

E-mail: Donald.McLean@act.gov.au

Abstract. In 2011 the International Commission on Radiological Protection (ICRP) lowered the occupational eye dose limit from 150 to 20 mSv/yr [1]. While international jurisdictions are in a process of adopting these substantial changes, medical physicists at the clinical level have been advising medical colleagues on specific situations based on dose measurements. Commissioned and calibrated TLDs mounted in commercially available holders designed to simulate the measurement of Hp(3), were applied to staff involved in x-ray procedures for a one month period. During this period clinical procedure data was concurrently collected and subject to audit. The use or not of eye personal protective equipment (PPE) was noted for all staff. Audits were conducted in the cardiac catheterisation laboratory, the interventional angiography rooms and the procedural room where endoscopic retrograde cholangiopancreatography (ERCP) procedures are performed. Significant levels of occupational dose were recorded in the cardiac and interventional procedures, with maximum reading exceeding the new limit for some interventional radiologists. No significant eye doses were measured for staff performing ERCP procedures. One outcome of the studies was increased use of eye PPE for operators of interventional equipment with increased availability also to nursing staff, when standing in close proximity to the patient during procedures.

1. Introduction

In 2011the ICRP made a statement lowering the occupational eye dose limit from 150 to 20 mSv/yr [1]. This significant change resulted from new evidence of the risk of development of lens opacities or cataracts from studies with greater sensitivity and longer follow up periods to the initial studies [2]. Analysis by the International Atomic Energy Agency (IAEA) [3] has identified the implications of this change in dose limit as having the largest impact on occupational dose in the medical sector [4] for staff working in close proximity to patients undergoing fluoroscopically or CT guided interventional procedures, staff involved in preparation of sources or radiopharmaceuticals, or in manual brachytherapy and staff working with cyclotrons.

In response to this new understanding the monitoring of occupational dose has gained increased importance along with the established work of optimization of patient dose. The extent of occupational risk needs to be determined at each institution. In our hospital three audits of occupational eye dose have been undertaken so far, involving staff in Cardiac catheterization, Angiographic interventional studies and the Gastroenterology & Hepatology unit when performing ERCP procedures.

The clinical context for each of the above three audits was distinct, although all included medical staff, essentially the operators of the equipment and in close proximity to the irradiated area of the patient, assisted by nursing staff who on occasion could also be close to the radiation field. Radiographic staff were in attendance (except in Cardiology) but typically remote from the radiation field, with anesthetic staff in attendance for the ERCPs, also remote from the radiation field.

A review of occupational dose results in the UK and elsewhere has been published for cardiac and angiographic procedures [5]. A number of reports on ERCP doses for both patients and staff, illustrating a wide range of determined eye doses have also been published [6-8]. The current work was then designed to determine the dose to the lens of the eye for occupational workers in the three environments and also to determine the eye dose per tube output, to assist in determining areas where dose could be reduced through a change in practice.

2. Methods and materials

2.1. TLD dosimetry

The dose to the lens of the eye was measured using TLDs of type-100H (Thermo Scientific, Ohio, USA) with dimension $3.2 \times 3.2 \times 0.38$ mm^3, which were selected because of their good relative energy response in the diagnostic x-ray energy range and enhanced sensitivity (uncertainty of +/- 5% from calibration energy). The TLDs were mounted in custom made holders (Eye-D™, RadCard, Krakow, Poland) that simulate the dose to the lens of the eye [9] designed for the measurement of Hp(3) doses [10]. Data from the manufacturers indicated that the relative angular response was extremely uniform (uncertainty of +/- 5% from normal to 80° incidence).

The TLDs were commissioned under ISO 12794 [11] using a Harshaw 5500 TLD reader (Thermo Scientific, Ohio, U.S.A.). The TLDs were calibrated free in air outside of the holders through the use of TLDs reserved for calibration and background correction. The calibration beam was generated with a medium frequency x-ray generator at tube voltage of 80 kV with a measure half-value-layer of 4 mm Al. The x-ray equipment was under regular quality control to ensure its performance complies with the regulatory requirements. The dose to the lens of the eye was calculated using the following formula:

$$\text{Eye lens dose} = [\mu_{en}/\rho(eye)/\mu_{en}/\rho(air)]RER.k.M$$

where RER is the relative energy response (held to unity), k is the TLD calibration factor, and M is the background subtracted TLD reading. Mass energy absorption coefficients [12] were for the effective energy of the calibrating beam. The associated uncertainty, taking into account the relative energy and angular response, the mass energy absorption coefficient variation across the possible energy range and TLD reading accuracy was estimated to be about 10% for readings above 1 mSv and about 14% for reading just above background levels.

2.2. Clinical measurement and data collection

For each of the three clinical environments (table 1) where eye dose measurements were made the same methodology was used. Each staff member present in the room during a fluoroscopic procedure was assigned an individual eye holder that was attached to an elastic head band, which allowed the eye holder to be placed near the corner of the left eye and outside the protection of the lead glasses if worn. This positioning was ascertained to typically receive the maximum dose.

Table 1. Clinical environments investigated.

Clinical situation	No. staff involved	procedures	average KAP (Gycm^2)	average procedure time (min)	No. X-ray units
Cardiac laboratory	17	192	34.8	32.8	2
Angiography Suite	18	93	242.3	67.9	2
ERCP procedures	14	34	5	20	1

The measurements were made over a one month period at each location. During this time, reference data was collected; including patient data such as age, height, weight; radiographic factors such as the tube output in terms of kerma area product (KAP) and cumulative air kerma (CAK), as well as the staff present for each procedure, their duty and location and whether they wore lead glasses to protect their eyes. This data was gathered for each procedure under the scrutiny of each laboratory coordinator before final review of data integrity.

2.3. Data analysis

For each clinical environment the staff's use of protective lead glasses and workload data was established by analysis of each staff member to determine the number of procedures they had been present for and also the cumulative KAP value for those particular procedures. The proportion of the total tube output that staff was protected by lead glasses was calculated. Further the measured eye doses were normalized by the cumulative KAP to give the eye dose rate for each individual.

3. Results

Analysis of the collected data revealed a significant spread in the eye doses received by staff within and also between clinical environments as seen in table 2. It was noted in cardiology that there were 2 consultants with recorded eye doses at 50% of the ICRP recommended limit. However the eye doses in angiography were of a higher magnitude, with three medial consultants having eye doses in excess of the ICRP limit (figure 1), requiring a report to be made to the local regulators. The doses in ERCP were low as is consistent with the nature of the radiation exposure used by this environment as seen in table 1.

Table 2. Statistical breakdown of the extrapolated annual eye dose (mGy) for both medical staff and nurses in each of the clinical environments.

	Cardiology		Angiography		ERCP	
	medical	nurse	medical	nurse	medical	nurse
minimum	0.68	0.49	2.40	0.65	0.21	0.17
1st quartile	1.41	0.78	12.16	1.13	0.30	0.18
median	1.54	1.32	18.05	6.06	0.37	0.21
3rd quartile	4.45	2.59	**26.55**	7.39	1.00	0.35
maximum	9.64	3.09	**59.01**	14.88	2.77	0.46

The cumulative protection from lead glasses to staff for each environment was 50% in cardiology, 30% in angiography and nil for ECRP procedures. The rate of eye dose as a function of tube output, as seen in Table 3, again showed a marked difference between the clinical environments. Of interest the readings from ERCP are the highest, despite the fact that these staff experienced the lowest total doses.

Table 3. Statistical breakdown of the eye dose/KAP (μGy/Gycm^2) for both medical staff and nurses in each of the clinical environments

	Cardiology		Angiography		ERCP	
	medical	nurse	medical	nurse	medical	nurse
minimum	0.13	0.05	0.19	0.04	0.96	0.26
1st quartile	0.25	0.08	0.56	0.06	1.24	0.29
median	0.34	0.13	0.93	0.10	1.51	0.36
3rd quartile	0.43	0.18	1.64	0.19	1.90	0.40
maximum	0.56	0.23	2.04	0.48	1.51	0.61

4. Discussion

The audits revealed that for each of the clinical environments, that the use of protective glasses was not thought a high priority. However, McVey *et al.* [13] have shown that the wearing of correctly

Journal of Physics: Conference Series **694** (2016) 012054 doi:10.1088/1742-6596/694/1/012054

fitted leaded glasses reduces dose to the eye by approximately 80%. In the case of the angiographic suite Figure 1 clearly demonstrates staff members that measured as over 50% (yellow) and 100% (red) of the ICRP limit. Staff awareness of this data has resulted in the almost universal wearing of lead glasses in this environment.

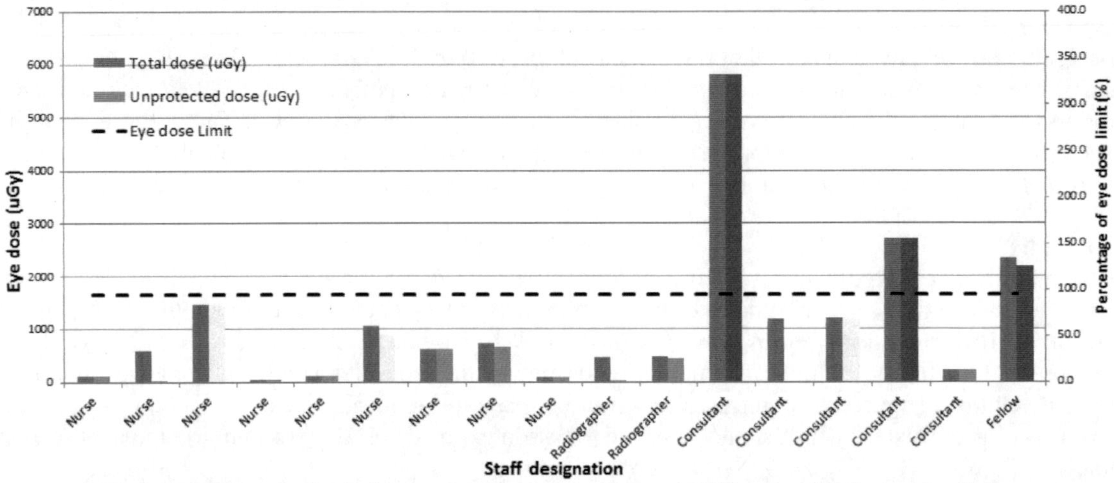

Figure 1. Range of eye doses measured over one month for the Angiography Suite.

Upon review of the audit results a number of components critical to dose reduction were identified. Comparison of typical tube output for similar procedure types in cardiology revealed that the older of the two x-ray units on average used 40% more radiation dose that would be expected to contribute directly to a similar increase in eye dose. This unit has subsequently been replaced and a recent dose audits reveals the new unit is operating at considerably lower dose to that of the previous units.

Conduct of the procedure has also been found to have a significant impact of occupational dose. It was noted that in some procedures that staff maintained positions very close to the patient during acquisition runs. While this might be considered necessary in some cases, the use of continuous collar radiation monitors (Rad Safe i2) has demonstrated real time the importance of considering the proximity to the patient in scatter dosage and as appropriate the importance of properly utilising protective lead glass screens. Further study is planned to quantify changes in eye dose due to heightened staff awareness.

One surprise of the audits was the relatively high dose rate found in ERCP. It must be stressed that the actual eye dose levels were low, however when normalised with tube output, there were indications that some changes in practice might be useful. It was further noted that in the room used for ERCP there is not overhead moveable shield available, which might have contributed to higher eye dose per tube output. It should be noted that the configuration of the room, utilising under table tube, was otherwise consistent with best practice [6].

One important outcome of the work was the realisation that nursing staff, while not the closest to the radiation source, often are present for a large number of procedures. Often they also may be in close proximity to the patient, but without the additional protection of a mobile lead screen. As seen in Figure 1 three of the highest six eye dose measurements in angiography were received by nurses. For this reason changes have been made to the relevant radiation management plans to ensure provision of lead glasses for nurses, should they elect to wear them during procedures.

5. Conclusion

The use of TLD dosimetry in combination with well-designed TLD holders that allow acceptable clinical implementation, can add substantially to knowledge on staff eye dose due to high dose rate fluoroscopy. The three audits undertaken had a number of beneficial outcomes on clinical practice, including the increased usage of lead glasses, examination of procedural technique to reduce both patient and staff dose, and a recognition that nursing staff need to be considered when assessing risk to the eyes from x-ray exposure during fluoroscopy.

References

[1] International Commission on Radiological Protection 2011 *Statement of Tissue Reactions*, ICRP ref 4825-3093-1464 (ICRP, Elsevier)

[2] International Commission on Radiological Protection 2012 *ICRP Statement on Tissue Reactions and Early and Late Effects of Radiation in Normal Tissues and Organs – Threshold Doses for Tissue Reactions in a Radiation Protection Context* (ICRP Publication 118, Elsevier)

[3] International Atomic Energy Agency, Implications for occupational radiation protection if the new dose limits for the lens of the eye, IAEA-TECDOC-1731, IAEA Vienna. http://www-pub.iaea.org/books/IAEABooks/10628/Implications-for-Occupational-Radiation-Protection-of-the-New-Dose-Limit-for-the-Lens-of-t

[4] International Commission on Radiological Protection 2000 *Avoidance of Radiation Injuries from Medical Interventional Procedures, ICRP Publication 85* (Oxford and New York: Pergamon Press)

[5] Martin C J and Magee J S 2013 Assessment of eye and body dose for interventional radiologists, cardiologists, and other interventional staff *J Radiol Prot* **33(2)** pp 445-60

[6] O'connor U, Gallagher A, Malone L and O'reilly G 2013 Occupational radiation dose to eyes from endoscopic retrograde cholangiopancreatography procedures in light of the revised eye lens dose limit from the International Commission on Radiological Protection *Br J Radiol* **86** 1022 pp 20120289

[7] Naidu L S, Singhal S, Preece D E, Vohrah A and Loft D E 2005 Radiation exposure to personnel performing endoscopic retrograde cholangiopancreatography, *Postgrad Med J* **81** 960 pp 660-2

[8] Health Protection Agency 2005 Doses to patients from radiographic and fluoroscopic X-ray imaging procedures in the UK - 2005 Review, HPA Rep. HPA-RPD-029 Chilton

[9] Gualdrini G *et al.* 2011 Eye lens dosimetry: task 2 within the ORAMED project *Radiat Prot Dosimetry* **144** 1-4 pp 473-7

[10] Borges F L S, Guimarães M C, Da Silva T A and Tavares M S N 2014 Calibration of an eye lens dosemeter in terms of Hp(3) to be used in interventional radiology *Radiation Physics and Chemistry* **104** pp 45–47

[11] International Organization for Standards, Individual Thermoluminescence Dosemeters for Extremities and Eyes Rep. ISO 12794-1 Geneva, Switzerland

[12] Hubbell J H and Seltzer S M NIST X-Ray Mass Attenuation Coefficients, Table 4, http://physics.nist.gov/PhysRefData/XrayMassCoef/tab4.html.

[13] McVey S, Sandison A and Sutton D G 2013 An assessment of lead eyewear in interventional radiology *J Radiol Prot* **33** 3 pp 647-59

Psychophysiological interaction between superior temporal gyrus (STG) and cerebellum: An fMRI study

A N Yusoff[1], X L Teng[1], S B Ng[1], A I A Hamid[2] and S Z M Mukari[1]

[1] Faculty of Health Sciences, Universiti Kebangsaan Malaysia, Jalan Raja Muda Abdul Aziz, 50300 Kuala Lumpur, Malaysia
[2] Otto-Von-Guericke University Magdeburg, Biomedical Magnetic Resonance Leipziger Str. 44, D-39120 Magdeburg, Germany

E-mail: nazlimtrw@ukm.edu.my

Abstract. This study aimed to model the psychophysiological interaction (PPI) between the bilateral STG and cerebellum (lobule VI and lobule VII) during an arithmetic addition task. Eighteen young adults participated in this study. They were instructed to solve single-digit addition tasks in quiet and noisy backgrounds during an fMRI scan. Results showed that in both hemispheres, the response in the cerebellum was found to be linearly influenced by the activity in STG (vice-versa) for both in-quiet and in-noise conditions. However, the influence of the cerebellum on STG seemed to be modulated by noise. A two-way PPI model between STG and cerebellum is suggested. The connectivity between the two regions during a simple addition task in a noisy condition is modulated by the participants' higher attention to perceive.

1. Introduction

The psychophysiological interaction (PPI) analysis was aimed to identify regions whose activity depended upon the interaction between physiological factors (the time course of a region of interest) and psychological factors (the tasks) [1]. The principal idea underlying PPI is that if two brain areas are interacting, the level of activity in those areas will correlate over time. In the situation where the correlation coefficient of the two interacting brain areas changes with any experimental factors, PPI can be assumed to exist. We investigated the influence of background (noise) on the connectivity between bilateral STG and cerebellum during the performance of simple addition task in quiet (AIQ) and in noise (AIN) using a functional magnetic resonance imaging (fMRI) technique. The objectives of the study were to determine the psychophysiological interaction (PPI) [1, 2] between the STG and cerebellum in each particular hemisphere and to investigate the influence of background noise in modulating the effective connectivity between STG and cerebellum during a simple addition task in quiet and in noise. If the activity in the STG correlates to that of the cerebellum and vice versa, we should see a linear relationship between the activities of the two regions [3]. If any of the two regions interacts with the experimental factor that exist (such as a noisy condition that distracts attention) in order to exert or to influence the activity in another region, we should see a change in the regression slope on the plot of the response of the second region vs. the activity of the first region [4].

In the present study, we investigated the PPI between STG and cerebellum in both hemispheres during the performance of simple addition task. Subtraction task is not included in this study because the results obtained are assumed to be similar to addition task.

2. Methods

Eighteen native, Malay-speaking, male adults participated in the present study. The average age and standard deviation of the participants were 23.2 ± 2.5 years (ranged between 20 to 28 years). All participants reported no history of the psychiatric or neurological disorder and no current use of any psychoactive medications. All participants were confirmed to be right handed with average laterality index (LI_{ave}) of 87 in the range of sixth to the seventh right. The fMRI data were analysed using statistical parametric mapping (SPM8) [5]. Details about the analysis of the fMRI data are given elsewhere [6]. Bilateral STG and cerebellum were chosen for PPI analysis due to their important roles as processing centre in a task involving auditory and cognitive functions [7]. The PPI analysis was undertaken separately for each hemisphere. The response in one region of interest (ROI), known as the target region, in terms of an interaction between the influence of another ROI, known as the source region and the background condition was investigated. A 4-mm radius sphere (VOI = volume of interest) with the point of maximum intensity as the centre was drawn from each ROI in both quiet and noisy backgrounds. The PPI response in the target region was then plotted as a function of the activity in the source region, assumed to be interacting with the experimental factor (noise) and thereby influencing the response in the target region. PPI is considered to exist if there is a significant change in the slope of the regression line (Friston *et al.* 1997), which is the plot of the response in a target region vs. the activity in the source region, either with or without the interaction with the experimental factor [1]. A simple linear regression analysis was performed using IBM SPSS Version 20 to investigate the interaction between the ROIs in quiet and noisy backgrounds at $p = 0.05$ (95% confidence interval).

3. Results

Figure 1(a) illustrates the response in the right cerebellum plotted against the activity in the right STG, while figure 1(b) shows the response in the right STG plotted against the activity in the right cerebellum, as obtained from the PPI analysis. The best-fit line equation, the R^2 value and the number of participants involved (N) are shown on each of the PPI figures. It can be seen from both plots that the response over the target region (vertical axis) increased linearly and significantly ($p < 0.05$) with the increase in the activity within the source region (horizontal axis) for AIQ and AIN. The activity in the right STG had a similar influence upon the response in the right cerebellum in both conditions as indicated by the same regression coefficient values (figure 1(a)).

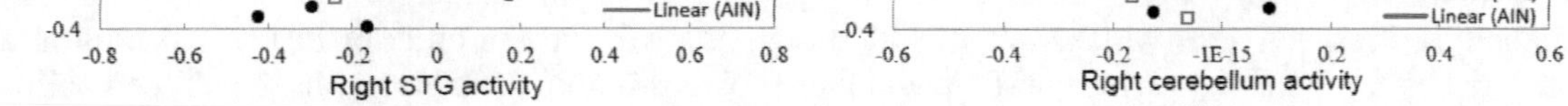

(a) (b)

Figure 1. Results of psychophysiological interaction (PPI) analysis between right hemisphere STG and cerebellum for (a) right STG as the source region and (b) right cerebellum as the source region.

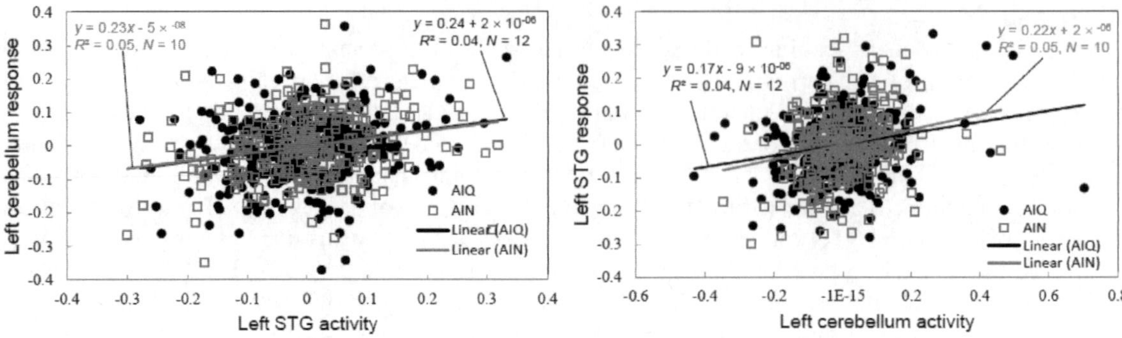

(a) (b)

Figure 2. Results of psychophysiological interaction (PPI) analysis during AIQ and AIN between left hemisphere STG and cerebellum for (a) left STG as the source region and (b) left cerebellum as the source region.

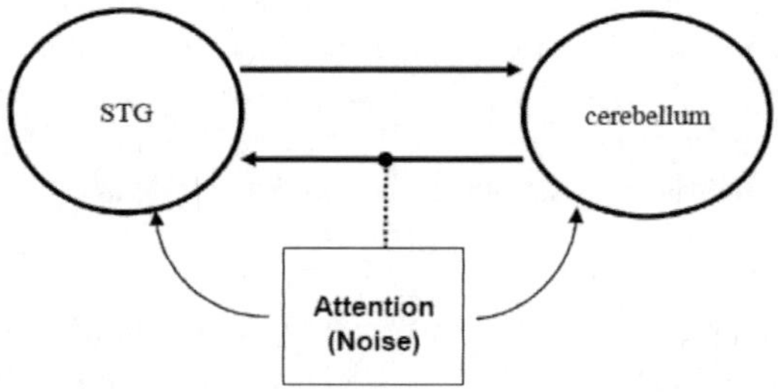

Figure 3. PPI models with attention (noise) to the stimuli as the experimental factor; connectivity from STG to cerebellum is not modulated while connectivity from cerebellum to STG is modulated (dotted line) by the noise.

However, the influence that the right cerebellum has on the right STG during the performance of a simple addition task seemed to increase in noise as compared to in quiet as shown by the steeper slope of the red regression line (figure 1(b)).The steeper slope of the red regression line suggested that noise increased the influence of the cerebellum on the right STG during simple addition task.

Similarly, figure 2(a) illustrates the response in the left cerebellum plotted against the activity in the left STG, while figure 2(b) shows the response in the left STG plotted against the activity in the left cerebellum. The response over the target region (vertical axis) again increased linearly and significantly ($p < 0.05$) with the increase in the activity within the source region (horizontal axis) for both AIQ and AIN conditions. The activity in the left STG also had a similar influence upon the response in the left cerebellum irrespective of whether simple addition was performed in quiet or in noisy condition as indicated by the same regression coefficient values shown in figure 2(a). However, the left cerebellum activity seemed to have some influence on the left STG response that was modulated by the experimental factor (noise) as shown by the steeper slope of the red regression line in figure 2(b).

Figure 3 shows the suggested PPI model that can be used to explain the effective connectivity between the STG and cerebellum during AIQ and AIN with noise as the experimental factor, modulating the influence the cerebellum has on STG but not on the influence the STG has on the cerebellum. Noise, however, had a direct influence on both the STG and cerebellum.

4. Discussion

The PPI between the two areas is presented graphically by plotting the PPI data obtained from the target region against the PPI data obtained from the source region. Figure 1 and 2 illustrate the PPI between right STG and right cerebellum (lobule VI) and left STG and left cerebellum (lobule VII), respectively. Both figures show that the response over the target region increases linearly with the increase in activity within the source region regardless of the task being performed in quiet (dark line) or in noise (red line). From figure 1 (right hemisphere), it can be seen that there is an increase in the slope (gradient) of the line from 0.22 for AIQ to 0.30 for AIN but only when the right cerebellum (lobule VI) acts as the source region. No change in the slope is observed when STG acts as the source region from which the slope remains at 0.20 for both the AIQ and AIN lines. Similarly, in the left hemisphere, there is an increase in the slope (gradient) of the line from 0.17 for AIQ to 0.22 for AIN, again only when the left cerebellum (lobule VII) acts as the source region. No change in the slope is observed when STG acts as the source region from which the slope remains at 0.24 and 0.23 for both the AIQ and AIN lines respectively. The slope on a line reflects the influence that the source region exerted over the target region. The change in the slope on a plot indicates the presence of PPI [1], with noise (increase in attention) as the experimental factor that modulates the influence that the activity within the source region has on the response over the target region. Thus, a steeper plot slope means that a small change in the activity of the source region results in a relatively larger response over the target region.

Figures 1(a) and 2(a) show that the slope of the AIQ line equals that of the AIN line (STG as the source region and cerebellum as the target region regardless of the hemisphere of the brain). Thus, PPI is not observable in the plots of the response of the right cerebellum (lobule VI) against the activity of the right STG and the response of the left cerebellum (lobule VII) against the activity of the left STG. These results indicate that the connectivity from STG to the cerebellum is not influenced by any experimental factors that exist, which is noise, in this case, see figure 3.

Figures 1(b) and 2(b) show interesting findings (cerebellum as the source region and STG as the target region). The slope of the AIN line is larger than that of the AIQ line for the right hemisphere regions as mentioned above. This is based upon the increase in the slope value when the task is performed in a noisy background (comparing the slope values depicted in figure 1(b)). Thus, PPI is said to be present in the plot of the response of the right STG against the activity of the right cerebellum (lobule VI). These results indicate that the connectivity from the cerebellum to STG in the right hemisphere is influenced or modulated by the experimental factor (noise) that exists, as shown in figure 3. In the presence of distraction such as in a noisy condition, participants' effort in accomplishing the task would increase to overcome the disturbance from the noise which is reflected by an increase in the right cerebellum activity. As a result, the activity within the source region (lobule VI) interacts with the experimental factor (an increase in attention due to noisy condition) and exerts a larger influence on the target region. In the left hemisphere, PPI also appears to exist in the plot of the response of the STG against the activity of the left cerebellum (lobule VII). The influence that the left cerebellum (lobule VII) has on the left STG is also enhanced by the presence of noise. This is based upon the increase in the slope value when the task is performed in noise (comparing the slope values depicted in figure 2(b)). It can be said that the activity within the source region (lobule VII) interacts with the experimental factor (noise) and exerts a larger influence on the target region. The suggested PPI model is shown in figure 3. The experimental factor also has a direct influence on the respective regions. In the absence of noise, participants would be able to listen to the given stimuli better without any needs to increase their attention to overcoming distraction, which is shown by the smaller slope value.

In the present study, bilateral STG and cerebellum were chosen to be the source and target regions (interchangeably) due to their important roles in arithmetic working memory [8]. On the one hand, STG is assumed to act as the input centre for auditory processing whereby auditory signals from sub cortical areas were received before being transmitted to other cortical areas for further processing [9]. In addition to that, STG has been well-known about its role in processing verbal (e.g. speech or word numbers) and non-verbal (e.g. pure tones and noise) stimuli [6]. On the other hand, earlier imaging studies have highlighted the role of the cerebellum in working memory processes and various other processes related to attention [10], in addition to its well-known role in coordinating motor movement [11].

In this study, one of the experimental factors to be considered is the 80-dB white noise in which the 83-dB stimuli are embedded. The presence of the noise during the task performance increases the difficulty for the participants to listen to the delivered stimuli thus forcing the participants to increase their efforts to overcome distraction in order to accomplish the task, e.g. paying more attention to the given stimuli [12]. Interestingly, the presence of noise only affects the connectivity from cerebellum to STG either in the right or left hemisphere (figure 3) but not the other way around. As mentioned above, the main role of STG is to process sound stimuli that are delivered through the two ears [9]. The information gathered in the STG is then transmitted to other brain areas depending upon the type of task being executed, regardless of whether the task is performed in a quiet or in a noisy background. In other words, the information that was sent to the cerebellum is not affected by the noise. These explain why the connectivity from STG to the cerebellum is free from being influenced or modulated by the interaction between the noise and the activity in the STG. In contrast, in solving the arithmetic addition problems in a noisy background, cerebellum comes into play due to its cognitive functions such as for attention and language [13]. Furthermore, the cerebellum is involved in attenuating noise and/or increasing attention to task performance and is connected to STG in the process [14]. Thus, it is quite reasonable to suggest that the information that was sent back to STG from cerebellum was modulated by the interaction between the experimental factor (noise) and the activity in the cerebellum itself.

5. Conclusion

PPI results indicated that the connectivity from STG to the cerebellum in both hemispheres is not influenced by the experimental factor. In contrast, the connectivity from the cerebellum to STG is modulated by the interaction between the activity of STG and the experimental condition, e.g. noise. This suggests that the persisting interaction when the cerebellum acts as the source region is modulated by the noise which could be due to the increase in attention in order for the participants to get a better perception of the stimuli in a noisy surrounding. These results strengthen previous findings on the involvement of the bilateral cerebellum in arithmetic working memory processing and might be useful in explaining brain performance in suppressing irrelevant information for patients with disordered brain function.

References

[1] Friston K J, Buechel C, Fink G R, Morris J, Rolls E and Dolan R J 1997 *NeuroImage* **6(3)** 218
[2] O' Reilly J X, Woolrich M W, Behrens T E J, Smith S M and Johansen-Berg H 2012 *SCAN* **7** 604
[3] Yusoff A N 2013 *Sains Malaysiana* **42 (6)** 765
[4] Yusoff A N, Manan H A, Mukari S Z M, Hamid K A and Franz E A 2013 *Mod. App. Sci.* **8 (5)** 97
[5] Friston K J 2002 Statistics I: Experimental Design and Statistical Parametric Mapping. In W T Arthur & C M John (Eds.), Brain Mapping: The Methods (Second Edition) (pp. 605-631). San Diego: Academic Press.
[6] Hamid A I A, Yusoff A N, Mukari S Z M and Mohamad M *Mal. J. Med. Sci.* **18 (2)** 3
[7] Abdul Manan H, Yusoff A N, Franz E A and Mukari S Z M 2013 *ISRN Neurology* **13**
[8] Zhou X, Chen C, Zang Y, Dong Q, Chen C, Qiao S and Gong Q 2007 *NeuroImage* **35(2)** 871
[9] Langers D R, van Dijk P and Backes W H 2005 *NeuroImage* **48(2)** 490
[10] Stoodley C J and Schmahmann J D 2009 *NeuroImage* **44(2)** 489
[11] Yusoff A N 2014 *Sains Malaysiana* **43(2)** 313
[12] Manan H A, Yusoff A N, Franz E A, Mukari S Z M 2013 *Neurology, Psychiatry and Brain Research* **19(4)** 207
[13] Wolf U, Rapoport M J and Schweizer T A 2009 *J. Neuropsychia. Clin. Neurosci.* **21(3)** 245
[14] Abdul Manan H, Franz E A, Yusoff A N and Mukari S Z M 2012 *Psychol. Neurosci.* **5(2)** 247

Acknowledgements

The authors would like to thank the Department of Radiology, UKMMC for the permission to use the MRI scanner and the Audiology Program, Faculty of Health Sciences UKM for conducting a hearing test on the participants. This work was supported by the research grant UKM-GUP-SK-07-020-205.

Journal of Physics: Conference Series **694** (2016) 012056 doi:10.1088/1742-6596/694/1/012056

Development of a software tool for an internal dosimetry using MIRD method

A Chaichana[1] **and C Tocharoenchai**[1]

[1] Department of Radiological Technology, Faculty of Medical Technology, Mahidol University, Bangkok, Thailand

E-mail: anucha.cha@student.mahidol.ac.th

Abstract. Currently, many software packages for the internal radiation dosimetry have been developed. Many of them do not provide sufficient tools to perform all of the necessary steps from nuclear medicine image analysis for dose calculation. For this reason, we developed a CALRADDOSE software that can be performed internal dosimetry using MIRD method within a single environment. MATLAB software version 2015a was used as development tool. The calculation process of this software proceeds from collecting time-activity data from image data followed by residence time calculation and absorbed dose calculation using MIRD method. To evaluate the accuracy of this software, we calculate residence times and absorbed doses of 5 Ga-67 studies and 5 I-131 MIBG studies and then compared the results with those obtained from OLINDA/EXM software. The results showed that the residence times and absorbed doses obtained from both software packages were not statistically significant differences. The CALRADDOSE software is a user-friendly, graphic user interface-based software for internal dosimetry. It provides fast and accurate results, which may be useful for a routine work.

1. Introduction

In nuclear medicine, radionuclides are used in variety of diagnostic and therapeutic procedures. Internal dosimetry is essential to evaluate the risk and benefit of any procedures. It is also used to optimize imaging techniques and develop new radiopharmaceuticals. Internal radiation dosimetry in nuclear medicine is usually performed using MIRD (Medical Internal Radiation Dose) method [1], which is based on measurement of the biokinetics data of serial image scans. There are three steps of calculation as follows: first, the time-activity data for each source organ is determined, then the residence time in each source organ is calculated and finally, the radiation absorbed dose in each target organ is determined using MIRD method. To manually perform internal dosimetry is time-consuming and errors can occur in each step leading to developing software tools to ease users.

There are many software packages available, however, many of them have limited functions. For example, OLINDA/EXM software [2] uses residence time to calculate absorbed dose per unit of administered activity. It also provides EXM tool to plot time-activity data and fit the data to multi-exponential function, which is then integrated to calculate residence time. Whereas, separate software such as ULMDOS [3] or SPRIND [4] is required to obtain the time-activity data that is used as input for OLINDA/EXM.

To include all three steps in one software tools, we have developed the ease-of-use, platform independent software named CALRADDOSE, which provides tools used for internal dose calculation using MIRD method in a single environment.

2. Materials and methods

2.1 Software development

Our software namely 'CALRADDOSE' was developed using MATLAB (MATrix LABoratory) program version 2015a. The development process covered the design of graphic user interfaces (GUIs), data structure, and calculation algorithms. The Graphic User Interface Development Environment (GUIDE) tool was used to create software GUIs. The dose calculation was performed using MIRD method. The calculation process in this software was divided into three major steps: time-activity data collection, residence time calculation, and dose calculation. Figure 1 showed a workflow of the software design including 'Image Processing UIs', 'Residence Time Calculation UI', and 'Dose Calculation UI', which provided functional units of the calculation process.

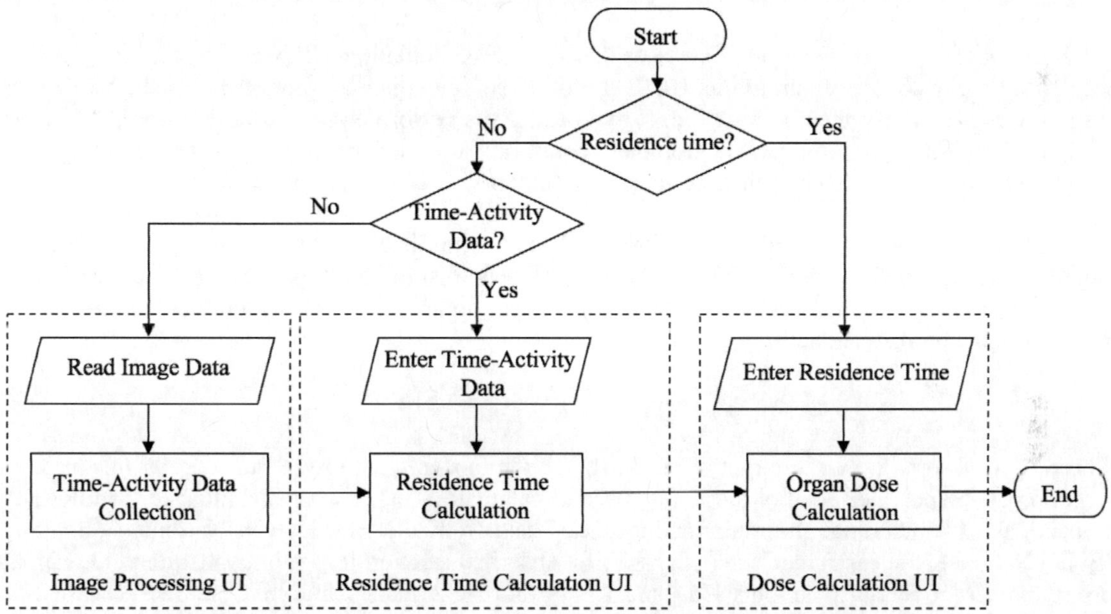

Figure 1. A flowchart of the calculation process of CALRADDOSE.

2.1.1 Time-activity data (TAD) collection. To collect time and activity from image data, the Image Processing UIs including 'Planar Image Processing UI' and 'SPECT Image Processing UI' were created to read DICOM file, draw ROI, obtain numbers of count and calculate count rate.

A simple screen shot of the 'Planar Image Processing UI' was presented in figure 2. Anterior and posterior whole-body images were read and displayed side-by-side. Study information including study description, study date-time, acquisition date-time, were automatically read from DICOM header. This UI provided a simple tool for drawing a region of interest (ROI) of source organ into image display. After an ROI was drawn on an anterior image, this ROI was copied to a posterior image, and also copied to series image at a time different. Counts from ROIs were background corrected [5] and averaged using geometrical mean [6], and finally, count rate was obtained.

For 'SPECT Image Processing UI', SPECT image data in transversal plane were read and used to generate images in coronal and sagittal plane. The user can determine the organ VOI by manual drawing ROIs combined with thresholding-based segmentation. Count rate was calculated by dividing total count in VOI with scanning time (frame time x numbers of frame).

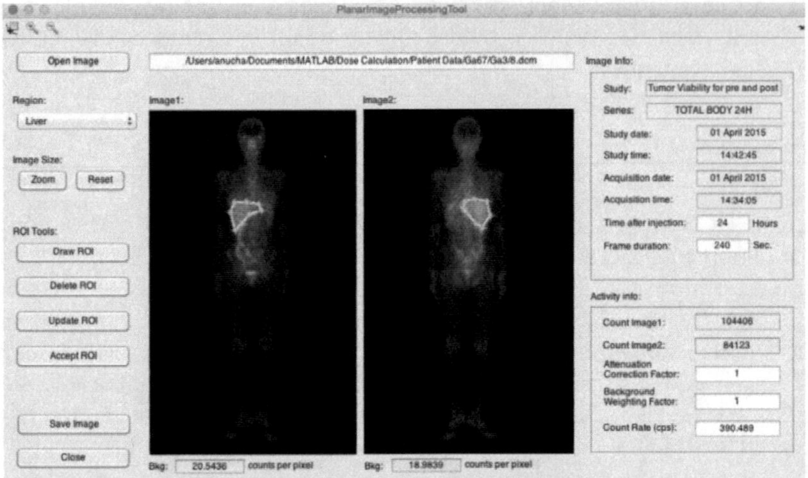

Figure 2. A simple screen shot of the 'Planar Image Processing UI' includes two planar images with liver ROIs.

2.1.2 Residence time calculation. The 'Residence Time Calculation UI' was created to calculate a residence time in source organ. In this UI, TADs of all source organ are plotted and fitted with either a single- or two-term exponential function. Curve fitting was performed using the Levenberg-Marquardt method [7, 8]. After the fitting was completed, count rate at each time point from zero to infinity was integrated and then divided by initial count rate of total body data to get residence time.

2.1.3 Dose calculation. The absorbed doses per unit administered activity in target organs were calculated using MIRD method. The general concept was to sum all the product of residence time for source organs (τ_h) and the S-values for the source-target organ $S(k \leftarrow h)$, which yielded the absorbed dose per unit administered activity as:

$$\frac{D_k}{A_0} = \sum_h \tau_h S(k \leftarrow h) \quad (1)$$

where D_k is absorbed dose to target organ k; A_0 is administered activity; h and k refer to source and target organ, respectively. All of these data used to calculate S-value were downloaded from RADAR website [9, 10] including: the radionuclide decay data from the Brookhaven National Laboratory's (BNL) National Nuclear Data Center, the specific absorbed fraction for various phantom [11, 12], and organ masses of Stabin phantoms [11] and Cristy and Eckerman phantoms [12, 13]. The software calculated S-value for each target organ by considering each emission in decay scheme. After calculation of S-value and residence time, the absorbed dose in target organ was calculated using equation (1). For photon emission, the absorbed dose was scaled with the cube root of organ mass for self-irradiation. Whereas alpha and beta emissions, the dose calculation needed to be linearly scaled with organ mass [2, 14].

2.2 Software evaluation

In this study, software evaluation was focused on the accuracy of residence time calculation and organ dose calculation by comparing with those from OLINDA/EXM. Planar images of five Ga-67 and five I-131 MIBG studies were used. The TAD were obtained from series of planar whole-body images at 24, 48 and 72 hours after injection for all the patients using 'Planar Image Processing UI'. The selected source organs were liver, spleen, and total body. These time-activity data were used to calculate residence time for all source organs and then organ doses with the CALRADDOSE and OLINDA/EXM. The differences of the results from both software were compared using a paired t-test with 95% confident interval (p-value<0.05).

3. Results and discussion

Table 1 shows residence times from 10 patients calculated from CALRADDOSE and OLINDA/EXM. The average percent difference was 0.76. A paired t-test was performed and there was no statistically significant difference in the residence time calculated by CALRADDOSE and OLINDA/EXM with p-value = 0.329 (>0.05). That meant the residence times obtained from both software were similar.

Table 2 shows organ doses from patient no.1 calculated from both software. The average percent difference was 2.26. The results showed that the organ dose calculated from both software also were not significant difference with p-value = 0.228 (>0.05). However, there were high percent differences due to difference calculation of S-value, which could be affected by many reasons. First, both software used

Table 1. A comparison of residence time from total body data calculated by CALRADDOSE and OLINDA/EXM.

Patient	Injected Nuclide	Calculated Residence Time (MBq-s/MBq)		Difference (%)
		CALRADDOSE	OLINDA/EXM	
1	Ga-67	94.49	94.00	0.52
2	Ga-67	82.67	83.10	0.52
3	Ga-67	81.22	81.10	0.15
4	Ga-67	92.01	92.30	0.31
5	Ga-67	78.43	79.40	1.23
6	I-131	26.49	26.50	0.05
7	I-131	40.19	40.10	0.22
8	I-131	41.74	41.80	0.14
9	I-131	64.11	64.00	0.17
10	I-131	63.64	64.00	0.56
Average				0.76

Table 2. A comparison of organ doses from patient no.1 calculated by CALRADDOSE and OLINDA.

Target	Calculated Dose per Administered Activity (mGy/MBq)		Difference (%)
	CALRADDOSE	OLINDA/EXM	
Adrenals	0.1261	0.1230	2.52
Brain	0.0785	0.0805	2.48
Breasts	0.0738	0.0745	0.94
Gall Bladder Wall	0.1479	0.1400	5.64
LLI Wall	0.1048	0.1070	2.06
Small Intestine	0.1051	0.1060	0.85
Stomach Wall	0.1104	0.1100	0.36
ULI Wall	0.1140	0.1140	0.00
Heart Wall	0.1127	0.1120	0.62
Kidneys	0.1126	0.1110	1.44
Liver	0.3021	0.2810	7.51
Lungs	0.1035	0.1030	0.49
Muscle	0.0870	0.0879	1.02
Ovaries	0.1075	0.1100	2.27
Pancreas	0.1333	0.1290	3.33
Red Marrow	0.0925	0.0878	5.35
Osteogenic Cells	0.1323	0.1360	2.72
Skin	0.0649	0.0659	1.52
Spleen	0.2480	0.2370	4.64
Thymus	0.0943	0.0958	1.57
Thyroid	0.0817	0.0836	2.27
Urinary Bladder	0.0997	0.1020	2.25
Uterus	0.1059	0.1080	1.94
Total Body	0.0963	0.0968	0.52
Average			2.26

different sources of radiation decay data, the OLINDA/EXM used the decay data that published by Stabin *et al.* [15] whereas the CALRADDOSE used that from the BNL's National Nuclear Data Center. Second, S-values in OLINDA/EXM were corrected when a source was in hollow organ (gall bladder, urinary bladder, GI tract) by estimating electron dose as the maximum dose at content/wall interface [14]. Third, the dose to source organs (liver, spleen) from CALRADDOSE was higher than that from OLINDA/EXM due to different methods for self-irradiation correction. Lastly, the dose to red marrow was high different because marrow model was not included in CALRADDOSE.

4. Conclusion

CALRADDOSE is a user-friendly, graphic user interface-based software. This software is imaging modality and analysis software vendor independent. It can be performed all steps of internal dosimetry within single environment lead to reducing calculation time and reducing possibility of error. CALRADDOSE also provides fast and accurate results which may be useful for a routine work.

References

[1] Bolch W, Eckerman K, Sgouros G and Thomas S 2009 *J. Nucl. Med.* **50** 477-48
[2] Stabin M, Sparks R and Crowe E 2005 *J. Nucl. Med.* **46** 1023-27
[3] Glatting G, Landmann M, Kull T, Wunderlich A, Blumstein N, Buck A and Reske S 2005 *Med. Phys.* **32** 2399-2405
[4] Visser E, Postema E, Boerman O, Visschers J, Oyen W and Corstens F 2007 *Eur. J. Nucl. Med.Mol.* **34** 413-421
[5] Buijs W, Siegel J, Boerman O and Corstens F 1998 *J. Nucl. Med.* **39** 2167
[6] Thomas S, Maxon H and Kereiakes J 1976 *Med. Phys.* **3** 253-255
[7] Levenberg K 1944 *Quarterly of Applied Mathematics.* **2** 164-168
[8] Marquardt D 1963 *SIAM. J. Appl. Math.* **11** 431-441
[9] Stabin M, Siegel J, Hunt J, Sparks R, Lipsztein J and Eckennan K 2001 RADAR: the radiation dose assessment resource *J. Nucl. Med.* **42**
[10] Stabin M, Xu X, Emmons M, Segars W, Shi C and Fernald M 2012 *J. Nucl. Med.* **53** 1807-13
[11] Stabin M, Watson E, Cristy M, Ryman J, Eckerman K, Davis J, Marshall D and Gehlen M 1995 Mathematical models and specific absorbed fractions of photon energy in the nonpregnant adult female and at the end of each trimester of pregnancy *Preprint* ORNL-TM/12907
[12] Cristy M and Eckerman K 1987 Specific absorbed fractions of energy at various ages from internal photon sources. VI. Newborn *Preprint* ORNL-TM/8381
[13] Cristy M 1980 Mathematical phantoms representing children of various ages for use in estimates of internal dose *Preprint* ORNL-NUREG-TM/367
[14] Stabin M 2008 *Fundamentals of nuclear medicine dosimetry* (New York: Springer Science+Business Media) pp 9-27
[15] Stabin M and da Luz C Q P L 2002 *Health Phys.* **83** 471-475

Acknowledgments

This study was supported by Postgraduate Student Exchange Program 2014 from Mahidol University. The authors also thank Faculty of Medical Technology for all facilities.

Assessment of cardiac function using myocardial perfusion imaging technique on SPECT with ^{99m}Tc sestamibi

M R A Gani[1], F Nazir[2], S A Pawiro[1] and D S Soejoko[1]

[1] Department of Physics, Faculty of Mathematics and Sciences, University of Indonesia, Depok, Indonesia
[2] National Nuclear Atomic Energy Agency, Jakarta, Indonesia

E-mail: m.roslan@sci.ui.ac.id

Abstract. Suspicion on coronary heart disease can be confirmed by observing the function of left ventricle cardiac muscle with Myocardial Perfusion Imaging techniques. The function perfusion itself is indicated by the uptake of radiopharmaceutical tracer. The 31 patients were studied undergoing the MPI examination on Gatot Soebroto Hospital using ^{99m}Tc-sestamibi radiopharmaceutical with stress and rest conditions. Stress was stimulated by physical exercise or pharmacological agent. After two hours, the patient did rest condition on the same day. The difference of uptake percentage between stress and rest conditions will be used to determine the malfunction of perfusion due to ischemic or infarct. Degradation of cardiac function was determined based on the image-based assessment of five segments of left ventricle cardiac. As a result, 8 (25.8%) patients had normal myocardial perfusion and 11 (35.5%) patients suspected for having partial ischemia. Total ischemia occurred to 8 (25.8%) patients with reversible and irreversible ischemia and the remaining 4 (12.9%) patients for partial infarct with characteristic the percentage of perfusion ≤50 %. It is concluded that MPI technique of image-based assessment on uptake percentage difference between stress and rest conditions can be employed to predict abnormal perfusion as complementary information to diagnose the cardiac function.

1. Introduction

Coronary heart disease is one of the causes number of death in Indonesia. The malfunction heart is caused by constriction of the coronary arteries. If the heart could not pump blood and distribute it's to all parts of the body, pain in the chest will be happen while heavy work or walk a rush. Heart disease has caused death 26.4% of all mortality in the country in 2001 according to the Health Survey Indonesia [1]. Nuclear cardiology as a diagnostic procedure is commonly used to evaluate coronary artery disease. Nuclear cardiology is dedicated for cardiac function assessment, metabolism and to observe the images of the left ventricle of the heart which is often used in clinical evaluation [2].

Myocardial perfusion imaging (MPI) techniques is an ideal approach to doing the assessment of myocardial perfusion. As non-invasive imaging modalities, it can examine the functions of the various segments of the heart muscle to reflect whether there is a malfunction on coronary arteries. In myocardial perfusion imaging studies, there are stress and rest sets of data [3]. The study of stress patient data was obtained from exercise using a treadmill or pharmacological stress agent. Furthermore, the study of rest patient data was gained with the patient in complete resting condition on the same or different day, and then it is injected after satisfying of these conditions [4]. Evaluation carried out by observing the distribution of myocardial activity shown on bull eyes that illustrate image 2D polar map [5].

Radionuclide Thallium 201 (^{201}Tl) and Technetium 99m (^{99m}Tc) have become the most widely used in MPI procedure [5]. Radionuclide ^{201}Tl that is produced in the cyclotron has long half-life ($t_{1/2}$ = 73 h) and low energy photon (60 to 83 keV) therefore, the image results have low resolution, great scatter and attenuation. Radionuclide ^{201}Tl has been abandoned for assessment of myocardial perfusion which it. In 1990, radionuclide ^{99m}Tc have been introduced of utilization for the assessment of myocardial perfusion [1]. Several characteristics of this radionuclide provided clear advantages over ^{201}Tl for gamma camera imaging. Compare to radionuclide ^{201}Tl, radionuclide ^{99m}Tc is produced through a generator which have higher energy (140 keV) and that is ideal for standard gamma camera imaging with lesser scatter and attenuation, and shorter half-life ($t_{1/2}$ = 6 h) [6]. Radionuclide ^{99m}Tc will be labeling pharmaceutical methoxyisobutyl isonitrile (sestamibi) that is commonly used to study perfusion myocardial [6]. Radiopharmaceutical ^{99m}Tc-sestamibi would accumulate in the myocardial tissue of the bloodstream. SPECT imaging is a tomographic technique that provide three-dimensional (3D) heart organ for given information of uptake radiopharmaceutical on each segment of the heart [4]. In this study will be informed evaluation the image of left ventricle to assess function cardiac-based on myocardial perfusion imaging technique. By quantifying of radiopharmaceutical uptake of ^{99m}Tc-sestamibi by every segment illustrated in the bull eyes, the percentage of myocardial perfusion was determined as described in figure

2. Materials and methods

In this research, 31 patient in records were examined with SPECT myocardial perfusion at the Department of Nuclear Medicine, Hospital Gatot Subroto Jakarta between February and June 2015. The patients were suspected of suffering from disorders of the heart muscle were used as a data sample in this work. Patient data were taken which used ^{99m}Tc sestamibi perfusion and the percentage obtained under conditions of stress and rest. SPECT imaging scans performed with dual head with LEGP collimator and detector NaI(Tl). Patient examination was done in 1 day with 2 phases of stress and rest protocol [5].

2.1. Protocol stress

Stress testing the heart organ in myocardial perfusion imaging techniques are through physical exercise (ergometric) and pharmacologic. Conditions stress will cause an increase in oxygen consumption in the heart muscle tissue [2]. Increased myocardial perfusion are characterized by heart rate, blood pressure and myocardial contraction, resulting in coronary artery vasodilation [7]. Myocardial perfusion can also enhance through the provision of pharmacologic Persantine. There is occurred differential capacity of dilation of blood vessels so that the tracer can be identified as cardiac dysfunction. When stress testing was done, it is necessary to observe the heart rate on the computer check the condition of the patient such as heart rate recovery in the initial conditions. Radiopharmaceutical injection can perform after the heart rate has reached 85 percent of the difference between 220 minus the age of the patient [7]. Radiopharmaceutical was injected with the range of dose between 259 to 333 MBq [3], and then patients were encouraged to consume a large amount of fat so that no distortion of the image of the heart of the surrounding organs.

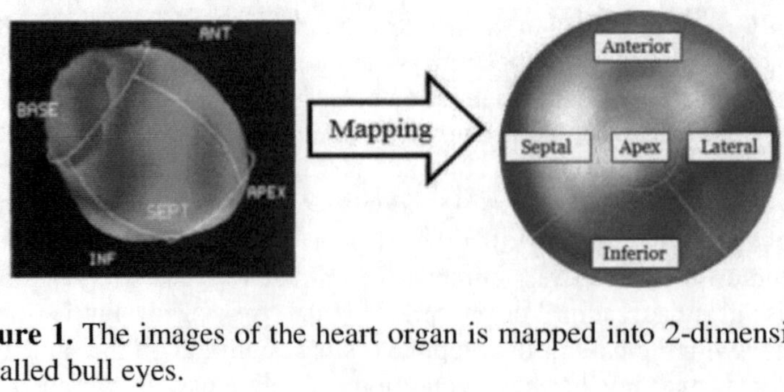

Figure 1. The images of the heart organ is mapped into 2-dimensional to called bull eyes.

2.2. Protocol rest

Testing rest was done after 2 to 4 hours of scanning images under stress completed [5]. The test was done by using radiopharmaceutical injection with doses greater than the stress condition in the range 629 to 814 MBq [3]. Afterward, the scanning will perform with the same parameters SPECT which was using in protocol stress.

2.3. Scanning and acquisition SPECT

The patient in the supine position with both arms placed on the head and the electrodes attached to the patient's body to observe the heart rate [8]. Scanning was done with dual head SPECT gamma camera parameters Cardiac Gated SPECT. This study were applied parameters zoom 1.46 x (40.9) cm, matrix size 64 x 64, the number of angles 64, time per angle 20 seconds, the saturation level of 32.767, the relative angle detector 90^0, the initial angle of 45^0, and the direction of rotation clockwise.

SPECT gamma camera detector was positioned on the patient's body left chest. The detector was rotated 180 degrees from the position of 45 degrees right anterior oblique (RAO) to 45 degrees left posterior oblique (LPO) [8]. Both detectors are perpendicular, with a position as close as possible to the body of the chest, which is located in the middle of the field of view (FOV) detector.

The result of scanning are a reconstruction of the left ventricle and then we did the quality control image acquisition with cine display [9]. The reconstruction was done with applications Philips Astonish Cardiac SPECT Auto Pro that can directly doing to correction of attenuation, scatter, and partial volume effect. This application provided a correction resolution, minimize noise and improved uniformity. Moreover, the scanned image would be reconstructed to determine the boundary around the heart to the organs. Automatic segmentation in cardiac image reconstruction program could make the slide image from the volumetric data. The evaluation of slice image was performed by Quantitative Perfusion/Gated SPECT (QPS/QGS) from Cedars-Sinai Medical Center, Los Angeles [4]. The volumetric image was segmented into three slices, namely the short axis, vertical long axis, and horizontal long axis for each condition of stress and rest as illustrated in figure 2 [2].

QPS method used ellipsoidal models to identify myocardial, heart can be assessed visually quantify 3D distribution approach in myocardial activity. It uses cylindrical and spherical coordinates so that it becomes a 2D map of the image is called bull eyes [4]. The bull eyes were divided into 5 regions of the left ventricle heart, there is anterior, lateral, inferior, septal and apex [5]. Tracer each area will be displayed in the interpretation of the percentage of perfusion in the color scale [10]. The color scale ranges from 0 to 100 percent, a higher percentage of myocardial perfusion showed good function, and conversely, a lower percentage indicates abnormal myocardial function.

Myocardial function was determined by observation of concentration of tracer uptake, and proportional to the percentage myocardial perfusion [2]. Tracer uptake on stress condition must be greater than rest condition. Conversely, it suspected abnormal myocardial perfusion function. Diagnosis of abnormal function can be determined from the defining difference between the percentage perfusion of stress and rest conditions. The percentage difference (Δ) of the segment of the heart wall has been used to assess the left ventricular function.

Observations percentage of cardiac perfusion of the left ventricle, infarction in coronary artery occurred if the percentage perfusion is less than 50 percent. Myocardial necrosis happens due to blood flow in the coronary arteries stalled consist of myocardial infarction viable (reduced perfusion between stress and rest is not significant in the range of 0 to 20 percent) and nonviable infarction (decrease stress and rest perfusion is in the range of 21 to 50 percent). Furthermore, we considered percentage difference (Δ) between stress and rest based on the percentage of perfusion in between 51 to 100 percent for each segment. If the percentage difference (Δ) is a positive value, the myocardial blood flow tends to normal, but if the percentage difference (Δ) is a negative value tend to ischemia symptom. Ischemia could be categorized in reversible (if the percentage difference between stress and rest more than 20 percent) and irreversible (if the percentage difference about of 0 to 20 percent) [2, 4, 10].

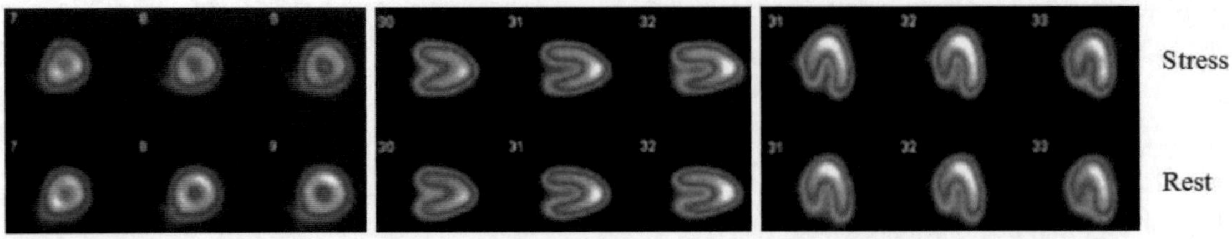

Figure 2. Slice images of segmentation: a) Short axis (left images) apex to base; b) Vertical long axis (middle images) septal to lateral; c) Horizontal long axis (right images) posterior to anterior.

3. Results and discussion

The average of patient age is 59 ± 11 years with a range of 42 to 72 years from 31 patients with 17 males and 14 females. Examination load procedure for the patient is expected to pass the activities ergo cycle to see a heart function, but there were 2 patients are unable to do it because 1 patient had a stroke and 1 patient a new surgery on the right foot. Therefore, two patients were given pharmacological Persantine to make heart muscle dilatation.

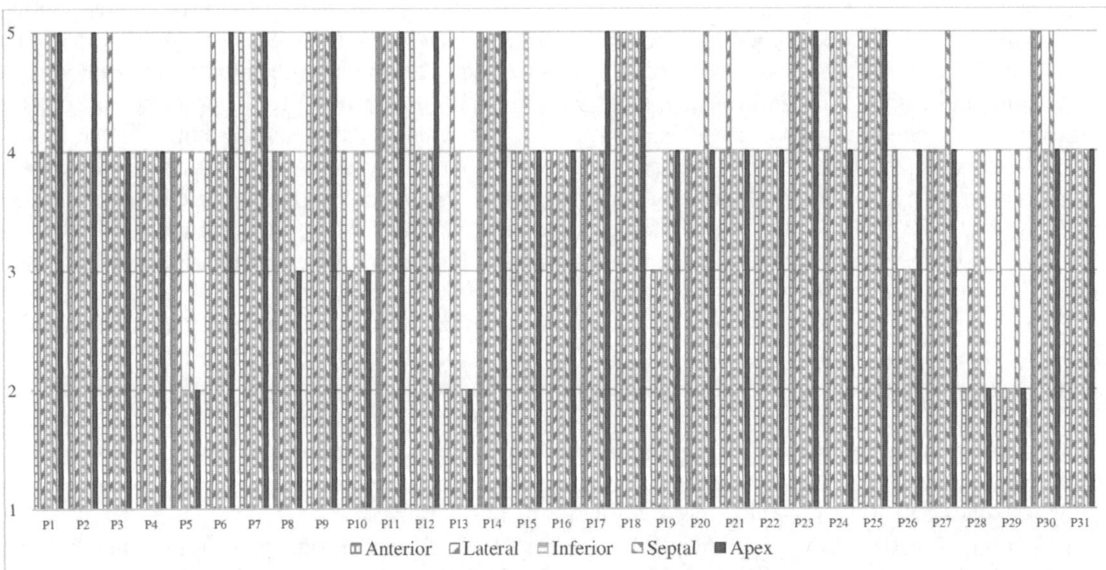

Figure 3. Graphic assessment of perfusion percentage of each segment to all patients, by parameter: 5 normal; 4 irreversible ischemia; 3 reversible ischemia; 2 nonviable infarct; 1 viable infarct

Patient data was evaluated by QPS program to obtain a percentage of left ventricular perfusion and divided into five segments; anterior, apex, lateral, inferior and septal as illustrated in figure 3. For anterior segment showed 10 normal patients and 21 patients with abnormal, there were 18 patients with ischemia irreversible (Δ = 2 to 10%), 1 patient ischemia reversible (Δ = 13%), and 2 patients infarction nonviable (percentage perfusion 26 and 45%). The lateral segment showed 12 normal patients and 19 patients with perfusion abnormalities, there were 14 patients with ischemia irreversible (Δ = 1 to 8%), 4 patients with ischemia reversible (Δ = 11 to 18%), and 1 patient infarction nonviable (percentage perfusion 40%). The inferior segment showed 10 normal patients and 21 patients with abnormal perfusion which are 18 patients with irreversible ischemia (Δ = 1 to 6%), 1 patient with ischemia reversible and 2 patients with infarct nonviable (percentage perfusion 41 and 45%). The septal segment showed 12 normal patients and 19 patients with abnormal perfusion which are 17 patients with irreversible ischemia (Δ = 1 to 8%), 1 patient reversible ischemia (Δ = 14%) and 1 patient infarction nonviable (perfusion percentage of 36%). And the last segment, apex segment showed 12 normal patients and 19 patients with perfusion abnormalities which are 13 patients with irreversible (Δ = 1 to

10%), 2 patients with reversible ischemia (Δ = 13 to 16%) and 4 patients with infarction nonviable (15 to 36%).

Assessment allegations with MPI technique can be followed by observing the five segments of each patient, the average assessment of each segment of the anterior, lateral, inferior, septal and apex show that allegation of 8 patients with normal perfusion (consisting of 4 to 5 segments of normal), 11 patients with partial ischemia (there are 2 to 4 ischemia reversible and there are at least 1 normal), 8 patients total ischemia (segment consists ischemia reversible and irreversible) and 4 partial infarction patients (there are 2 to 3 nonviable infarction). Abnormal perfusion is resulting that created to allegations of partial ischemia, total ischemia, and partial infarction because of tracer uptake stress condition less than rest condition cardiac. In stress condition heart is increased of perfusion blow flood on left ventricle of cardiac. It is demonstrated on the image and be evidenced of percentage perfusion every segment of left ventricle.

If the myocardial wall segment is associated with coronary artery, segments of the anterior and lateral indicating the left coronary, inferior segment indicating the right coronary and lateral segment indicating the circumflex arteries [9]. Most likely, the abnormality of heart due to blockage in a coronary artery or branches left artery [2]. Five segments of the left ventricle, anterior has the biggest risk factor in the process of perfusion because the location of anterior is front of the heart muscle and very active with the function to pump blood. All segments of myocardial wall showed nearly equal amounts in the diagnosis that means less in perfusion, and then it will be discussed by a cardiologist for further medical action.

4. Conclusion

Myocardial Perfusion Imaging (MPI) techniques can be used to diagnose the function muscle heart of the review perfusion blow flood cardiac. Based on the uptake segment of the left ventricle, it shows the scanning image described the percentage perfusion. The assessment uptake of percentage perfusion is predicting abnormal perfusion, for example, ischemia or infarct.

References

[1] Delima, Mihardja L and Siswoyo H 2009 *Bulletin of Health Research* **37** 142-159
[2] Dilsizian V 2003 *Atlas of Nuclear Cardiology* (Philadelphia: Current Medicine)
[3] Strauss W, Miller D D, Wittry M D, Cerqueira M D, Garcia E V, Iskandrian A S, Schelbert H R, Wackers F J, Balon H R, Lang O and Machac J 2002 *Society of Nuclear Medicine Procedure Guideline for Myocardial Perfusion Imaging* **3** 9-17
[4] Khalil M M 2011 *Basic Sciences of Nuclear Medicine* (London, New York: Springer)
[5] Berman D S, Kiat H, Train K V,Garcia E, Friedman J and Maddahi J 1991 *Seminars in Nuclear Medicine* **21** 190-212
[6] Saha G B 2010 Fundamentals of Nuclear Pharmacy Sixth Edition (London: Springer)
[7] Lindner O, Rusche H, Schafers M, Schober O and Burchert W 2007 *Dtsch Arztebl* **104** 1-9
[8] Hansen C L *et al.* 2006 *Journal Nuclear Cardiology* **13** e39-e60
[9] Bushberg J T, Seibert J A, Leidholdt E M and Boone J M 2002 *The Essential Physics of Medical Imaging* (Philadelphia : Lippincott Williams & Wilknis)
[10] Dvorak R A, Brown R K J and Corbett J R 2011 RadioGraphics 31 2041-2057

Acknowledgements

The authors wish to thank the head and medical staffs of Department Nuclear Medicine, Gatot Soebroto Hospital for their help of this study. We also thank Laboratory of Medical Physics and Biophysics for supporting to attend the SEACOMP 2015 under Hibah Pascasarjana 2015.

Journal of Physics: Conference Series **694** (2016) 012058 doi:10.1088/1742-6596/694/1/012058

Survival outcome of radioiodine therapy in post thyroidectomy thyroid carcinoma patients: Outcome of long term follow up

F Haque[1], N Nahar[1], S Sultana[1], F Nasreen[1], Z Jabin[1], A S M M Alam[2]

[1] National Institute of Nuclear Medicine and Allied Sciences, Dhaka, Bangladesh
[2] Dhaka Community Medical College and Hospital

E-mail: farhanahaque2013@yahoo.com

Abstract. The overall prognosis of patients with thyroid carcinoma is excellent whenever managed following best practice guidelines. Objective: To calculate sex and age group affected by thyroid cancer; to compare between single or multiple dose of radio ablation needed after thyroidectomy and to determine the percentage of patients become disease free during their follow up. Methods: This was a retrospective study done in NINMAS, Bangladesh on 687 patients from 1984 to 2004. In all cases total or near total thyroidectomy was done before commencing radioiodine therapy. Patients TG level, neck ultrasonography, thyroid scan, whole body I^{131} scans, neck examination were done every six monthly/yearly. Results: Among 687 patients, female were more sufferers (68.1%) and female to male ratio was 2:1. Age group 19-40 years was mostly affected (57.8%). Most common type seen was papillary carcinoma (81.8%). After ablation 100 patients did not follow-up. Total 237 patients discontinued within 4 years. Remaining 450 patients undergone regular follow-up for 5 years and more, 394 were disease free (87.6%). Total recurrence of metastasis was 23 and 12 patients expired at different times. Conclusions: Long-term regular follow-up is necessary after radioiodine ablation to become free of disease.

1. Introduction

Differentiated thyroid cancer (DTC) accounts for 98% of thyroid cancer, with neoplasms arising from the follicular cells (papillary, follicular and Hurthle cell thyroid cancer) and parafollicular cells (medullary thyroid cancer). Thyroid cancer is the most common endocrine malignancy [1]. Management of DTC includes total or near total thyroidectomy followed by radio-iodine ablation. Ablation is given to destroy the remnant thyroid tissue as well as residual cancer tissue. Ablation facilitates follow-up of these patients by whole body I^{131} scan and estimation of serum TG level. Long term follow-up is necessary after radio-ablation to detect early relapse and distant metastasis as well as evaluate the effectiveness of TSH suppression [2]. DTC is slowly growing cancer and good prognosis with appropriate management including radio iodine ablation after total / near total thyroidectomy and thyroid hormone suppression [1, 2, 6]. Radioactive Iodine (RAI) is administered as adjuvant therapy to reduce risk of recurrence and to facilitate future cancer surveillance. RAI ablation should be completed within 4-6 weeks of thyroidectomy [3, 4].

The aim of this study was to calculate the sex and age group mostly affected by thyroid cancer; to compare between single and multiple dose of radio ablation needed after thyroidectomy and to determine the percentage of carcinoma patients become disease free apparently during their follow-up.

2. Method

This study was a retrospective study done in National Institute of Nuclear Medicine and Allied Sciences (NINMAS), Bangladesh reviewing the medical records of the 687 DTC patients over the 21-year period from 1984 to 2004. Data on patients' status at half-yearly / yearly follow-up was collected and included those patients who received radio-active iodine and then lost without any follow-up, those who continued for follow-up for one to twenty-eight years till 2015. The patients' data contained age at diagnosis, gender, carcinoma types, state of metastasis at diagnosis, number and type of surgery, therapy details (1st dose, number of doses, total doses), duration of follow-up and their conditions at last follow-up on individual basis. We analyzed the data using SPSS version 19.1.

3. Results

Figure 1 shows the distribution of the 687 patients by their year of registration in NINMAS. It was found that only 3 patients registered for therapy in 1984-85 and only 7 patients in 1986-90. Then number of patients increased gradually in the successive years up to 2004.

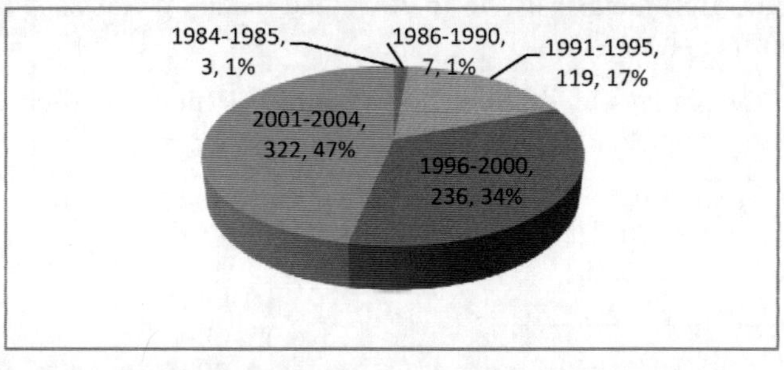

Figure 1. Distribution of the patients by their year of registration in NINMAS

Figure 2 shows the distribution of the patients by their age. Among the total of 687 patients, 468 (68.12%) were female and 219 (31.88%) were male resulting in a female to male ratio almost 2:1. Patients mostly affected were 397 (57.8%), at the age group of 19-40 years. Only 52 patients (7.5%) were children and adolescents (belong to 1-18 years) and only 43 (6.2%) patients from elderly group (belong to 61-94 years).

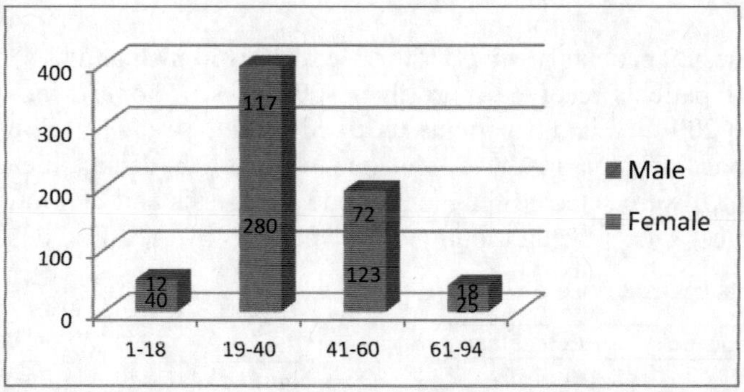

Figure 2. Distribution of the patients by their age and gender.

Table 1. Distribution of the patients by the type of their carcinoma.

Carcinoma type	Frequency	Percentage
Papillary Thyroid Carcinoma (PTC)	563	81.80
Follicular Thyroid Carcinoma (FTC)	108	15.72
Others	16	2.48

Table 1 shows the distribution of the patients by the type of their carcinoma. Of the 687 patients with DTC, 563 (81.80%) had papillary thyroid carcinoma (PTC) including 68 of follicular variant of papillary thyroid carcinoma, 108 (15.72%) had follicular thyroid carcinoma (FTC), among the other 16 (2.43%) patients 5 (0.73%) had medullary thyroid carcinoma (MED) and 2 (0.29%) had Hurthle cell carcinoma. The data about the type of the rest 9 patients was unavailable.

In this study, most common type of surgery was found as total / near total thyroidectomy which had been done on 510 (74.23%) patients. Two to three different types of surgeries (hemi-thyroidectomy, sub-total thyroidectomy, lobectomy and completion thyroidectomy) were done in 85 (12.37%) patients. Hemi-thyroidectomy had been done in 26 patients, subtotal thyroidectomy in 26 and lobectomy in 11. Hospital documents of operation notes of 29 patients about their surgery were not available.

Among 687 patients, 485 (70.60%) were found without metastasis at the time of diagnosis, 156 (22.71%) had lymph node (LN) metastasis, 46 (6.70%) had distant metastasis. Distant metastasis sites were bone, lung and multiple sites.

Table 2. Distribution of the patients by the site of metastasis at the time of diagnosis.

Site of metastasis	Frequency	Percentage
Without Metastasis	485	70.60
Lymph Node (LN)	156	22.71
Distant	46	6.70

Table 3. Patients by their duration of follow-up.

Years of follow-up	Frequency	Percentage
Lost after ablation	100	14.56
1-4	137	19.51
5-9	182	26.49
10-14	181	26.35
15-19	64	9.32
More than 20 years	11	1.75
Expired	12	1.75

Table 3 shows the distributions of patients by their duration of follow-up. Of the total of 687 patients 100 patients did not come for follow-up after receiving RAI ablation. 137 patients came for regular follow-up for 1-4 years only and subsequently lost. Rest of the patients visited for more than 5 years. Follow-up was at least 5 years after initial therapy in 182 (26.49%), 10 or more years in 181 (26.35%), 15 or more years in 64 (9.32%) patients, 20 or more years in 11 patients. About 78.5% patients followed for 2 or more years. Among the total cases 12 expired at different times of follow-up.

Among the total cases 29 patients received low dose of I^{131} (50 mCi and less), 554 patients received 51-100 mCi of I^{131}, 81 patients received more than 100 doses up to 150 mCi as the first dose, 20 patients received up to 200 mCi and 3 patients received up to 250 mCi. Below 100 mCi of I^{131} was administered for remnant ablation in those who had no local or distant metastasis at the time of diagnosis. Up to 150 mCi were selected for lymph node metastasis and 200 mCi and above doses for distant metastasis. Table 4 shows the distribution of the patients by their first dose of I^{131}.

Table 4. No. of Patients by first dose of I^{131}.

Amount in mCi	Frequency	Percentage
≤50	29	4.22
51-100	554	80.64
101-150	81	11.79
151-200	20	2.91
201-250	3	0.44

Table 5. No. of Patients by No. of doses of I^{131}.

Number of doses	Frequency	Percentage
1	364	52.98
2-5	305	44.40
6-10	18	2.62

Maximum number of 364 (52.98%) patients received single dose of I^{131}, 305 (44.40%) patients received 2-5 doses. Only 18 patients were found to receive 6-10 doses. Table 5 shows the distribution of the patients by their number of doses of I^{131}. Cumulative doses up to 500 mCi were given to 617 (89.81%) patients. Only 9 (1.30%) patients were given cumulative doses of more than 1500 mCi. Cumulative dose of 501-1500 mCi was given to 61 patients (8.88%)

Table 6. Distribution of the patients with no metastasis at the time of diagnosis.

Type of Carcinoma	Number at diagnosis	Follow-up for less than 5 years	Follow-up for 5 years & more			
			Normal	TG High	Metastasis	Expired
PTC	406	122	265	4	LNM-11	4
FTC	70	35	29	2	MBM-1	3
Others	9	4	4	0	0	1

Table 6 shows distribution of the patients having no metastasis at all at the time of diagnosis. Total number of patients without metastasis was 485 among which 161 patients discontinued follow-up within 4 years. Remaining 324 patients undergone regular follow-up more than 5 years and 298 among them were found disease free. About 92% patients became disease free on regular follow-up for more than 5 years. Out of 324 patients, lymph node metastasis occurred in 11 patients and only 1 patient developed multiple bone metastases. Therefore, the rate of recurrence is only 3.70% in this group.

Table 7. Distribution of the patients with lymph node (LN) metastasis at the time of diagnosis.

Type of Carcinoma	Number at diagnosis	Follow-up for less than 5 years	Follow-up for 5 years & more			
			Normal	TG High	Mets	Expired
PTC	140	51	75	6	LNM-7	1
FTC	11	2	7	2	0	0
Others	5	2	2	0	LNM-1	0

Table 7 shows distribution of the patients with lymph node (LN) metastasis at the time of diagnosis. Total number of patients with lymph node (LN) metastasis was 156 among which 55 patients discontinued follow-up within 4 years. Remaining 101 patients undergone regular follow-up for more than 5 years and 84 among them were found disease free. Of them 83% patients became disease free on regular follow-up for more than 5 years. Out of 156 patients, lymph node metastasis occurred in 8 patients. Therefore, the rate of recurrence is only 7.9% in this group.

Table 8. Distribution of the patients with distant (D) metastasis at the time of diagnosis.

Type of Carcinoma	Number at diagnosis	Follow-up for less than 5 years	Follow-up for 5 years & more			
			Normal	TG High	Mets	Expired
PTC	17	9	8	0	0	0
FTC	27	11	4	6	D-3	3
Others	2	1	0	1	0	0

Table 8 shows distribution of the patients with distant (D) metastasis at the time of diagnosis. Total number of patients with distant (D) metastasis was 46 among which 21 patients discontinued follow-up within 4 years. Remaining 25 patients undergone regular follow-up for more than 5 years and 12 (48%) among them were found disease free. Of the 25 patients, distant metastasis occurred in 3 patients. Therefore, the rate of recurrence is only 12% in this group.
Among the total 687 patients 12 died in different times of their follow-up.

4. Discussion

The ratio of female to male in the study is almost 2:1 which is comparable to other's study. In studies of Catharina Ihre Lundgren 1.9:1, Mazzaferri *et al.* 1.6:1, Mazzaferri and Jhiang 2.2:1, and Caron N. R 3:1 respectively [7][6][2][1]. The most of the patients affected in thyroid carcinoma (DTC) in our study belongs to 19-40 years age group (57.8%). Children and adolescent population is only 7.5%

which is comparable to the study of Caron N. R. which is less than 10% [1]. In this study out of 687 patients 81.80% were PTC and only 15.72% were FTC. It is comparable to the study of Catharina *et al.* (PTC 70%) [7]. 81% of PTC in Mazzaferri *et al.* [6]. 77% PTC in Hundahl *et al.* [8]. Patients came for RAI therapy mostly after total or near total thyroidectomy. In our study total thyroidectomy is 74.23%. This finding is similar with the study of Hundahl *et al.* Among the 324 patients having no metastasis at the time of diagnosis and who had undergone regular follow-up for 5 years and more, 298 (92)% became disease free, recurrence of 11 cases with lymph node metastasis and 1 with multiple bone metastases were found resulting in a rate of recurrence only 3.70%. Again, among 101 patients with lymph node metastasis at the time of diagnosis under follow-up of 5 years or more, 84 (83%) patients became disease free, recurrence was found in 8 (7.9%) patients with lymph node metastasis. In the distant metastasis group, 25 patients out of 46 who undergone follow-up for more than 5 years, 12 (48%) patients became disease free and recurrence of distant metastasis occurred in 3 patients (12% only).

5. Conclusions

It is clear that the post-thyroidectomy patients those have received long term and regular follow up after radioiodine therapy had good prognosis.

References

[1] Caron N R and Clark O H 2004 Well differentiated thyroid cancer *Scandinavian journal of Hand surgery* **93** 261-271
[2] Mazzzaferri E L and Jhiang S M 1995 Differentiated thyroid cancer long term impact of initial therapy. Transactions of the American Clinical and Climatological Assoc. 106151-170.
[3] Cooper D S, Doherty G M, Haugen B R and Kloos R T 2009 Revised American thyroid association management guidelines for patients with thyroid nodules and differentiated thyroid cancer *Thyroid* **19(11)** pp1167-1214
[4] Sultana S, Nahar N, Begum F, Alam F and Hasan M 2015 Management of Patients with Differentiated Thyroid Carcinoma – SNMB Guidelines *Bangladesh Journal of Nuclear Medicine* **18(1)** 73-84
[5] Schneider A B and Ron E 2005 *Carcinoma of follicular epithelium Epidemiology and pathogenesis* 9th edition (Philadelphia: Lippincott williams and Wilkins) pp 889-906
[6] Mazzaferri E L, Young R L and Oertel J E 1977 Papillary Thyroid Carcinoma: The Impact of therapy in 576 patients *Medicine (Baltimore)* **56(3)** pp171-196.
[7] Lundgren C I, Hall P, Dickman P W and Zedenius J 2005 Clinically Significant Prognostic Factors for Differentiated Thyroid Carcinoma. American Cancer Society. Published online December 2005 www.interscience.wiley.com
[8] Hundahl S A, Fleming I D, Fremgen A N, Menck H R 1998 A National Cancer Data Base Report on 53,856 cases of Thyroid Carcinoma treated in U.S., 1985-1995. American Cancer Society. December 1998 **83** pp 2638-48

Organ doses from hepatic radioembolization with ^{90}Y, ^{153}Sm, ^{166}Ho and ^{177}Lu: A Monte Carlo simulation study using Geant4

N A A Hashikin[1,2], C H Yeong[1,2], S Guatelli[3], B J J Abdullah[1,2], K H Ng[1,2], A Malaroda[3], A B Rosenfeld[3] and A C Perkins[4]

[1] Department of Biomedical Imaging and [2] University of Malaya Research Imaging Centre, Faculty of Medicine, University of Malaya, Kuala Lumpur, Malaysia
[3] Centre for Medical Radiation Physics, Faculty of Engineering and Information Sciences, University of Wollongong, Wollongong, New South Wales, Australia
[4] Radiological and Imaging Sciences, Medical Physics and Clinical Engineering, Medical School, University of Nottingham, Nottingham, United Kingdom

E-mail: chyeong@um.edu.my

Abstract. ^{90}Y-radioembolization is a palliative treatment for liver cancer. ^{90}Y decays via beta emission, making imaging difficult due to absence of gamma radiation. Since post-procedure imaging is crucial, several theranostic radionuclides have been explored as alternatives. However, exposures to gamma radiation throughout the treatment caused concern for the organs near the liver. Geant4 Monte Carlo simulation using MIRD Pamphlet 5 reference phantom was carried out. A spherical tumour with 4.3cm radius was modelled within the liver. 1.82GBq of ^{90}Y sources were isotropically distributed within the tumour, with no extrahepatic shunting. The simulation was repeated with ^{153}Sm, ^{166}Ho and ^{177}Lu. The estimated tumour doses for all radionuclides were 262.9Gy. Tumour dose equivalent to 1.82GBq ^{90}Y can be achieved with 8.32, 5.83, and 4.44GBq for ^{153}Sm, ^{166}Ho and ^{177}Lu, respectively. Normal liver doses by the other radionuclides were lower than ^{90}Y, hence beneficial for normal tissue sparing. The organ doses from ^{153}Sm and ^{177}Lu were relatively higher due to higher gamma energy, but were still well below 1Gy. ^{166}Ho, ^{177}Lu and ^{153}Sm offer useful gamma emission for post-procedure imaging. They show potential as ^{90}Y substitutes, delivering comparable tumour doses, lower normal liver doses and other organs doses far below the tolerance limit.

1. Introduction

Liver cancer is the second most common cause of death from cancer worldwide [1]. Radioembolization is widely used for palliative treatment of hepatocellular carcinoma (HCC), which is the most common type of liver cancer. The procedure involves intra-arterial administration of embolic Yttrium-90 (^{90}Y) microparticles directly to the tumour, delivering all the radiation in-situ. Two commercial radioembolic agents are available, the glass-based TheraSphere® (Nordion, Canada) and the resin-based SIR-Spheres (SIRTex, Australia) microspheres. ^{90}Y decayed 99.98 % via beta emissions, thus considered as a pure beta emitter. Since gamma emission is absent, verification of ^{90}Y-microsphere distribution following each procedure is a challenge. The post-procedure imaging is crucial for patient monitoring, in case there is presence of extrahepatic deposition of the microparticles. Bremsstrahlung imaging can be utilised, but the images are very poor in resolution.

Several theranostic (therapy and diagnostic) radionuclides have been suggested as ^{90}Y alternative [2-5]. These include Samarium-153 (^{153}Sm), Holmium-166 (^{166}Ho), and Lutetium-177 (^{177}Lu). The physical characteristics are tabulated in table 1. Due to the presence of gamma emissions throughout the length of the treatment, the exposure from gamma radiation towards the organs in the proximity of the liver has been a concern. Moreover, in order to achieve similar therapeutic effect as ^{90}Y, higher activities may be required, increasing the gamma exposure to these organs. The aim of this study was to estimate the absorbed dose to these organs from the exposure of gamma radiations by the alternative radionuclides, as compared to ^{90}Y for hepatic radioembolization. The Geant4 Monte Carlo code was used to simulate the treatment.

Table 1. Physical characteristics of suggested radionuclides for radioembolization [6].

Radionuclides	^{90}**Y**	^{153}**Sm**	^{166}**Ho**	^{177}**Lu**
Physical half-life, $T_{1/2}$ (days)	2.67	1.93	1.12	6.65
Maximum E_{β^-} (keV) (*% yield*)	2279.8 (*99.98*)	807.6 (*19.5*)	1854.5 (*48.2*)	498.3 (*79.3*)
Principal E_γ (keV) (*% yield*)	-	103.2 (*29.2*)	80.6 (*6.6*)	208.4 (*10.4*)
Maximum beta range (mm) in soft tissue	11.4	4.0	9.3	2.5

E_{β^-}: energy of beta; E_γ : energy of gamma

2. Materials & Methods

2.1. Simulation with ^{90}Y

Monte Carlo simulation using Geant4.9.6.p03 advanced example *human_phantom* was carried out. The 70 kg, 1.74 m tall human phantom adopted from MIRD Pamphlet 5, and a spherical tumour of radius r = 4.3 cm, simulated within the liver volume are shown in figure 1 (a) and (b), respectively. The tumour size was chosen since it was the largest spherical tumour able to be constructed within the liver, without interceptions between the tumour and liver boundaries. The tumour and liver volumes are 333 and 1833 cm^3, respectively, which correspond to 18.2 % tumour involvement.

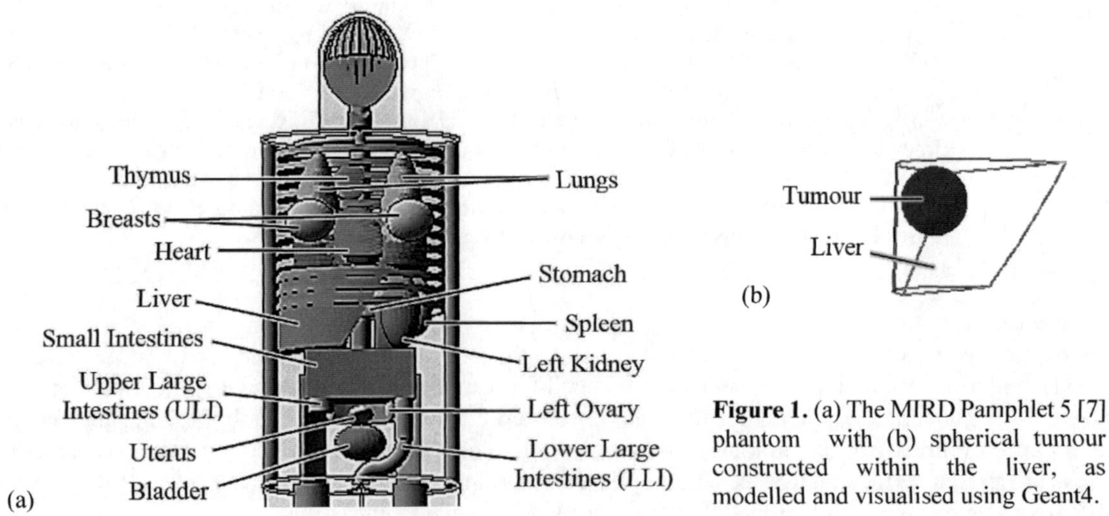

Figure 1. (a) The MIRD Pamphlet 5 [7] phantom with (b) spherical tumour constructed within the liver, as modelled and visualised using Geant4.

The total activity, A of 1.82 GBq ^{90}Y was prescribed to the tumour, as determined using equation (1) [8], which incorporates the Body Surface Area (BSA) value calculated from equation (2) [9]. The ^{90}Y sources were assumed to be fully taken by the tumour, with no lung and extrahepatic shunting. The number of events, N corresponding to the total ^{90}Y activity was 6.07 x 10^{14} events, as determined using equation (3) where decay constant, λ of ^{90}Y is 3.00 x 10^{-6} s^{-1}.

$$A\ (GBq) = BSA\ (m^2) - 0.2 + (\ \%\ tumour\ involvement\ /\ 100) \tag{1}$$

$$BSA\ (m^2) = 0.20247 \times height^{0.725}\ (m) \times weight^{0.425}\ (kg) \tag{2}$$

$$A\ (Bq) = \lambda\ (s^{-1})\ N \tag{3}$$

The electromagnetic interactions of particles were modelled using the Low Energy Livermore and Radioactive Decay Physics List. The threshold of production of secondary particles was fixed to 1 mm. The Geant4 General Particle Source was used to generate the decay of the radionuclide in the tumour. ^{90}Y point-like sources were isotropically distributed within the tumour volume with randomised direction of emissions. Approximately 6.07 × 10^7 events were generated in each simulation to obtain a statistical accuracy of less than 1 % in the results. The energy deposited (MeV) to all organs were multiplied to 10^7, converted to Joules (J), and divided by each organ mass (kg) to obtain the total absorbed dose (Gy).

2.1.1. Comparison between Geant4 and the partition model. The tumour absorbed dose calculated by means of Geant4 was compared to the tumour dose estimated using the MIRD based partition model [10] as shown in equation (4). This was done to quantify the agreement between the two models for internal dosimetry.

$$Absorbed\ dose\ (Gy) = \frac{49670 \times A\ (GBq)}{mass\ (g)} \tag{4}$$

2.2. Simulations with ^{153}Sm, ^{166}Ho and ^{177}Lu.

2.2.1. Determination of dose factor. The activities corresponding to 10^7 events calculated using equation (3), are shown in table 2 for each radionuclide. A pilot run simulation with 10^7 events, uniformly distributed throughout the liver volume were carried out for each radionuclide. The energy deposited (J) to the liver were converted to absorbed dose (Gy) and divided by the corresponding activity to obtain the dose per activity (Gy/GBq) for each radionuclide. The dose per activity value was then normalised to that of ^{90}Y to obtain the dose factor.

Table 2. Radionuclide activity corresponding to 1 × 10^7 events.

Radionuclides	**Decay constant, λ (s^{-1})**	**Corresponding activity, A (Bq)**
^{90}Y	3.00E-6	30
^{153}Sm	4.16E-6	41.6
^{166}Ho	7.18E-6	71.8
^{177}Lu	1.21E-6	12.1

2.2.2. Simulations with other radionuclides as compared to ^{90}Y. The total ^{90}Y activity of 1.82 GBq were divided by each dose factor to determine the activity needed for each radionuclide to deliver identical tumour dose as ^{90}Y. These activities were then applied into the tumour and the simulation setup as in section 2.1 were repeated for each radionuclide. The dose to the tumour, normal liver and other organs were recorded and compared with ^{90}Y.

3. Results and discussion

The activity of each alternative radionuclide used to achieve the same tumour dose of (262.9 ± 0.6) Gy, are shown in table 3. The dose estimated using the partition model was 271.5 Gy. The 3 % dose difference between the two methods can be explained due to presence of cross-over radiation at the tumour edge, which escaped from the tumour volume and recorded by the Geant4 simulation. This fraction of radiation was neglected by the partition model because of the assumption that beta particles are fully absorbed by the source volume, due to the small beta penetration depth.

Table 3. Radionuclide activities to deliver 263 Gy tumour dose.

Radionuclides	Dose per activity (Gy/GBq)	Dose factor	Equivalent activity (GBq)
^{90}Y	27.62	1	1.82
^{153}Sm	6.04	0.219	8.32
^{166}Ho	8.62	0.312	5.83
^{177}Lu	11.30	0.409	4.44

The remaining radiation that escaped from the tumour was absorbed by the normal liver. As shown in table 4, the other radionuclides contribute to lower normal liver doses which are advantageous as it allows for normal liver tissue sparing. This was due to the shorter beta ranges (lower beta energies), and thus, only a small fraction of these energies reaching the normal liver volume as compared to ^{90}Y.

Table 4. Normal liver dose (Gy) as a result of 263 Gy tumour dose.

Radionuclides	Absorbed dose (Gy)
^{90}Y	2.92 ± 0.05
^{153}Sm	2.30 ± 0.01
^{166}Ho	2.20 ± 0.06
^{177}Lu	1.59 ± 0.02

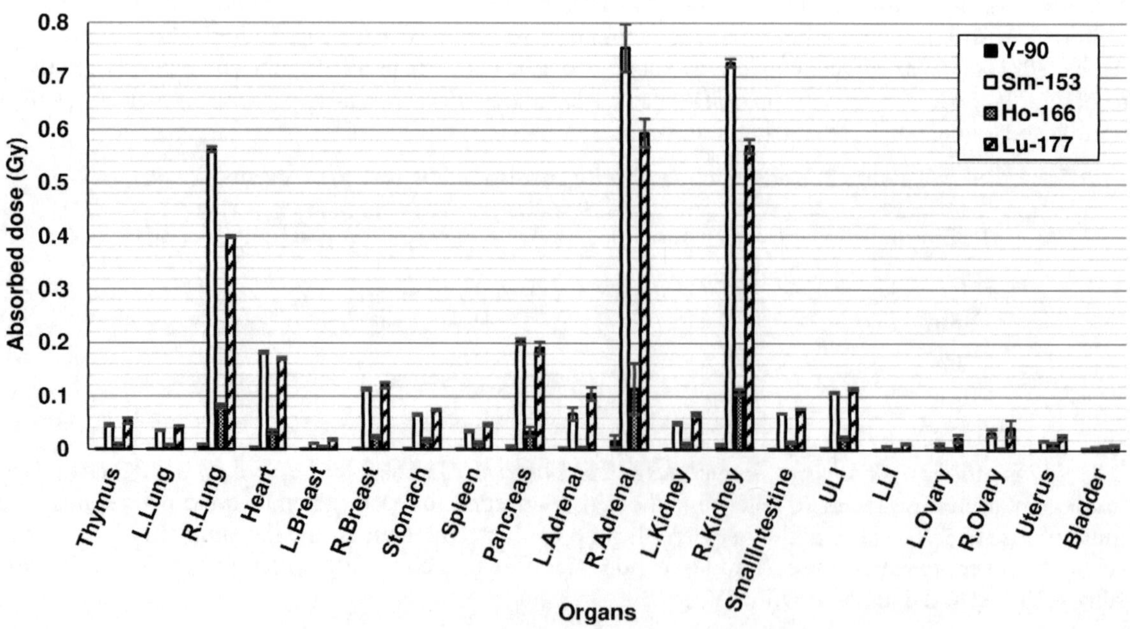

Figure 2. Organ doses as a result of 263 Gy tumour dose from ^{90}Y, ^{153}Sm, ^{166}Ho and ^{177}Lu. L: left, R: right, ULI: upper large intestines, LLI: lower large intestines.

The absorbed doses to the other organs from the various radionuclides are shown in figure 2. It was observed that the highest organ doses were from ^{153}Sm followed by ^{177}Lu, however all doses were still far below the maximum tolerance limit, i.e. 60, 23 and 17.5 Gy for adrenal glands, kidneys and lungs, respectively. The absorbed doses to these organs were found to be less than 1 Gy, and were affected by the energy of the gamma emissions, the half-life of the radionuclides, the distance from the liver and the organs masses.

Even though the absorbed doses to organs found in this study were lower than 1 Gy, it should be noted that they may be higher when more complicated cases are being simulated i.e. larger tumour hence higher activity, as well as the presence of normal liver uptake and lung or extrahepatic shunting. Further work to simulate more complex situations should be carried out and the maximum limit for treatment eligibility should be determined.

4. Conclusions

^{166}Ho, ^{177}Lu and ^{153}Sm offer the useful gamma emissions for post-procedure imaging. They show potential as ^{90}Y substitutes, delivering comparable tumour doses and lower normal liver doses, with other organ doses far below the maximum tolerance limit.

References

[1] Ferlay J, Soerjomataram I, Dikshit R, Eser S, Mathers C, Rebelo M, Parkin D M, Forman D and Bray F 2015 *Int. J. Cancer* **136** E359-86

[2] Hashikin N A A, Yeong C H, Abdullah B J, Ng K H, Chung L Y, Dahalan R and Perkins A C 2015 *PLoS One* **10** e0138106

[3] Hashikin N A A, Yeong C H, Abdullah B J J, Ng K H, Chung L Y, Dahalan R and Perkins A C 2015 Samarium-153 labelled microparticles for targeted radionuclide therapy of liver tumor *World Congress on Medical Physics and Biomedical Engineering, June 7-12, Toronto, Canada (IFMBE Proc.)* ed Jaffray DA (Switzerland: Springer International Publishing) **51** pp. 471-4

[4] Mumper R J, Ryo U Y and Jay M 1991 *J Nucl Med.* **32** 2139-43

[5] Poorbaygi H, Reza Aghamiri S M, Sheibani S, Kamali-Asl A and Mohagheghpoor E 2011 *Appl. Radiat. Isot.* **69** 1407-14

[6] Laboratoire National Henri Becquerel 2014 France http://www.nucleide.org/DDEP_WG/DDEPdata_by_Z.htm

[7] Snyder W S, Ford M R and Warner G G 1978 *MIRD Pamphlet No. 5 Revised: Estimates of Specific Absorbed Fractions for Photon Sources Uniformly Distributed in Various Organs of a Heterogeneous Phantom* (New York: Society of Nuclear Medicine)

[8] Sirtex 2004 [package insert] (Lane Cove, Australia: Sirtex Medical)

[9] Dubois D and Dubois E 1916 *Arch. Int. Med.* **17** 863-71

[10] Ho S, Lau W Y, Leung T W, Chan M, Ngar Y K, Johnson P J and Li A K 1996 *Eur. J. Nucl. Med.* **23** 947-52

Acknowledgements

This study was funded by the Ministry of Science, Technology and Innovation (MOSTI) Science Fund SF011-2014 and University of Malaya Postgraduate Research Fund PG104-2013B.

The standardization methods of radioactive sources (^{125}I, ^{131}I, ^{99m}Tc, and ^{18}F) for calibrating nuclear medicine equipment in Indonesia

G Wurdiyanto[1] **and H Candra**[1]

[1] Center for Technology of Radiation Safety and Metrology (PTKMR)
National Nuclear Energy Agency of Indonesia (BATAN)
Jalan Lebak Bulus Raya, No. 49, Jakarta Selatan, Indonesia 12440

E-mail: gatot_w@batan.go.id

Abstract. The standardization of radioactive sources (^{125}I, ^{131}I, ^{99m}Tc and ^{18}F) to calibrate the nuclear medicine equipment had been carried out in PTKMR-BATAN. This is necessary because the radioactive sources used in the field of nuclear medicine has a very short half-life in other that to obtain a quality measurement results require special treatment. Besides that, the use of nuclear medicine techniques in Indonesia develop rapidly. All the radioactive sources were prepared by gravimetric methods. Standardization of ^{125}I has been carried out by photon-photon coincidence methods, while the others have been carried out by gamma spectrometry methods. The standar sources are used to calibrate a Capintec CRC-7BT radionuclide calibrator. The results shows that calibration factor for Capintec CRC-7BT dose calibrator is 1,03; 1,02; 1,06; and 1,04 for ^{125}I, ^{131}I, ^{99m}Tc and ^{18}F respectively, by about 5 to 6 % of the expanded uncertainties.

1. Introduction

The utilization of nuclear technology in all fields has grown rapidly both in the international as well as in Indonesia. In their application, it is required the standard sources that meets the technical requirements so that their products have a sufficient level of quality. Some laboratories use standard to calibrate owned measuring apparatus, as well as the guarantor of quality for their products. One of the requirements that are used to obtain a qualified standard reference material will require an adequate level of homogeneity. Based on the duties and functions as a laboratory of national reference in the field of the measurement of radioactivity, then Center for Technology of Radiation Safety and Metrology (PTKMR-BATAN) must be able of provide a liquid standard source with various types that qualify as standard reference materials so that results of the measurement and testing of samples has a value of accurate, precise and traceable to the International System.

To solve that problems, it is required some procedures to calibrate the radionuclide calibrators that are used by hospitals for routine measurement of activity of ^{125}I, ^{131}I, ^{99m}Tc and ^{18}F that was administered to patients. Although ^{137}Cs is used to monitor the radionuclide calibrator, standards of ^{125}I, ^{131}I, ^{99m}Tc and ^{18}F are required to determine the calibration factors for this radionuclide.

Iodine-125 has a half-life of 59.388 (28) days, decay by electron capture and emits X-ray photon at 27 keV (112.5%), 31 keV (20.9%), 32 keV (4.5%), as well as γ-rays at 35.5 keV (6.6%) and Auger and conversion electrons between 22 and 35 keV (Chiste and Be, 2011). Iodine-131 has a half-life of 8.0233 (19) days, disintegrates through emission to the excited levels of ^{131}Xe, the isomeric state ^{131m}Xe included, and γ-ray at 364.46 keV (82.6%), 636.97 keV (7.12%), and 722.89 keV (1.78 %),

(Chiste and Be, 2004). Technicum-99m decays with a half-life 6.0067(10) hours, mainly by isomeric transition to ^{99}Tc. The most intense gamma transitions are 2.1726 keV (99.0%) and 140.511 keV (99.0%) (Chiste and Be, 2004). Fluorine-18 disintegrates to the ground state of ^{18}O by positron emission with an intensity of 96.86(19) % and electron capture with an intensity of 3.14(19)%. The half-life of ^{18}F is 1.8288 (3) hours. The positron has a maximum energy of 633.5 keV and an average energy of 249.3 keV. The 511 keV annihilation radiation following the positron emission has an intensity of 193.72 %. The decay data are those given by the International Bureau of Weights and Measures, or BIPM (BIPM, 2004).

This paper describes the procedures used to standardize the activity of ^{125}I, ^{131}I, ^{99m}Tc and ^{18}F, and the results of the measurements used to calibrate the nuclear medicine equipments in Indonesia. The volume of ^{18}F solutions used was about 16 ml because this is the volume most often used in hospitals. Vials that are used are made by Wheaton, USA and are made of borosilicate glass, with a volume of 20 ml, height of 55 mm, outer diameter of 30 mm, and thickness of 1 mm.

The PTKMR-BATAN Secondary Standard Ionization Chamber is a Capintec CRC-7BT radionuclide calibrator that is used as a working chamber for the routine dissemination of activity standards for photon emitting radionuclides. The calibrator is checked routinely to ensure constancy and relative accuracy using standard sources from national metrology institutes (NMIs). This apparatus is calibrated using standard sources that are calibrated by the gamma-ray spectrometry method for ^{131}I, ^{99m}Tc and ^{18}F radionuclides, but for ^{125}I is measured by absolute photon–photon coincidence counting methods. The activity concentrations of radioactive solutions are calibrated in the point source geometry first, and then standard sources are prepared in other geometries by careful gravimetric transfer of the calibrated solution into ampoules or vials.

2. Methodology

2.1. Preparation of sources

Source solutions of ^{125}I and ^{131}I in the form of potassium iodide (KI) in H_2O was diluted by a factor of 10.25 for ^{125}I and 6.53 for ^{131}I. From that solution, point sources were prepared by droping about 10 – 25 mg onto a thin polyethilene plastic backing. After dispensing the droplet of ^{125}I and ^{131}I onto the source support, a drop of $AgNO_3$ solution was added immediately to precipitate the volatile iodium anion (I^-) as AgI. In addition, the reminder of the original source and the dilution source solution of ^{125}I were transferred into five PTKMR type glass ampoules. The masses of all samples, about 2 g, were determined gravimetrically using KERN ABT 220-5DM type balance, traceable to SI through the Calibration Laboratory of Indonesia.

^{99m}Tc solution is obtained from molydenum generator was prepared in the form of ampoules and thin film layer of mylar. Preparation was carried out by gravimetric method using a calibrated semi-micro balance device because it is more accurate than other methods. 5 pieces of point sources and 3 ampoules were prepared. All of the point sources were dripped on mylar buffer and then dried using infra-red lamp. The dried sources were covered with a mylar layer and coded and ready to be measured.

^{18}F is obtained through the reaction ^{18}O(p, n)^{18}F using a cyclotron that is situated in a hospital in Jakarta. The chemical form of ^{18}F that is used is FDG (fluoro-deoxyglucose) without carrier solution. Three sets of measurements were made and, for each, between 3 and 5 point sources were prepared together with a single 16 ml liquid source in a vial. The ^{18}F point source and the solution in the vial originated from the same master solution.

The point sources had a sample diameter of about 4 mm, and the weight varied between 10 and 15 mg. After drying under a heat lamp for about 30 minutes, they were then covered with a film of mylar with a ± 25 μg/cm^2 thickness. The samples was measured in the activity range of 10^3 Bq to 4.0 10^5 Bq. Preparation of the vials is done by gravimetric transfer of a volume of about 16 ml. The density of the solution was measured as 1.033 g/ml.

2.2. Activity measurement

2.2.1. Photon-photon gamma coincidence measurements. All of the ^{125}I point sources were measured absolutely by a photon-photon coincidence system that was constructed in PTKMR-BATAN. The system using two NaI(Tl) detectors, crystal size of 76 mm diameter x 6 mm thickness with 0.5 mm aluminium window, connected to coincidence unit. The measurements were carried out twice with 25 replicates. Their activity was measured using source to detector distances ranging from 2 to 100 mm, to obtain varying detection efficiencies. Measurements were taken in the energy region between 13 and 100 keV. The duration for individual measurements varied between 100 s at low distance and 500 s at larger distances. Three counting rates were corrected for background, dead time and resolving time for the coincidence count rate. The activity of 125I was calculated using the equation for photon photon coincidence counting by Schrader and Walz (1987) and Schrader (1990, 2006).

2.2.2. γ-spectrometry measurements. The activities of all of the point sources were measured by gamma spectrometry. The detector is an HPGe model GC1018 (Canberra, USA), which has a relative efficiency of 10.3% with an energy resolution of 1.69 keV FWHM at 1332.5 keV. The detector is equipped with a model 2002CSL pre-amplifier, and a Canberra model 2020 amplifier and operates at a bias voltage of +4500 V. Signals from the detector are processed by the Canberra gamma spectrum analysis system using GENIE 2000 software (Canberra Industries, USA). The source-to-detector distance was 25 cm. As described above, the gamma-ray spectrometry system was first calibrated using standard sources of ^{152}Eu, ^{60}Co and ^{137}Cs that have traceability to the SI. Three sets of measurements were made with a counting time of 10 minutes in each case. Impurities were also checked using gamma spectrometry.

2.2.3. Measurement with ionization chamber. All of the liquid solutions in ampoule or vial glass were measured using the Capintec radionuclide calibrator ionization chamber in two ways, firstly with the predefined calibration factor corresponding to the each (^{125}I, ^{131}I, ^{99m}Tc and ^{18}F) button on the activity calibrator and secondly using the dial setting determination method. Preset push-buttons are provided for quick selection of the calibration number of the most often used radioisotopes. Thus, the calibration settings for each radionuclide and geometry should be determined individually.

The determination of the appropriate dial settings in this work was achieved in the manner described by Zimmerman and Cessna (2000). The ratio, R, of the activity measured with the special button on the standard activity, is determined after the gamma spectrometry activity results are known. The ratio R is given by R = Astd / Acap, where Astd is the activity of standard source and Acap is the activity measured by the Capintec radionuclide calibrator. This R value can be used directly as a calibration factor for the ^{18}F and ^{99m}Tc when the special button or calibration setting number is used.

3. Results and discussion

The efficiency (ε_1, ε_2) of the detector varied between 0.12 to 0.05. This values is determined by variation of the detector to source distance between 2 mm to 100 mm. The absolute activities of ^{125}I source could be obtained from a plot of the activity distribution versus $x = f(\varepsilon_1, \varepsilon_2)$, as shown in figure 1. The relative standard deviation (counting statistics) of the measurement result for individual sources varied from 0.4% to 0.5%. The average activity concentration of the 125I solution derived from photon-photon coincidence measurement results was (441.7 ± 4.0) kBq g^{-1}. The relative standard uncertainty components are presented in table 1.

The result of the efficiency calibration for the HPGe detector using 152Eu, 60Co and 137Cs standard sources is shown in figure 2. The energies used to produce the efficiency calibration curve were 244.7, 344.3, 661.657, 778.9, 964.1, 1112.1, 1173.228, 1332,492 and 1408.0 keV, resulting in a curve described by the equation efficiency at energy E = 0.1598 e-0.98 with a correlation coefficient of 0.9998. The uncertainty of the interpolated efficiency at any point is 1.5 % at k = 1. This value was

Journal of Physics: Conference Series **694** (2016) 012060 doi:10.1088/1742-6596/694/1/012060

determined from the residuals between the true efficiencies and the efficiencies obtained from the efficiency curve.

There was no significant impurity detected above 0.05% of the ^{125}I, ^{131}I, ^{99m}Tc and ^{18}F activity. This value was determined when the Capintec radionuclide calibrator was calibrated. The counting time was 10 minutes, and the time difference between the gamma spectrometry and Capintec radionuclide calibrator measurements was about 8 half-lives of ^{131}I, ^{99m}Tc and ^{18}F, respectively. The results of the activity determinations of the ^{125}I, ^{131}I, ^{99m}Tc and ^{18}F solution are shown in table 2. The corrections on standardization of radioactive sources are impurity, dead time, background counts, half-life, and decay on process of counting.

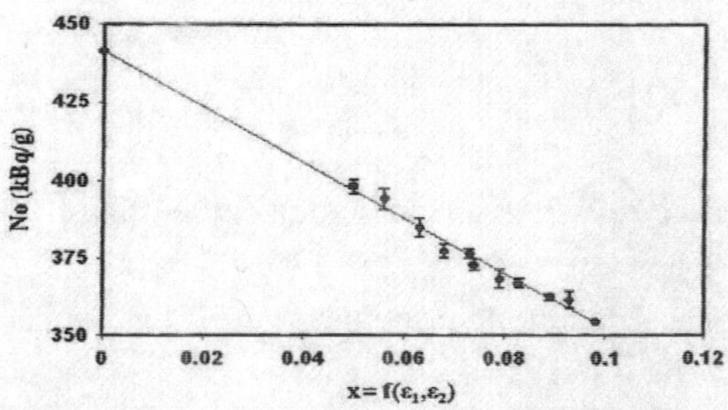

Figure 1. An example of activity distribution versus x = f(ε_1,ε_2).

Tabel 1. Relative uncertainty components of the 125I activity concentrations (in %).

Components	U,(%)	Type
Statistics counting	0.54	A
Counting time	0.035	B
Weighing	0.2	B
Dead time	0.05	B
Background	0.35	B
Resolving time	0.1	B
Decay parameters	0.13	B
Impurity	0	-
Extrapolation	0.5	B
Combined Uncertainty (k=1)	0.86	

Tabel 2. The result of activity measurement base on gamma spectrometry and photon-photon coincidence system.

Nuclide	Activity (MBq)	Methods	Dose calibrator (MBq)	Calibration Factor
^{125}I	7.30	coincidence	7.11	1.03
^{131}I	109.23	spectrometry	107.22	1.02
^{99m}Tc	214.67	spectrometry	201.67	1.06
^{18}F	744.82	spectrometry	715.23	1.04

Table 3, shows some types of components used to determine the value of the measurement an expanded uncertainty that used the gamma spectrometer equipment, dose calibrator and the coincidence system. Several components have the same uncertainty value, this is because the standard sources used are the same, namely ^{152}Eu, ^{137}Cs, and ^{60}Co. The experimental results showed that if the specific button is used for about 16 ml solutions, the correction factor, R, would be 1.03, 1.02, 1.06, and 1.04 for ^{125}I, ^{131}I, ^{99m}Tc and ^{18}F, respectively; with an expanded uncertainty of 4.9, 5.7, 5.6, and 5.7 % with 95 % of confidence level at coverage factor, k=2. ^{125}I has an expanded uncertainty smallest because it used absolute methods for standardizing that nuclide.

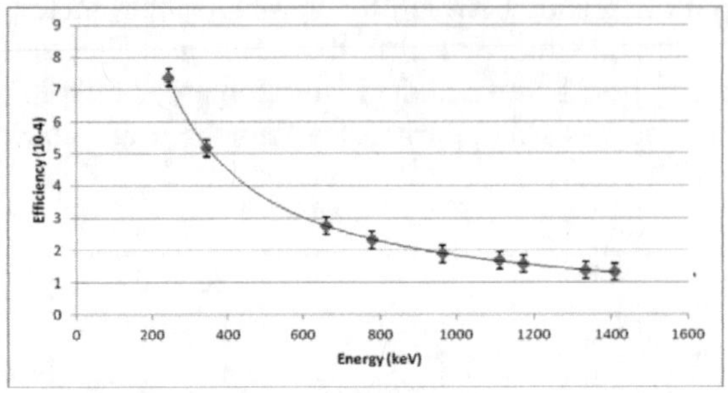

Figure 2. Efficiency calibration curve using ^{152}Eu, ^{60}Co and ^{137}Cs standard sources.

Table 3. Uncertainty components for the R-value determined by gamma-ray spectrometer and dose calibrator for ^{125}I, ^{131}I, ^{99m}Tc, and ^{18}F.

Source of uncertainty	Standard uncertainty components (%)				
	^{125}I	^{131}I	^{99m}Tc	^{18}F	Type
Gamma-ray Spectrometer:					
Standard Source	-	0.5	0.5	0.5	B
Half- life of standard	-	0.12	0.12	0.12	B
Efficiency	-	1.5	1.5	1.5	B
Intensity of sample	-	0.6158	0.2695	0.1962	B
Half life of sample	-	0.0237	0.0166	0.0164	B
Area of sample	-	0.56	0.56	0.67	A
Dead time	-	0.05	0.05	0.05	B
Photon-photon coincidence	*0.86*	-	-	-	*Comb.*
Ionization chamber :					
Half-life of sample	0.0471	0.0237	0.0166	0.0164	B
Statistics of counting	0.60	0.50	0.48	0.6	A
Detector response	1.155	1.155	1.155	1.155	B
Accuracy of reading	1.732	1.732	1.732	1.732	B
Repeatability	0.577	0.577	0.577	0.577	B
Non-linearity	0.35	0.35	0.35	0.35	B
Mass	0.05	0.05	0.05	0.05	B
Combined standard uncertainty	2.45	2.87	2.82	2.86	
Expanded Uncertainty (k = 2)	4.9	5.7	5.6	5.7	

4. Conclusion

Standardization of radioactive sources ^{125}I, ^{131}I, ^{99m}Tc, and ^{18}F have been successfully carried out in PTKMR – BATAN. The use of standard sources that has been standardized by PTKMR can well be used to calibrate the dose calibrator Capintec CRC 7 BT. Calibration factor values of dose calibrator are 1,03; 1,02; 1,06; 1,04 respectively for the source of ^{125}I, ^{131}I, ^{99m}Tc and ^{18}F, with the value of the expanded uncertainty of a stretch of 4.9, 5.7, 5.6, and 5,7 for a confidence level of 95 % at coverage factor, k = 2. PTKMR-BATAN can calibrate the equipment in field of nuclear medicine using a short half-life of ^{125}I, ^{131}I, ^{99m}Tc and ^{18}F standard source.

References

[1] BIPM 2004 *Table of Radionuclides* Monographie BIPM-5. Bureau International des Poids et Mesures, Pavillon de Breteuil Sevres

[2] Chiste V and Be M M 2011 ^{125}I. *Table of Radionuclides* (Vol. 6-A = 22-242), Monographie BIPM-5, BIPM, Sevres, pp. 37-41.

[3] Schrader H 1990 Standardization of ^{129}I by tracer method with photon-photon coincidences from decay of ^{125}I *Appl. Radiat. Isot.* **41(4)** 417 – 421

[4] Schrader H 2006 Photon-photon coincidences for activity determination: ^{125}I and other radionuclides *Appl. Radiat. Isot.* **64** 1179 – 85

[5] Schrader H and Walz K. F 1987 Standardization ^{125}I by Photon-photon coincidences counting and efficiency extrapolation *Appl. Radiat. Isot.* **41** 417

[6] Zimmerman B E and Cessna J T 2000 Experimental determinations of commercial 'dose calibrator' setting for nuclides used in nuclear medicine *Appl. Radiat. Isot.* **52** 615 – 619

Acknowledgments

This activity is funded by DIPA budget of PTKMR-BATAN in 2013-2014 fiscal. The author would like to thank the Government of Indonesia, especially the leaders of all ranks BATAN which has provided the opportunity to successfully complete this study.

Radiation exposure to nuclear medicine staffs during ^{18}F-FDG PET/CT procedures at Ramathibodi Hospital

T Donmoon[1], W Chamroonrat[1] and M Tuntawiroon[2]

[1] Medical Physic School, Division of Nuclear Medicine, Faculty of Medicine Ramathibodi Hospital, Mahidol University, Bangkok, Thailand

[2] Division of Nuclear Medicine, Faculty of Medicine Siriraj Hospital, Mahidol University, Bangkok, Thailand

E-mail: tinnagorn_d@yahoo.co.th

Abstract. The aim of this study is to estimate the whole body and finger radiation doses per study received by nuclear medicine staff involved in dispensing, administration of ^{18}F-FDG and interacting with radioactive patients during PET/CT imaging procedures in a PET/CT facility. The whole-body doses received by radiopharmacists, technologists and nurses were measured by electronic dosimeter and the finger doses by ring dosimeter during a period of 4 months. In 70 PET/CT studies, the mean whole-body dose per study to radiopharmacist, technologist, and nurse were 1.07±0.09, 1.77±0.46, μSv, and not detectable respectively. The mean finger doses per study received by radiopharmacist, technologist, and nurse were 265.65±107.55, 4.84±1.08 and 19.22±2.59 μSv, respectively. The average time in contact with ^{18}F-FDG was 5.88±0.03, 39.06±1.89 and 1.21±0.02 minutes per study for radiopharmacist, technologist and nurse respectively. Technologists received highest mean effective whole-body dose per study and radiopharmacist received the highest finger dose per study. When compared with the ICRP dose limit, each individual worker can work with many more ^{18}F-FDG PET/CT studies for a whole year without exceeding the occupational dose limits. This study confirmed that low levels of radiation does are received by our medical personnel involved in ^{18}F-FDG PET/CT procedures.

1. Introduction

Positron Emission Tomography accompanied with Computed Tomography (PET/CT) has played a major role in tumor imaging. Fluorine-18 FDG as a positron emitter radiopharmaceutical is the most widely and clinically utilized. It has been successfully used for assessing primary tumor and metastasis, to indicate prognosis, treatment planning and monitoring therapy as well as early detect recurrent tumor growth [1]. The 0.511 MeV annihilation photons associated with positron decay are much higher energy than other diagnostic radiations. The exposure rate constant from unshielded ^{18}F source in air (1.49×10^{-4} mSv. m^2/MBq.h) is approximately 7 times greater than ^{99m}Tc (2.06×10^{-5} mSv.m^2/MBq.h) which is commonly used for general Nuclear Medicine studies [2].

Fluorine-18 FDG PET/CT is being used increasingly to evaluate tumor response in addition to diagnosis and staging of tumors. Although PET scanner has been used widespread since about the year 2000, a dedicated PET/CT scanner at our institution was just installed in December 2010. Clinical PET/CT service has been started and increasing since May 2011. The average monthly PET/CT studies are almost double in 2012. More than ninety percent of our ^{18}F-FDG PET/CT studies are for oncologic purpose.

Increasing number of ^{18}F-FDG PET/CT studies and the high penetrating ability of ^{18}F raised the issue of radiation doses to medical staffs while performing their duties with ^{18}F-FDG PET/CT procedures i.e. preparation, handling and administration of the radiopharmaceuticals as well as patient positioning and monitoring during routine PET/CT examinations [3,4,5]. Workers do not have separate dosimeters for variety of tasks. The radiation doses from ^{18}F-FDG PET/CT procedures have assessed together with routine nuclear medicine procedures. The lack of separated radiation doses from ^{18}F-FDG PET/CT studies lead to our proposed study. In addition, the newly available optically stimulated luminescence ring type dosimeter (OSLRD) allows us to measure equivalent dose to fingers.

The objectives of this study are to assess the whole-body and finger radiation doses received by radiopharmacists, technologists and nurses while performing ^{18}F-FDG PET/CT procedures for tumor imaging at Ramathibodi hospital, Bangkok, Thailand, and to compare these occupational burdens with the dose limits recommended by the International Commission on Radiological Protection (ICRP Publication No. 103) [6].

2. Materials and methods

2.1 Dosimeter and readout system

Electronic dosimeter (ED) (ALOKA model PDM-112) was used to measure the Hp(10) from 40 keV gamma and above in a range from 1 to 9999 μSv. The energy dependence is within ±30% in the gamma range between 50 keV to 3MeV. The dosimeters were calibrated with a ^{137}Cs (662 keV) source in the Secondary Standard Dosimetry Laboratory (SSDL) of the Office of Atomic for Peace (OAP), Thailand. The new ring dosimeters developed by Hideaki Miyauchi *et al.* [7] was used for monitoring finger doses. The OSLD consists of two OSL elements and beta cut filter between the OSL elements (nanoDots) produced by Landauer Inc. The first nanoDot is used to measure the Hp(0.07) from gamma and beta rays. The second nanoDot is used for gamma rays only. To maintain dosimetry standard at national level, all measurements were monthly evaluated at the SSDL of the Division of Radiation and Medical Devices, Department of Medical Sciences, Ministry of Public health.

2.2 Exposed personnel

Properly trained radiopharmacists, technologists and nurses in handling radioactive sources who rotated to PET/CT center are groups of worker potentially exposed to ionizing radiation while carrying out tasks associated with ^{18}F-FDG PET/CT procedures. All of them were informed about the method for measuring their doses. In each ^{18}F-FDG PET/CT study, one radiopharmacist, one nurse and two technologists worked together.

Radiopharmacists perform 3 steps of task in the preparation of ^{18}F-FDG dose. The first step is measurement of the total activity of the ^{18}F-FDG multi-dose syringe and drawing up of each 185 MBq (5 mCi) ^{18}F-FDG dose in a syringe and mounted in a tungsten shield. The multi-dose ^{18}F-FDG is sent from the Thailand National Cyclotron and PET Center. The dose withdrawal is always performed within hot laboratory, being drawn up behind a bench-mounted lead shield with a lead equivalent glass insert. The second step is transportation of ^{18}F-FDG dose from the hot laboratory to the injection room. The last step is measurement of the post-injection residual dose in the syringe. The total radiation dose received by each radiopharmacist was timed from the first step and ending when the radiopharmacist threw the syringe into radioactive trash container.

The nurses perform 3 steps in the administration of radiopharmaceutical. The first step is injection the tracers into the patient through pre-canalized intravenous (IV) line. The second step is flushing of normal saline into the IV line. The last step is removal of IV line. The nurses were stand behind a bench-mounted lead shield with a lead equivalent glass insert while performing all steps. The total radiation dose received by each nurse was measured from the first step and ending when leaving the injection room.

The technologists perform all PET/CT imaging tasks. These include escorting the patient to void in a reserved bathroom. After voiding the patient is positioned on the tomograph for the procedure. These include positioning, acquiring images and helping the patient during and until the study is completed. During the time of camera operation, patients were viewed via lead glass window between the scanner and console room. The total radiation dose received by technologist was measured from the time the patient was escorted to void to the time when patient is released to the waiting area. The acquiring time was 60 minutes after injection. Patient was kept in a quiescent state during an uptake phase in a designated room. For PET/CT (Philips Gemini TrueFlight) imaging, spiral CT is firstly performed from the level of the skull vertex to the level of upper thighs by using a scout view with 30 mA and 120 kVp, followed by a spiral CT scan with 50 mA and 120 kVp. This is followed by 3D PET acquisition with 2 minutes per bed positions.

2.3 Whole-body dose [$H_p(10)$] measurement

The ED was placed on the upper left pocket of the coat. Each individual worker carried one dosimeter throughout a FDG PET/CT study. The radiation doses received by each worker were read directly from the dosimeter and recorded at the end of each procedure. The dose was multiplied by calibration factor obtained from the SSDL of the OAP and calculated as the mean value. The time spent of each procedure was recorded. This study was conducted over 4-months period.

2.4 Finger dose [$H_p(0.07)$] measurement

Two OSLRDs were placed on the base of index finger of both hands and noted the dominant hand. The finger dose was calculated as the mean value, corrected for environmental background and the differences in photon energy between the radiopharmaceuticals and reference radiation. The overall finger doses to each personnel were measured for 4 months. OSLRDs were worn during the procedures only otherwise, they were kept in the shielded box.

3. Results

3.1 Effective whole-body dose

Over 4-months period, a total number of 70 ^{18}F-FDG PET/CT studies were included. The ^{18}F-FDG injection activity per patient was between 3.68 to 7.20 mCi (5.89±0.60 mCi). The mean time spent and effective dose per study were reported in Table 1.

Table 1. Effective whole-body dose [$H_p(10)$] to nuclear medicine staffs while performing duties with ^{18}F-FDG PET/CT procedures.

Medical staff	N	Time spent (minutes)/study	Effective dose (μSv)/study
Radiopharmacist 1	30	5.90 ± 2.88	1.01 ± 0.84
Radiopharmacist 2	40	5.85 ± 2.60	1.13 ± 0.93
Mean ± SD		5.88 ± 0.03	1.07 ± 0.09
Nurse 1	34	1.20 ± 0.29	Not detectable
Nurse 2	36	1.22 ± 0.24	Not detectable
Mean ± SD		1.21 ± 0.02	Not detectable
Technologist 1	20	38.45 ± 7.63	1.87 ± 1.05
Technologist 2	8	35.75 ± 8.65	1.31 ± 0.75
Technologist 3	25	39.28 ± 8.95	1.94 ± 1.04
Technologist 4	13	38.54 ± 8.13	1.92 ± 0.83
Technologist 5	25	39.52 ± 8.86	2.20 ± 1.25

Technologist 6	6	42.50 ± 13.52	1.73 ± 0.54
Technologist 7	10	37.40 ± 7.09	1.35 ± 0.99
Technologist 8	12	39.75 ± 6.76	1.13 ± 0.82
Technologist 9	21	40.38 ± 9.32	2.58 ± 1.66
Mean ± SD		39.06 ± 1.89	1.77 ± 0.46

3.2 Finger doses

Finger radiation doses were reported in Table 2. The measurement was performed in all radiopharmacists and nurses similar to the whole-body measurement. However, we were able to measure finger doses from only 2 technologists who rotated to PET/CT center in the last month. The mean finger doses per study were 265.65±107.55, 19.22±2.59 and 4.84±1.08 μSv, for radiopharmacists, nurses and technologists, respectively.

Table 2. Finger doses [$H_p(0.07)$] to nuclear medicine staffs while performing duties with ^{18}F-FDG PET/CT procedures.

Medical staff	N	Time spent minutes/study	Finger dose, μSv/study	
			Right hand	Left hand
Radiopharmacist 1	30	5.90 + 2.88	229.10	511.30
Radiopharmacist 2	40	5.85 ± 2.60	150.09	172.05
Mean ± SD		5.88 ± 0.03	189.59 ± 55.87	341.70 ± 239.92
Nurse 1	34	1.20 ± 0.29	17.06	11.31
Nurse 2	36	1.22 ± 0.24	17.71	30.79
Mean ± SD		1.21 ± 0.02	17.38 ± 0.46	21.05 ± 13.76
Tech 5	25	39.52 ± 8.86	3.76	4.86
Tech 9	21	40.38 ±9.32	4.40	5.91
Mean ± SD		39.95 ± 0.61	4.08 ± 0.46	5.60 ± 0.53

4. Discussion

As shown in table 1, unsurprisingly, the mean effective whole-body dose per study received by technologist, 1.77±0.46 μSv was higher than other personnel which are likely to depend on time spent in proximity to a patient (35 to 43 minutes). High radiation doses may also imply that their patients need more attention, therefore, more contact time. The total radiation dose incurred by a technologist during a PET/CT studies is the product of dose rate and exposure time. Then, trying to spend unnecessary around the patient will minimize radiation dose. In contrast, the steps of the procedures for radiopharmacists and nurses are shorter and could be performed behind a bench mounted lead shield. Undetectable whole body dose means that radiation dose received by nurses were below the detection threshold of the ED.

In table 2, undoubtedly, the higher mean finger dose per study than other personnel received by radiopharmacist was from high amount of radioactivity being handled and from interaction with vials and syringes containing ^{18}F-FDG. Practicing dose dispensing and calibration techniques with nonradioactive materials will reduce the chance of high exposure. The experienced one would be able to reduce the time for each step of the procedure.

There were several reports on whole-body dose per study in the literature; such as 8.9 μSv by Zeff *et al.* [8], 8.5 μSv by Chiesa *et al.* [3], 6.5 by Benetar *et al.* [9], and 7.2 μSv by Biran *et al.* [10]. In our study, if one personnel was to perform all duties it can be assumed that the combined dose from 1 radiopharmacist, 1 nurse and 2 technologists, 4.6 μSv would be delivered per study. It is difficult to compare these doses between institutes because of the variability in the condition factors in each

Journal of Physics: Conference Series **694** (2016) 012061 doi:10.1088/1742-6596/694/1/012061

individual PET/CT facility, such as the patient doses, the procedure, the staff performance and shielding devices.

Finger doses were found to be within permissible limits. The highest finger dose to radiopharmacists are likely from handing of the ^{18}F-FDG multi-dose syringe, transferring the dose to the injection room and measuring the post-injection residual dose in the syringe. The nurse performed shorter part with ready-made individual radiopharmaceutical syringe and pre-canalized IV line for administration. Although, technologists spent the maximal time per study, however, they have the lowest finger dose because they did not directly handle the radioactive material.

In addition, our study reported the left hand majorly received higher dose than the right hand except one nurse while all staff is right handed. We assumed that it is an inconvenience handling anything in the non-dominant hand. Or simply, the non-dominant hand would be the one holding the radioactive material while the dominant hand would be moving, dividing, injecting etc. Therefore, the dominant hand received less finger dose than the non-dominant one. However, in order to decrease the finger dose, especially high volume PET/CT center, an automatic dose dispenser would be incorporated in the hot laboratory to draw ^{18}F-FDG multi-dose.

When compared to the ICRP occupational dose limits [6], 20 mSv per year for whole-body and 500 mSv per year for fingers. If each of our staffs continues to work with their maximum capacity e.g. 1,200 PET/CT studies during a whole year, the whole body doses would be about 1.28, and 2.12 mSv for radiopharmacists and technologists, respectively. The finger doses would be about 410.04, 25.26 and 6.72 mSv for radiopharmacists, nurses and technologists, respectively. All would be very minimal and far below the limit.

5. Conclusion

This study confirmed that our staff can work safety within recommended safety levels of occupational radiation doses to which they are routinely exposed.

References

[1] Thientunyakit T 2007 The ^{18}F-FDG PET scan in oncology *Songkla Med J.* 549-59

[2] Smith D S and Stabin M 2012 Exposure rate constants and lead shielding values for over 1,100 radionuclides *Health Physics.* **102** 271–91

[3] Chiesa C, De Sanctis V, Crippa F 1997 Radiation dose to technicians per nuclear medicine procedure: comparison between technetium-99m, gallium-67, and iodine-131 radiotracers and fluorine-18 fluorodeoxyglucose *Eur J. Nucl Med.* **24** 1380-89

[4] Guillet B *et al.* 2005 Technologist radiation exposure in routine clinical practice with^{18}FDG PET *J. Nucl Med Technol.* **33** 175-79

[5] Fiona O *et al.* 2005 Radiation dose to PET technologists and strategies to lower occupational exposure *J. Nucl Med Techol.* **33** 44-47

[6] International Commission on Radiological Protection 2007 Recommendations of the International Commission on Radiological Protection ICRP Publication 103

[7] Miyauchi H, Takahashi M, Yoshitomi H 2011 Development of ring type dosimeter using OSL elements *JAEA-Technology* 2010-50

[8] Zeff B W and Yester M V 2005 Patient self-attenuation and technologist dose in positron emission tomography *Med Phys.* **32** 861-5

[9] Benatar N A, Cronin B F and O'Dohery M J 2000 Radiation dose rates from patients undergoing PET: implication for technologists and waiting areas *Eur J. Nucl Med.* **27** 583-9

[10] Biran T, Weininger J, Malchi S, Marciano R and Chisin R 2004 Measurements of occupational exposure for a technologist performing 18F-FDG scans *Health Phys.* **87** 539-44

Acknowledgment

We are grateful to the Secondary Standard Dosimetry Laboratory (SSDL) of the Division of Radiation and Medical Devices, Department of Medical Sciences, Ministry of Public health for providing the

OSLRDs for this study. We also wish to thank radiopharmacists, technologists and nurses for their co-operation to make this study possible.

Journal of Physics: Conference Series **694** (2016) 012062 doi:10.1088/1742-6596/694/1/012062

The correlation between effective renal plasma flow (ERPF) and glomerular filtration rate (GFR) with renal scintigraphy ^{99m}Tc-DTPA study

D Ratnasari[1], F Nazir[2], L O H Z Toresano[1], S A Pawiro[1] and D S Soejoko[1]

[1] Department of Physics, Faculty of Mathematics and Natural Sciences, Universitas Indonesia, Depok, Indonesia
[2] National Nuclear Energy Agency, Jakarta, Indonesia

E-mail: diana.ratnasari@sci.ui.ac.id

Abstract. The prevalence of chronic renal diseases in Indonesia has an increasing annual trend, because it is frequently unrecognized and often co-exists with other disease. GFR and ERPF are parameters currently utilized to estimate renal function at routine renal scintigraphy $^{99m-}$Tc DTPA study. This study used $^{99m-}$Tc DTPA to measure GFR and ERPF. The purpose of this study was to find the correlation between ERPF and GFR, for ERPF analysis with Schlegel's method, and GFR analysis with Gate's method, as well as to find correction factor between both variables. Analysis of renal scintigraphy has been performed at Department of Nuclear Medicine Pertamina Center Hospital to thirty patient images acquired from 2014 to 2015 which were analyzed retrospectively data, using gamma camera dual head with counting method from renal scintigraphy $^{99m-}$Tc DTPA study. The calculation was executed by means of both display and manual calculation. Pearson's statistical analysis resulted on Positive Correlation for all data, with ERPF and GFR (display) showing Strongly Positive Correlation ($r = 0.82$; *p-value* < 0.05). Standard deviation was found to be 27.58 and 107.64 for GFR and ERPF (display), respectively. Our result indicated that the use of ^{99m}Tc-DTPA measure ERPF was not recommended.

1. Introduction

The glomerular filtration rate (GFR) is the best way to estimate renal function [1]. GFR is amounts of filtrate which formed per minute at both of kidneys. For male, filtration rate around of 125 ml/minute, and for female, filtration rate around of 110 ml/minute [2]. The concept of measurement of effective renal plasma flow (ERPF) only describes a fraction of plasma flow through a nephron [3]. Normal value of ERPF in the range of 491 to 817 ml/minute for male, and in the range of around 439 to 745 ml/minute for female [4].

Two kinds of radiotracer used to renal study, there are radiotracer which is cleaned from plasma by glomerular filtration, and the others would be cleaned by tubular secretion [3]. In nuclear medicine field, ^{99m}Technetium-diethylene triamine pentaacetic acid (DTPA) is often used to routine renal scintigraphy and to estimate GFR, because this radiotracer would be cleaned by glomerular filtration. Whereas radiotracer which is used to estimate ERPF was cleaned by tubular secretion, such as I-123 and 1-131 OIH Orthoiodohippurate or 'Hippuran', ^{99m}Tc-EC, and ^{99m}Tc-MAG3 [5].

In clinical study (commonly Hospital in Indonesia), radiotracer or radiopharmaceutical which is commonly used to routine renal scintigraphy is ^{99m}Tc-DTPA, because of the limited supply of I-131 Hippuran, ^{99m}Tc-EC, and^{99m}Tc-MAG3. Additionally, in order to know the relative renal function using ^{99m}Tc-DTPA only need 21 minutes, so that radiotracer is the correct choice for clinical use. Actually, the estimation of renal function is not only GFR, but ERPF was too important especially for hypertension and diabetes mellitus.[6] Concerned to this situation, it would be a necessary to do the study the use of ^{99m}Tc-DTPA could be implemented to estimate ERPF or not by using Schlegel's[7] and Gate's[8] method in gamma camera.

2. Materials and methods

In this study, data was collected at Department of Nuclear Medicine Pertamina Center Hospital between 2014 and 2015. The collecting and processing data at least 4 months between January and April 2015. We use retrospective data in total of 30 adults, with 18 male and 12 female. ^{99m}Tc - DTPA dynamic images were acquired with the patient in the supine position (until 21 minute totally after injection) and the detector of gamma camera took placed at the posterior plane. The gamma camera Skylight ADAC [Philips, USA] were equipped with low energy general purpose collimator and 59 PMTs Na(I)Tl crystal. For dose measurements, we used the dose calibrator which was calibrated on March 2015. All patients were injected in the dose range of 300 to 500 µCi (at least 3-5 mCi) of ^{99m}Tc - DTPA. Furthermore, dynamic images were recorded in a 512 × 512 matrix format per second for 1 minute which is known by pre injection and post injection counts. Relative renal function was measured and analysed in a composite image acquired at 2 minutes after injection of radiotracer.[9] Renal and semilunar background regions of interest (ROIs) were drawn manually. Relative renal function was measured using the ROIs of each kidney in the posterior image.

The correlation between ERPF and GFR measurement using ^{99m}Tc-DTPA renal scintigraphy was performed by Pearson Correlation (table 1) [10]. The correlation results were expressed with Schlegel – Gate analysis, ERPF using Schlegel's analysis, which need physical data of patients such as weight and height because it is the important to calculate renal depth [11]. Comparison of relative renal function measurement using ERPF and GFR parameter was performed with display (Gamma camera integrated program calculation) and manual (Microsoft Excel calculation). The statistical analysis was performed using Statistica Software Version 10.

Table 1. Meaning of Pearson correlation coefficient value (*r*).

Correlation Coefficient Value (*r*)	Direction and Strength of Correlation
-1	Perfectly negative
-0.8	Strongly negative
-0.5	Moderately negative
-0.2	Weakly negative
0	No association
0.2	Weakly positive
0.5	Moderately positive
0.8	Strongly positive
1	Perfectly positive

The calculation in related to dosimetry, radiation counts of gamma camera detector has information about activity per unit times. Conversion factor from radiation counts to activity is much needed because we need to convert to international unit (MBq). In order to calculate the conversion factor, the detector was placed at the anterior plane upon the source ^{99m}Tc-Pertechnetat with distance between detector and source at about 7 centimeters. The acitivity of source is 1.2 mCi which normalized 1 mCi, counting was carried out 1 minutes for N = 12 of measurement. The calibration factor was found to be 6010 cps/mCi.

3. Results and discussion

The hospital recorded 30 adults (18 male and 12 female) patients with kidney function would be analyzed. The age of patient was in the range of 17 to 81 with mean of patient age of 54 ± 15years. Furthermore, the data was grouped by gender between male and female. The mean age of male was 55 ± 15 years, and was 53 ± 16 years for female.

Pearson correlation was founded with the analysis of ERPF Schlegel and GFR Gates with parameter *r* and *p-value*. Data presented was divided into three groups, include all patients, male, and female without grouping on the type of their kidney diseases as indicated in figure 1–8. The correlation of ERPF Schlegel based on display and manual calculation, data of all patients was resulted on positive correlation ($r = 0.5350; p < 0.05$), include male ($r = 0.363; p > 0.05$), and female patients ($r = 0.7545; p < 0.05$). The correlation of ERPF for all patients had significant positive correlation, with standard deviation of ERPF display and ERPF manual was found to be 107.64 and 2529.47, respectively. If the data was grouped by a gender, there was a shifted in *r* and *p-value*, which is correlation of ERPF for male had not Significant Positive Correlation ($p > 0.05$). Then, the correlation of GFR Gates based on display and manual calculation, data of all patients was resulted on Positive correlation ($r = 0.8333; p < 0.05$), include male($r = 0.8039; p < 0.05$), and female patients ($r = 0.8991; p < 0.05$). GFR measurement for both calculation (display and manual) was resulted on significant positive correlation, with standard deviation of GFR display and GFR manual was found to be 27.58 and 14.61, respectively. In outline, the correlation between ERPF and GFR would be discussed according to the parameters of Pearson. Firstly, based on display calculation was resulted on significant and strongly positive correlation ($r = 0.8181; p < 0.05$). Secondly, based on manual calculation was resulted on not significant and weakly positive correlation ($r = 0.2030; p > 0.05$). As a correction factor, standard deviation which is obtained from GFR was recommended for clinical use, but for ERPF was not recommended as a correction factor, because it was out of the range. Some values were out of the range was because these were mostly related to non-uniform kidney disease with different renal function measured by ^{99m}Tc-DTPA in renal scintigraphy. Discussed about formula used on the ERPF and GFR calculation, there was a difference in body surface area (formula used European's anatomical sizes) to measure the renal depth which is invented by Tonnesen [12]. He measured the renal depths of 55 subjects by ultrasonography with the patient in the sitting position and the probe positioned at an oblique angle to the kidney. It turned out the Tonnesen's method which is couldn't be used in Indonesia.

Discussed about the correlation between ERPF and GFR, we must know that these parameters have the different measurement at the part of kidney. GFR was occured at glomerular, but ERPF was at tubular. ^{99m}Tc-DTPA acts like a plasma, hence it can measure flow system or 'Perfusion' [5]. ^{99m}Tc-DTPA emitting gamma-ray with mean energy 140 keV, which is this radiotracer was suitable using at routine renal scintigraphy. Discussed about ERPF, it was more suitable using radiotracer worked at tubular like I-131 or I-123 OIH which is known by 'Hippuran'. It was emitting beta particle with mean energy 193 keV, and it was not suitable at routine renal scintigraphy. I-131 Hippuran has a beta radiation that could be penetrate larger than gamma radiation. At the ERPF measurement, ^{99m}Tc-DTPA acts like a plasma which is filtrated by glomerular frequently in all nephrons. The tubular was existed after glomerular as a plasma flow in kidney. Therefore, ^{99m}Tc-DTPA which is flew at the tubular, occured only a few amounts for 2 minutes images acquired (renal uptake GFR and ERPF was at 2-3 minutes after injection) [7, 8].

In the study related to this work, Celik *et al* was compared the use of ^{99m}Tc-DTPA and ^{99m}Tc-DMSA for relative renal function in pediatric patients, their concluded that for patients who needs renogram curve and GFR calculations ^{99m}Tc-DTPA could be a choice for the calculation of relative renal function although ^{99m}Tc-DMSA is the gold standard method for the calculation of relative renal function [13]. On the other study, Suapang *et al* who estimated of GFR and ERPF in dynamic renal scintigraphy, the clearance of ^{99m}Tc-DTPA and ^{99m}Tc-MAG3 were the relatively index of GFR and ERPF in the clinical evaluation of patients with acute and chronical renal failure using camera-based methods [11]. This study used^{99m}Tc-DTPA for measure GFR, and ^{99m}Tc-MAG3 for measure ERPF.

Journal of Physics: Conference Series **694** (2016) 012062 doi:10.1088/1742-6596/694/1/012062

Equals with our study, they use ERPF Schlegel and GFR Gates for the analysis. However our study using an original formula of Schlegel's analysis without R (the predicted 32 minutes return of the injected radionuclide) because our examination time was performed at 21 minutes only.

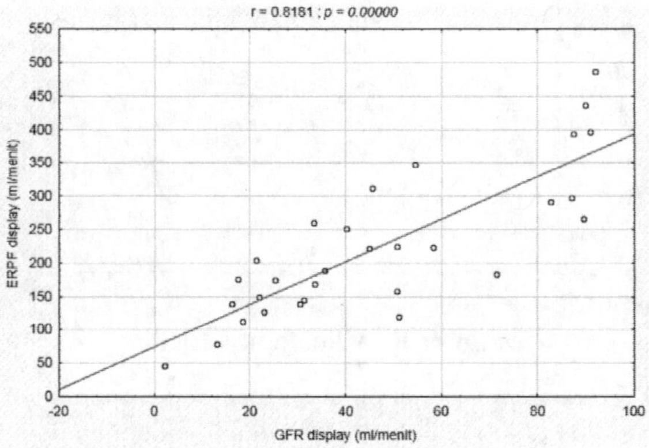

Figure 1. ERPF and GFR (display).

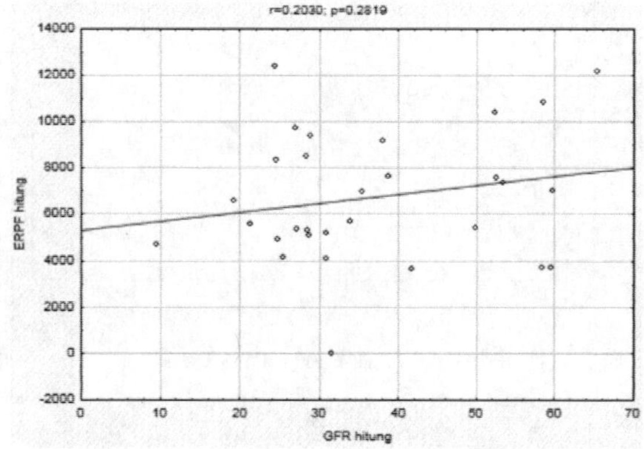

Figure 2. ERPF and GFR (manual).

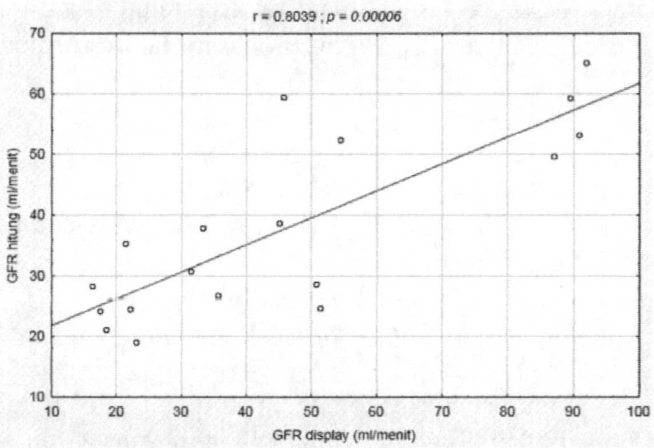

Figure 3. Male's GFR.

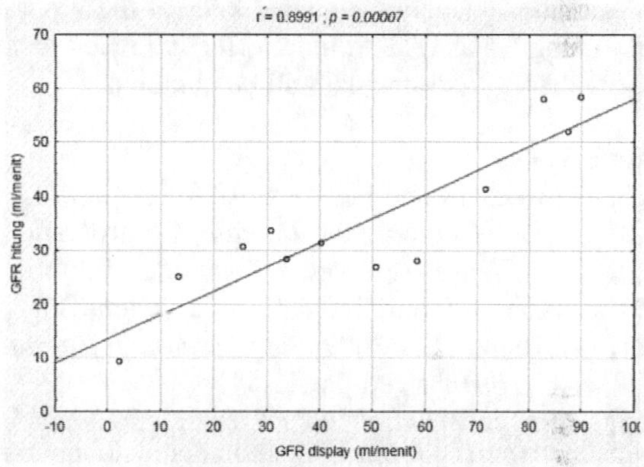

Figure 4. Female's GFR.

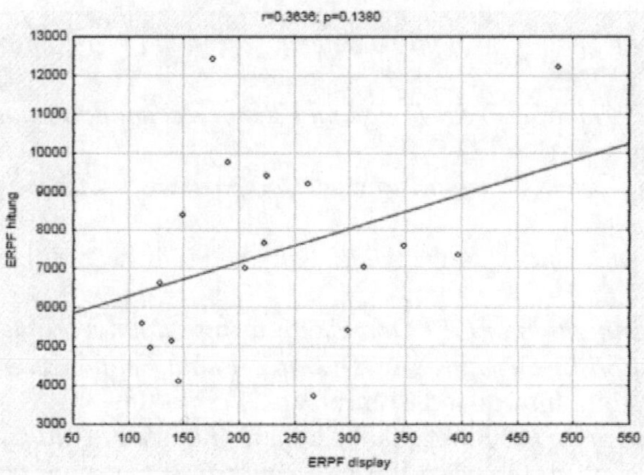

Figure 5. Male's ERPF.

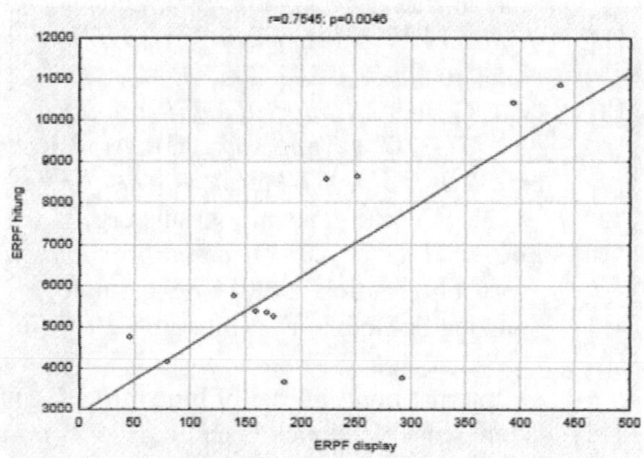

Figure 6. Female's ERPF.

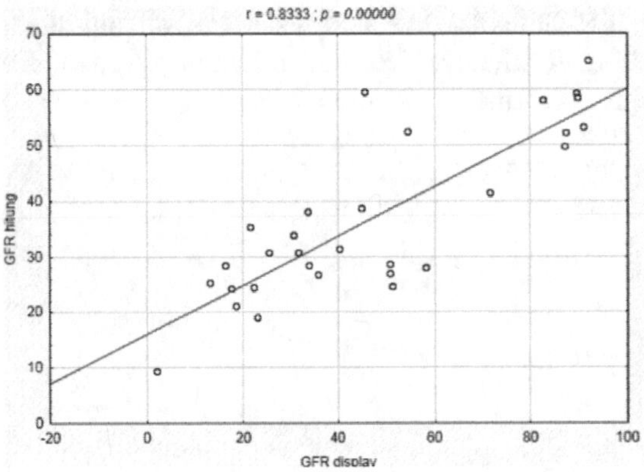

Figure 7. All patient's GFR.

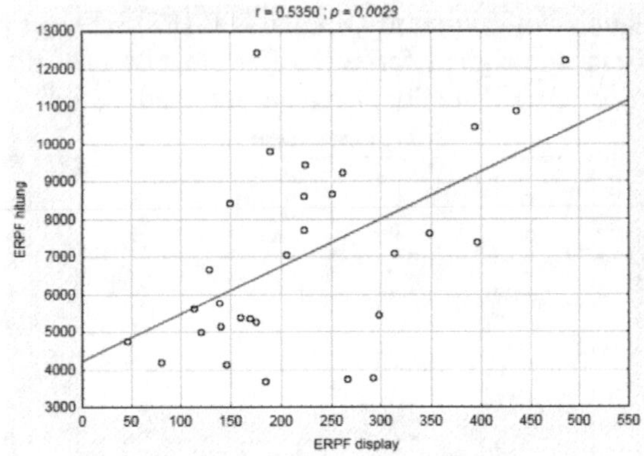

Figure 8. All patient's ERPF.

4. Conclusion

The correlation between ERPF and GFR with renal scintigraphy^{99m}Tc-DTPA study showed Positive Correlation, which means that both of these parameter was related to estimate relative renal function. By using ^{99m}Tc-DTPA, more effective result to measure GFR than ERPF based on correlation formed from display and manual will be obtained.

References

[1] Prigent A and Piepsz A 2006 *Functional Imaging in Nephro-Urology* (UK: Springer)

[2] CDC (Centers for Disease Control and Prevention) 2010 *Kidney Disease: Key Facts and Figures (*Ireland: Association of Public Health Observations)

[3] Baert A L and Sartor K 2006 *Diagnostic Nuclear Medicine* 2nd rev(London: Springer)

[4] Masjhur J S 2009 *Buku Pedoman Tatalaksana Diagnostik dan Terapi Kedokteran Nukir. In: Nuklir K*, ed. Pp 41-54

[5] Jurgen BH *et al.* 2007 *Clinical Nuclear Medicine* (London: Springer)

[6] CDC (Centers for Disease Control and Prevention) 2011 *Chronic Kidney Disease Surveillance System*. Atlanta, GA: Centers for Disease Control and Prevention, US Dept of Health and Human Services;. http://www.cdc.gov/ckd.

[7] Schlegel J U and Hamway S A 1976 *Individual renal plasma flow determination in two minutes* J Urology 116:282-285.

[8] GatesG F1982 *Glomerular Filtration Rate: Estimation From Fractional Renal Accumulation of ^{99m}Tc-DTPA (Stannous)*. AJR Am J Roentgenol 138:565-570.

[9] Fine E J *et al* 1987 *Measurement of Effective Renal Plasma Flow: A Comparison of Methods* (New York: jnm.snmjournals.org)

[10] Zou K H *et al* 2003 *Correlation and Simple Linear Regression*, RSNA. Published online 10.1148/radiol.2273011499 Radiology 227:617–628.

[11] Suapang P, Dejhan K, Yimman S 2015*The Estimation of GFR and ERPF using Adaptive Edge-Based Active Contour for The Segmentation of Structures in Dynamic Renal Scintigraphy* International Journal of Innovative Computing, Inform and Control Vol. 11, No. 11.

[12] Tonnesen K J, Munck O, Hald T 1976 *Influence on the renogram of variation in skin to kidney distance and the clinical importance thereof.* Presented at the International Symposium on Radionuclides in Nephrology, Berlin, (cited by Schlegel JU, Jammway SA: Individual renal plasma flow determination in 2 minutes) *JUROL* **116** 282

[13] Çelik T, Yalçin H, Günay EC, Özen A and Özer C 2014 Comparison of the relative renal function calculated with ^{99m}Tc-diethylenetriaminepentaacetic acid and ^{99m}Tc-dimercaptosuccinic acid in children *World J Nucl Med* **13** 149-53

Acknowledgements
I would like to express my sincere thanks to Duta Kamesworo and Chavid Varuna of Department of Nuclear Medicine, Pertamina Center Hospital, for their helps to preparing manuscript data. We also thanks to Laboratory of Medical Physics and Biophysics for supporting to attend SEACOMP 2015 though Hibah PUPT Directorate General of Higher Education, Ministry of Education and Culture, Republic Indonesia 2015.

Integration of soft tissue model and open haptic device for medical training simulator

G F Akasum[1], L N Ramdhania[2], Suprijanto[2] and A Widyotriatmo[2]

[1] Graduate Program of Instrumentation and Control, Faculty of Industrial Technology, Institute of Technology Bandung, Indonesia
[2] Instrumentation and Control Research Group, Faculty of Industrial Technology, Institute of Technology Bandung, Indonesia

E-mail: gina.akasum@gmail.com

Abstract. Minimally Invasive Surgery (MIS) has been widely used to perform any surgical procedures nowadays. Currently, MIS has been applied in some cases in Indonesia. Needle insertion is one of simple MIS procedure that can be used for some purposes. Before the needle insertion technique used in the real situation, it essential to train this type of medical student skills. The research has developed an open platform of needle insertion simulator with haptic feedback that providing the medical student a realistic feel encountered during the actual procedures. There are three main steps in build the training simulator, which are configure hardware system, develop a program to create soft tissue model and the integration of hardware and software. For evaluating its performance, haptic simulator was tested by 24 volunteers on a scenario of soft tissue model. Each volunteer must insert the needle on simulator until rearch the target point with visual feedback that visualized on the monitor. From the result it can concluded that the soft tissue model can bring the sensation of touch through the perceived force feedback on haptic actuator by looking at the different force in accordance with different stiffness in each layer.

1. Introduction

Minimally invasive surgery (MIS) is a surgical technique in which short, trocars are inserted through small incisions. Compared with open surgery, MIS has some advantages, such as cause less damage to muscle tissue, creating a faster recovery for the patient. The application of MIS on the clinical situation require advanced technology and surgeons must be trained to operate MIS devices since MIS procedure required particular eye-hand coordination that totally different with open surgery procedure. Needle insertion is one of simple MIS procedure. Its procedures are often used to take the sample of soft tissue (biopsy) or give the local drug with epidural anaesthesia. Before the needle insertion technique used in the real situation, it essential to train this type of medical student skills. Naik *et al.* conducted a research using the learning curve cumulative summation method and obtained that a novice can attains proficiency of epidural anaesthesia after 75 attempts [1]. Thus, the training simulator for medical students is urgently needed.

The application of minimally invasive surgery can be found at many hospitals in Indonesia. The surgeons use this technique in various operation with the limitation of sight, so that camera laparoscope is used to guide the operation procedures. This competence is essential for surgeon so that the simulator used in the practical period must be completed by visualization device system as the guidance during the simulation process.

Many types of research and development on medical training simulator have been conducted in the past few decades. Some research was trying to found the appropriate model for friction force along the needle [2] while the others observed the lost of resistance effect [3] or application of haptic technology in many cases of surgery [4]. In this research, the simulator is developed to train the medical students the needle insertion procedure through the liver tissue.

2. Theoretical backgrounds

In this work, the simulator is built to facilitate medical students practice the needle insertion procedure on liver tissue. To provide the realistic feel corresponding the sensation in the real surgery, haptic technology is used in this simulator.

2.1. Haptic

Haptic is the term that have been used to explain the interaction between human and computer through the virtual object [5]. The sensation of touch is the human perception of the force felt by a tactile sensor at his hand. Haptic works in bidirectional direction, it can sense and responds the human's action as shown in figure 1.

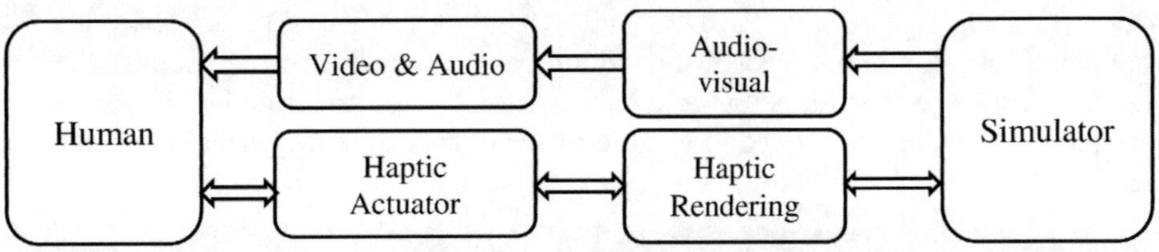

Figure 1. Interaction between human and simulator through haptic rendering.

Haptic can provide the realistic feel to human's hand since it completes the audio and visual sensation along the needle insertion procedure. The delivery process of touch between human and computer is called haptic rendering. In this simulator, human create the change of needle tip position (in mm) and receive the feedback from the computer as force (in N) that perceived as touch sensation.

2.2. Liver

The liver is the largest gland in the human body that located in the upper right abdomen, just below the lungs. Liver filled the upper right abdomen from the front to the rear so that the medical action of the liver is usually done from the right side of the patient's body. In the transverse cross-section, the body viewed from the direction of the foot or head, liver located under the skin. Each layer between the skin and the liver have different characteristics. One of the characteristics that affect the sensation of touch is stiffness. This difference will give a different sensation when the needle inserted to the body. Some research [6] found that the characteristics of each layer are given in table 1.

Table 1. The characteristic of layers of skin and liver.

Layer	Young Modulus
Stratum corneum	333 MPa
Epidermis	17 MPa
Dermis	1,7 kPa
Liver	2,3-5,9 kPa

Young Modulus represents the degree of elasticity of a layer material. The value of stiffness (k) can be calculated from the relationship with Young Modulus as follows.

$$k = \frac{AE}{l} \tag{1}$$

where E is the Young's Modulus of the tissue (N/m^2), A is the area of the tissue (m^2) and l is the thickness of the tissue (m).

In this simulator, the characteristic creates the force feedback as the response according to the following equation.

$$F = k(y_2 - y_1) \tag{2}$$

where y_2 is the actual position, y_1 is the previous position, and (y_1-y_0) is displacement in position of the virtual needle in a model of soft tissue.

3. Simulator design

The research has developed an open platform of needle insertion simulator with haptic feedback that providing the medical students a realistic feel encountered during the actual procedures. There are three main steps in build the training simulator, which are configured hardware system, develop a program to create soft tissue model and the integration of hardware and software.

3.1. Hardware

The hardware of training simulator consists of the dummy of human body (mannequin), the lever system for coupling force from a hand to an actuator, a haptic actuator that integrated with computer, a camera and a screen that displays visual feedback to guide user during inserting needle through the body as shown in figure 2.

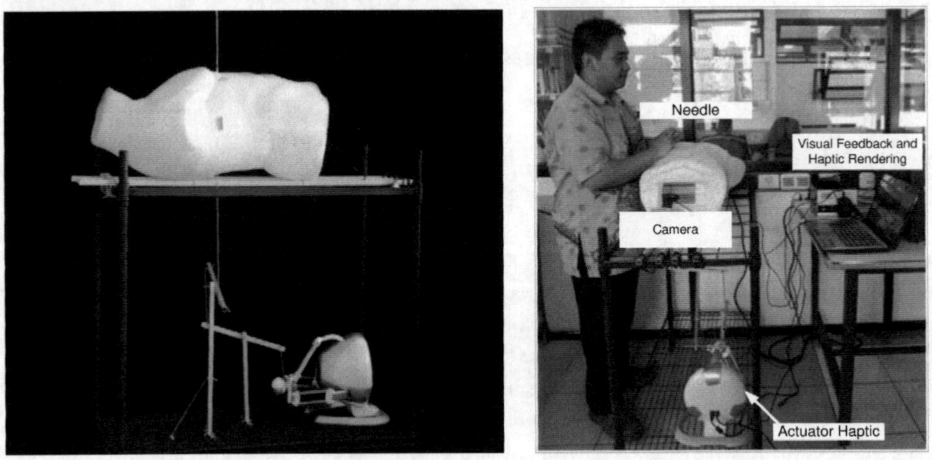

Figure 2. The configuration of training simulator.

In this haptic-based simulator, the mannequin is not filled by medium, but it is left empty. A background picture has mounted in the mannequin to give the visual sensation along needle insertion procedure. The camera is located in the mannequin for user guidance while needle insertion occurred. It capture the needle's movement and display it on the screen. Furthermore, the haptic actuator is connected to the computer where the computation of force feedback is calculated in it.

3.2. Software

The software of haptic simulator consists of stiffness model of resemble skin and liver tissues and module of realtime communication with the haptic actuator. Figure 3 shows the model for the layers of skin and liver. The stiffness and thickness are different for each layer and the value is adapted with the haptic actuator's limited work field. The change of needle tip position made by the user will be detected in this model and the force feedback is calculated using equation (2).

Suppose there are two conditions, the initial condition is called condition (*n*) with the initial position y_1 and next condition $(n + 1)$ with position y_2. In each condition the displacement $\Delta y = y_2 - y_1$ position is constantly changing as user change the position of state (*n*). The force feedback is formed into an equation, namely

$$F_m(n+1) = -F_{offset} - k[y(n+1) - y(n)] \quad (3)$$

where *m* is the model layer (layer I-V), F_{offset} is the force on the haptic device, *n* is the state position, and *k* is a constant stiffness layer.

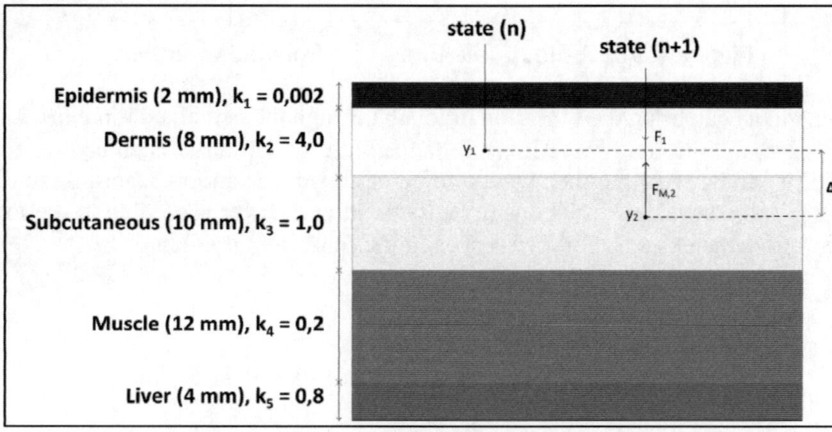

Figure 3. The model of skin and liver layers.

3.3. *The integration of hardware and software*

As input for the model, a needle position is detected based on the encoder of the haptic actuator and a model calculate feedback force that different for each layer of skin and send it to the haptic actuator. The force feedback from the haptic actuator and then reach the user's hand through the lever system so the user can feel the sensation of touch.

4. Evaluation

For evaluating its performance, the haptic simulator was tested by 24 volunteers on a scenario of soft tissue model. Each volunteer is not allowed to try the simulator so that all the volunteers have the same baseline. They must insert the needle on the simulator until reaching the target point with visual feedback that visualized on the monitor. To see the needle tip on the monitor, the user can estimate the progress of needle insertion procedure on the target. Users have a constraint in completing needle insertion procedure that is a time limit. Each trial must be completed in less than 15 seconds.

5. Result and discussion

Users have four times trial to perform needle insertion. The result of the evaluation from one volunteer is shown in figure 4. The time required to reach the target is shorter with the increasing number of trial. Each trial shows the different pattern in the depth changes. It means that sensation of touch can be felt by the user so the reactions of the user are different too. It shows that the training simulator works well.

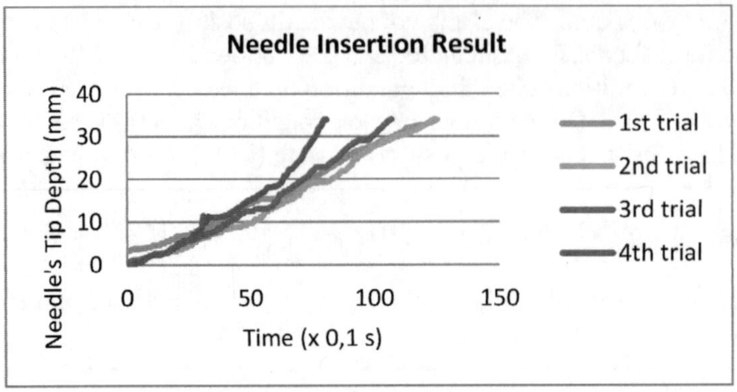

Figure 4. The result of four times trial from the volunteer.

The soft tissue model can bring the sensation of touch through the perceived force feedback on haptic actuator by looking at the different force in accordance with modelling in each layer. From the result shown in figure 5, it can be seen that the average force of each layer differs according to differences in stiffness constants. There are large variations in the style of each layer according to the modelling that has been made to prove the existence of differences in the sensation of touch.

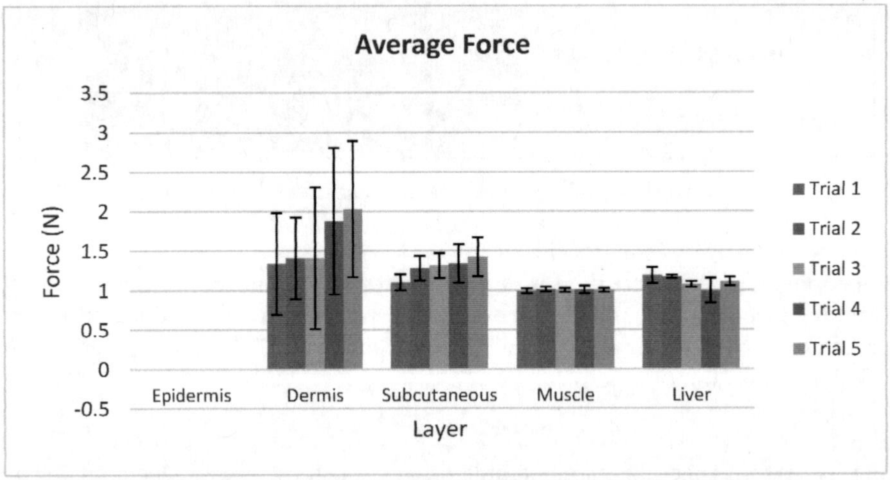

Figure 5. The average force perceived when volunteer evaluate the simulator.

From the results, we can show that the simulator provided the different force in each layer that perceived by the human skin as the sensation of touch. This different force is being an input to the tactile sensor in human's hand and made users give the force feedback to insert the needle. The force in each layer can be different because of the different displacement made by the user, so in each trial, the sensation is different too. This condition requires the user to keep practicing in order to familiarize the sensation of touch.

6. Conclusion

The results show that the different force feedback of each layer can provide the good perception of the surgeon in distinguishing different layers of skin. The time required by the needle to reach the target is increasingly shorter when the trial is repeated. This is consistent with the function of the simulator as a training tool, which is to familiarize the user perform needle insertion procedure. Therefore, the proposed simulator with haptic feedback could be used for basic training of medical students, particularly in simulate needle insertion.

References

[1] Robert D C, Kent G H, David J B and Daniel S J P 2014 Student attainment of proficiency in a clinical skill: the assessment of individual learning curves *PLoS One* PMC3930528

[2] Allison M O 2004 Force modeling for needle insertion into soft tissue *IEEE TRANSACTIONS ON BIOMEDICAL ENG* **51** No. 10

[3] Dogliotti A M 1993 Research and clinical observations on spinal anesthesia: with special reference to the peridural technique *Anesth Analg* **12**:59–65

[4] Yongyao Y 2011 Epidural injection simulator using haptic techniques *A Project of eResearch SA Summer Scholarships 2010/2011*

[5] Alejandro G 2008 Haptic deformable shapes using open source libraries *Thesis, Department of Computing Imperial College London, UK*

[6] Hendriks F M, Brokken D, Oomens C W J, Baaijens F P T and Horsten J B A M 2000 Mechanical properties of different layers of human skin *Philips Research Laboratories Eindhoven*

Influence of bone morphological properties on a new expandable orthopaedic fastener

M Oldakowski[1], I Oldakowska[1], T B Kirk[1], C T Ford[1], T B Sercombe[2], P Hardcastle[1] and R E Day[2,3]

[1] School of Civil and Mechanical Engineering, Curtin University, Kent St, Bentley, WA 6102, Australia
[2] School of Mechanical and Chemical Engineering, University of Western Australia, 35 Stirling Hwy, Crawley, WA 6009, Australia
[3] Medical Engineering and Physics, Royal Perth Hospital, Wellington St, Perth, WA 6000, Australia

E-mail: intan.oldakowska@postgrad.curtin.edu.au

Abstract. Previous studies have demonstrated that bone morphological properties are a significant determinant of orthopaedic fastener fixation strength. The authors previously tested a new design of unthreaded expandable fastener (UEF) prototype against screws and demonstrated a significant increase in pull-out strength. However the effect of bone morphology on the pull-out strength of the UEF and expandable fasteners in general is unknown. This study assessed the correlation between failure force and maximum force against five microstructural parameters. The failure force of the UEF was correlated to the trabecular bone volume fraction, as with screws. Unlike screws, however, the maximum force of the UEF has an inverse relationship with cortical volume. No correlation was found between failure force and the Structural Model Index (SMI). Additionally the critical volume of interest (VOI) for the UEF is around the bottom of the fastener where the expansion occurs, whereas for the screw a full height VOI is critical. Furthermore, we observed that screw mechanical performance may be affected more by bone morphological parameters that are associated with poorer quality bone. Therefore the UEF may perform better than screws in low quality osteoporotic bone.

1. Introduction

Orthopaedic screws have long been used to affix orthopaedic implants to bone in order to stabilise fractures and fuse joints. However screw loosening and pull-out are still significant problems [1, 2]. In order to address these problems, the authors are developing a new design of expandable fastener to provide superior fixation strength and reduce the incidence of loosening.

Pull-out strength is a critical measurement of the performance of orthopaedic fasteners. Low fastener pull-out strength can lead to screw loosening in–vivo, causing injury and potentially requiring revision surgery. Often the cause of this failure is low quality bone due to osteoporosis [1, 2]. In a previous study [3], the authors have demonstrated a 41% higher failure strength for a new design of unthreaded expandable fastener (UEF) prototype compared to conventional orthopaedic screws of equivalent length and diameter in a non-osteoporotic ovine bone model. King and Cebon have also demonstrated favourable results for expandable fasteners in osteoporotic bone but not in normal bone, indicating that expandable fasteners may perform better relative to screws in poor quality bone [4].

Many ex–vivo studies have investigated the correlation between Bone Mineral Density (BMD) and screw pull-out force with varied success [5]. More recently researchers have begun to assess the effects of trabecular bone morphology on screw pull-out strength. In 2010, Poukalova *et al.* investigated how the trabecular bone structure immediately adjacent to screws affected their pull-out strength and demonstrated significant correlations with some microstructural properties [6]. Ab-Lazid *et al.* investigated the effect of bone microstructure, using a volume of interest (VOI) centred on the screw and demonstrated a strong relationship between most microstructural properties [7]. The diameter of the VOI for their study was taken from the deformed area during pull-out loading in the FEA study by Wirth *et al.* [8].

This paper aimed to quantify the effects of bone quality on the pull-out strength of a new design of UEF and compare this to standard orthopaedic screws. Furthermore the study investigated the VOIs in which the bone properties are most critical, allowing future studies to use the most appropriate VOI.

2. Method

The pull-out testing data and CT scans analysed in this study were taken from a previous study by the authors [3]. Briefly, experimental square unthreaded expandable fasteners were manufactured using a Realizer SLM100 machine from a low–modulus pure–beta titanium alloy Ti2448 (Ti-24Nb-4Zr-8Sn, all in wt%). The parts were produced using a standard contour and vector fill scanning strategy, with the direction of the fill vectors rotated 90° between layers. The laser power was 120W for the contour and 175W for the fill, while the scan speed, scan spacing, beam compensation and layer thickness were 1000 mm/s, 100 µm, 60 µm and 50 µm, respectively. The particle size of the powder was 45 µm.

The expandable fasteners were 10mm long, 4mm wide and expanded to approximately 6.2mm. The most critical novel design feature of the UEF is the expansion sides having dedicated bending sections that have a thin rectangular profile with minimal area moment of inertia. This allows a much larger expansion than currently available expandable fasteners with purely elastic expansion (see figure 1). An ISO 5835 cancellous orthopaedic screw was also tested for comparison purposes. The screws were self-tapping and had a 4 mm outer diameter, were 40 mm long, had a pitch of 1.9 mm and thread depth of 0.5 mm (see figure 1). Both the fasteners and screws were inserted at 10mm depth and tested in pull-out using an ovine vertebral body.

The bones were CT scanned using the Skyscan 1176 micro-CT scanner at the Centre for Microscopy, Characterisation and Analysis, at the University of Western Australia. Scanning was performed after the pilot holes were created at 18 microns resolution, using a rotation step of 0.5°, 2 frame averaging, a source voltage of 80 kV, an exposure time of 285 ms and Cu+Al filter. A 3D median filter was applied to remove noise and cortical segmentation was performed using a modified method based on the work of Buie *et al.* [9]. Pull-out testing was performed using an Instron 8874 with a quasi-static 5mm/minute loading rate and failure force and maximum force were recorded. Failure force was defined as the highest force before significant non-linearity in the force-displacement curve and maximum force was defined as the highest force anytime during the pull-out test.

Bone property measurements were calculated from CT scans of the bone samples in the VOI created by centring a cylinder on the fastener location, with varying diameters and heights measured from the bottom of the fastener (figure 2). We created VOIs from 6mm to 16mm diameter in 2mm increments and from 2mm height to 10mm height in 2mm increments.

Bone properties were calculated using FIJI [10], an open source image analysis program and the BoneJ [11] plugins. Trabecular volume (Tb.V) and cortical volume (Cor.V) were calculated by voxel counting. The trabecular bone volume fraction (Tb.V/TV) was calculated as the quotient of the trabecular bone volume and total bone volume, where the total bone volume was calculated by applying a 15 iteration 3D voxel dilation/erosion to create a solid bone volume. Trabecular length (Tb.L) was calculated using a 3D skeletonisation procedure [12]. Trabecular Bone surface area (Tb.SA) was calculated by creating a surface mesh using the marching cubes algorithm and summing the area of the triangles making up the mesh [13]. Structural Model Index (SMI) was calculated using the voxel dilation method [14]. All calculations were performed at 35 microns resolution.

Correlation analysis of sample bone properties to failure force and maximum force were performed based on paired Pearson coefficients with a high degree of correlation defined as r>0.6 or r<-0.6. The probability of significance of the correlation was calculated using a two-tailed test with a correlation having p<0.05 considered to have a high probability of significance.

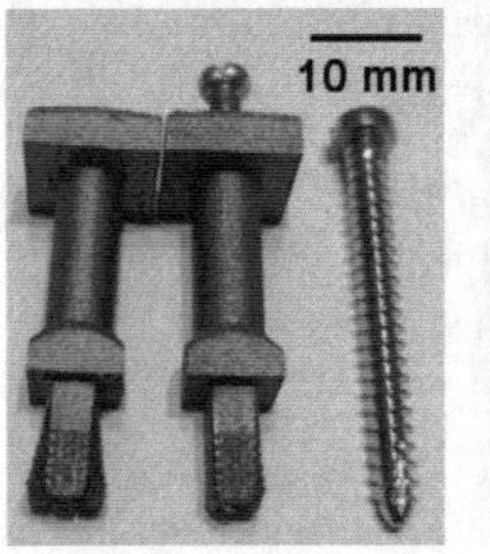

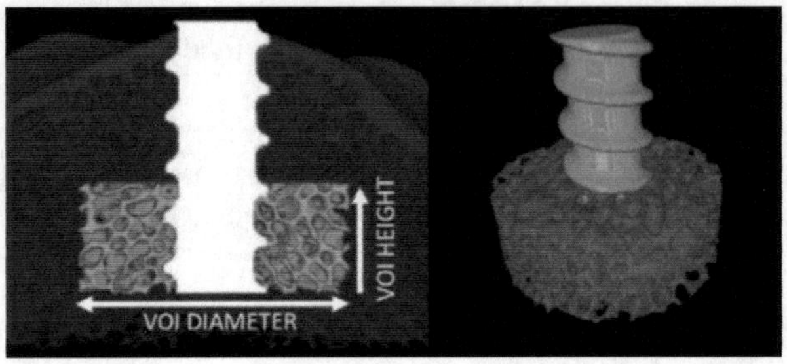

Figure 1. (left) Tested UEF prototype and screw [3].

Figure 2. (right) VOI defined by the VOI diameter and VOI height.

3. Results

Table 1 reports the correlation coefficients (r-values) and the corresponding probability of its significance (p-values) for the relationships between mechanical performance and the bone morphological properties. Only at the VOI diameter and height combination with the lowest p-value is reported to highlight the critical VOI for the screws and expandable fasteners. For screws, the maximum force is equal to the failure force.

Table 1. Correlation between mechanical performance and bone morphology.

		Tb.V/TV	Tb.L	SMI	Tb.SA	Cor.V
EF - Failure Force	r-value	0.740	0.870	-0.471	0.769	-0.538
	p-value	0.018	0.002	0.201	0.015	0.135
	VOI Diameter (mm)	8	14	10	14	8
	VOI Height (mm)	2	2	4	2	-
EF - Maximum Force	r-value	0.526	0.765	-0.343	0.838	-0.857
	p-value	0.146	0.016	0.366	0.005	0.003
	VOI Diameter (mm)	8	16	10	16	14
	VOI Height (mm)	2	2	4	2	-
Screw – Failure/ Maximum Force	r-value	0.710	0.414	-0.685	0.580	0.222
	p-value	0.032	0.268	0.042	0.102	0.566
	VOI Diameter (mm)	8	6	8	6	16
	VOI Height (mm)	10	10	10	10	-

Trabecular bone volume fraction demonstrated a high degree of positive correlation with failure force for the screw over all full height VOI diameters (figure 3). The highest correlation (r=0.710) with a high probability of significance (p=0.032) was demonstrated at full (10mm) height and 8mm diameter. For the UEFs a high degree of positive correlation was only demonstrated for failure force over a small range of VOI diameters (6-8mm) and VOI heights (2-4mm). The highest correlation (r=0.740) was found at 2mm height and 8mm diameter, and is considered to have a high probability of significance (p=0.018).

Cortical volume demonstrated a high degree of negative correlation with maximum force for the UEFs over all VOIs tested. The highest correlation (r=-0.857) with a high probability of significance (p=0.003) was found for a 16mm diameter VOI for cortical volume (figure 4). This relationship was negative, indicating a higher pull-out strength for thinner cortices. Only a weak correlation was found between the cortical volume of the screw set and the failure force.

Trabecular length demonstrated a high degree of positive correlation (r=0.870) with UEF failure force (p=0.002) and trabecular bone surface area also demonstrated a high degree of positive correlation (r=0.838) with UEF maximum force (p=0.005). Both correlations having a high probability of significance. A weak correlation was demonstrated between screw failure force and trabecular bone surface area. SMI demonstrated a high degree of negative correlation with screw failure force for VOIs between 6mm and 10mm diameter with a full height VOI, however no significant correlation was found for the UEFs.

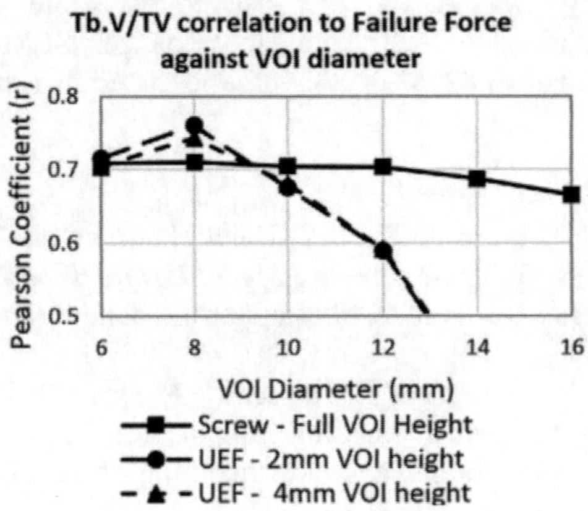

Figure 3. Correlation coefficient between Tb.V/TV and failure force against VOI diameter.

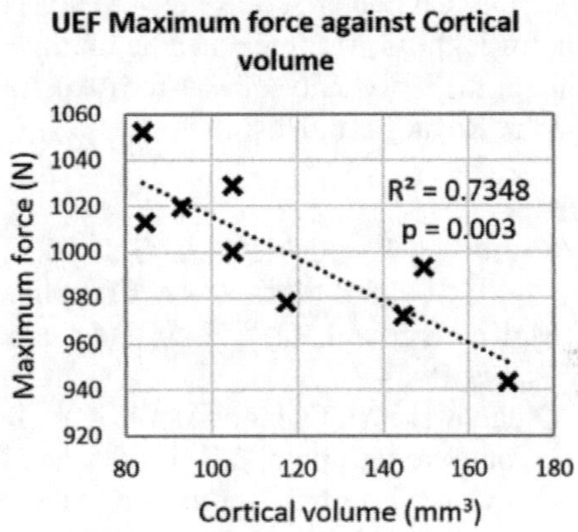

Figure 4. Correlation between UEF maximum force and cortical volume.

4. Discussion and conclusion

Correlation of fastener performance to bone properties was found to vary with the VOI over which the bone properties are measured. By taking an arbitrary VOI or else by including the entire bone volume a significant relationship can be obscured, which may explain some previous inconclusive attempts to correlate bone properties to pull-out strength.

Previous studies reported on the significance of correlations between pull-out force and up to nine microstructural properties without correcting for multiple comparisons [6, 7]. In this study, not correcting for multiple comparisons would be problematic as we were making multiple comparisons across different VOIs to determine the effect of VOI size in addition to multiple microstructural properties. Consequently we have reported p-values and r-values without commenting on significance.

It was found that the critical VOI for screws covered the entire length of the screw with the highest correlation at 8mm diameter, reducing in degree of correlation in both directions (figure 3). This relationship (critical VOI twice the diameter of the screw) is similar to the deformed area calculated with FEA by Wirth *et al.* [8] and the experimental set up of Ab-Lazid *et al.* [7].

Trabecular bone volume fraction, trabecular length and trabecular surface area only demonstrated a high degree of correlation with UEF failure load when only the bone around the bottom of the fastener is considered in isolation. The authors hypothesise that this is because the UEF fails due to slipping and so only the bone that contributes to increased pressure at the expansion zone is critical. This hypothesis would indicate that the expandable fasteners will be less sensitive to fastener length.

These results confirm previous studies with trabecular SMI and trabecular bone volume fraction being the most highly correlated parameters. Some deviation is inevitable, as previous studies use the epiphysis of human long bones where the current study used ovine vertebral bone. Contrary to epiphysis of human long bones where the current study used ovine vertebral body bone. Contrary to previous studies in long bones [6, 7] no correlation between cortical volume and screw failure force was found. The authors hypothesise that this was because, in the vertebral body, the cortex was thinner and the trabecular bone denser than in the epiphysis of the long bones and therefore the influence of the cortical properties was reduced.

Osteoporotic bone can be characterised by a reduction in bone volume fraction, cortical thinning, and a deterioration of plate-like trabeculae into rod-like trabeculae [15]. Compared to screws, UEFs demonstrated a similar dependence on trabecular bone volume fraction but no relation to SMI and an inverse relationship with cortical thickness, indicating potential suitability for osteoporotic bone.

Acknowledgements

The authors acknowledge the facilities of the National Imaging Facility at the Centre for Microscopy Characterisation & Analysis, a facility funded by the University, State and Commonwealth Governments. The authors also acknowledge the use of the Realizer SLM100 machine at the University of Western Australia and the Instron 8874 servo-hydraulic materials testing machine at Royal Perth Hospital.

References

[1] Katonis P, Papadakis S A, Galanakos S, Paskou D, Bano A, Sapkas G and Hadjipavlou A G 2011 Lateral Mass Screw Complications: Analysis of 1662 screws *J Spinal Disord Tech* **24** 6

[2] McKoy B E and An Y H 2001 An expandable anchor for fixation in osteoporotic bone *J Orthop Res* **19** 545-7

[3] Oldakowski M, Oldakowska I, Kirk T B, Ford C T, Sercombe T B, Hardcastle P, Day R E Pull-out strength comparison of a novel expanding fastener against an orthopaedic screw in an ovine verterbal body: An ex-vivo study *J Med Eng Tech.* Forthcoming 2016. DOI: 10.3109/03091902.2015.1127441

[4] King T S and Cebon D 1993 An alternative to screws for plating osteoporotic bone *J Biomed Eng* **15** 79-82

[5] Thiele O, Eckhardt C, Linke B, Schneider E and CA L 2007 Factors affecting the stability of screws in human cortical osteoporotic bone: a cadaver study *J Bone Joint Surg Br* **89** 5

[6] Poukalova M, Yakacki C M, Guldberg R E, Lin A, Saing M, Gillogly S D and Gall K 2010 Pullout strength of suture anchors: Effect of mechanical properties of trabecular bone *J Biomech* **43** 1138-45

[7] Ab-Lazid R, Perilli E, Ryan M K, Costi J J and Reynolds K J 2014 Pullout strength of cancellous screws in human femoral heads depends on applied insertion torque, trabecular bone microarchitecture and areal bone mineral density *J Mech Behav Biomed Mat* **40** 354-61

[8] Wirth A J, Goldhahn J, Flaig C, Arbenz P, Müller R and van Lenthe G H 2011 Implant stability is affected by local bone microstructural quality *Bone* **49** 473-8

[9] Buie H, Campbell G, Klinck R, Macneil J and Boyd S 2007 Automatic segmentation of cortical and trabecular compartments based on a dual threshold technique for in vivo micro-CT bone analysis *Bone* **41** 505-15

[10] Schindelin J, Arganda-carreras I, Frise E, Kaynig V, Longair M, Pietzsch T, Preibisch S, Rueden C, Saalfeld S, Schmid B, Tinevez J-y, White D J, Hartenstein V, Eliceiri K, Tomancak P and Cardona A 2012 Fiji: an open-source platform for biological-image analysis *Nat Methods* **9** 676-82

[11] Doube M, Kłosowski M M, Arganda-Carreras I, Cordelières F P, Dougherty R P, Jackson J S, Schmid B, Hutchinson J R and Shefelbine S J 2010 BoneJ: Free and extensible bone image analysis in ImageJ *Bone* **47** 1076-9

[12] Lee T C, Kashyap R L and Chu C N 1994 Building Skeleton Models via 3-D Medial Surface Axis Thinning Algorithms *CVGIP: Graphical Models and Image Processing* **56** 462-78

[13] Lorensen W E and Cline H E 1987 Marching cubes: A high resolution 3D surface construction algorithm *Comp Graphics (ACM)* **21** 163-9

[14] Skyscan AB Structural parameters measured by the SkyScan™ CT-Analyser software

[15] Stauber M and Muller R 2006 Age-related changes in trabecular bone microstructures: global and local morphometry *Osteoporosis International* **17** 616-26

Image analysis for dental bone quality assessment using CBCT imaging

Suprijanto[1], L Epsilawati[2], M S Hajarini[1], E Juliastuti[1] and H Susanti[1]

[1] Instrumentation and Control Research Group, Faculty of Industrial Technology, Institut Teknologi Bandung, Indonesia
[2] Department of Dentomaxilofacial Radiology, Faculty of Dentistry, Padjadjaran University, Bandung Indonesia

E-mail: supri@tf.itb.ac.id

Abstract. Cone beam computerized tomography (CBCT) is one of X-ray imaging modalities that are applied in dentistry. Its modality can visualize the oral region in 3D and in a high resolution. CBCT jaw image has potential information for the assessment of bone quality that often used for pre-operative implant planning. We propose comparison method based on normalized histogram (NH) on the region of inter-dental septum and premolar teeth. Furthermore, the NH characteristic from normal and abnormal bone condition are compared and analyzed. Four test parameters are proposed, i.e. the difference between teeth and bone average intensity (**s**), the ratio between bone and teeth average intensity (**n**) of NH, the difference between teeth and bone peak value (**Δp**) of NH, and the ratio between teeth and bone of NH range (**r**). The results showed that n, s, and **Δp** have potential to be the classification parameters of dental calcium density.

1. Introduction

Recent development in technology of medical imaging has a new perspective in medical diagnosis and therapy planning. The X-ray based imaging equipments are also employed in the cases of dental treatment for functional and aesthetic purposes. The periapical and panaromic dental imaging are the type of 2D imaging. For 3D imaging, cone beam computed tomography (CBCT) is recently widely used for dental maxillofacial radiography application. CBCT uses much lower radiography percentage, faster scan time, and safer to use in maxillofacial region than conventional multidetector CT scan. For assesment of bone quality, CT image has relatively straightforward because a gray value is proportional with energy attenuation of bone material. Unlike CT scan, CBCT image has problem on variability of gray values due to limited field size, relatively high amount of scattered radiation and limitations of currently applied reconstruction algorithms. However, to support pre-operative dental implant planning, 3D CBCT data has potential to be applied if the problem on variability of gray values can be suppressed. To overcome this problem, we proposed an image analysis scheme, based on relative comparison between quantitative data on the different location of region of interest (ROI) on the same slice orientation. The dental expert defined the locations of ROI that are predicted as normal and abnormal condition. On the 3D CBCT data, an image plane could be selected based on the slice that used for further analysis and diagnosis.

2. Material and Methods

2.1. Material

In this work, sponges of alveolar bone in the lower jaw is selected as ROI. These selected teeth are shown in 3D reconstruction CBCT image in figure 1(a). Alveolar bone that used as ROI was inter septal bone that located between two teeth as described by blue circle in figure 1(b). The alveolar bone is consist of spongus and cortical, where spongus bone calcium density was lower than cortical bone calcium density.

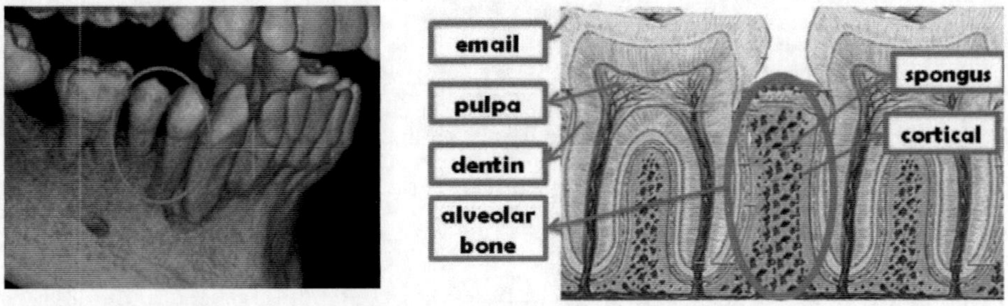

(a) (b)

Figure 1. (a). Right lower jaw premolar teeth. The blue circle represent the permolar teeth (b). Teeth and alveolar bone structure.

In the research, ten sets of 3D CBCT data was acquired using CBCT (EPX-Impla, E-WOO Tech-Department of Dentomaxilofacial Radiology, Faculty of Dentistry, Padjadjaran University). Each set of CBCT data contain 256 axial slices of human. Each slice is a 12 bit DICOM (Digital Imaging and Communication in Medical) image. These sets were used to build 3D image and another orthogonal slice such as coronal and sagittal slices by *EPX-IMPLA* software. ROI were recognized by identifying dental structure from 3D images and orthogonal slices. The dentist makes preliminary diagnosis in order to classify the set of 3D CBCT data. The dentist also decides that five sets of 3D CBCT data are normal and the rest of data sets are abnormal.

2.2. Methods

Based on 3D CBCT reconstruction, the slice orientation that is proper for the representation of the premolar teeth and interseptal bone image was selected. Sagittal orientation was chosen for image plane (see figure 2-left). Furthermore, the image area that is used for further processing is selected. On the pre-processing step, contrast stretching is applied to normalize the contrast range of images.

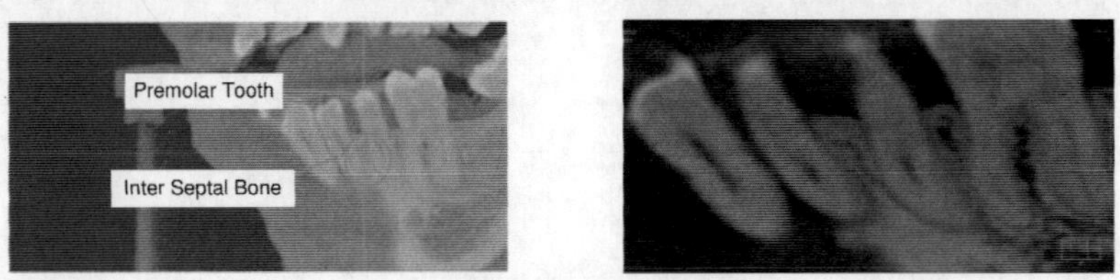

Figure 2. Sagittal plane of premolar teeth and inter septal bone (left) and the pre-processing image for further analysis (right).

The bone condition could be diagnosed based on the calcium density of bone that is represented as grayscale on the images. We propose comparison method based on normalized histogram (NH) on the ROI inter-dental septum and premolar teeth. Under the guidance of the dentist, ROI is selected and marked by several points around the targeted area. Then, based on selected points, a close curve is reconstructed using spline interpolation (see figure 3).

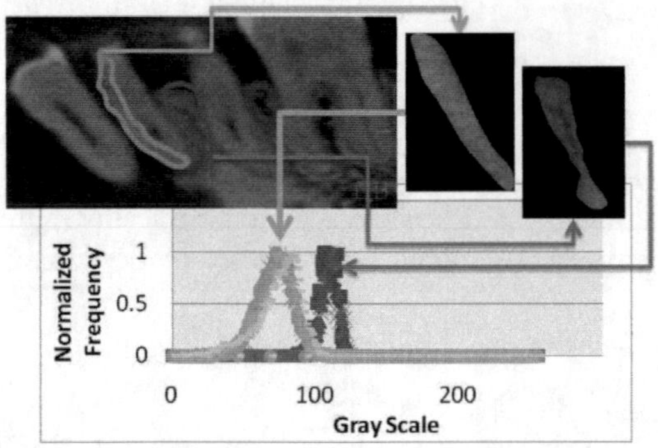

Figure 3. Proposed scheme of image analyis bone condition.

The probalistic density function of each selected ROI is determined based on NH, that is represented as

$$h(k) = \frac{p(I_k)}{p(I_k)_{\max}} \tag{1}$$

where $p(I_k)$ was number of pixel on one grayscale value, k was a grayscale value. Each histogram was normalized by its maximum value to minimize CBCT inconsistency effect that influences the data. Furthermore, the NH characteristic from normal and abnormal bone condition are compared and analyzed. Five sets of NH of normal data and five sets of NH of abnormal data are used on the research. Each set of data contains four sagittal slices of premolar tooth and alveolar bone that located between two premolar teeth.

Four test parameters are used to analyze the prediction of bone condition based on two NH distributions (figure 3). The first test parameter is the difference between teeth and bone peak value (**Δp**) of NH. This variable was also recognized as the difference between bone and tooth gray scale mode.

The second parameter is the ratio between range of NH tooth and NH bone **(r).** NH range is calculated based on the difference between maximum and minimum of grayscale intensity in the NH. The third parameter is the difference between average of grayscale distribution on NH tooth and NH bone **(s)**. The fourth parameter is the ratio between bone grayscale intensity averages and tooth grayscale intensity averages **(n)**.

Sets of four test variables ({Δp, r,s,n}) are analyzed with averaging four axial slice histogram values in each data. Analysis was conducted by comparing five normal and five abnormal data to observe the difference between groups. This comparison was done by two methods, directly and analyzing its variance.

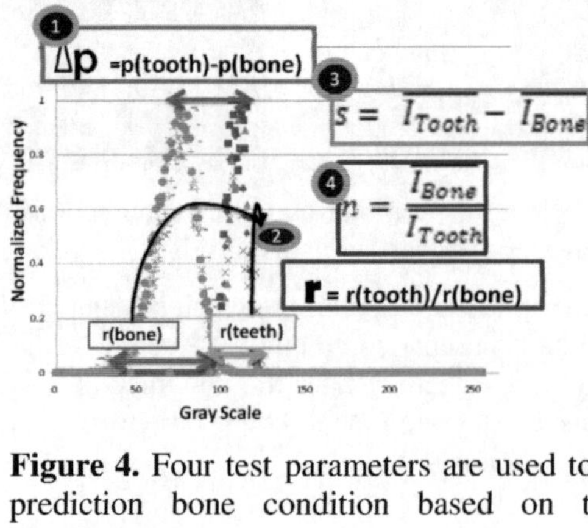

Figure 4. Four test parameters are used to analyze prediction bone condition based on two NH distribution.

Analysis of variance (ANOVA) was a general method to overlook the variance of data inside group and inter groups. If the variance inside the group is significantly smaller than inter group, then it means we can observe the difference between groups clearly. F-test was used to evaluate the diference between group, where F value determined by ratio between intergroup variability (Va) and group variability (Vg). The significant of four test variables to determine bone condition is evaluated based on comparing F value of data and F critical (F_{crit}) value.

3. Results

Based on the sets of CBCT jaw image data (five normal and five abnormal), the range of **Δp** value from left and right jaws data was shown in Figure 5. These results directly showed that **Δp** value could differentiate normal and abnormal teeth clearly.

Figure 5. The range of Δp value from right and left jaws.

ANOVA of this data Δp showed that F of this data is 22.93, which is much larger than its F critical of 3.96. These mean that the first variable, **Δp**, has potential to be a test variable for teeth classification. From their trends, normal teeth **Δp** values tend to be smaller than abnormal teeth **Δp** values. The same procedure to determine range of **Δp**, the range of **r** value from left and right jaws data was shown in figure 6.

Figure 6. The range of **r** value from right and left jaws.

The result of the left jaws **r** values showed that r could differentiate normal and abnormal teeth. But the result of the right jaws **r** values showed that normal teeth **r** area covers all abnormal **r** area. That fact showed **r** could not differentiate normal and abnormal teeth **r** group. ANOVA of this data **r** showed that F of this data is 0.69, which is much smaller than its F critical of 3.96. These mean that the second variable, **r**, has no potential to be a test variable for teeth classification. Furthermore, the third parameter, **s**, left and right jaws showed in figure 7. These results directly showed that **s** value could differentiate normal and abnormal teeth on both jaws.

Figure 7. The range of **s** value from right and left jaws.

ANOVA of this data **s** showed that F of this data is 53.01, which is much larger than its F critical of 3.96. Those two facts have give us indication that the third variable, **s**, has potential to be a test variable for teeth classification. From its trend, normal teeth **s** values tend to be smaller than abnormal teeth **s** values. Finally, the fourth parameter,**n**, left and right jaws showed in Figure 8. Calculation of the fourth variable, **n**, left and right jaws showed in figure 8.

Figure 8. The range of **n** value from right and left jaws

These results directly showed that **n** value could differentiate normal and abnormal teeth as two different groups. ANOVA of this data n showed that F of this data is 70.11, which is much larger than its F critical of 3.96. Those two facts have give us indication that the fourth variable, **n**, has potential to

Journal of Physics: Conference Series **694** (2016) 012065 doi:10.1088/1742-6596/694/1/012065

be a test variable for teeth classification. From its trend, normal teeth n values tend to be bigger than abnormal teeth **n** values.

4. Conclusion

The proposed image analysis method has potential for the assessment of dental bone quality using CBCT images. Four test parameters are proposed, i.e. the difference between teeth and bone average intensity **(s)**, the ratio between bone and teeth average intensity **(n)** of NH, the difference between teeth and bone peak value (**Δp**) of NH, and the ratio between teeth and bone of NH range **(r)**. The results showed that **n**, **s**, and **Δp** have potential to be the classification parameters of dental calcium density.

References

[1] Shokri, Eskandarloo, Noruzi-Gangachin and Khajeh 2015 Detection of root perforations using conventional and digital intraoral radiography, multidetector computed tomography and cone beam computed tomography *Restor Dent Endod* **40(1)** pp 58-67

[2] Kafieh R, Shahamoradi M, Hekmatian E, Foroohandeh M and Emamidoost M 2012 Removing distortion of periapical radiographs in dental digital radiography using embedded markers in an external frame *J Med Sign Sens* **2** pp 219-24

[3] D'Addazio P S, Campos C N, Özcan M, Teixeira H G, Passoni R M and Carvalho A C 2011 A comparative study between cone-beam computed tomography and periapical radiographs in the diagnosis of simulated endodontic complications *Int Endod J* **44(3)** pp 218-24

[4] Grondahl H G and Huumonen S 2004 Radiographic manifestations of periapical inflammatory lesions. How new radiological techniques may improve endodontic diagnosis and treatment planning *Endodontic Topics* **8** pp 55–67

[5] Tanomaru-filho M, Gonc¸ Alves M and Tanomaru J M G 2008 Detection of periapical lesion development by conventional radiography or computed tomography *Oral Surgery, Oral Medicine, Oral Pathology, Oral Radiology and Endodontics* **10** pp 56–61

[6] Lee T, Jeong H, Kim D and Lee B 2009 Quantitative analysis of endodontic treatment for periapical lesions in intraoral radiographs *IEEE International Conference on Bioinformatics and Bioengineering*, Taichung

[7] William C S and Allan G F 2007 Cone beam computed tomography : A paradigm shift for clinical densitry *Australian Dental Practice* pp 102-110

[8] Ayuningtiyas A, Putra N K, Suprijanto, Juliastuti E and Epsilawati L 2013 Quantitative image analysis of periapical dental radiography for dental condition diagnosis *3rd International Conference on Instrumentation, Communications, Information Technology, and Biomedical Engineering (ICICI-BME)*

13th South-East Asian Congress of Medical Physics 2015 (SEACOMP) IOP Publishing
Journal of Physics: Conference Series **694** (2016) 012066 doi:10.1088/1742-6596/694/1/012066

Dental panoramic image analysis for enhancement biomarker of mandibular condyle for osteoporosis early detection

Suprijanto[1], Azhari[2], E Juliastuti [1], A Septyvergy[3] and N P P Setyagar[3]

[1] Instrumentation and Control Research Group, Faculty of Industrial Technology, Institut Teknologi Bandung, Indonesia
[2] Department of Dentomaxilofacial Radiology, Faculty of Dentistry, Padjadjaran University, Bandung, Indonesia
[3] Engineering Physics, Faculty of Industrial Technology, Institut Teknologi Bandung, Indonesia

E-mail : supri@tf.itb.ac.id

Abstract. Osteoporosis is a degenerative disease characterized by low Bone Mineral Density (BMD). Currently, a BMD level is determined by Dual Energy X-ray Absorptiometry (DXA) at the lumbar vertebrae and femur. Previous studies reported that dental panoramic radiography image has potential information for early osteoporosis detection. This work reported alternative scheme, that consists of the determination of the Region of Interest (ROI) the condyle mandibular in the image as biomarker and feature extraction from ROI and classification of bone conditions. The minimum value of intensity in the cavity area is used to compensate an offset on the ROI. For feature extraction, the fraction of intensity values in the ROI that represent high bone density and the ROI total area is perfomed. The classification will be evaluated from the ability of each feature and its combinations for the BMD detection in 2 classes (normal and abnormal), with the artificial neural network method. The evaluation system used 105 panoramic image data from menopause women which consist of 36 training data and 69 test data that were divided into 2 classes. The 2 classes of classification obtained 88.0% accuracy rate and 88.0% sensitivity rate.

1. Introduction

Degenerative bone disease which is indicated by the occurrence of low bone mineral density (BMD) may occur in older people. This disease can increase the fragility, fractures risk and lowering mechanical force on supporting body normal activity [1]. Actually, osteoporosis can develop in both men and women, but women postmenopausal have more significant risk. The decrease of bone mineral density in postmenopausal woman is a direct result from the loss of estrogen. Osteoporosis is also known as a silent disease because people may not know that they have osteoporosis until their bones become so weak that a sudden strain, bump, or fall causes a hip to fracture or a vertebra to collapse.

The gold standar of measurement bone mass, bone mineral density or other aspect related to bone structure is dual energy X-ray absorptiometry (DXA). In the developing countries, the number of DXA apparatus is relatively limited, therefore early osteoporosis detection using DXA becomes not easy to perform [1, 2].

Usually older people may have greater opportunity to visit dentists for dental treatment than visit medical expert to measure bone density using DXA. Previous researcher states that there is a correlation between condition of hip and mandibular bone [1] . Therefore, information of mandibular bone that is recorded using panoramic radiography is becoming alternative to be researched as early detection of osteoporosis [3-5]. Various studies have been proposed to measure the width of mandibular cortical bone or pattern of trabecular bone that is imaged on panoramic image as osteoporosis prediction. In patients with postmenopausal osteoporosis, trabecular number decreased due to reduced estrogen hormones that regulate the transport of calcium into bone [2, 4, 5]. The architecture of trabecular bone changes due to osteoporosis could be imaged in panoramic radiography images. However, a spesific processing scheme must be development to extract the pattern of trabecular bone image in the spesific region.

This work reported alternative scheme, that consists of the determination of the Region of Interest (ROI) in the image as a biomarker and feature extraction from ROI and classification of bone conditions. ROI was taken on the left and right condyle mandibular. The classification will be evaluated from the ability of feature and its combinations for the BMD detection in 2 classes (normal and abnormal) with the artificial neural network method.

2. Material and methods

2.1. Material

The database of panoramic images for supporting this research was prepared by Department of Dentomaxilofacial Radiology, Faculty of Dentistry, Padjadjaran University. Each patient data consist of panoramic image and BMD information that was determined from DXA. The data were acquired from 50-85 year old postmenopausal women. Two panoramic radiograph machines were used, i.e. orthopantomograph E-Woo EPX-Impla at 8 mA and the voltage of 66 kV and panoramic radiograph Rotograph Evo at 13,8 mA and the voltage of 74 kV. The total samples were 105 patients of postmenopousal women.

2.2. Methods

The mandibula bone of patient was imaged using extra-oral imaging where the arrangement of the teeth was projected on the panoramic image plane. In this study, the Region of Interest (ROI) was taken on the dental panoramic radiograph image that shows the area of the jaw bone. Figure-1A illustrates the process of panoramic imaging. The film or digital detector and also X-ray source will be rotated in several steps, until covering the whole mandibula bone. The results of the process is panaromic images (figure 1b).

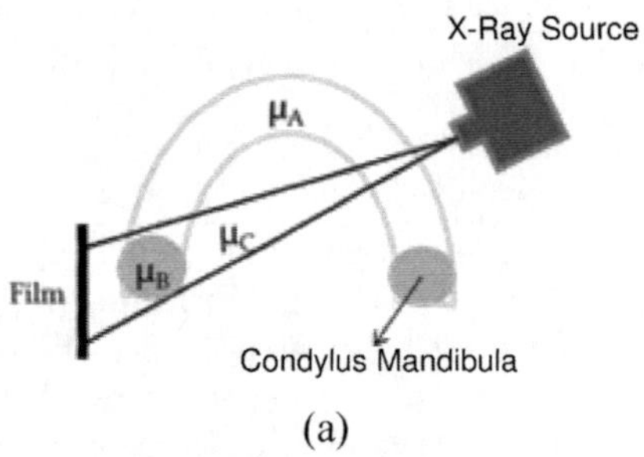

(a)

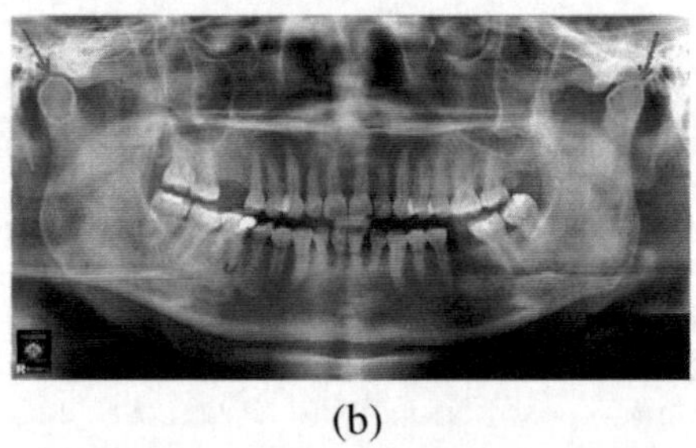

(b)

Figure 1. (a). Illustration process of panarmic imaging. (b). Panaromic image and ROI region on the condyle marked with “red arrow”.

For the extraction of digital biomarker, ROI was taken on the left and right *condyle mandibular* (lower jaw bone ends). Due to the condyle shape, the specific method is used to define the ROI with follow the shape of the condyle bone. The ROI region is marked with “red arrow” in the Figure 1B. Condyle mandibular is selected as the ROI for biomaker because its region is less affected by local factor such as infection.

Ideally, the image intensity value on the ROI (I_{ROI}) is only proportional to the attenuation factor, only on the condyle area (μ_B). However, a superimpose problem occurs in the panaromic imaging. Due to this problem, the intensity on the condyle area is modelled as:

$$I_{ROI} = I_S \exp(\mu_A + \mu_B + \mu_C)\, x \tag{1}$$

where I_s is intensity of source, μ_A attenuation factor of soft-tissue and μ_C is attenuation factor of air cavity and x is thicknes of tissue. The factor of μ_A and μ_C are assumed as background offset intensity on the I_{ROI}. To minimize the problem of background offsets, it was determined that the minimum of variance value in the cavity is that had the minimum attenuation X-ray in the surrounding condyle area. If the background offset is represented by I_{offset}, then the compensated I_{ROI-C} is modelled as:

$$I_{ROI-C} = I_{ROI} - I_{offset} \tag{2}$$

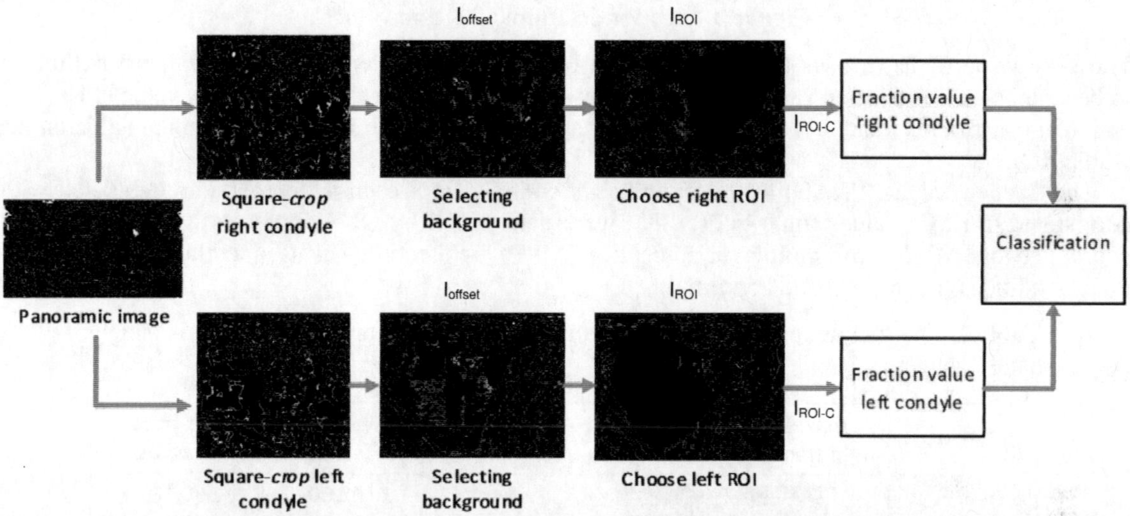

Figure 2. Proposed scheme of image analysis for osteoporosis early detection.

The steps of compensation process are shown in Figure 2. The feature extraction is processed on I_{ROI-C}, in order to obtain quantification value of bone condition prediction or digital biomarker. We assume that, on I_{ROI-C}, the gray-value is proportional with a bone density on the condyle. The pattern of gray scale variation represents the pattern of trabecular bone in condyle. Based on the assumption, segmentation was done on I_{ROI-C} to determine the lowest value of bone density (marked with 0), based on threshold that is chose based on isodata algorithm. Next, if the gray value on I_{ROI-C} is higher than the threshold, the gray value on I_{ROI-C} is marked with 1. After the binary image of I_{ROI-C} is obtained using segmentation process, the final step determines $I_{FEATURE}$ that is formulated as:

$$I_{FEATURE} = I_{ROI-c} \text{ x (biner of } I_{ROI-c}) \tag{3}$$

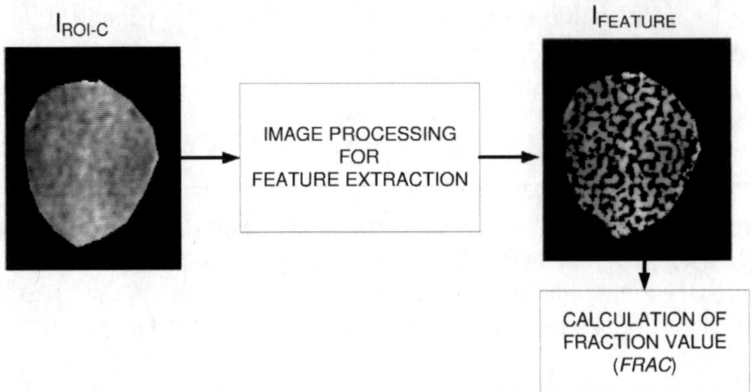

Figure 3. Step for determine fraction value.

The gray value of $I_{FEATURE}$ is the representation of digital biomarker of bone density prediction on trabecular bone. The fraction value is determined by the ratio between total sum of gray value in $I_{FEATURE}$ and total sum of gray value in I_{ROI-C}, if assumed that the gray value on I_{ROI-C} has maximum value on the whole ROI.

The fraction values (FRAC) from left and right condyle from each patient data are tabulated with correspond to BMD values (on T-score) that were determined by DXA. For classification, artificial neural network (ANN) with multilayer perceptron (MLP) architecture was used to classify 2 classes of bone conditions (i.e. normal and abnormal).

Table 1. The sample results of fraction value from 2 normal and 2 osteoporosis patient data.

Patient	Image of right fraction	Image of left fraction	Result of DXA	Right fraction value	Left fraction value
Patient 1	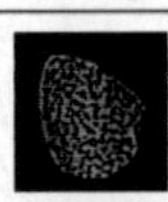		Normal	1. 0,180 2. 0,178 3. 0,178	1. 0,181 2. 0,181 3. 0,178
Patient 2			Osteoporosis	1. 0,018 2. 0,021 3. 0,015	1. 0,049 2. 0,050 3. 0,054
Patient 3	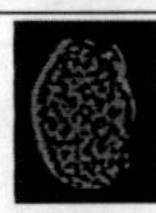		Normal	1. 0,102 2. 0,102 3. 0,098	1. 0,024 2. 0,018 3. 0,018
Patient 4		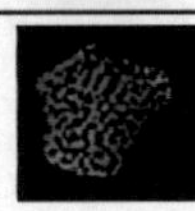	Osteoporosis	1. 0,051 2. 0,052 3. 0,053	1. 0,075 2. 0,077 3. 0,079

3. Results

The sample results of fraction value is tabulated on Table 1. These results are determined from 2 normal and 2 abnormal patient data. For patient with less bone density, the value of abnormal FRAC is less than normal FRAC. However, from 105 patient data, the range of FRAC values are often overlap between normal and abnormal condition. The overlap of FRAC value is usually only on the left or right of condyle. To overcome these problem, the set of FRAC pattern related with T-score value are trained using ANN with 4 layer-MLP. The evaluation system used 105 panoramic image data from menopause women and from different image devices which consist of 36 training data and 69 test data that were divided into 2 classes proportionally.

Table 2. Uncertainty matrix.

Actual	**Prediction**	
	Normal	Osteoporosis
Normal	*True Positive* (TP)	*False Negative* (FN)
Osteoporosis	*False Positive* (FP)	*True Negative* (TN)

Testing of methods and systems that have been designed are performed using accuracy and sensitivity [3]. The accuracy and sensitivity value of classification can be defined using the uncertainty matrix elements as follows:

$$Accuracy\,(\%) = \frac{TP + TN}{TP + TN + FP + FN}\,x100 \tag{4}$$

$$Sensitivity(\%) = \frac{TP}{TP + FN}\,x100 \tag{5}$$

where the TP, TN, FN and FP are tabulated on table 2. The proposed scheme has performance of 88.0% in accuracy rate and 88.0% in sensitivity rate.

4. Conclusion

From the results, it was concluded that the image analysis scheme of dental panoramic radiograph image can be used as a basis of computer aided diagnosis for dentist in providing early information about osteoporosis. Feature extraction of fraction value on the condylus areas are promoted as digital biomarker. The rate of classification based on 69 test data is 88.0% in accuracy rate and 88.0% in sensitivity rate. The representation of the bone mineral density condition has not been able to achieve the results of 100%. It may be caused the quality of dental panoramic image is varied due to size of the patient's head and noise from X-ray detector To improve the ability of the system, it is needed many test data of the BMD value based on DXA value at the distal radius and femur area.

References

[1] Oliveira M L, Pedrosa E F N C, Cruz A D, Haiter-Neto F, Paula F J A and Watanabe P C A 2013 Relationship between bone mineral density and trabecular bone pattern in postmenopausal osteoporotic Brazilian women *Clin. Oral Investig.* **17(8)** pp. 1847–53

[2] Taguchi A, Sanada M, Krall E, Nakamoto T, Ohtsuka M, Suei Y, Tanimoto K, Kodama I, Tsuda M and Ohama K 2003 Relationship between dental panoramic radiographic findings and biochemical markers of bone turnover *J. Bone Miner. Res.* **18(9)** pp. 1689–94

[3] Suprijanto, Juliastuti E, Diputra Y, Mayantasari M and Azhari 2013 Dental panoramic image analysis on mandibular bone for osteoporosis early detection, in *Instrumentation Control and Automation (ICA), 2013 3rd International Conference on*, 2013, pp. 138–143

[4] A Katsumata and H Fujita 2014 Progress of computer-aided detection/diagnosis (CAD) in dentistry CAD in dentistry *Jpn. Dent. Sci. Rev.* **50(3)** pp. 63–68

[5] Arifin A Z, Asano A, Taguchi A, Nakamoto T, Ohtsuka M, Tsuda M, Kudo Y and Tanimoto K 2006 Computer-aided system for measuring the mandibular cortical width on dental panoramic radiographs in identifying postmenopausal women with low bone mineral density *Osteoporos. Int.* **17(5)** pp. 753–759

Model-based assessment of probe placement criteria in cancer therapy using RF ablation

E W Tanotogono[1], G F Akasum[1], Suprijanto[1] and J J Sudirham[2]

[1] Instrumentation and Control Research Group, Faculty of Industrial Technology, Institute of Technology Bandung, Indonesia
[2] Engineering Physics Research Group, Faculty of Industrial Technology, Institute of Technology Bandung, Indonesia

E-mail: evan.wijayat@gmail.com

Abstract. Radiofrequency ablation has been developed as a minimally-invasive method for cancer therapy. Nevertheless, the unfeasibility of direct observation during ablation process sometimes becomes a challenge for practitioners, particularly those constrained by the absence of a proper monitoring system. Thus, aiming to develop a prudent cancer therapy planning, this research develops a 3D model that enable practitioners to predict the tissue damage resulted by a simulated ablation before a real ablation is executed. The model, developed using finite element method, is made to mimic real human liver tissue by simulating its physical properties as temperature-dependent functions. Three probe placement cases, representing three different approaches, are analysed to study the effect of probe placement configuration on tissue damage formed during a time-dependent ablation process. The three placement cases are surface-perpendicular placement, misaligned placement, and relatively accurate placement. It can be concluded that the accuracy of a probe placement configuration can be assessed by quantifying two major parameters: average tissue damage in the target domain and accumulated damage resulted in complementary tissue domain. Optimum ablation duration can also be determined by considering those parameters.

1. Introduction

Radiofrequency (RF) ablation is one of minimally invasive techniques that have been developed for various purposes including cancer therapy. In many cases, without the presence of proper monitoring system, Indonesian practitioners are faced with observation unfeasibility during ablation process. This condition may challenge the execution of an optimum ablation due to incomplete ablation process or inaccurate probe positioning. In addition to ineffectiveness, inaccuracy in probe placement planning could also lead to damage resulted in normal tissue surrounding the cancer.

This research applies numerical simulation method to develop a 3D model of bipolar RF ablation. It enables practitioners to predict damage formed in liver tissue before performing the actual therapy. As an outcome, practitioners can assess the effectiveness of each configuration and the corresponding time duration in order to eliminate cancer tissue with certain position, shape, and size. Thus, an ablative cancer therapy is planned quantitatively.

2. Theoretical backgrounds

Applied across the tissue, high-frequency AC electricity generates heat (Q_{heat}) proportional to the product of current density and electric field produced in the tissue [1]. The generated heat is then substituted into Pennes' equation (1948), which has been modified by Yang *et al.* [2] by considering heat for tissue water evaporation process, to be

$$\rho_{ti}\left[C_{ti} - \frac{\alpha}{\rho_{ti}}\frac{\partial \rho_{water}}{\partial T_{ti}}\right]\frac{\partial T_{ti}}{\partial t} = \nabla . k_{ti}\,\nabla T_{ti} + \rho_{bl}C_{bl}W_{bl}(T_{bl} - T_{ti}) + Q_{heat} + Q_{meta} \tag{1}$$

Variables ρ_{ti}, C_{ti}, T_{ti}, and k_{ti} are consecutively the density, specific heat, temperature, and thermal conductivity of liver tissue. Variables ρ_{bl}, C_{bl}, W_{bl}, and T_{bl} are consecutively the density, specific heat, perfusion rate, and temperature of blood, while α and ρ_{water} are water evaporation heat and tissue water density. Q_{meta} represents the generated metabolic heat. The level of tissue damage resulted by the ablation can be approximated by the Arrhenius equation as

$$\Omega(t) = \ln\left[\frac{c(0)}{c(t)}\right] = \int_0^t A.\exp\left[-\frac{\Delta E}{RT}\right]dt \tag{2}$$

A is the frequency factor, ΔE is the activation energy and R is the universal gas constant. Biological tissue is assumed to be dead (99% damage) when $\Omega(t)$ reaches 4.6 [3].

3. Model development

This model is developed using finite element method (FEM) in COMSOL Multiphysics 5.0. Multiphysical coupling is designed between Time-harmonic Electric Current and Time-dependent Bioheat Transfer modules. Bipolar RF ablation is modeled as in the prior research of Tanotogono *et al.* [4], yet only 280 kHz 121.8 Volt AC electricity is used in this paper. Targeted cancer tissue was modeled as a separated ellipsoid domain (figure 1) and the remaining tissue domain that is not covered by cancer domain is labeled as complementary domain.

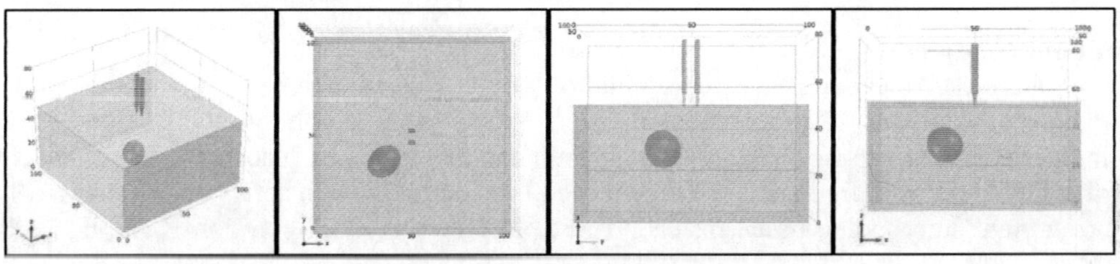

Figure 1. Geometry of ablation model with targeted cancer domain.

Whilst the model developed in prior Tanotogono's model [4] are based on *ex-vivo* experiment, some improvements are added to this model to mimic tissue activities in *in-vivo* condition, such as blood perfusion phenomenon and metabolic heat generation, as well as setting the initial tissue temperature to 37^{0}C. Blood perfusion rate, W_{bl}, is defined as a piecewise function of tissue damage fraction [5] as

$$W_{bl} = \begin{cases}(1 + 25\Omega - 260\Omega^2)W_{bl_0} & 0 < \Omega \leq 0{,}1\\ (1-\Omega)W_{bl_0} & 0.1 < \Omega \leq 1\end{cases} \tag{4}$$

Density and heat capacity of the blood is assumed 1,060 kg m^{-3} and 3840 J kg^{-1}K^{-1} [1]. W_{bl0}, as the blood perfusion rate in fully active tissue, is approximated to be 1.5833 x 10^{-5} m^{3}kg^{-1}s^{-1} [6]. On the other hand, metabolic heat generation [7] is defined as

$$Q_{meta} = \begin{cases}33800(1 + 0.1(T-37)) & ,\ 37 \leq T \leq 50\\ 0 & ,\ T > 50\end{cases} \tag{5}$$

In this model, the characteristic of cancer tissue is not differentiated from the surrounding normal tissue.

Probe placement configuration, shown in figure 2, is defined by six parameters that determine probe position and direction. Those six parameters are: x and y, as 2-dimensional cartesian coordinates of probe insertion point at tissue surface; θ and φ, as angles in spherical coordinates to determine the direction of insertion; r, as depth of insertion, which represents how far probe is moved along its axial; and α, as probe rotation angle towards its own axial.

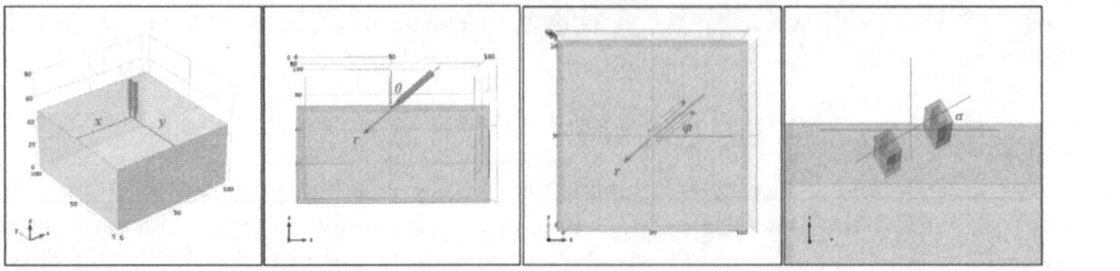

Figure 2. Six probe placement parameters.

Three test configurations (table 1) were subjectively chosen for predictive assessment using the developed model. Configuration A represents a condition in which practitioners are able to locate the cancer position and insert the probe just in perpendicular to tissue surface. Next, Configuration B shows a condition of inaccurate probe placement. Lastly, parameters in Configuration C were carefully determined by considering the position, direction, and size of the target domain to ensure relatively effective therapy.

Table 1. Test configurations subjected for assesment in this paper.

Parameters	Configuration A	Configuration B	Configuration C
x	36.5 mm	42.5 mm	40.5 mm
y	36.5 mm	40 mm	40.5 mm
θ	0^0	30^0	20^0
φ	0^0	30^0	42.5^0
α	-50^0	90^0	90^0
r	22.5 mm	25 mm	23 mm

4. Result and analysis

Images in figure 3 show isosurface representations of tissue damage resulted by the three probe configurations after 60 and 120 seconds of ablation. From the three configurations, Configuration B, shown in figure 3(b), depicts how the least ideal result can be yielded by inaccurate probe placement. It is clear that ablation in Configuration B keeps some of the target domain active and destroys a sizeable volume of normal surrounding tissue instead. Thus, placement in Configuration B can be said as not optimum.

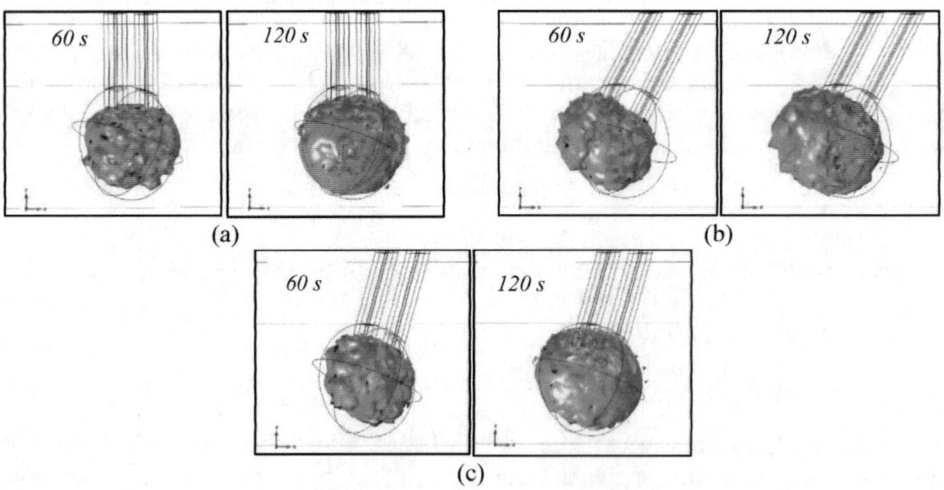

Figure 3. Simulation result of tissue damage using configuration (a) A (b) B (c) C after 60 and 120 seconds of ablation.

At overall, Configuration A in figure 3(a) is more optimal than Configuration B. Nevertheless, in addition to the fact that a part of the target domain still remains, a sizeable amount of complementary tissue is affected by the damage. Finally, figure 3(c) shows that Configuration C is the most effective probe placement since the shape and position of damage are almost identical to the target domain. Therefore, the damage of complementary tissue around the target is minimized and cancer is eliminated effectively.

Spatially averaged tissue damage ratio in target domain could serve as a parameter to analyze the level of completion during the ablation. It exposes the ratio of cancer cell damaged by the ongoing ablation. Graphic on figure 4 below shows a range of value between 0 and 1 along y-axis since it represents the ratio of killed cancer cells (*c(0)-c(t)*) over initial cancer cells (*c(0)*) spatially averaged over the target domain. Ideally, an optimum probe configuration should bring this value up to 1 in the shortest duration possible. We can see from the plot that Configuration B is not able to reach total cancer killing even after 120 seconds of ablation. This shows that inaccurate probe positioning is not effective in killing cancer. On the other hand, Configuration C is the most effective ablation, followed by Configuration A.

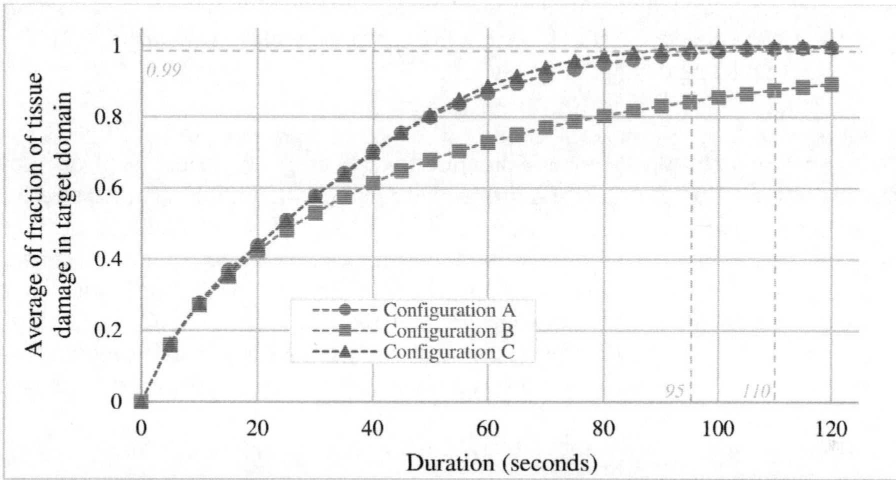

Figure 4. The average fraction of tissue damage in target domain.

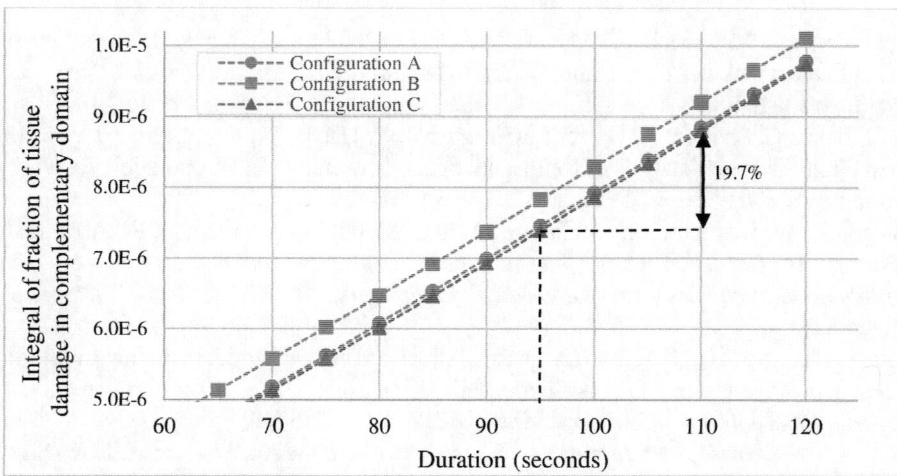

Figure 5. Integral fraction of the tissue damage in complementary domain.

Measuring the effectiveness of probe placement configuration should also consider the damage of normal surrounding tissue as a constraint. In figure 5, the complementary domain is being taken into concern by calculating the integral of tissue damage ratio all over the domain. This exposes how much damage on complementary tissue already generated by an ablation process over time. It is clearly visible that at all time, more complementary tissue damage is produced by Configuration B, compared to those resulted by Configuration A and C. This gives idea that inaccurate probe placement, in addition to its ineffectiveness in killing target domain, could lead to the damage of normal surrounding tissue. Configuration C, on the other hand, has the lowest impact on complementary domain, followed by Configuration A.

By first glimpse at figure 4, Configuration A and C appear to have nearly the same effectiveness, yet whether or not they are all optimum is still questionable. Let us assume that target domain will be dead when the spatial average of tissue damage is 0.99. Configuration C only needs 95 seconds of ablation to reach a spatial average of 0.99 ($\Omega = 4.6$) in target domain. Nevertheless, Configuration A can only reach that level of completion no earlier than 110 seconds. This 15-second difference definitely brings substantial impact on complementary tissue damage. At figure 5, it is shown that complementary damage resulted by Configuration A after 110 seconds is 19,7% bigger than that resulted after 95 seconds of ablation using Configuration C. It can be inferred that Configuration A damages 19.7% more non-targeted tissue compared to Configuration C. Conclusively, despite the similar effectiveness, converting from surface-perpendicular insertion to a more careful placement could prevent more surrounding tissue from being damaged.

5. Conclusion

Optimum RFA ablation probe placement and duration for cancer therapy could be planned by conducting model-based assessment. The accuracy of a probe placement configuration can be tested by quantifying two major parameters: average tissue damage in target domain and accumulated damage resulted in surrounding tissue. In this research, Configuration C, which is carefully determined by considering cancer position, direction, and size, is the most optimum ablation: yielding maximum tissue damage spread in target domain while producing minimum damage in complementary domain. Configuration C optimally eliminates target after 95 seconds. Due to shorter required duration compared to surface-perpendicular insertion, Configuration C generates 19.7% less damage in surrounding tissue compared to Configuration A.

6. References

[1] Andersson P 2008 *Radio Frequency Thermal Treatment of Liver Tumours* Master thesis of Linköping University

[2] Yang D, Converse M C, Mahvi D M, and Webster J G 2007 Expanding the Bioheat Equation to Include Tissue Internal Water Evaporation During Heating *IEEE Transactions on Biomedical Engineering* **54(8)** pp 1382-1388

[3] Long G, Bakos G, Shires P K, Gritter L, Crissman J W, Harris, J L, and Clymer J W 2013 Histological and Finite Element Analysis of Cell Death due to Irreversible Electroporation *Technology in Cancer Research and Treatment* 31 Agustus 2013 pp 1-10

[4] Tanotogono E W, Suprijanto, and Sudirham J J 2015 Model Development for RF Ablation Using Bipolar Forceps in Cancer Therapy *International Conference on Instrumentation, Communication, Information Technology and Biomedical Engineering (ICICI-BME) 2015 Proceeding* p P3C-6 (Preprint)

[5] Abraham J P and Sparrow E M 2006 A Thermal-ablation Bioheat Model Including Liquid-to-vapor Phase Change, Pressure-and-necrosis-dependent Perfusion and Moisture-dependent Properties *International Journal of Heat and Mass Transfer* **50** pp 2537-44

[6] Duck F A 1990 *Physical Properties of Tissue: A Comprehensive Reference Book* (San Diego USA: Academic Press Ltd)

[7] Zhu Q, Shen Y, Zhang A, and Xu L X 2013 Numerical Study of The Influence of Water Evaporation on Radiofrequency Ablation *Biomedical Engineering Online* **12** p127

Preliminary study of Alzheimer's Disease diagnosis based on brain electrical signals using wireless EEG

N Handayani[1,2], Y Akbar[1], S N Khotimah[1], F Haryanto[1], I Arif[1] and W P Taruno[3]

[1] Department of Physics, Faculty of Mathematics and Natural Science, Institut Teknologi Bandung, Jl. Ganesa No. 10, Bandung 40132, Indonesia
[2] Department of Physics, Faculty of Sciences and Technology, UIN Sunan Kalijaga, Jl. Marsda Adisucipto No. 1, Yogyakarta 55281, Indonesia
[3] Neuroscience Laboratory, CTech Labs PT Edwar Technology, Jl. Jalur Sutera Kavling Spektra Blok 23 BC No. 10-12 Alam Sutera, Tangerang, Banten, Indonesia

E-mail: nitahandayani@yahoo.com

Abstract. This research aims to study brain's electrical signals recorded using EEG as a basis for the diagnosis of patients with Alzheimer's Disease (AD). The subjects consisted of patients with AD, and normal subjects are used as the control. Brain signals are recorded for 3 minutes in a relaxed condition and with eyes closed. The data is processed using power spectral analysis, brain mapping and chaos test to observe the level of complexity of EEG's data. The results show a shift in the power spectral in the low frequency band (delta and theta) in AD patients. The increase of delta and theta occurs in lobus frontal area and lobus parietal respectively. However, there is a decrease of alpha activity in AD patients where in the case of normal subjects with relaxed condition, brain alpha wave dominates the posterior area. This is confirmed by the results of brain mapping. While the results of chaos analysis show that the average value of MMLE is lower in AD patients than in normal subjects. The level of chaos associated with neural complexity in AD patients with lower neural complexity is due to neuronal damage caused by the beta amyloid plaques and tau protein in neurons.

1. Introduction

Alzheimer's Disease (AD), a type of dementia, is a progressive neurodegenerative damage that affects many people aged 65 years and above. The disease is generally characterized by declining memory function caused by damages in nerve cells (neurons). Neuronal damage is caused by the accumulation of beta-amyloid plaque attached to the nerve cells and neurofibrillary tangled inside the nerve cells [1]. Early diagnosis of AD is a very important step in order to be able to immediately provide an effective treatment when symptoms appear early. Doctors usually perform diagnosis based on several tests such as physical and neurological examination, neuropsychological tests, blood and spinal fluid tests, mental status tests [2].

In the last few years, significant developments in the diagnosis of dementia have been conducted through biomarkers. Biomarkers are molecular markers concentrated in the brain, physiological or anatomical variables that can be used to diagnose and assess disease progression in therapy response. There are several types of biomarkers that can be used to help diagnose: biochemical marker, genetics biomarkers, neuroimaging biomarkers and neurophysiological biomarkers [3]. Neuroimaging biomarkers, structural and functional, are widely used in detecting AD. Structural imaging such as computed tomography (CT) and magnetic resonance imaging (MRI) can detect the type of athropy and morphological changes in the damaged areas. Meanwhile, functional imaging such as positron emission tomography (PET), single photon emission computed tomography (SPECT) and functional magnetic resonance imaging (fMRI) can detect AD based on the parameters of brain metabolism [4]. Despite having a high spatial resolution, this technique has drawbacks such as low temporal resolution, high prices and the effects of radiation exposure to the patient.

Currently, the development of diagnostic technology is directed towards non-invasive, rapid detection, inexpensive and reliable method. This technique includes the category neurophysiological biomarkers based on observations on neural activities in the brain that produce electrophysiological signals such as action potential and post-synaptic potential [5]. One of the neuroimaging modalities which can detect neural changes is electroencephalography (EEG). EEG is capable to read brain by electrical signals generated by neurons measured through scalp. In this paper, we will assess the use of EEG for the study of AD diagnosis based on the brain's electrical signals.

2. Methods

2.1. Participants

Subjects of this study consisted of five patients with Alzheimer's Disease and five healthy subjects as control, aged between 50-70 years. Ascertained control subjects were healthy, have not experienced neurological and physiological disorders and have never experienced brain injuries. The recording process was done in a relaxed condition (eyes closed) for a duration of 3 minutes.

2.2. EEG Measurements

EEG recording using Emotiv Epoc (BCI & Scientific Contextual EEG) 14 channels plus 2 channel references that have a sampling frequency of 128. Based on the results of previous research, there are three characteristics that play recorded brain electrical signals from the AD patient: a shift in the spectrum of low brain waves (delta and theta frequency), low levels of chaos and low synchronization signals between brain regions and others. In this study, EEG's data is processed using power spectral analysis, brain mapping and chaos analysis with MMLE (Moving Maximum Lyapunov Exponents). The selected power spectrum method is periodogram Welch that has a level of precision better than the other methods. MMLE chaos level calculation is developed by using Rosenstein techniques [6].

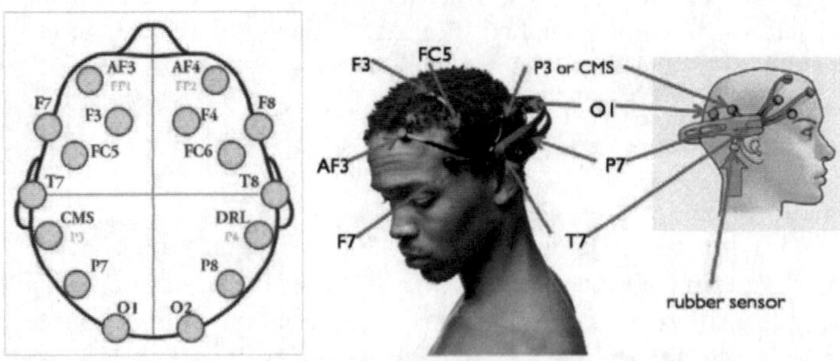

Figure 1. The position of the electrode on the Emotiv Epoc 16 channels.

3. Results

EEG signals in patients with AD are found to be slower, which means that the EEG power spectrum is concentrated at lower frequencies. The charts in figure 1 show (a), an overall power spectral in the delta frequency band (1-4 Hz) and theta (4-7 Hz) in patients with AD. Conversely, a decrease occurs in the power spectral in alpha (7-13 Hz) and beta (13-30 Hz) frequency band. Delta waves which have a voltage amplitude range between 20 μV to 200 μV occur in a deep sleep condition and the people who have the disease (abnormality) of the brain [7]. For adults, these waves are recorded on the frontal part of the brain. The graph shows that F8 has the highest amplitude, followed by F7 and P7. Theta waves which have a voltage amplitude range between 5 μV to 10 μV are recorded on the temporal and parietal areas [7]. This wave spreads on the scalp corresponding to drowsiness, alertness, cognitive impairment and dementia.

EEG's signals in normal subjects under relaxed conditions are dominant at alpha frequency (7-13 Hz). Alpha brain waves appear when under the conscious condition, relax and closed eyes. The voltage range for the alpha waves is between 20 μV to 200 μV [7]. Alpha waves are mostly observed in the posterior area of the head. The chart in figure 1(b) shows that the electrodes O2, P8 and O1 have a high amplitude corresponding to the occipital and parietal lobes (located in the posterior part of the head).

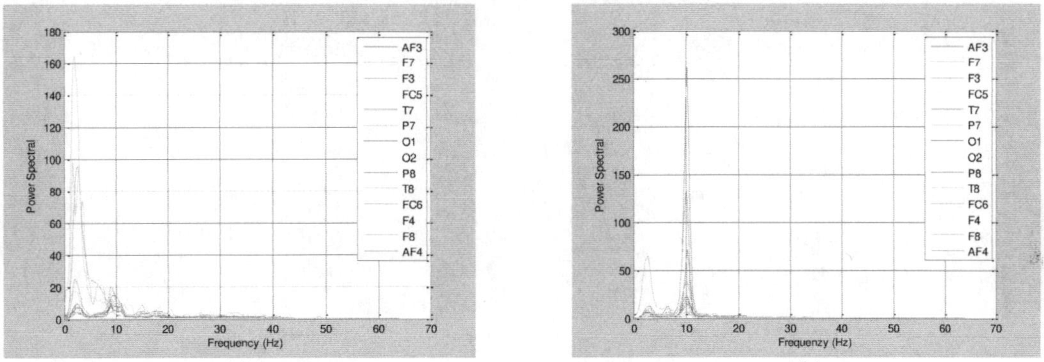

(a) (b)

Figure 2. Graph of EEG's power spectral in (a) AD patients and (b) normal subjects.

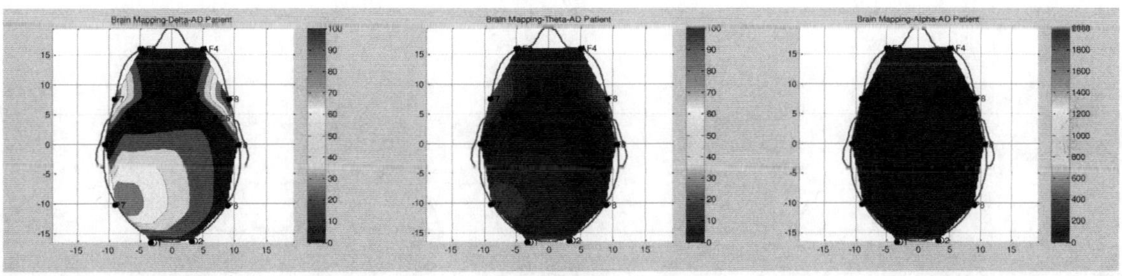

(a) (b) (c)

Figure 3. Brain mapping of power spectral (a) delta wave (b) theta wave and (c) alpha wave in AD patients.

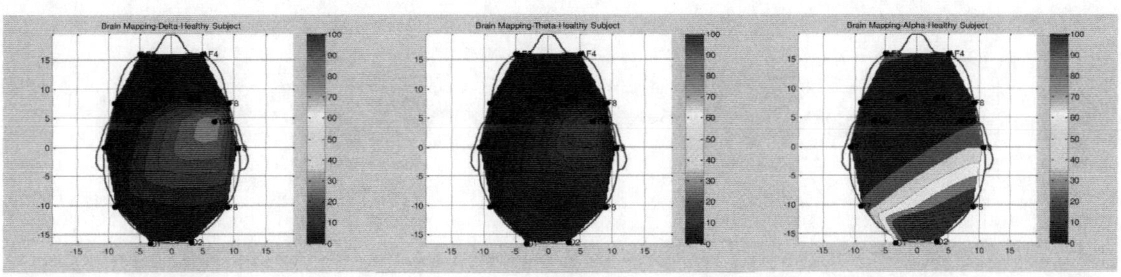

(a) (b) (c)

Figure 4. Brain mapping of power spectral (a) delta wave (b) theta wave and (c) alpha wave in normal subjects.

The result of brain mapping show that in patients with AD (figure 3), brain waves are dominant at the delta frequency, while in the normal subjects, brain waves are dominant at alpha frequency.

Table 1 shows the comparison of average MMLE values between normal subjects and AD patients. It shows that the average MMLE normal subjects on channel AF3, F7, FC5, O1, P8, T8, F4, F8 and AF4 has a higher value than the average MMLE in AD patients. The examples of EEG raw data and estimation MMLE chart are shown in figure 5.

Table 1. Average MMLE value for normal subjects and patients with AD

Channel	MMLE of each subject	
	AD	Normal
AF3	0,069	0,115
F7	0,035	0,100
F3	0,098	0,014
FC5	0,033	0,107
T7	0,165	0,032
P7	0,117	0,051
O1	0,042	0,064
O2	0,044	0,036
P8	0,026	0,121
T8	0,008	0,029
FC6	0,231	0,172
F4	0,030	0,052
F8	0,037	0,107
AF4	0,069	0,135
Average MMLE	0,071	0,081

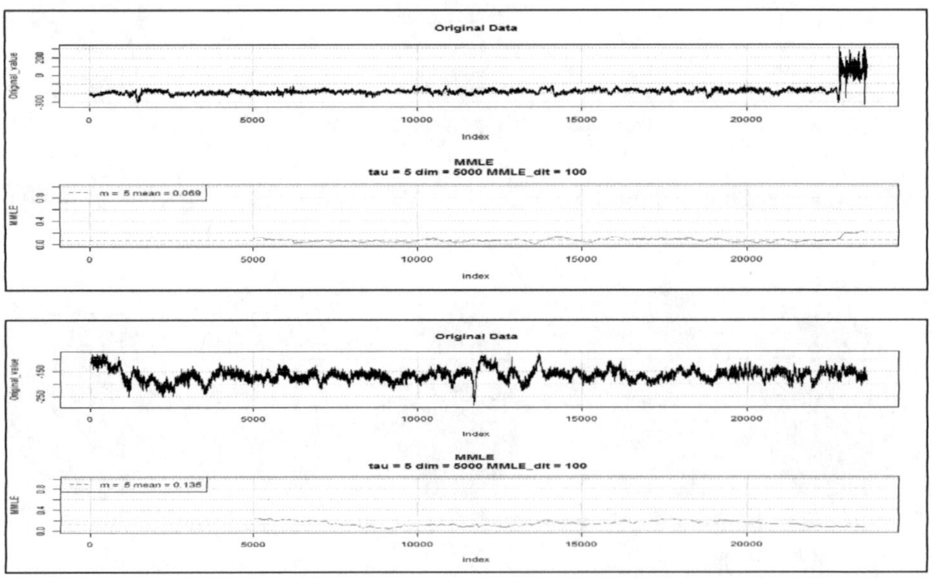

Figure 5. Raw data EEG and estimated MMLE graph at channel AF4 for AD patients (top) and normal subjects (below).

Based on theoretical review, the chaos level of normal subjects is higher than that of patients with brain abnormalities, including patients with AD. Brain signals are the sum of the electrical potential of a number of nerve cells in the cerebral cortex which are recorded through the scalp. Brain signals show macroscopic characteristics of brain function. For patients with AD, there was atrophy in the prefrontal areas related to the ability of concentration, the ability of thinking, decisions-making, and the emotional personality. The temporal lobe is associated with memory function.

4. Discussion

AD patients suffered damage or dysfunction in the cerebral cortex. Abnormality of electric potential recorded by EEG in the cortex is directly related to the pathological changes of the structure and function of cortical layers in AD [8]. Currently, the analysis of EEG abnormalities of AD patients based analytic techniques can be divided into three techniques, namely (a) time-frequency analysis, (b) wavelet analysis, and (c) chaos analysis. The marker most observed in EEG for AD patients is the reduction of high-frequency content which is called EEG slowing. This is the result of the increase in power at a low frequency band and a decrease in power at a high frequency band. Power on the frequency band represents activities on the frequency band. AD patients are characterized by the increase of theta activity and the decrease of alpha and beta activities. The increase of delta activity occurs at an advanced stage. The increase of EEG slowing rate is reported by some researchers where the increase occurs in the right postero-temporal region, the occipital region, the left frontal and the temporal lobes [9]. The result of this research found an increase of delta activity in the frontal lobe and an increase of theta activity in the left parietal area. The dominance of alpha activity in the posterior area is weak in AD patients.

Analysis of chaos can be used to identify certain characteristics of the EEG which are not visible to the naked eyes. Many researchers focus on quantitative measures of EEG complexity, as another marker of abnormality. The result has reported a decline of average MMLE in AD patients compared to the normal subjects. This neuropathology is related to cognitive deficits, which reduces neural complexity due to the decreased ability in processing information. This damage is associated with neuronal damages due to the presence of beta-amyloid plaques and tau protein in the nerve cells.

5. Conclusion

EEG's brain signals can be used as a marker for brain abnormalities and distinguish the abnormalities from normal conditions. Analytical techniques such as time-frequency analysis and chaos analysis can be used for data processing of EEG. Results of the analysis of the data show the shift of the frequency of the EEG in AD patients in comparison with normal subjects. Power spectral shifts occur at lower frequencies (delta and theta) under relaxed condition, while the results of the chaos analysis show that the complexity of neural chaos in AD patients is low compared to the normal subjects represented by average MMLE. This is consistent with the characteristics of AD where the deterioration of cognitive abilities occurs due to damage nerve cells in the brain.

References

[1] Jeffrey R P, Edward C R and Murral D P 2003 *Radiology* **226** pp 315-36
[2] Justin D, Francois V and Andrzej C 2010 *Current Alzheimer Research* **7 (6)** pp 487-505
[3] Tinu V, Sheelakumari R and Jija S J 2013 *Neurology Asia* **18 (3)** pp 239-248
[4] Keith A J, Nick C F, Reisa A S and William E K 2012 *Cold Spring Harb Perspect Med* **2** 1-23
[5] Noor K A, Sawal H, Ali, Siti A A, Kalaivani C, Shabiul I and Javier E 2014 *The Scientific World Journal*
[6] Hiroki O, Ken U, Dwi H, Muhammad F I, Rizky M, Panji N, Habib R and Warsito P T 2015 *Advanced Science, Engineering and Medicine* **7**
[7] John G W 1998 *Medical Instrumentation Application and Design* (New York: John Wiley & Sons)
[8] Jaeseung J 2004 *Clinical Neurophysiology* **115** 1490-1505

[9] Hojjat A 2010 *Automated EEG Based Diagnosis of Neurological Disorders: Inventing The Future of Neurology* (Boca raton:CRC Press Taylor & Francis Group) chapter 11 pp 207-220

Acknowledgements
The authors would like to thank Rizky Maharani, the researcher of the CTech Labs Edwar Technology and Hiroki Okada, Kyoto University, for their assistance in the chaos test. This work was partially supported by RIK ITB 2016 (006p/I1.C01/PL/2016) and PUPT Ristekdikti 2016.

Journal of Physics: Conference Series **694** (2016) 012069 doi:10.1088/1742-6596/694/1/012069

Two dimensional finite element method simulation to determine the brain capacitance based on ECVT measurement

S H Sirait[1], W P Taruno[2], S N Khotimah[1] and F Haryanto[1]

[1] Physics Department, Faculty of Mathematics and Natural Sciences, Institut Teknologi Bandung, Jalan Ganesa 10 Bandung 40132, Indonesia
[2] Center of Medical Physics and Cancer Research, CTECH Laboratories, Indonesia

E-mail: syarif.hussein.sirait@gmail.com

Abstract. A simulation to determine capacitance of brain's electrical activity based on two electrodes ECVT was conducted in this study. This study began with construction of 2D coronal head geometry with five different layers and ECVT sensor design, and then both of these designs were merged. After that, boundary conditions were applied on two electrodes in the ECVT sensor. The first electrode was defined as a Dirichlet boundary condition with 20 V in potential and another electrode was defined as a Dirichlet boundary condition with 0 V in potential. Simulated Hodgkin-Huxley -based action potentials were applied as electrical activity of the brain and sequentially were put on 3 different cross-sectional positions. As the governing equation, the Poisson equation was implemented in the geometry. Poisson equation was solved by finite element method. The simulation showed that the simulated capacitance values were affected by action potentials and cross-sectional action potential positions.

1. Introduction

In the field of tomography, electrical capacitance volume tomography (ECVT) is a relatively new modality for direct 3D imaging permittivity distribution [1]. This modality was developed based on 2D permittivity imaging modality so called ECT which usually used in industry processes. ECVT utilize several capacitance sensors to collect capacitance data from its objects (forward problem) and reconstruct permittivity distribution image using appropriate reconstruction algorithms (inverse problem). ECVT has been used in multiphase flow systems include bubble columns, gas-solid fluidized beds and consider as potential imaging modality in industry [2].

Recently, ECVT has been developed for brain tumor detection [3] and study of the brain's activity using ECVT has been proposed by Warsito *et al.* [4] [5]. They showed that ECVT was available to observe the activity of the brain during different kinds of simulation and motoric movement. It is possible that due to the depolarization in neuron the impedance will change [6] then the change in the impedance will possibly change its capacitance. However, it is still lack of confirmation on what specific parameters in neuron which affect the capacitance measurements.

In this simulation we used simulated action potential as electrical activity. Cross-sectional positions of action potential were varied on geometry. This paper aimed to study the effects of action potential on brain's capacitance using finite element simulation based on 2 electrodes ECVT measurement, butthis simulation is still need to be verified using another experiment. However, the result can be used to improve ECVT's data acquisition system for the measurement of brain activities.

2. Methodology

2.1. Head design

The head was formed from five different layers with different relative permittivity (ε_r) values and was assumed to be isotropic [7]. We used relative permittivity values at 2.5 MHz in corresponding to real ECVT measurement which applied 2.5 MHz in frequency. Those five layers were white matter, grey matter, CSF, skull, and skin with its permittivity values were 308, 604, 109, 93, and 799 [8], respectively, as shown in figure 1.

2.2. Sensor design

Two electrodes ECVT sensor were designed and followed the real 2 electrodes brain ECVT sensor. It consists of vessel, excitation electrode, detection electrode, axial screen, and grounded screen with its relative permittivity values are 3, 1, 1, 1, and 1, respectively as shown in Figure 1. The electrodes had 5cm in length and width. Vessel is usually used in ECVT as part of the sensor so the measurement system becomes non-invasive, and non-destructive. Axial screen and grounded screen were applied in ECVT so unwanted electrical field would not affect the capacitance measurement. Next, the head design and the sensor design were combined as shown in figure 1.

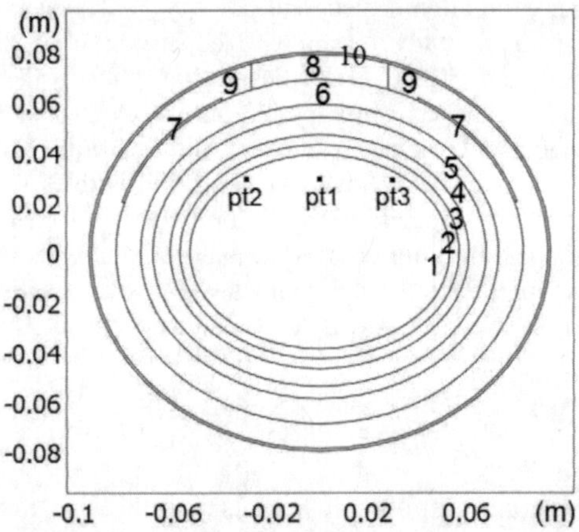

Head layer
1. White matter
2. Grey matter
3. CSF
4. Skull
5. Skin

ECVT sensor part
6. Vessel
7. Excitation electrode (left), detection electrode (right)
8. Axial screen
9. Air
10. Grounded screen

Figure 1. Simulated 2D design of coronal head and ECVT sensor.

2.3. Electrical activity

Simulated Hodgkin-Huxley action potentials were applied as brain's electrical activity. One cycle of 6 ms action potential was used and had 1×10^{-5}s in sampling interval. We assumed that action potential was occurred at a point in the geometry. This action potential can be viewed in figure 2. Cross-sectional position was defined as distance from the center and in this case the center was point 1. Action potential subsequently was put on 3 different cross-sectional positions with point 1 on (0; 0.03), point 2 on (-0.03; 0.03), and point 3 on (0.03; 0.03). The positions of these points can be viewed in figure 1 above as pt1, pt2, and pt3.

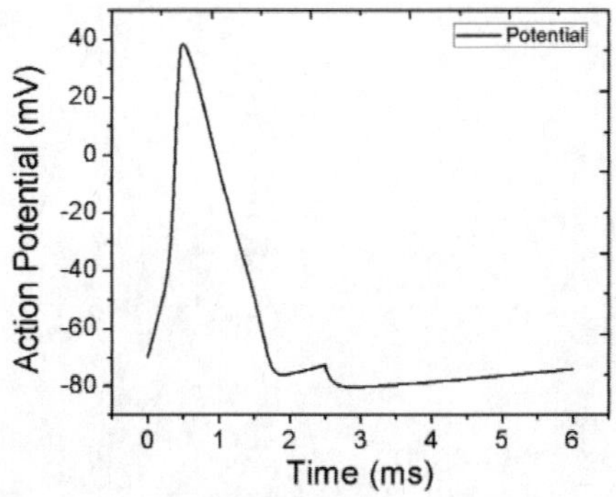

Figure 2. Simulated Hodgkin-Huxley action potential.

2.4. Simulation setup

Poisson equation was implemented as the governing equation to estimate the physical phenomena in geometry like it was described in ECT [9] [10] [11] and ECVT [1] [2]. Excitation electrode was defined as a Dirichlet boundary condition with static 20 V in potential and detection electrode was defined as a Dirichlet boundary condition with 0 V in potential. Static potential was used rather than time-dependent potential to simplify the equation into electrostatic approximation. In future, time dependent potential with f=2.5 MHz could be applied but with different estimation. Axial screen and grounded screen were defined as ground. Relative permittivity in the head design and the sensor design were assumed to be isotropic. Simulated action potential was defined as a point potential and was put subsequently on point 1, point 2, and point 3 in geometry. Free charges were assumed that it did not exist in the simulation domain. Poisson equation and the boundary condition hold:

$$\nabla \cdot \varepsilon \nabla \phi(x, y) = 0 \tag{1}$$

$$\phi|_{\Gamma_i} = V\,; \quad \phi|_{\Gamma j} = 0$$

where $\phi(x, y)$ were potential distribution in geometry, with V = 20 volt. Γ_i and Γ_j were Dirichlet boundary condition for excitation electrode and detection electrode, respectively. Finite element method was used to solve Poisson equation for finding potential distribution. Finite element method worked and subsequentialy changed the Poisson equation into its weak formulation.

$$-\int_{\Omega} \nabla w \cdot \mathbf{D}\mathrm{dA} = -\oint_{\Gamma} w\mathbf{D} \cdot \mathbf{n}\,\mathrm{dl} \tag{2}$$

$$\mathbf{D} = \varepsilon \nabla \phi(x, y) \tag{3}$$

w as residual weight and **D** as electric displacement. After potential distributions had been found, the capacitance was calculated using this capacitance equation.

$$C_{ij} = -\frac{1}{\Delta\phi_{ij}} \oint_{A_j} \varepsilon \nabla \phi(x, y) \cdot \mathbf{n}\mathrm{dA} \tag{4}$$

3. Results and discussion

The entire domain was solved by finite element method. Every simulation was performed for 6 milliseconds long potential action. The simulation was run for every potential action of position in geometry. Head and sensor design were discretised using triangle element, as can be viewed in Figure 3. Meshing process were good and produced 24948 elements.

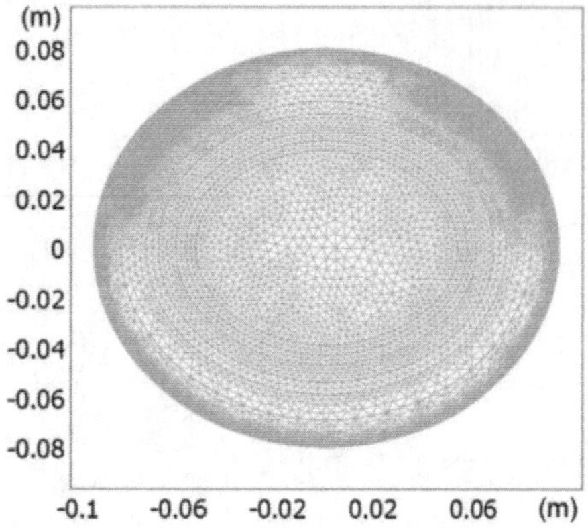

Figure 3. Simulated meshing result.

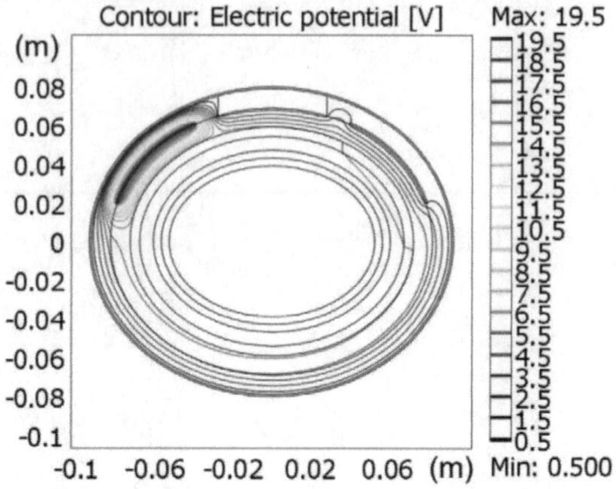

Figure 4. Simulated potential distribution without action potential.

As can be viewed from figure 5, high potential was blocked and accumulated around the excitation electrode due to a high relative permittivity (ε_r =799) value in skin layer. A potential of about 4.5 V from excitation electrode was successfully penetrated into the deepest layer and achieved the detection electrode. Hence, this potential was adequate to be used in excitation electrode. Axial screen and grounded screen were effective to block unwanted electrical field because there was no potential penetration.

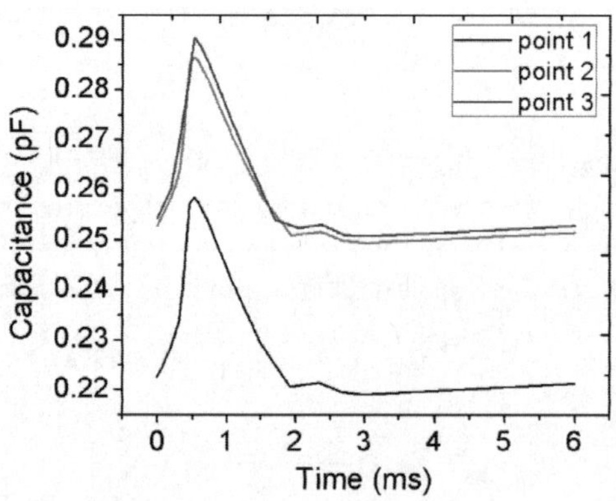

Figure 5. Simulated brain's capacitance in head geometry affected by action potential which were put on 3 different cross-sectional position.

Figure 5 shows that the pattern of capacitance values followed the pattern of action potential value which is shown in figure 2. Action potential from different cross-sectional position affected simulated brain's capacitance value. Brain's capacitance affected by action potential on point 1 had peak 0.258pF and valley 0.219pF, affected by point 2 had peak 0.286pF and valley 0.249pF, and affected by point 3 had peak 0.290pF and valley 0.250pF. All capacitance had the range of peak and valley for about 0.04pF. Capacitance that affected by action potential on point 1 had a difference for about 0.03pF with capacitance affected by action potential on point 2 and point 3 due to the difference in the cross-sectional position. Capacitance affected by action potential on point 2 had slightly difference with capacitance affected by action potential on point 3 due to the similarities in cross-sectional positions but in opposite direction.

4. Conclusions

We have studied the action potential effect to capacitance value in ECVT measurement by using finite element simulation. It showed that the capacitance values were changed simultaneously with the action potential. These simulation results a suggestion that the action potential possibly could be detected using ECVT due to its effect to the capacitance value. It also showed that the action potentials which were put on different cross-sectional positions made difference in simulated capacitance value. These differences in capacitance were happened due to the cross-sectional positions that showed a potential for ECVT as imaging modality for brain's activity.

References

[1] Warsito W, Marashdeh Q and Fan L S 2007 *IEEE Sensor Journal* **7(4)** pp 525-35
[2] Marashdeh Q, Fan L S and Warsito W 2008 *Ind. Eng. Chem* **47** pp 3708-37
[3] Taruno W P, Baidillah M R, Sulaiman R I, Ihsan M F, Yusuf A, Widada W and Aljohani M 2013 *6th Annual International IEEE EMBS Conference on Neural Engineering* pp 743-6
[4] Taruno W P, Baidillah R M, Sulaiman R I, Ihsan M F, Fatmi S E, Muhtadi A H, Haryanto F and Aljohani M 2013 *IEEE 10th International Symposium on Biomedical Imaging* pp 1006-9
[5] Taruno W P, Ihsan M F, Baidillah M R, Tandian T, Mahendra M and Aljohani M 2014 *Middle East Conference on Biomedical Engineering* pp 147-50
[6] Aristovich K Y, Packham B C, Koo H, Santos G S D, McEvoy A and Holder D S 2016 *NeuroImage* **124** pp 204-13
[7] Miranda C Pedro, Mekonnen A, Salvador R and Basser P J 2014 *Phys. Med. Biol* **59** pp 4137-47
[8] An internet resources for the calculations of dielectric properties of body tissues. http:niremf.ifac.cnr.it/tissprop. [Accessed February 2015]
[9] Alme K J and Mylvaganam S 2006 *IEEE Sensor Journal* **6(5)** pp 1256-66
[10] Banasiak R, Wajman R and Soleimani M 2009 *Insight* **51** pp 36-9

Acknowledgment

The authors would like to thank CTech Labs in Tangerang, Indonesia for providing collaboration in ECVT research. This work was partially supported by RIK ITB 2016 (006p/I1.C01/PL/2016) and PUPT Ristekdikti 2016.

Spectral and brain mapping analysis of EEG based on Pwelch in schizophrenic patients

Y Akbar[1], S N Khotimah[1] and F Haryanto[1]

[1] Department of Physics, Faculty of Mathematics and Natural Sciences, Institut Teknologi Bandung, Jalan Ganesa 10, Bandung 40132, Indonesia

E-mail: yudiansyahakbar1@gmail.com

Abstract. The aim of this study is to investigate and analyze the differences of power spectral distribution in various frequency bands between healthy subjects and schizophrenic patients. Subjects in this study were 8 people consisting of 4 schizophrenic patients and 4 healthy subjects. Subjects were recorded from 12 electrodes with Electroencephalography (EEG). EEG signals were recorded during a resting eye-closed state for 4-6 minutes. Data were extracted and analyzed by centering and filtering, then performed using Welch Periodogram technique for the spectral estimation with a Hamming window. The results of this study showed that delta power spectral in schizophrenic patients increased ten times from healthy subjects; theta power spectral in schizophrenic patients increased three times from healthy subjects; alpha power spectral in schizophrenic patients decreased with an increase of one third of healthy subjects. These results were confirmed by Kolmogorov-Smirnov test showing there were significant differences between schizophrenic and healthy subjects on delta, theta and alpha brain wave. Based on the results of Brain Mapping analysis showed that there was significant increasing in the activity of delta waves and theta waves in frontal lobe of schizophrenics, whereas the alpha waves indicated a decrease in the occipital lobe in all schizophrenic patients.

1. Introduction

Schizophrenia is conceptualized as a psychotic disorder with symptoms such as delusions (the belief in something that actually never happened or not real) and hallucinations (the belief that people see or hear us something that does not exist or is not real) [1]. One of the modalities that can observe the brain abnormalities in schizophrenia is EEG. EEG signals have characteristics to change each time (non-stationary) and random because of a very dynamic activity in the brain. Although the EEG signals is dynamic, but it can be classified based on certain characteristics, one of them through the power spectral analysis of several brain waves frequency range.

In the previous study, the power spectral analysis was used for classification of patients with brain damage asymmetric signal [2], the classification of EEG signals for automatic detection of epileptic seizure [3] also early detection of mild cognitive impairment and Alzheimer's disease [4]. In this study, EEG signals of schizophrenics analyzed and quantified using power spectral analysis through Welch periodogram approach with Hamming window, in an attempt to identify the components of the delta, theta, alpha, beta and gamma wave. Power spectral distribution of schizophrenia patients compared to power spectral distribution of normal subjects to analyze power spectral differences in both conditions. The representation of the power spectral distribution was displayed with brain mapping.

2. Method

The method used in this study was the experimental method with a direct measurement in normal subjects and patients with schizophrenia. Subjects in this study were 8 people consisting of 4 patients with schizophrenia (hereinafter referred SZ1, SZ2, SZ3, SZ4) who had been diagnosed based on the criteria of the Diagnostic and Statistical Manual of Mental Disorder, 4th edition (DSM-IV) and 4 healthy subjects (hereinafter referred to as H1, H2, H3, H4) as a control. Subjects were recorded from 12 electrodes with Electroencephalography (EEG) with a sampling frequency of 128 rps. EEG recording in schizophrenic patients was performed at the Mental Hospital of West Java province, Indonesia. While recording for healthy subjects was conducted at the Laboratory of Biophysics, Bandung Institute of Technology. The method of research that was conducted was depicted in the figure 1 below:

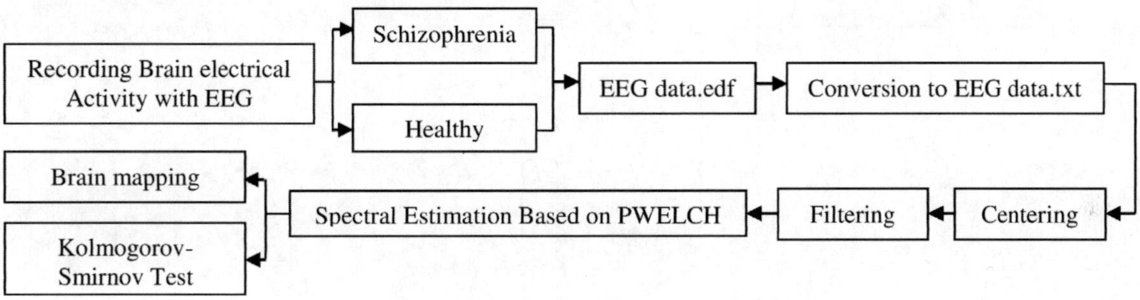

Figure 1. A block diagram of data processing.

Data recording results were in European Data Format (EDF) format. This data then were converted into American Standard Code for Information Interchange (ASCII) using TestBench software v.1.5.1.2. EEG data extraction was done by centering in order to remove the DC offset of each electrode. The principle of centering the data was done in order to averaging all the data and then each of the data was reduced by the average data. This was done so the the data which have a value above the average will have positive results and data which has a value below the average will have negative results. The data were then filtered from the noise and artifacts. This study used a band-pass filter that passes a signal at a frequency range above the lower limit frequency (2 Hz) and below the upper limit frequency (45 Hz).

Welch power spectral estimation was done by segmenting the data into the P part, the overlap segment length was D and one segment to the other segment displaced along the S Data ($S \leq D$). Thus, the maximum length of P and the data obtained were [5]:

$$x^{(p)}(k) = w(k)x(k + pS), \quad 0 \leq k \leq D-1,\ 0 \leq p \leq P-1 \tag{1}$$

Furthermore periodogram estimate for each data-p to [5]:

$$U = T\sum_{k=0}^{D-1} w^2(k)\tilde{P}_{xx}^{(p)} = \frac{1}{UDT}\left|T\sum_{K=0}^{D-1} x^{(p)}(k)\exp(-j2\pi fkT)\right|^2 \tag{2}$$

Kolmogorov-Smirnov test (KS test) was applied to justify the power spectral differences between schizophrenic patients and healthy subjects. KS test principle showed compatibility between two cumulative distributions by taking into account the deviation between the two cumulative distributions [8]. Both sets of data were from the same distribution or there were no significant differences between both sets of data if the value of $p > 0.05$ at the 5% significance level. The representation of the power spectral distribution of each brain wave was displayed in 2D Brain mapping. The algorithm performed in conducting Brain mapping was as follows:

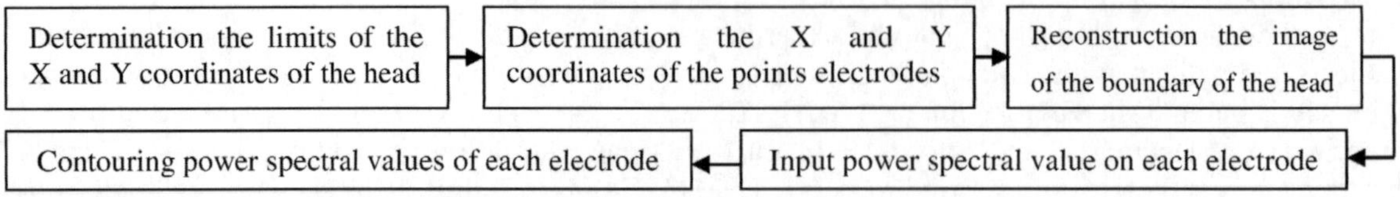

Figure 2. The algorithm of brain mapping.

3. Results

3.1. The results of spectral estimation

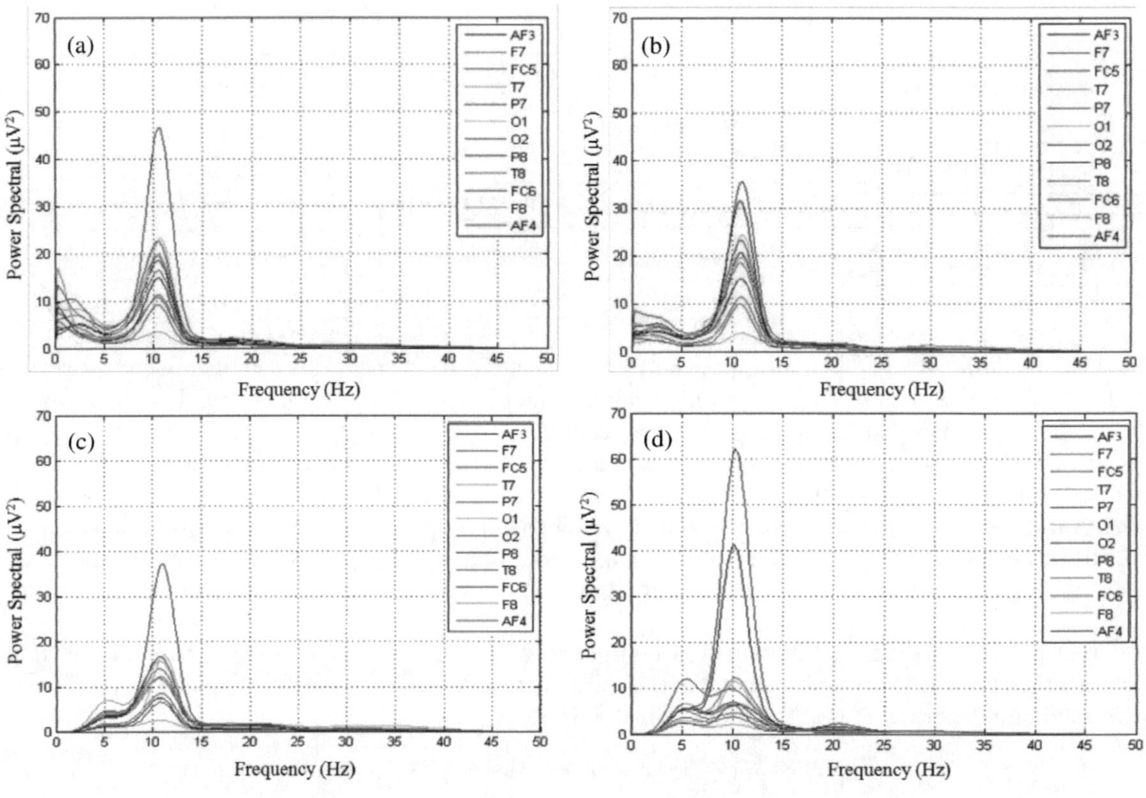

Figure 3. Results of spectral analysis in healthy subjects, (a) H1 (b) H2 (c) H3 (d) H4.

Figure 3 shows the results of spectral analysis using the Welch method in normal subjects with a resting state condition. The power spectral for normal subjects had varying values. The dominant power spectral distribution in healthy subjects was in the frequency 8-12 Hz. The highest value of power spectral was 62.11 μV^2 at the O2 electrode in the frequency 10.25 Hz. This indicated that the right occipital lobe had the highest alpha activity when recording a resting state. The results of the power spectral analysis for schizophrenics were shown in figure 4. The power spectral for schizophrenic patients had varying value. The dominant power spectral distribution in healthy subjects was in the range frequency of delta and theta. The highest value of power spectral was 263.3 μV^2 at the AF4 electrode in the frequency 3 Hz. This indicated that the right anterior frontal lobe experienced the highest alpha wave activity when recording the resting state.

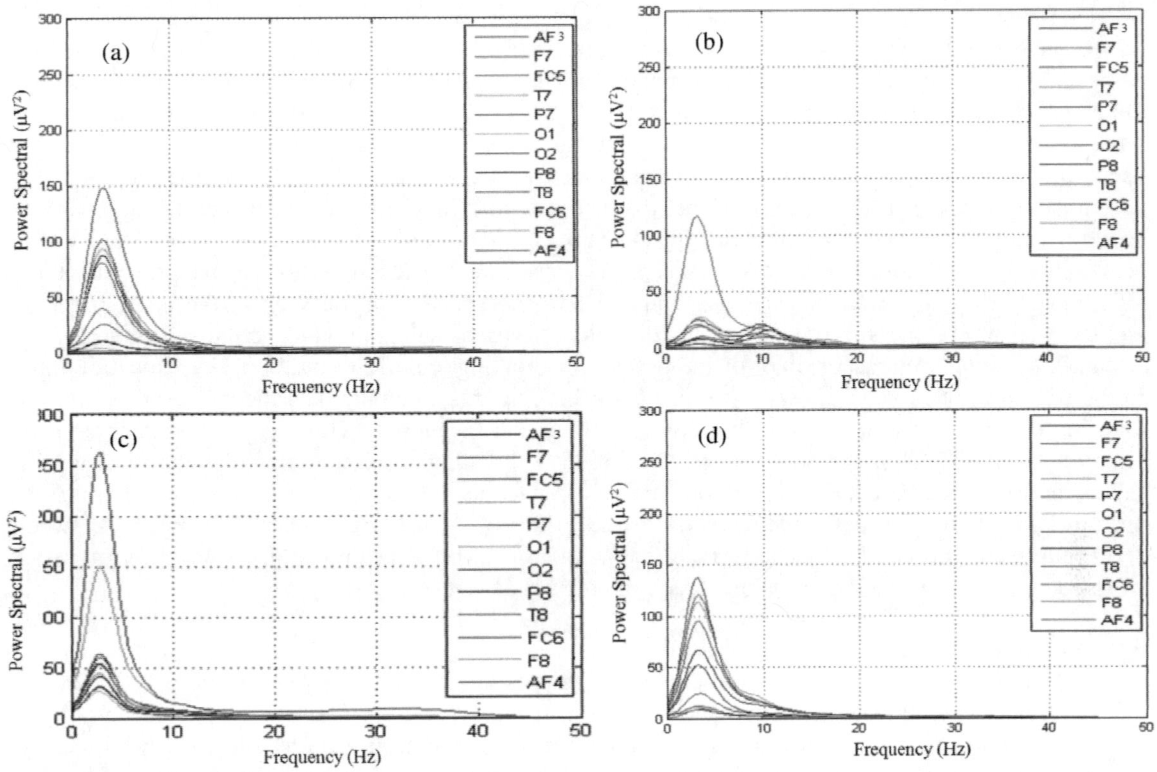

Figure 4. Results of spectral analysis in schizophrenic patients, (a) SZ1 (b) SZ2 (c) SZ3 (d) SZ4

These results were confirmed by Kolmogorov-Smirnov test which showed that there were significant differences between schizophrenic and healthy subjects on the delta ($p = 0.007$), theta ($p = 0.021$) and alpha ($p = 0.021$).

3.2. The results of brain mapping

Figure 5 shows the results of Brain mapping in healthy subjects and schizophrenic patients on the delta, theta, alpha beta and gamma. Based on these images it could be seen that there were significant differences between schizophrenic and normal subjects in the delta waves in the frontal lobe, theta waves in frontal lobe and in the alpha waves in the occipital lobe. There were no significant differences between schizophrenic and normal in the beta and gamma waves.

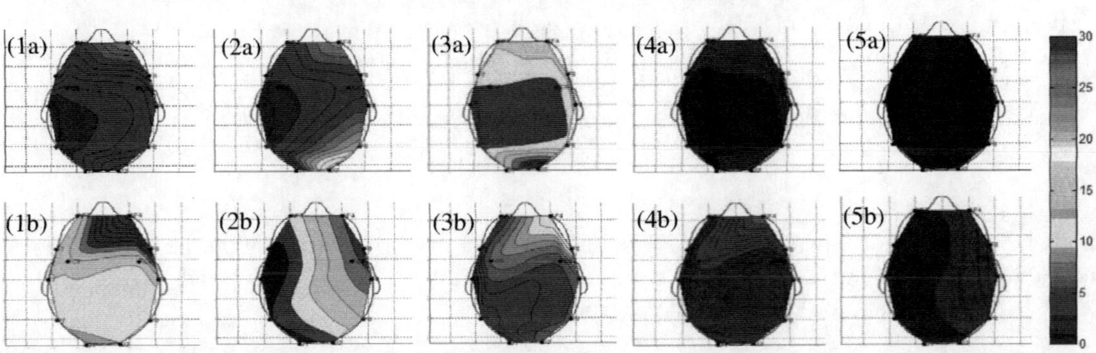

Figure 5. Results of brain mapping, (1) delta, (2) theta, (3) alpha, (4) beta, (5) gamma; (a) healthy subject, (b) schizophrenic patients

4. Discussion

Based on spectral analysis with Welch method, it could be seen some differences between healthy subjects and schizophrenic patients. The spectral power for normal subjects had the highest value on O2 electrode in the frequency range 10.25-11 Hz. The highest power spectral values of schizophrenics patients were contained in the AF4 electrode with a frequency range of 3 - 3.5 Hz. The lowest power spectral values were contained in the electrode electrodes O1. Electrodes O1 describe the electrical activity that occured in the occipital lobe which has a role in vision and visual processing. Power spectral value in the occipital lobe was low because the recording was done on resting state where schizophrenic conditioned at an eye-closed state. Based on the results of brain mapping, the most significant differences were in the delta and theta waves. This could be seen in the contour of the frontal lobe where the spectral power in the delta waves of schizophrenic patients increased. The frontal lobe is the area responsible for the process of thinking, emotions and behavior. Spectral value in the frontal of delta wave showed low levels of thinking in schizophrenic patients. It was suspected of being the cause of low levels of thinking and emotional responses. This finding was supported by Khan *et al.* (1993) [6], and Schellenberg *et al.* (1990) [7]. Unlike the case with delta and theta waves, the alpha waves decrease the occipital lobe activity in patients with schizophrenia. The contours of the spectral power in the beta and gamma waves did not show significant differences between schizophrenic and normal subjects. It could happen because the recording subjects were not in a state of full consciousness and did not experience high mental activity.

5. Conclusions

Spectral analysis of brain waves showed an increasing in delta power spectral (2-4 Hz) in schizophrenic patients with ten times increase from healthy subjects. The theta power spectral (4-8 Hz) in schizophrenic patients increased with three times increase from healthy subjects. Additionally, alpha power spectral (8-12 Hz) in schizophrenic patients decreased with one third increase of healthy subjects. Based on the results of Brain Mapping analysis showed that there were significant increasing in the activity of delta waves and theta waves in frontal lobe of schizophrenics, whereas alpha waves indicated a decreasing in the occipital lobe in all patients with schizophrenia.

References

[1] Tandon R *et al.* 2013 *J. Schizo Research* **150** 3-10
[2] Davey M P, Victor J D and Schiff N D 2000 *J. Clin. Neurophysiology* **111** 1949-1954
[3] Naderi M A and Mahdavi N 2010 *Proc. Int. Conf. of 17th Iranian Conference of Biomedical Engineering (ICBME)* (Isfahan: IEEE) pp 1 - 4
[4] McBride J C, Zhao X, Munro N B, Smith C D, Jicha G A , Hively L, Broster L S, Schmitt FA, Kryscio RJ and Jiang Y 2014 *J. Com. Methods and Programs in Biomed* **114** 153–163
[5] Tjokronegoro H A 2004 *Digital Spectral Analysis* (Bandung: ITB)
[6] Engmann S and Cousineau D 2011 *J. of Appl. Quant. Research* **6** 4
[7] Michael K, Richard D W, Richard C, Harold S K and Katherine S 1993 *J. Biol psychiatry* **4** 284-290
[8] Schellenberg R, Knorr W, Schindler M, Kropf S and Beyer H 1990 *J. Schizo research* **3** 357-359

Acknowledgements

The authors would like to thank Mental Hospital of West Java province, Indonesia which have collaborated in recording data. This work was partially supported by RIK ITB 2016 (006p/I1.C01/PL/2016) and PUPT Ristekdikti 2016.

Journal of Physics: Conference Series **694** (2016) 012071 doi:10.1088/1742-6596/694/1/012071

Analysis of respirable particulate exposure and its effect to public health around lead smelter and e-waste processing industry in West Java, Indonesia

M Marselina[1], D Roosmini[1], I R S Salami[1], M Ayu A[1] and W Cahyadi[1]

[1] Department of Environmental Engineering, Faculty of Civil and Environmental Engineering, Institut Teknologi Bandung, Indonesia, Jalan Ganesa 10, Bandung 40132, Indonesia

E-mail: indahrss@yahoo.com

Abstract. Respirable particulate exposure strongly affects human health, especially for children who lived around industrial area. This study was conducted to evaluate the effect of respirable particulate exposure to lung capacity of children. Study location in this study was Parung Panjang District, area of lead smelter industry and also in Astana Anyar District, area of e-waste processing industry. Thirty children were involved in Astana Anyar District and also thirty children in Parung Panjang District. The control groups were also studied in both areas. Predicted average daily intake (ADD) of respirable particulate was estimated and lung or respiration condition of children was measured by using spirometer. The lung condition of respondents was estimated by $FEV_{1.0}$ and FVC values. As the result, the predicted ADD of children in lead smelter area is 3 times higher than the predicted ADD of children in e waste processing area. It was correlated positively with the higher $PM_{2.5}$ concentration in Parung Panjang District than the $PM_{2.5}$ concentration in Astana Anyar District. Metals concentration in Parung Panjang was also measured with X-Ray Fluorescence (XRF) in this study and it was clearly state that metals concentration in location study were higher than metals concentration in control area.

1. Background

Numerous studies had been performed to associate respirable particulate exposure with respiratory diseases such as respiratory illness, chronic obstructive pulmonary disease (COPD), asthma, fibrosis, and even lung cancer as well. Dyosi (2007) stated that ambient lead concentrations in the areas of refining and smelting in used batteries industries have exceeded the Occupational Exposure Limit (OEL) and can exposed the worker. Slaughter (2010) declared that there is a positive relationship between the high concentration of respirable particulates, formaldehyde, and carbon monoxide from smoke to the lung capacity of fire-fighters which declared by the decrement of $FEV_{1.0}$ value. The research was conducted in the western of United States. Jain (2007), that $FEV_{1.0}$, FVC, $FEV_{1.0}$/FVC may be less sensitive to factors that associated with respiratory tract alteration and can detect the severity of respiratory tract obstruction.

This study discussed the respiratory health risk of children living around industrial area in two separate studies in Indonesia. The first study was conducted around e-waste processing activity in Astana Anyar District, Bandung Regency, West Java and the second one was in children who living around lead smelter industry in Parung Panjang District, Bogor Regency, West Java.

Journal of Physics: Conference Series **694** (2016) 012071 doi:10.1088/1742-6596/694/1/012071

Astana Anyar District was selected because there is some school that is located very near to e-waste processing activity. Study in Parung Panjang District was obtained to analyze the effect from lead smelter industry to children around that area. Both of this study will discussed and compare the impact of those industrial activity on children respiratory health risk.

2. Materials and method

The measurement of respirable particulate was done by using personal sampling pump with flow rate of 1.7 L/min that is placed for 4 hours on each respondent. This research used Mixed Cellulose Esther (MCE) with a diameter of 25 mm and pore size is 8 μm as filter. According to Occupational Safety and Health Administration (OSHA), MCE is a good filter to catch metal particulates. The filter was weighed in a room with 21°C temperature and 55% humidity.

The measurement was performed on boy and girl respondents. The respondents were group of children aged 10-14 years old that goes to school near the lead smelter industry in Parung Panjang and also the e-waste processing activity in Astana Anyar. There were 15 boy respondents and 15 girl respondents in both location study. Every respondent has been selected based on their teaching-learning activities in an open window classroom. The sampling process is carried out for 4 hours.

Respiratory function was measured using spirometer tool Minato AS-500 from Japanese manufacturers. Lung function values expressed in $FEV_{1.0}$ (Forced expiratory volume in one second), is forced expiratory volume in one second and FVC (Forced Vital Capacity) is the forced vital capacity (Ward, 2008) The measurement of $FEV_{1.0}$ was conducted on respondents who have no smoking habit that can reduce the lung health. Metals concentration in Parung Panjang was also measured with X-Ray Fluorescence (XRF) analysis.

3. Results and discussion

The measurement of respirable particulate was done by using personal sampling pump with flow rate of 1.7 L/min that is placed for 4 hours on each respondent. This research used Mixed Cellulose Esther (MCE) with a diameter of 25 mm and pore size is 8 μm as filter. According to Occupational Safety and Health Administration (OSHA), MCE is a good filter to catch metal particulates. The filter was weighed in a room with 21°C temperature and 55% humidity.

3.1. Measurement of $PM_{2.5}$ and predicted average daily intake (ADD)

$PM_{2.5}$ is particulate less than 2.5 micrometers in diameter ($PM_{2.5}$) that referred as "fine" particles and believed to pose the greatest health risks. The concentration of $PM_{2.5}$ in area of e-waste processing industry showed the average of $186.75 \pm 125.1 \mu g/m^3$. Furthermore, in area around lead smelter industry, the concentration of $PM_{2.5}$ was average of 225.9 ± 214.9 μg/m^3. Table 1 below showed that the concentration of $PM_{2.5}$ in Parung Panjang District was higher than those in Astana Anyar measurement.

Table 1. Measurement of $PM_{2.5}$ (μg/m^3).

Location of Children	Astana Anyar District	ParungPanjang District
Study Area	186.75 ± 125.1	225.9 ± 214.9
Control Area	80.5 ± 42.6	75.11 ± 39.7

Predicted average daily intake (ADD) of respirable particulate was estimated for both area of study (table 2). Children living around e-smelter processing was estimated to have ADD of 32.14 ± 28.42 μg/kg.day. Whereas in Parung Panjang, the ADD was 42.36 ± 39.66 μg/kg.day higher than those inAstana Anyar. For both studied area, the concentration of PM2.5 in area of e-waste processing industry showed the average of 186.75 ± 125.1 μg/m^3. Furthermore, in area around lead smelter industry, the concentration of $PM_{2.5}$ was average of 225.9 ± 214.9 μg/m^3.

Table 2. Predicted ADD of Respirable Particulate Intake (μg/kg.day)

Location of Children	Astana Anyar District	Parung Panjang District
Study Area	32.14 ± 28.42	42.36 ± 39.66
Control Area	10.78 ± 7.89	15.42 ± 8.36

3.2 Respiratory function analysis

Table 3 described the $FEV_{1.0}$ result of the children. It was observed that the lung capacity of boys in Parung Panjang was always lower than the lung capacity in Astana Anyar. This is the same pattern with the lung capacity of girls in Parung Panjang was always lower than the lung capacity of girls in Astana Anyar. For both study area, $FEV_{1.0}$ and FVC value in location study were always higher than in control area. The proportion of fine particle in the air quality had affected the lung capacity of respondents. The air pollution in Parung Panjang was also considered worsen as the lead smelter activity generated other pollutant than particulate matter.

Table 3. $FEV_{1.0}$ and FVC measurement results of sampled children.

Location	Children	Study Area		Control Area	
		$FEV_{1.0}$ (l)	**FVC (l)**	**$FEV_{1.0}$ (l)**	**FVC (l)**
Astana Anyar	Boys	1.41 ± 0.19	1.23 ± 0.22	1.51 ± 0.24	1.79 ± 0.20
	Girls	1.49 ± 0.16	1.09 ± 0.28	1.52 ± 0.20	1.78 ± 0.21
Parung Panjang	Boys	1.18 ± 0.23	1.68 ± 0.25	1.21 ± 0.22	1.24 ± 0.12
	Girls	1.06 ± 0.25	1.77 ± 0.14	1.52 ± 0.18	1.19 ± 0.12

Figure 1 and 2 showed that age of the children differentiated the lung function and condition. Figure 1 showed comparison of $FEV_{1.0}$ and FVC condition related to age of the children in Astana Anyar District. Respondents around e-waste processing industry exhibited lower $FEV_{1.0}$ and FVC at the age of 10-12 years. The value of $FEV_{1.0}$ and FVC in location study in area of e-waste processing industry was lower than the value in control area, it was caused by the higher pollution of respirable particulate in location study than in control area.

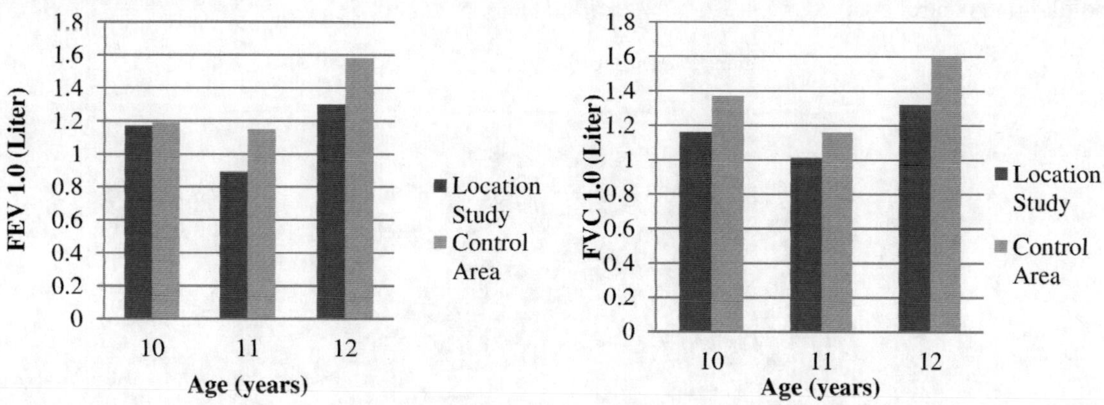

Figure 1. Comparison of $FEV_{1.0}$ and FVC in Astana Anyar District

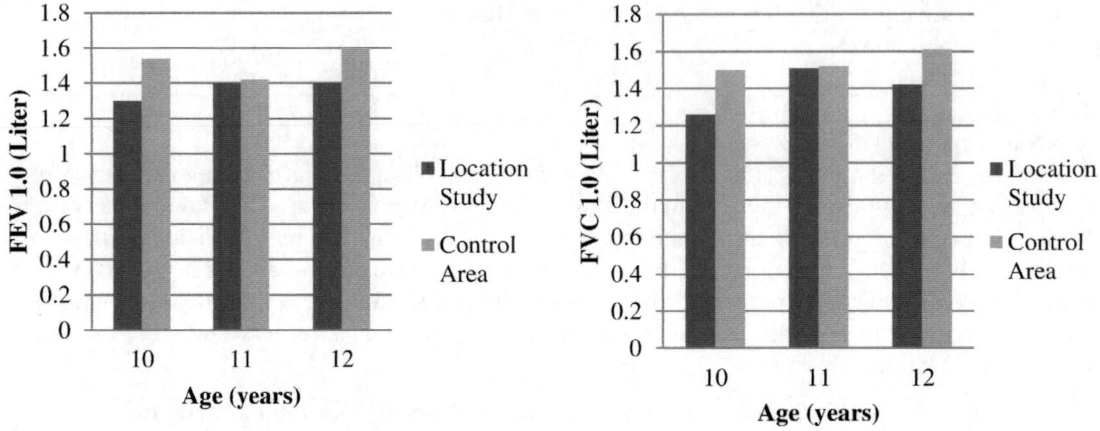

Figure 2. Comparison of $FEV_{1.0}$ and FVC in Parung Panjang district.

Figure 2 showed the lung condition of respondents in Parung Panjang District, area around lead smelter industry. The value of $FEV_{1.0}$ and FVC in location study showed a lower value than the value in control area. It was also noted that the age of 11 years in respondents around Parung Panjang District did not perform different $FEV_{1.0}$ and FVC values in both location study and control area. Besides that, resulted measurement of $FEV_{1.0}$ and FVC showed that lung condition of respondents at Parung Panjang was lower compare with lung condition of respondents at Astana Anyar District.

3.3. Metal concentration in respirable particulate

Based on figure 3 below, it can be seen that the concentrations of metals in samples of respirable particulate in Parung Panjang was higher than the concentrations of metals in samples of respirable particulate in control area. This shows that the air pollution in Parung Panjang is higher than the air pollution in control area.

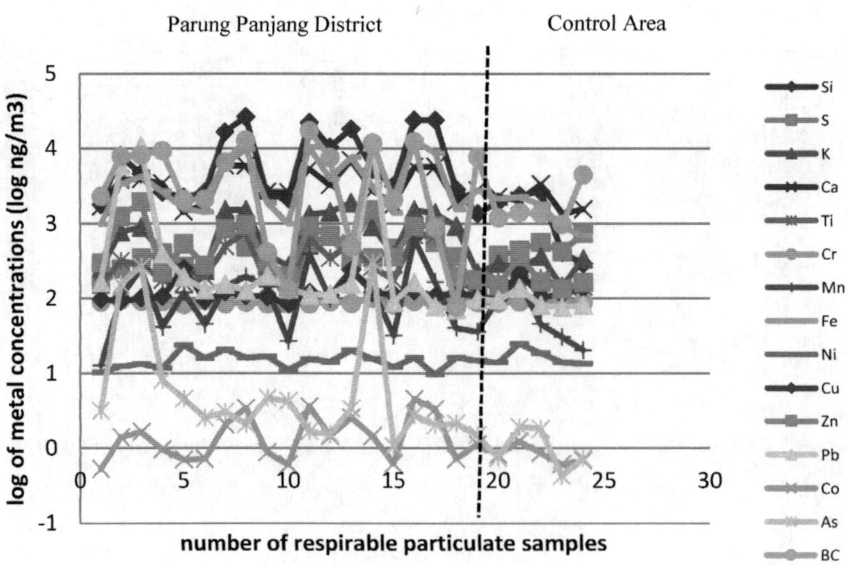

Figure 3. Metals concentration in respirable particulates in Parung Panjang and control area.

4. Conclusion

The concentration of $PM_{2.5}$ in area of e-waste processing industry showed the average of 186.75 ± 125.1µg/m3. Furthermore, in area around lead smelter industry, the concentration of $PM_{2.5}$ was average of 225.9 ± 214.9µg/m^3. Whereas in Parung Panjang, the ADD was 42.36 ± 39.66 µg/kg.day higher than those in Astana Anyar. It was observed that the lung capacity of boys in Parung Panjang was always lower than the lung capacity in Astana Anyar. This is the same pattern with the lung capacity of girls in ParungPanjang was always lower than the lung capacity of girls in Astana Anyar. This is positive relevant with the concentration of $PM_{2.5}$ dan ADD (average daily dose) of respirable particulate in Parung Panjang District was higher than Astanta Anyar District.

References

[1] Chen and Ching-Hwa 2010 Effect of uniformities of deposition of respirable particles on filters on determining their quartz contents by using the direct on filter X-ray diffraction (DOF XRD) method *Journal of Hazardous Materials* **176** pp 389-394

[2] Dyosi and Sindiswa 2007 Evaluation of Preventive and Control Measures for Lead Exposure in a South African Lead-Acid Battery Recycling Smelter *Journal of Applied Occupational and Environmental Hygiene* **4(10)** pp 762-769

[3] Farfel M R, Lees P S J, Bannon D, Lim B S and Rohde C A 2011 Comparison of Two Cyclone-Based Collection Devices for the Evaluation of Lead-Containing Residential Dusts *Applied Occupational and Environmental Hygiene* **9(3)** 212-217

[4] Jain 2007 Determinants of Forced Expiratory Volume in 1 Second ($FEV_{1.0}$), Forced Vital Capacity (FVC), and $FEV_{1.0}$/FVC in Chronic Spinal Cord Injury *American Parapelgia Society, September 7-9, 2004, Las Vegas, NV.*

[5] Maharani 2014 Analysis of Exposure Effect of Electronic Waste (E-Waste) Treatment on Lung Health of Elementary School Children in Astana Anyar District, Bandung, Bachelor Thesis, Department of Environmental Engineering, Faculty of Civil and Environmental Engineering, ITB, Bandung, Indonesia.

[6] Salami I R *et al.* 2014 Respiratory health risk assessment of children living close to industrial areas in Indonesia *Rev Environ Health* **29(1-2)** 139-142

[7] Slaughter and James C 2010 *Association. Between Lung Function and Exposure to Smoke Among Firefighters at Prescribed Burns*. Journal of Applied Occupational and Environmental Hygiene **1(1)** pp 45-49

[8] Wang and Eugene Y 2011The Relationship Between the Dust Lead Concentration and the Particle Sizes of Household Dusts Collected *Journal of Applied Occupational and Environmental Hygiene* **11(3)** pp 199-206

Modelling in conventional electroporation for model cell with organelles using COMSOL Multiphysics

M Y Sulaeman[1] **and R Widita**[1]

[1] Department of Physics, Faculty of Mathematics and Natural Sciences, Institut Teknologi Bandung, Jalan Ganesa 10, Bandung 40132, Indonesia

E-mail: ysulaeman2010@gmail.com

Abstract. Conventional electroporation is a formation of pores in the membrane cell due to the external electric field applied to the cell. The purpose of creating pores in the cell using conventional electroporation are to increase the effectiveness of chemotherapy (electrochemotherapy) and to kill cancer tissue using irreversible electroporation. Modeling of electroporation phenomenon on a model cell had been done by using software COMSOL Multiphysics 4.3b with the applied external electric field with intensity at 1.1 kV/cm to find transmembrane voltage and pore density. It can be concluded from the results of potential distribution and transmembrane voltage, it show that pores formation only occurs in the membrane cells and it could not penetrate into inside the model cell so there is not pores formation in its organells.

1. Introduction

Conventional electroporation is a formation process of temporary hydrophilic pores on the cell membrane surfaces. Electroporation phenomenon occurs when cells are exposed to external electric field and experiencing an increase of transmembrane voltages to a certain point because of electric field accumulation on cell membrane. The increased transmembrane voltages will trigger formation of pores at lipid double layer and allowing ions and big molecules to enter the cell. Conventional electroporation capabilities to form pores on cell membrane became a rapid-growth research topic, specifically at medicine, genes transportation, chemotherapy treatment, and tumor tissues ablation [1-4].

There are some researches in this area, but yet still a lot of unknown aspects from the electroporation phenomenon need to be investigated. In this research, cell model were constructed with generally electric parameter for electroporation method [2, 4, 5] and numerically interpreted by applying finite-element method supported by COMSOL Multiphysics 4.3b software. Thus, the increasing of transmembrane voltages and the density of pores which formed at cell membrane and organelles inside the cells on the conventional electroporation method can be observed.

2. Methodology

In this research, the cell models will be constructed on extracellular liquid (figure1a), with organelles such as cell nucleus (figure 1b), endoplasmic reticulum (Fig. 1c), and mitochondria (figure 1d). Electric field is exposed from the left-hand side of the model and the electric-potential distribution at cell model determined by equation (1):

$$-\nabla\left(\sigma\nabla\Phi\right)-\frac{\partial}{\partial t}\left(\nabla\varepsilon\left(\nabla\Phi\right)\right)=0 \tag{1}$$

where σ and ε are respectively conductivity and permittivity at each region of the cell model. Derived from eq. (1), another equation which determined transmembrane voltage at cell membrane and organelle membrane can be obtained equation (2):

$$\Psi_{\mathrm{tm}}=\Phi(R+h)-\Phi(R) \tag{2}$$

with R and h are respectively cell or organelle radius and membrane thickness.

Pore formation process at cell or organelle membrane approximated by asymptotic model equation, in equation. (3):

$$\frac{dN(t)}{dt}=\alpha\exp\left(q\left(\Psi_{\mathrm{tm}}(t)/\phi_{\mathrm{ep}}\right)^2\right)\left[1-\frac{N(t)}{N_0}\exp\left(-q\left(\Psi_{\mathrm{tm}}(t)/\phi_{\mathrm{ep}}\right)^2\right)\right] \tag{3}$$

where α, q, Ψ_{tm}, ϕ_{ep} , and N_0 are respectively pore formation rate density, pore formation rate, transmembrane voltage determined by equation (2), membrane threshold voltage, and pore density at equilibrium. To solve equation (1) and (3) COMSOL Multiphysics 4.3b software will be applied. Combining two modules simultaneously – which are electric current to determine potential distribution and transmembrane voltage and weak form PDE to determine pore density. The shape of applied external electric field for conventional electroporation that used in this research can be seen at figure 2.

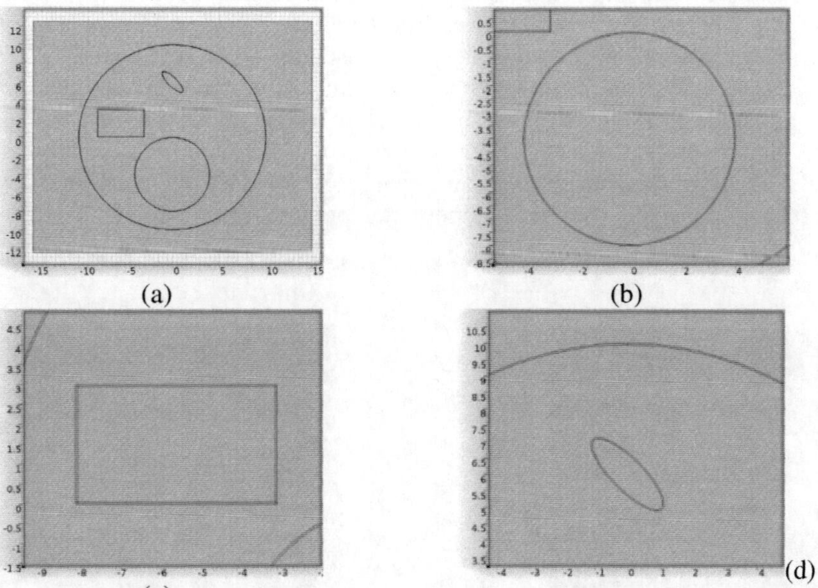

Figure 1. (a) cell model used in this research, (b) cell nucleus, (c) endoplasmic reticulum, and (d) mitochondria.

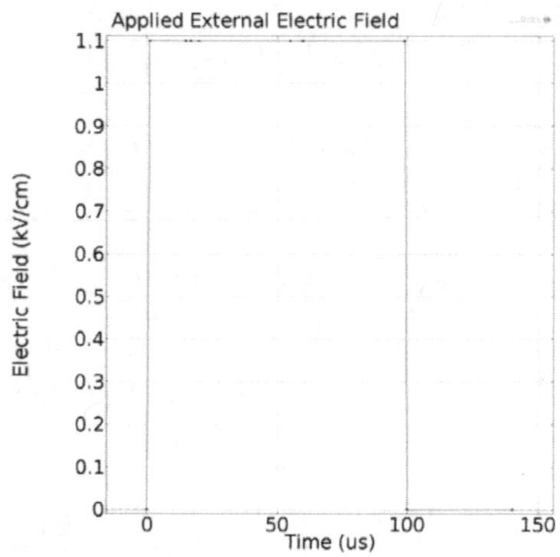

Figure 2. Electric field for this research (pulse width of 100 microseconds, rise and fall time 1 microsecond, and amplitude of 1.1 kV/cm).

3. Results and discussions

Potential distribution in model cell when applied external electric field were given can be seen in figure 3. It could be observed at the beginning of the exposure, t = 1 microsecond (figure. 3a) and at the near end of exposure, t = 89 microseconds (figure. 3b), the electric field were unable to penetrate the cell. Potential distribution values inside the model cell constantly at range 1.5 to 2 volt, even though the applied electric field at maximum intensity (1.1 kV/cm). Thus, on conventional electroporation the membrane of organelle inside the cell did not experienced an increasing of transmembrane voltage. Consequently, electroporation did not occur at nucleus cell membranes, endoplasmic reticulum, and mitochondria. Increasing transmembrane voltage only occurred at cytoplasm membrane (figure. 4a) causing formation of pores only occurred at these membranes (figure. 4b).

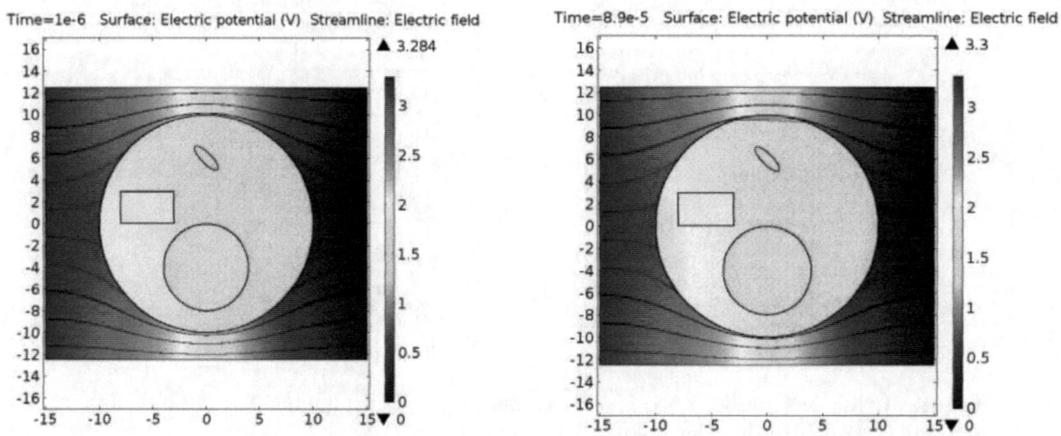

(a) (b)

Figure 3. Electric potential distribution for exposure of applied external electric field at, (a) t = 1 microsecond and (b) t = 89 microseconds.

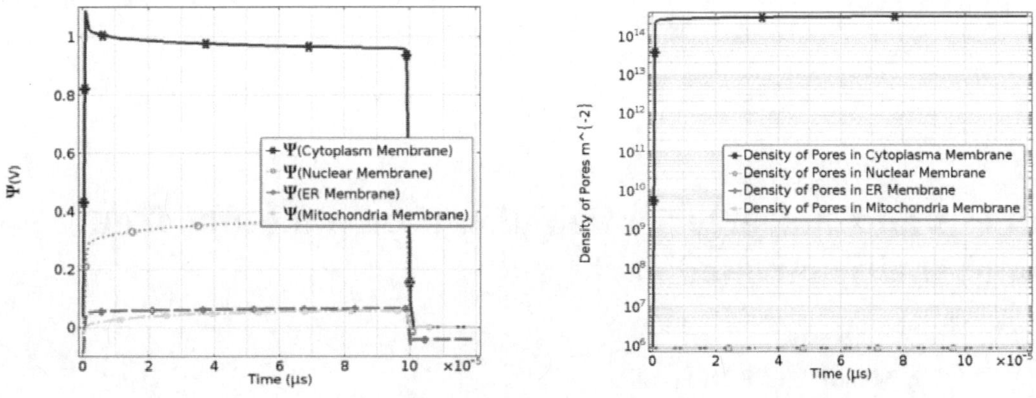

Figure 4. Result for (a) transmembrane voltage and (b) pore density after being exposured with applied external electric field.

4. Conclusions

Using numerical intreperation supported by COMSOL Multiphysics 4,3b to solve the finite-element method for electroporation phenomenom is essential to understanding the electric potential distribution and pores formation in the model cell. The model cell used in this research includes representations of several organelles, such as mitochondria, endoplasmic reticulum, and nuclear cell, so it will be very usefull for study the electroporation phenomenom and for this research, it can be seen in figure 3 and figure 4, exposure of applied external electric field to the model cell for conventional electroporation could only trigger pores formation on cytoplasm membranes. The applied external electric field is not strong enough to penetrate into inside the model cell to create secondary electroporation in the membrane of organelles.

References

[1] Kee S 2011 *Clinical aspects of electroporation* (New York: Springer)

[2] Gowrishankar T R 2006 Microdosimetry for conventional and supra-electroporation in cells with organelles *Biochemical and biophysical research communications* **341(4)** p 1266-1276

[3] Rubinsky B 2010 *Irreversible electroporation*. Series in biomedical engineering (Berlin: Springer) p 312

[4] Salimi E 2011 *Nanosecond pulse electroporation of biological cells: The effect of membrane dielectric relaxation* (Canada: University of Manitoba)

[5] Sulaeman M Y and R Widita 2015 *Modeling in conventional and supra electroporation for model cell with organelles.* AIP Conference Proceedings **1677** p 100003

Synthesis and characterization of carbon nanotube from coconut shells activated carbon

A Melati[1] and E Hidayati[1]

[1] Physics Department, Faculty of Science and Technology, UIN Sunan Kalijaga Yogyakarta, Jl. Maksda Adisucipto No. 1, Yogyakarta, Indonesia

E-mail: asih.melati@gmail.com

Abstract. Carbon nanotubes (CNTs) have been explored in almost every single cancer treatment modality, including drug delivery, lymphatic targeted chemotherapy, photodynamic therapy, and gene therapy. They are considered as one of the most promising nanomaterial with the capability of both detecting the cancerous cells and delivering drugs or small therapeutic molecules to the cells. CNTs have unique physical and chemical properties such as high aspect ratio, ultralight weight, high mechanical strength, high electrical conductivity, and high thermal conductivity. Coconut Shell was researched as active carbon source on 500 – 600^{0}C. These activated carbon was synthesized becomes carbon nanotube and have been proposed as a promising tool for detecting the expression of indicative biological molecules at early stage of cancer. Clinically, biomarkers cancer can be detected by CNT Biosensor. We are using pyrolysis methods combined with CVD process or Wet Chemical Process on 600^{0}C. Our team has successfully obtained high purity, and aligned MWCNT (Multi Wall Nanotube) bundles on synthesis CNT based on coconut shells raw materials. CNTs can be used to cross the mammalian cell membrane by endocytosis or other mechanisms. SEM characterization of these materials have 179 nm bundles on phase 83^0 and their materials compound known by using FTIR characterization.

1. Introduction

The international agency for research on cancer (IARC), the specialized cancer agency of the World Health Organization released new version of IARC's online database. Global burden rises to 14.1 million new cases and 8.2 million cancer deaths in 2012. GLOBOCAN 2012 estimated 14.1 million new cancer cases and 8.2 million cancer-related deaths occurred in 2012, compared with 12.7 million and 7.6 million, respectively, in 2008. Prevalence estimates for 2012 show that there were 32.6 million people (over the age of 15 years) alive who had had a cancer diagnosed in the previous five years. The most commonly diagnosed cancers worldwide were those of the lung (1.8 million, 13.0% of the total), breast (1.7 million, 11.9%), and colorectal (1.4 million, 9.7%). The most common causes of cancer death were cancers of the lung (1.6 million, 19.4% of the total), liver (0.8 million, 9.1%), and stomach (0.7 million, 8.8%) [1]. Carbon nanotubes (CNTs), due to their unique physicochemical properties, have become a popular tool in cancer diagnosis and therapy. They are considered one of the most promising nanomaterial with the capability of both detecting the cancerous cells and delivering drugs or small therapeutic molecules to these cells. Over the last several years, CNTs have been explored in almost every single cancer treatment modality, including drug delivery, lymphatic targeted chemotherapy, thermal therapy, photodynamic therapy, and gene therapy. A holy grail in cancer therapy is to deliver high doses of drug molecules cancer sites for maximum treatment efficacy while minimizing side effects to normal organs. For example, Liu *et al.* compared the efficacy and the toxicity of doxorubicin-loaded CNTs with that of doxorubicin-loaded nanoliposomes [2]. We are making CNT based on activated carbon from coconut shells. Indonesia enriched coconut trees product, contain coconut shell product to. Indonesia, Philippines and Sri Lanka are the main supplier market product of coco-based on activated carbon, Indonesia has 3810 Ha field produced 15.429.000 of coconut [3].

2. Coconut shells activated carbon

Coconut shells made by furnace process at temperature of 500 °C. Most carbonaceous materials do have a certain degree of porosity and internal surface area in the range of 10-15 m^2 / gram (m^2/g). Activation of carbon controlled oxidation of carbon atoms is done by steam at high temperature, enhances, the internal surface of the carbon to 1000-1200 m^2/g. The activation process generates a network of minute openings (pore or cavities) of different diameters on the carbon surface which become the path for water to access the extended internal surface created by the activation. The pore diameters are usually grouped: Micro-pores (< 4 nm), Meso-pores (4-500 nm) and Macro-pores > 500 nm (typically 500-2000 nm). The behavior of the bond structure and dynamic interactions between graphene layers produced strong electrical conductivity properties which functions as a lubricant [5]. The heating process by using various methods has conducted to accelerate the adsorption of electron [6]. The mechanism can be as the simple as removal of surface impurities, however it may also involve the removal of oxide and even the structural rearrangement [7]. The using temperature from 1000 °C to 1720 °C results in the increasing of ratio of graphite layer in the crystallites from 4.5 to 33 and their diameter growing bigger from 1.8 to 6.3 nm. Composite carbon can be obtained by carbonizing coconut shell, and polyvinyl alcohol (PVA) in a thermal decomposition process [8-9]. The main composition of coconut shell consists of cellulose, lignite, and hemicelluloses with a formula C, O, H, and N. These organic materials contain functional groups such as hydroxyl (R-OH), alkanes (R-(CH2) nR'), carboxyl (R-COOH), carbonyl (R-CO-R'), ester (R-COOR'), linear and cyclic clusters ether (ROR') with a variation of carbon [10].

3. Physicochemical properties and functionalization of CNT

Carbon nanotubes are a huge cylindrical large molecules consisting of a hexagonal arrangement of sp^2 hybridized carbon atoms (C-C distance is about 1.4 Å). Carbon nanotubes are a huge cylindrical large molecules consisting of a hexagonal arrangement of sp2 hybridized carbon atoms (C-C distance is about 1.4 Å). The wall of CNTs is single or multiple layers of graphene sheets. The wall formed by rolling up of single sheet are called single-walled carbon nanotubes (SWNTs) and formed by rolling up of more than one sheets are called multi-walled CNTs (MWNTs). Both SWNTs and MWNTs are capped at both ends of the tubes in a hemispherical arrangement of carbon networks called fullerenes warped up by the graphene sheet. SWCNTs have a better defined wall, whereas MWCNTs are more likely to have structural defects, resulting in a less stable nanostructure, yet they continue to be featured in many publications due to ease of processing.

4. Methods

In conducting this research, we are using pyrolysis methods and CVD or Wet Chemical Process. The coconut shells were cleaned with deionized water and dried at 110 °C for 48 h to reduce the moisture content. The dried samples were then crushed and sieved to a size range of 1–2 mm. Subsequently, coconut shells were carbonized gas up to the temperature of 500 °C and held for 2 h. After carbonization, these samples were mixed with ferrocene and benzene in a stainless steel beaker with the weight ratio of benzene/sample equal to 1:2 (CSC-A) and this mixtures were pyrolysis process at 700^0C with Ar gaseous for 4 h, and The products were cooled to room temperature and washed with HNO3 65% and deionized water until the pH of the washing solution reached 6–7. Aquadest was used

Journal of Physics: Conference Series **694** (2016) 012073 doi:10.1088/1742-6596/694/1/012073

to purify composite. We accomplished the characterization of these materials using SEM, EDX and FTIR analysis.

5. Results and discussion

Our materials characterization CNT using SEM, EDX, and FTIR analysis. Our Research has successfully obtained high purity and aligned MWCNT (Multiwall Nanotube).

5.1. SEM (scanning electron microscopy)

SEM is one of the most powerful tools in nanotechnology and plays an important role in the research of nanowires and CNTs. Figure 1 and 2 shows SEM micrographs taken at consecutive time intervals during the 15 KV heat treatment of the MWCNTs. Figure 1 shows the MWCNTs as dispersed. The nanotubes follow the morphology of the substrate closely because of the van der Waals attraction between the nanotubes and the surface. The characterization of these materials have 222 nm (θ=30°), 179 nm (θ=83°), 384 nm (θ=62°), 670 nm (θ=5°) and 187 nm (θ=72°), we found graphene MWNT bundles on 700°C. These size are nanometer so it can absorb in mammals cells.

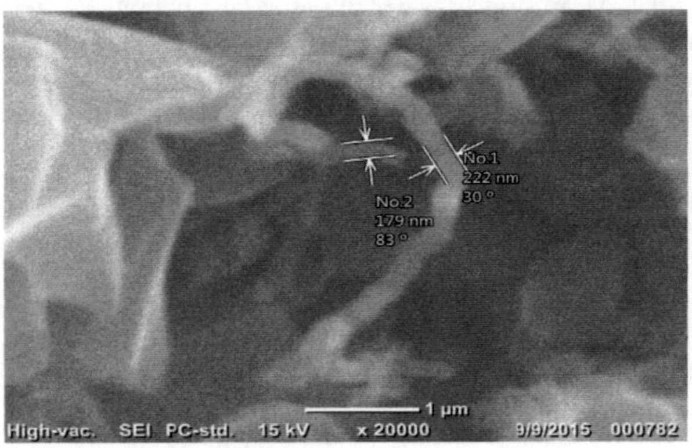

Figure 1. SEM CNT (another side).

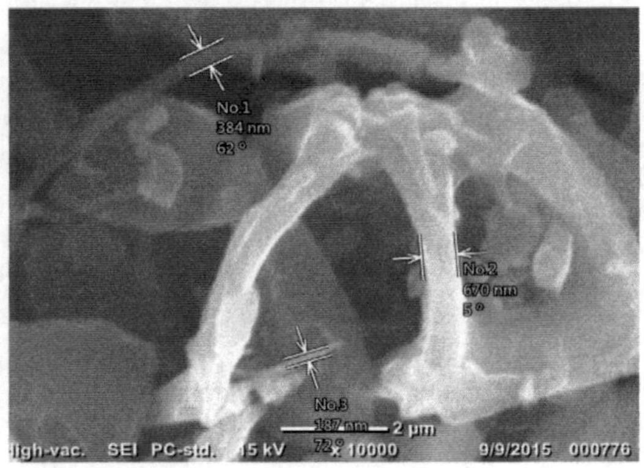

Figure 2. SEM CNT (another side).

5.2. FESEM (field emission scanning electron micrograph) and EDX analysis

Figure 3 show. FESEM image of the graphene exfoliated at 700°C in hydrogen atmosphere clearly reveals the formation of ~30 nm thick graphene flakes whereas graphene form. The atomic elements present in the as gown and purified samples has been characterized by Energy dispersive analysis of X-rays (EDX). Figure 4 shows the EDX spectra of graphite oxide which confirms the presence of carbon and oxygen observed at 0.265 eV (K_α) and 0.514 eV (K_α). The weight percentage of the oxygen in graphite oxide is ~33% due to the strong oxidation of the graphite structure by the oxidizing agents.

5.3. FTIR spectroscopy

FTIR spectroscopy present identify organic functional groups on a CNT's surface by measuring characteristic vibrational modes. However, IR does not provide a quantitative measure of functional group concentrations, and peaks are often hard to distinguish from background features. FTIR used to detect the carboxylic groups and hydroxylic groups where attached on CNT surface. Figure 5 present absorbs wavenumber in 3413, 2380,82 but in this case we have not found effectiveness of this characterization because this CNT have not absorb significantly the source of optical light.

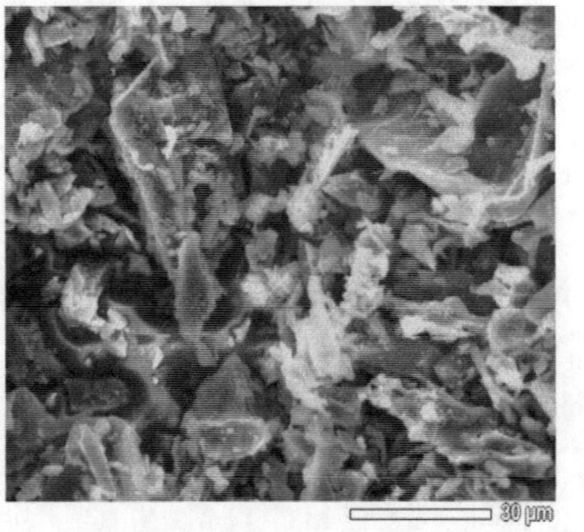

Figure 3. FESEM CNTs.

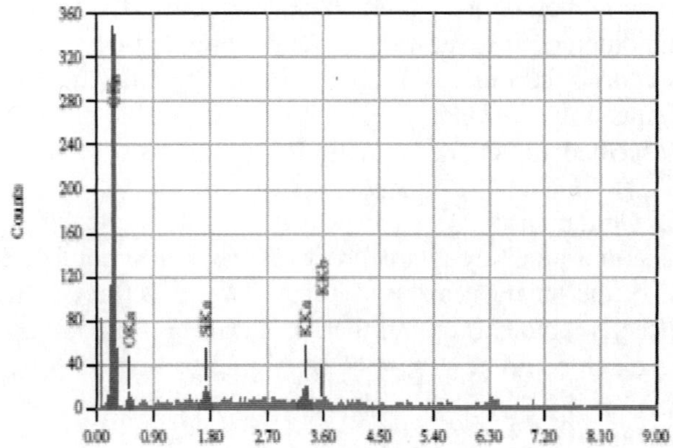

Figure 4. EDX analysis.

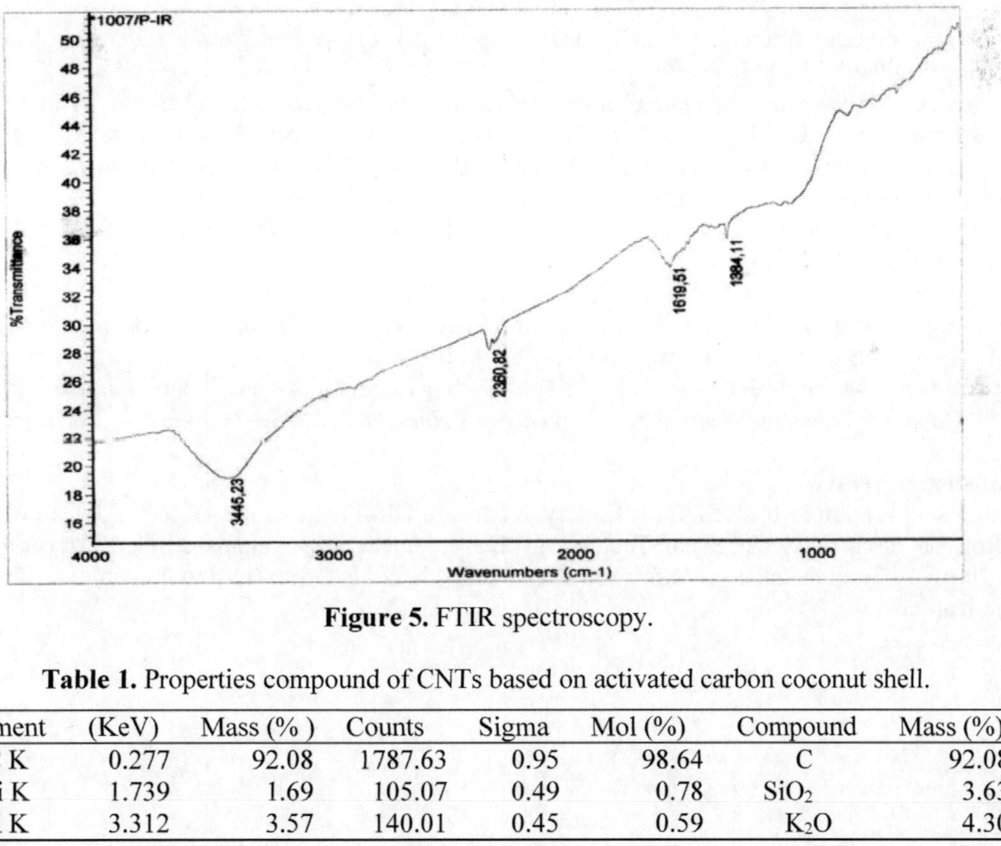

Figure 5. FTIR spectroscopy.

Table 1. Properties compound of CNTs based on activated carbon coconut shell.

Element	(KeV)	Mass (%)	Counts	Sigma	Mol (%)	Compound	Mass (%)
C K	0.277	92.08	1787.63	0.95	98.64	C	92.08
Si K	1.739	1.69	105.07	0.49	0.78	SiO_2	3.63
K K	3.312	3.57	140.01	0.45	0.59	K_2O	4.30

Table 1 present this CNT truly contains Carbon 92,08% which is 0,277 KeV and another material compound are Silika and Kalsium. MWCNTs have strong tendency to bundle together in ropes as a consequence of attractive van der Waals forces. Bundles contain many nanotubes and can be considerably longer and wider than the original ones from which they are formed. CNTs exist in different forms depending upon the orientation of hexagons in the graphene sheet and possess a very high aspect ratio and large surface areas. The available surface area is dependent upon the length, diameter, and degree of bundling. Theoretically, discrete SWCNTs have special surface areas of approximately 1300 m^2/g, whereas MWCNTs generally have special surface areas of a few hundred square meters per gram. The markedly CNTs have various lengths from several hundreds of nanometers to several micrometers and can be shortened chemically or physically for their suitability for drug carriers. To obtain desirable dispersion: (1) surfactant-assisted dispersion, (2) solvent dispersion, (3) functionalization of side walls, and (4) biomolecular dispersion [11]. Among the above described approaches, functionalization has been the most effective approach.

6. Conclusion

Coconut shells recommended for carbon nanotubes MWNT for graphene at 700°C, unhomogenous size, the smallest size we have 179 nm. Although it is need smaller size until 0,34 nm to becomes effective biomarker. Furthermore, Carbon Nanotubes based on Coconut Shell for cancer diagnosis research could be supported by higher technology and medical instruments.

References

[1] IARC press release, June 2013
[2] Liu Z, Fan A C, Rakhra K, Sherlock S, Goodwin A, Chen X, Yang Q, Felsher D W and Dai H 2006 Supramolecular stacking of doxorubicin on carbon nanotubes for in vivo cancer therapy. *Angew Chem Int Ed Engl* 2009 **48(41)** 7668-72
[3] Romulo N and Arancon J R 2013 *Market and trade of coconut Product in Asian and Pacific* (Thailand: Thailand Press)
[4] Inagaki M 2000 *New Carbon: Control of Structure and Functions* (Amsterdam: Elsevier)
[5] Hirose T, Zhao B, Okabe T and Yoshimura M 2002 *J. Mater. Sci.* **37(16)** 3453–58
[6] Sahajwalla V A S, Mehta A S and Khanna R 2004 *Metall. Mater. Trans.* **B35(1)** 75–83
[7] Elsayed M A, Hall P J and Heslop M J 2007 *Adsorption* **13(3-4)** 299–306
[8] Seok-Jin S, Jung-Sik C, Ko-Yeon C, Sun-Day S, Savithri V and Tae-Hwan K 2005 *Korean J. Chem. Eng.* **22(2)** 291–297
[9] Tae-Hwan K, Savithri V, Seok-Jin S and Dong K J 2002 *J. Porous Mater.* **9(4)** 279–286
[10] Van der Marrel H W and Beutelspacher M H 1976 *Atlas of Infrared Spectroscopy of Clay Minerals and Their Admixtures* (New York: Elsevier)
[11] Foldvari M and Bagonluri M 2008 Carbon nanotubes as functional excipients for nanomedicines: I. Pharmaceutical properties *Nanomedicine* **3** 173-182

Acknowledgements

This work is supported by Religion Ministry Indonesia DIKTIS (Islamic Higher Education) for the Funding Research on Science and Technology Development UIN Sunan Kalijaga Decision Rector UIN Sunan Kalijaga Number: 171.53 years 2015 and Materials Group Discussion study club of UIN Sunan Kalijaga.

Dosimetric impact of interplay effect in lung IMRT and VMAT treatment using in-house dynamic thorax phantom

S Julia[1], Nurlely[1] and D S Soejoko[1]

[1] Department of Physics, Faculty of Mathematics and Natural Sciences, Universitas Indonesia, Depok 16424, Indonesia

E-mail: s.julia2805@gmail.com

Abstract. The objective of this research was to determining of protein from connective tissue (protein) in the inner layer of organic and broiler chicken gizzard. The presence of collagens was observed by submersion with 0,1 M NaOH base pH 8 and without submersion. The physiochemical properties and characteristic of collagen from inner layer chicken gizzard such as FTIR spectra, element content, morphology and SDS-PAGE pattern were evaluated. FTIR spectrum in the sample without submersion shown collagen components had wave number that are in the range of wave number (3289-1078 cm^{-1}) bovine collagen. While, FTIR spectrum in the sample with submersion exhibited removal of molecular functional groups and decreasing intensity of infrared absorption. The element content and morphology of two samples indicated that the elements found in the samples were recognized as the elements generally found in proteins and had little structure at SEM micrograph. SDS-PAGE pattern demonstrated that collagen from inner layer chicken gizzard had >120 kDa on molecular unit weight indicating that type 1 collagen is a major component of inner layer chicken gizzard collagen. This study demonstrated the existence of collagen from exploited bio-waste to be an alternative possible source for collagen production.

1. Introduction

Collagen is a material which has a ranged strength and fibrous shape structure. Collagen is the most abundant animal protein with the amount about 30% of total protein in the body [1,2,3]. Nearly, a third of the protein in the body of vertebrates are collagen [2]. Collagen presents in all organ showing strength and stiffness. Collagen is a major fibrous component in bones, teeth, cartilage, dermis and tendons. This structure consists of collagen fibril that looked like transverse lines, which are organized by biological functions of a cell. Collagen fibrils consist of repetitive polypeptide subunit with a triple helix shaped and are called tropocollagen. In one piece of collagen, chains consist of more than 1000 amino acid residue [4].

Collagen has many gene types which involved in the biological function of a cell such as survival and differentiation [4]. Each type of collagen has different gene types which shown in the composition of the polypeptide chain. Due to its unique characteristic, collagen has been used in various fields such as medical and biotechnology [1, 2]. Collagen has been used as a raw material of cosmetics and pharmaceutical industries. Non-denatured collagens can be used as the component of food, cosmetics, biomedical, and pharmaceutical substances. Denatured-collagen known as gelatin is commonly used in food and many biomedical substances. Biomedical and pharmaceutical substances consisting of collagen can be used in various treatment such as hypertension, urinary incontinence and pain associated with osteoarthritis and also in tissue engineering for implants in humans [5].

Recently, most commercial collagens come from skin and bone from bovine and porcine [1, 2, 3, 5]. Due to the outbreak of bovine spongiform encephalopathy (BSE) and foot and mouth disease (FMD) [1, 2, 3, 6, 7, 8], so that the use of cow collagen has a risk of contracting this disease. Scientists have found that skin, bone, fin and scales of freshwater and marine fishes, chicken skin, marine sponge, and bullfrog skin can be used as alternative collagen source, which have no ethnic or safety-related consumer concern [1].

A greater amount of chicken consumption and a large number of chicken cuts produce wastes. Although, there is an attempt to decrease the waste, the quantity of the waste produced is increasing annually [5]. Facing of the above problems, recently, there has been much interest in investigating the most effective use of under-utilized resources wastes. Based on the results of research conducted by Kiruthiga *et al.* proved that the waste from the broiler chicken can be beneficial as a source of collagen production [6]. Additionally, Bonifer *et al.* also said that there are about 3% of collagen in the chicken skin that may provide the potential for increasing the value of products [9]. In this study, an alternative collagen production has been explored from the inner layer of chicken gizzard. The result indicated that chicken gizzard collagen was almost similar with bovine collagen.

2. Methodology

The chemicals that were used in this study were NaOH base 0,1 M pH 8, tris buffer HCl 0,5 M pH 6,8 and 200 mg KBr pellets. All chemicals used in this study were analytical grade and are from Merck-Germany. The samples were the inner layer of organic and broiler chicken gizzard and obtained from the traditional market.

2.1. Sample treatment

Sample from organic chicken (sample I) and those from broiler chicken (sample II) are divided into four groups, sample I without submersion (sample I.1) and with submersion (sample I.2), sample II without submersion (sample II.1) and with submersion (sample II.2).

All samples were initially cleaned and heated at the low temperature at about 60^0C. Sample I.2 and II.2 were soaked in NaOH base 0,1 M pH 8 for 16 hours and changed the solvent twice a day. The resultant solvent was washed with aqua dest for 8 hours until sample in the neutral condition. All of the samples (sample I.1, I.2, II.1 and II.2) were ground to be a powder with 50 mesh and 120 mesh particle size. Samples with 120 mesh particle size were further called sample I.1a and sample II.1a and the others with 50 mesh were called sample I.2b and II.2b.

2.2. Fourier transform infrared (FTIR) spectroscopy

Molecular functional group and its modification in the sample were observed with ABB FTIR MB3000 in Material Analysis Laboratory, Physics Department – FMIPA IPB. For infrared spectroscopy, 2 mg of samples and 100 mg KBr were pressed into a pellet. Measurements were performed in the range of wavenumber 4000 – 400 cm^{-1}.

2.3. Scanning electron microscopy (SEM) and energy dispersive X-ray (EDX)

Surface morphology was studied by SEM equipped with EDX (SEM-EDX EVO 50 ZEISS) in Puslitbang, Forest Departement, Gunung Batu - Bogor. SEM was used to examine the morphology and the microstructure of the collagen in four types of samples, and EDX identified the elements. The magnifications were 100x, 250x, 500x and 1000x.

2.4. Sodium dodecyl sulphate-polyacrylamide gel electrophoresis (SDS-PAGE)

The examination of the sample unit weight protein was performed by SDS-PAGE ATTO Electrophoresis Page Run R WSE-110 in LIPI Biotechnology-Biological Research Centre. The band was observed using polyacrylamide gel. Gel consist of 3.9% stacking gel and 8% separating gel. For this examination 0.25 gram of sample was dissolved in 4 ml of 0.1 M NaOH base and kept in -10^{0}C temperature. The sample was inserted into the plate wells. The plate containing the samples was connected to the cathode and the anode which was connected to the Tris buffer HCl. This was observed for 120 minutes and 300 mA electrical current. Electrophoresis results can be visualized by staining gel. This staining using a staining solution with 0.1% Coomassie Brilliant Blue R-250.

3. Results and discussion

Sample with submersion (sample I.2 and II.2) were clearer and thinner than sample without submersion (sample I.1 and II.1). The submerged sample had a larger size may be because of increasing fibre size. FTIR spectra of samples in Figure 1 exhibited the characteristic peaks of amides A, B, I, II and III (amide A at wavenumber 3367 cm^{-1}, amide B at wavenumber 2924 and 2854 cm^{-1}, amide I at wavenumber in the range of 1600-1700 cm^{-1}, amide II at wavenumber 1529 cm^{-1} and amide III at wavenumber 1232 and 1234 cm^{-1}). It was found that IR spectra of the sample I.1 and II.1 had similar wavenumber peaks that are in the range of wave number bovine collagen [8]. The absorption characteristics of amide A commonly associated OH-bond in hydroxyproline. Amide B peaks, representing the asymmetrical stretch of CH_2, for sample I.1 and II.1, were in the wave number of 2924 and 2854 cm^{-1}, whereas bovine collagen was at 2920 and 2853 cm^{-1} [8], and fish collagen was at the wavenumber of 2926 and 2928 cm^{-1} [2]. Amide I is mainly associated with CO-stretch vibration in the chain peptide group that are owned by amino acids in collagen. Amide I that attributed groups with C- or O- stretching vibration along the polypeptide backbone, is a sensitive marker of the peptide secondary structure [2]. Amide II is associated N-H bending vibration and C-N stretching vibration. The amide III are in the range of wave number from cow collagen (1260 cm^{-1}) [2] and fish collagen (1242 and 1244 cm^{-1}) [8]. This is associated with NH bending corresponded to glycine residue.

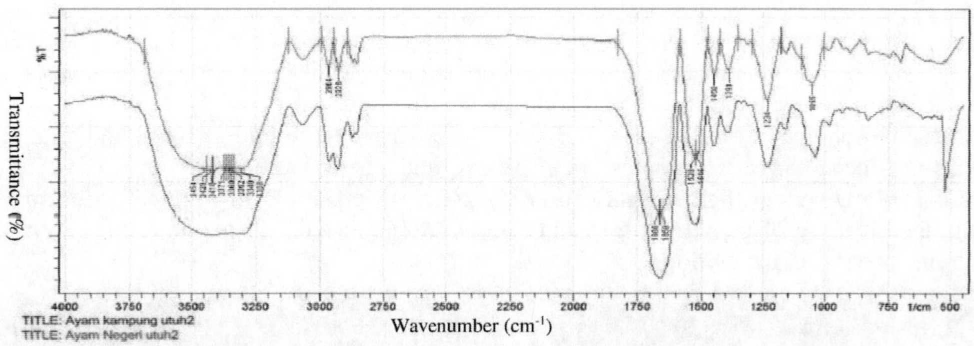

Figure 1. FTIR for sample without submersion (sample I.1 and II.1).

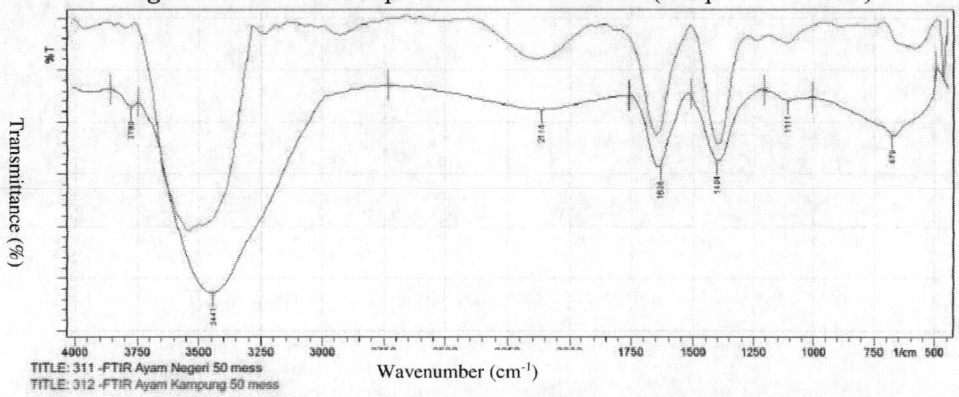

Figure 2. FTIR for sample with submersion (sample I.2 and II.2).

FTIR for the sample with submersion (sample I.2 and II.2) in figure 2 exhibited removal of molecular functional groups and decreasing intensity of infrared absorption. Both of samples has lost amide A. Moreover, amide B for sample II.2 did not appear but still occur on sample II.1 at the wavenumber of 2931 cm^{-1}. Differ with the sample without submersion, this sample shown amide II peaks at different wave number (1404 cm^{-1}). This wave number is similar to the amide II from fish collagen at the wavenumber of 1403 cm^{-1} that originated from COO- stretching symmetry. While, amide III only existed in sample II.2 with skeletal stretching. Amide lost from these samples was predicted due to the submersion in 0,1 M NaOH base and temperature treatment. Many studies that avoid amide lost were conducted at 4^0C in treatment process [1, 2, 5, 7, 8, 10, 11].

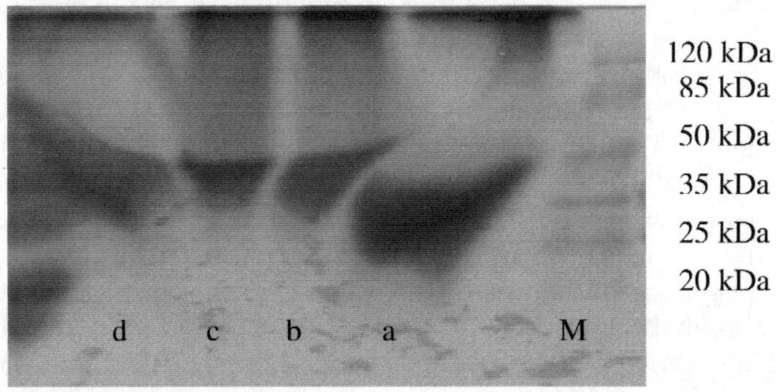

Figure 3. SDS-PAGE of the sample without and with submersion. M is high molecular weight marker, a is sample I.2, b is sample I.1, c is sample II.2 and d is sample I.2.

The SDS-PAGE pattern in figure 3 shown that samples had molecular weight more than 120 kDa. Based on P Sivakumar's study that type I collagen had large molecular weight with α1 chain [10, 11]. Therefore, collagen from samples had type I collagen.

The surface morphology of the samples in figure 4 was observed with low (100x) and high (1000x) magnification. Morphology of samples were fibres and there were pores in 250x and 500x magnification. Moreover, in high magnification shown was performed in Figure 5 had small fiber structure in the presence of granules which indicate as the impurities. These impurities were related with the drying process in the outdoors.

Figure 4. SEM of sample I.1a and II.1 a with 100x magnification.

Figure 5. SEM of sample I.1a and II.1a with 100x magnification.

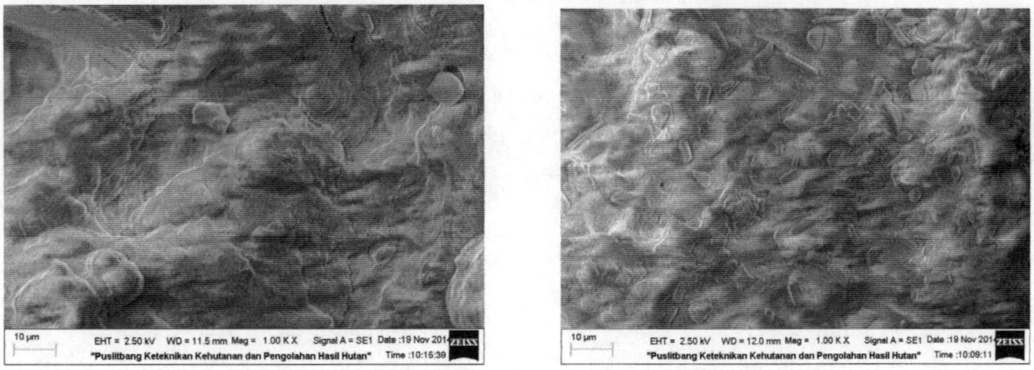

Figure 6. SEM of sample I.2b and II.2b with 1000x magnification.

4. Conclusions

Collagen samples isolated from the inner layer of organic and broiler chicken gizzards were successfully observed and there existed no significant difference between the contents of the two sample groups. The characteristic of collagen samples was also similar with bovine collagen from FTIR analysis. However, samples submerged in NaOH base had lost functional amide group shown in the range of 3300-3400 cm^{-1} of wave number. The existence of both samples is evidence that major collagen from inner layer chicken gizzard is type I collagen. The result of the present study reveals the existence of collagen and it may be concluded that the inner layer of chicken gizzard as waste may become an additional and alternative possible source for alternative sources of collagen production.

References

[1] Choi J H, Benham S H and Kim S M 2013 *J. Agr. Sci,Tech.* **15** pp 293-302
[2] Prabjeet S, Soottawat B, Maqsood S and Kishimura H 2011 *Elsevier Food Chemistry* **124** pp 97-105
[3] Liu D C, Lin Y K and Chen M T 2001 *Asian-Australian Journal* **14** pp 1638-1644
[4] Lehninger 1982 *Dasar-Dasar Biokimia Terjemahan* (Jakarta: Erlangga)
[5] Veeruraj A, Arumugam M and Balasubramanian T 2013 *Elsevier Process Biochemistry*. **48** pp 1592-1602
[6] Kiruthiga N, Sujithra S and Kumaresan R 2013 *International Journal of Institutional Pharmacy and Life Science* **3** pp 53-60
[7] Takeshi N, Nobutaka S, Yasuhiro T and Norihisa K 2012 *Food and Nutrition Sciences* **3** pp 72-79

[8] Vairamani S, Pasiyappazham R, Namasivayam S, Sadhasivan S, Palaniappan S, Meivelu M, Jayalakshmi K, Annaian S and Alagiri S 2012 *African Journal of Biotechnology* **11** pp 14326-14337
[9] Cliche S, Amiot J, Avezard C and Gariepy C 2003 *Poultry Science* **82** pp 503-509
[10] Hashim P, Mohd Ridwan M S and Bakar J 2014 *International Journal of Agricultural, Biosystems Sciences and Engineering* **8** pp 50-54
[11] Peck L K, Mashitah M D 2013 *Focussing on Modern Food Industry (FMFI)* **2** pp 123-128

Journal of Physics: Conference Series **694** (2016) 012075 doi:10.1088/1742-6596/694/1/012075

Study on the effect of nanoparticle bimetallic coreshell Au-Ag for sensitivity enhancement of biosensor based on surface plasmon resonance

Widayanti[1,2] and Abraha K[2]

[1] Department of Physics, Faculty of Sciences and Technology, UIN Sunan Kalijaga Yogyakarta, Indonesia

[2] Department of Physics, Faculty of Mathematics and Natural Sciences, Universitas Gadjah Mada Yogyakarta, Indonesia

E-mail: salma_wida@yahoo.com

Abstract. Bimetallic Au-Ag core-shell, a type of composite spherical nanoparticle consisting of a spherical Au core covered by Ag shell, have been used as active material for biomolecular analyte detection based on surface plasmon resonance (SPR) spectroscopy. SPR technology evolved into a key technology for characterization of biomolecular interaction. In this paper, we want to show the influence of nanoparticle bimettalic Au-Ag coreshell for optic respon of LSPR biosensor through attenuated total reflection (ATR) spectrum. The method consist of several steps begin from make a model LSPR system with Kretschmann configuration, dielectric function determination of composite bimetallic coreshell nanoparticle using effective medium theory approximation and the last is reflectivity calculation for size variation of core and shell bimetallic nanoparticle. Our result show that, by varying the radius of core and shell thickness, the peak of the reflectivity (ATR spectrum) shifted to the different angle of incident light and the addition of coreshell in SPR biosensor leads to enhancement the sensitivity.

1. Introduction

Recently, sensor based on localized surface plasmon resonance (LSPR) have attracted considerable attention because of its revolutionary technology [1]. The advantages brought about by curret LSPR technology include label free analysis, monitors the interaction real time, non-destructive technique, large tuneability from the visible into the infrared (IR) spectrum region, high sensitivity and selectivity. LSPR is a kind of electromagnetic resonances that exist when there is an interface between metal and dielectric material and this system have been use for sensing various biomolecules [2, 3]. When the biomolecules comes in contact to the metal film surface, that biomolecules get adsorbed on its surface and increasing the refractive index resulting in the change in the angle resonance. This LSPR properties are very sensitive to size, shape and the surrounding medium refractive index of metal nanoparticle or the medium that kept contact with metal. Tuning of the LSPR spectrum is the key issue in application such as in cancer treatment, bio medical diagnosis, enzyme detection, food safety and cellular imaging [4]. The Au-Ag bimetallic coreshell nanoparticle is one of the metal that have attracted extensive interest due to the addition of a second metal shell, changing the surface plasmon band, and enhancing the stability [5, 6]. The LSPR of Ag nanoparticle is located around 400 nm, whereas the LSPR of Au nanoparticle is located around 520 nm. The combination of that two

Journal of Physics: Conference Series **694** (2016) 012075 doi:10.1088/1742-6596/694/1/012075

metal is similar to combine the plasmon feature of each metal become the new plasmon feature that depend on the size and shape of core and shell that coreshell. The performance of the biosensor base on the LSPR depends on the adsorption of the biomolecule and enhancement of the field intensity of the excitation light. It has been reported that the addition other medium over metal film improves the sensitivity of the biosensor [5]. In this latter, we have been investigate the ATR spectrum of four and five mutilayer biosensor based on LSPR with Au-Ag coreshell addition and DNA as sensing analyte.

2. Biosensor System base on LSPR

2.1. *Kretschmann configuration with 5 layers*

In our study, we use the theory and computational approximation to determinate effective permittivity of composite AuAg coreshell and calculate reflectivity in attenuated total reflection (ATR) spectrum. The LSPR system using the Kretschmann configuration [7] with 5 layers, prism/Ag/coreshell/DNA/DNA+water, is shown in figure 1. θ_i and θ_r are the incident and the reflection angle , k_z is the wave vector component at *z* axis, and *d* is the thickness of the each layer.

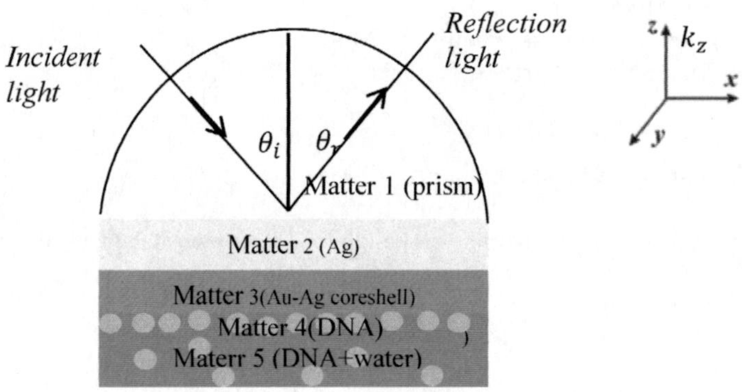

Figure1. Biosensor base on LSPR with Au-Ag coreshell

In this simulation have been carried out for wavelength 628,5 nm and the refractive indices of the SF10 glass prism used is 1,7230. The refractive indices of silver is *n*=0,13455 and extinction *k*=3,98651, the refractive indices of the gold is *n*=0,19715; *k*=3,0899 [8] , for DNA is *n*=1,335; *k*=0 and the DNA+water is *n*=1,330,; *k*=0 [9]. The thickness of Ag film is *d*=40 nm, and for DNA *d*=30 nm Finally, the reflectivity R is given by by Fresnel equation [10]

$$R = |r_{ijk}|^2 = \left| \frac{r_{ij}+r_{jk}e^{2ik_{jz}d_j}}{1+r_{ij}r_{jk}e^{2ik_{jz}d_j}} \right|^2 \tag{1}$$

with

$$r_{ij} = \frac{k_i\varepsilon_j - k_j\varepsilon_i}{k_i\varepsilon_j + k_j\varepsilon_i} \tag{2}$$

r_{ij} is the surface reflectivity between medium *i* and medium *j*. k_{ij} is the wave vector component perpendicular to the surface, k_x is the wave vector component parallel to the surface, whereas d_j and ε_i are the thickness *j*-th layer and the *i*-th medium dielectric constant

2.2. *A spherical coreshell Au-Ag*

We choose a spherical coreshell model like figure 2. The bimetallic composite model consists of a metallic Au core of radius a_2 coated by a metallic Ag of thickness $a_1 - a_2$. The dielectric constant of metallic Au, Ag and the embedding medium are ε_2, ε_1 and ε_0 respectively.

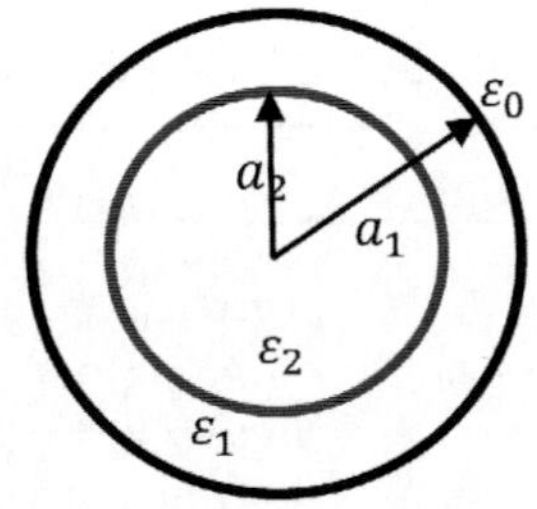

Figure 2. The Au-Ag coreshell model

In this calculation the complex form of ε_1 and ε_2 can be used from Johnson and Cristy literature [8].

2.3 Effective permittivity of coreshell
For calculating the effective permittivity of Au-Ag composite, with neglecting the correlation between the shell and core, can be employ using the Rayleigh mixture formula [11].

$$\frac{\varepsilon_{\text{eff}}-\varepsilon_0}{\varepsilon_{\text{eff}}+2\varepsilon_0} = (f_1 + f_2)\left(\frac{\varepsilon_{\text{incl}}-\varepsilon_0}{\varepsilon_{\text{incl}}+2\varepsilon_0}\right) \tag{3}$$

with

$$\varepsilon_{\text{incl}} = \frac{1-2G}{1+G}\varepsilon_1 \tag{4}$$

where

$$G = \frac{\varepsilon_1-\varepsilon_2}{2\varepsilon_1+\varepsilon_2}\left(\frac{a_2}{a_1}\right)^3 = \left(\frac{\varepsilon_1-\varepsilon_2}{\varepsilon_1+2\varepsilon_2}\right)\left(\frac{f_2}{f_1+f_2}\right) \tag{5}$$

ε_{eff} is the effective permittivity of coreshell, ε_0 is the vacuum permittivity, $\varepsilon_{\text{incl}}$ is the inclusion permittivity, ε_1 is the shell permittivity, ε_2 is the core permittivity, f_1 and f_2 are the volume fraction of the core and shell and $\frac{a_2}{a_1}$ is the rasio of core and shell radius.

2.4 Biosensor sensitivity
The calculation of the sensitivity of biosensor base on the LSPR is written [12]

$$S = \frac{\Delta\theta_{\text{sp}}}{\Delta n} = \frac{\theta_{\text{sp}}(\varepsilon_d+\Delta\varepsilon_d)-\theta_{\text{sp}}(\varepsilon_d)}{\Delta n} \tag{6}$$

Where $\Delta\theta_{sp}$ is the difference of the angle of SPR and Δn is the change in refractive index. While, the enhancement sensitivity value of LSPR biosensor with Au-Ag coreshell compare with the conventional SPR is written [13]

$$\Delta S = \frac{S_{\text{graph}}-S_{\text{kon}}}{S_{\text{kon}}} \times 100\% \tag{7}$$

3. Result and discussion
The changes of the radius of core and the thickness of shell, leads to the change of rasio $\boldsymbol{a_2/a_1}$ of the composite Au-Ag coreshell and then change the effective permittivity of composite. The increasing of effective permittivity of composite Au Ag coreshell, leads to the enhancement of the

sensitivity of this biosensor. Table 1 shows the effective permittivity of composite coreshell Au Ag that can obtained from equation (4) and depends on the rasio of radius core and shell.

Table 1. The effective permittivity of Au-Ag coreshell for the $\boldsymbol{a_2/a_1}$variation

a_2/a_1	$\varepsilon_{\text{incl}}$
0.50	-14.9742 + 1.1041 i
0.90	-11.0608 + 1.1991 i
0.75	-12.9660 + 1.1612 i
1.00	-9.5086 + 1.2183 i
1.20	-5.8433 + 1.2220 i
1.30	-3.8063 + 1.1987 i

The calculated ATR spectra (reflectivity) from biosensor based on LSPR in Kretschmann configuration are shown in figure 3 and figure 4. For the layer consist only the thin film of metal, the peak of reflectivity at the angle 55 deg (*blue line*). When the coreshell composite was deposited on the surface of the metal thin film, the peak of reflectivity shifted to the large angle. From figure 3. for the a_2/a_1 of the composite is fixed at 0.9 and the thicknesses of the composite is variated from 10 nm to 40 nm, the large shift of peak is obtained for thicknesses is 40 nm but the reflectivity is small . The figure shows that at 20 nm is the best performance of the peak. From figure 4, for the thicknesses of the composite is fixed at 20 nm and the a_2/a_1 is variated from 0.5 to the 1.3, the peak shifts from the 55 deg to the 66 deg. The large shiftiness is obtained for the a_2/a_1 is 1,3 but the reflectivity is smallest from the other. Table 1 shows that for a_2/a_1=0.9 resulting the effective permittivity $\varepsilon_{\text{incl}}$ =-11.0608 + 1.1991 i. Therefore at 20 nm thicknesses and ε_{incl} =-11.0608 + 1.1991 i, the shiftiness of angle is from 55 deg (*blue line*) to 57 deg (*red line*), whereas the change the permittivity is from *n*=0.13455; *k*=3.98651 (*blue line*/Ag thin film) to -11.0608 + 1.1991 i (*red line*) or *n*=0.180012 ; *k*=3.330616 . Therefore from equation (7) , calculation of the sensitivity resulting the sensitivity increase 3.636% compare with sensitivity of SPR biosensor without Au-Ag coreshell addition.

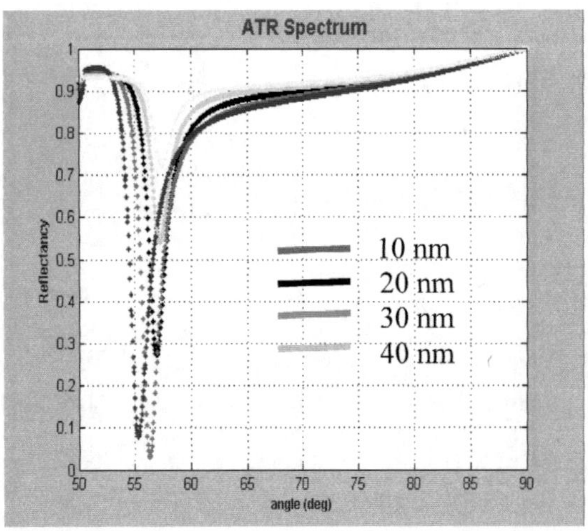

Figure 3. The reflectivity of ATR spectrum for the a_2/a_1 of the composite is fix at 0.9 and the thicknesses is variated from 10 nm to the 40 nm.

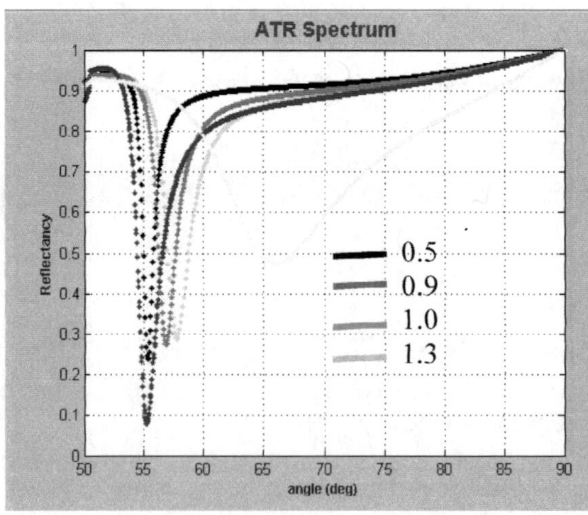

Figure 4. The reflectivity of ATR spectrum for the thicknesses of the composite is fixed at 20 nm and the a_2/a_1 is variated from 0.5 to the 1.3.

4. Conclusion

The enhancement of sensitivity of biosensor based on LSPR have been studied via the addition nanoparticle Au-Ag coreshell in LSPR 5 layer system, By variated the radius of core and shell, the refractive index of coreshell changes and leads to change the position of the peak of reflectivity spectrum to the large angle. The LSPR dips were shifted when Au-Ag coreshell were deposited on the surface of metal film (Ag). A large shift in the dip angle by that deposition suggest the potential for application in highly sensitive biosensor, in this case sensing biomolecules as analyte. The sensitivity enhancement of biosensor can be obtained from the rasio of the difference of the angle of the SPR peak with the change in refractive index.

References

[1] Kim D 2010 *Optical Guided-wave Chemical and Biosensor I* (Berlin: Springer-Verlag)

[2] Stuart D A, Haes A J, Yonson C R, Hicks E M and Van Duyne R P 2005 Biological application of localised surface plasmonic phenomenae *IEE Proc-Nanobiotechnol* **152(1)**

[3] Jain P K, Huang X, El-Sayed I and El-Sayed M 2007 Review of some interesting surface plasmon resonance-enhanced properties of noble metal nanoparticles and their application to biosystem *Plasmonics* **2** pp 107-118

[4] Ligler F S, Taitt C R, Shriver-Lake L C, Sapsford K E, ShubinY and Golden J P 2003 Array biosensor for detection of toxins *Anal. Bioanal. Chem.* **377** 469– 477

[5] Zhu J 2009 Surface Plasmon Resonance from Bimetallic Interface in Au–Ag Core–Shell Structure Nanowires *Nanoscale Res Lett* **4** 977–981

[6] Som T and Karmakar. B 2009 Coreshell Au-Ag Nanoparticle in Dielectric Nanocomposite with Plasmon-Enhanced Fluoresence: A New Paradigm in Antimony Glasses *Nano Res* **2** 607-616

[7] Kretschmann E and Raether H 1968 Radiative decay of non-radiative surface plasmons excited by light *Zeitschrift Naturforsh* **23** A2135-2136

[8] Johnson P B and Christy R W 1972 Optical constants of the noble metals *Phys Rev* B **6** No 12

[9] Wu L, Chu H S, Koh W S and Li E P 2010 Highly sensitive graphene biosensors based on surface plasmon resonance *Optics Express* **18(14)** 14395

[10] Rather H 1986 *Surface Plasmons on Smooth and Rough Surfaces and on Gratings* (Berlin: Springer-Verlag)

[11] Chen L F, Ong C K and Tan B T G 1998 Effective permittivity of layered dielectric *Journal of Material Science* **33** 5891–94.

[12] Verma R, Gupta B D and Jha R 2011 Sensitivity enhancement of a surface plasmon resonance based biomolecules sensor using graphene and silicon layers *Sensor and Actuators B: Chemical* **160** 623-631

[13] Adhib M and Abraha K 2011 Enhancement of sensitivity in Surface Plasmon Resonance Biosensor Using graphene materials: A theoretical prediction P*roceeding of the 3rd International Conferences And Workshop On Basic And Applied Sciences* Bandung. *Preprint*

Journal of Physics: Conference Series **694** (2016) 012076 doi:10.1088/1742-6596/694/1/012076

Nonlinear dynamics of specific DNA-protein interactions

D Dwiputra[1], W Hidayat[1,2], R Khairani[1] and F P Zen[1,2]

[1] Departement of Physics, Faculty of Mathematics and Natural Sciences, Institut Teknologi Bandung Jl. Ganesha 10, Bandung 40132, Indonesia
[2] Indonesian Center of Theoretical and Mathematical Physics (ICTMP), Indonesia

E-mail: donny.dwiputra@s.itb.ac.id

Abstract. Interactions between DNA binding protein and specific base pairs of nucleic acid is critical for biological process. We propose a new model of DNA-protein interactions to depict the dynamics of specific DNA-protein interactions. Hydrogen bonds (H-bonds) are, among the other intermolecular interactions in DNA, the most distinctive in term of specificity of molecular bonds. As H-bonds account for specificity, we only consider the dynamics affected by H-bonds between DNA base pairs and H-bonds connecting protein side chains and DNA. The H-bonds are modelled by Morse potentials and coupling terms in the Hamiltonian of coupled oscillators resembling a coupling between planar DNA chain and a protein molecule. In this paper we give a perturbative approach as an attempt for a soliton solution. The solution is in the form of nonlinear travelling wave having the amplitudes satisfying coupled nonlinear Schrödinger equations and is interpreted as the mediator for nonlocal transmittance of biological information in DNA.

1. Introduction

The dynamics of deoxyribonucleic acid (DNA) and its interaction to proteins is one of the most fascinating conundrum in biophysics because it constitutes the fundamental basis of life. The information transfer between proteins in the process of DNA transcription and replication has to be highly efficient and lossless to maintain the macroscopic life around us over the ages. However, an intact dynamical modelling is difficult due to the complexity of the role of regulatory proteins in DNA such as RNA polymerases. Thus, here we are simplifying the problem by examining only the significant aspects of DNA-protein binding.

The idea of nonlinear solitonic excitations in DNA has established in the past decades [1–3], the notion of the soliton is due to the ability of DNA to maintain a lossless transfer of information along vast distances—which cannot be satisfied with simple linear waves. Of interest here is the breather type of excitation which correspond to DNA denaturation bubbles by Peyrard and Bishop (PB) [2], they developed a statistical approach of the local base pair opening which precede the denaturation process. Later the dynamics and thermal effects of PB breather in presence of external potential are investigated [4, 5], giving the picture on how DNA and external perturbations behave. Regulatory protein would increase the breather amplitude as shown in [6] via nonequilibrium statistical calculations. A further effect of the oscillatory mode of the protein itself is examined on a thermal bath by [7].

Origins of the specificity in DNA-protein recognition rely on the unique chemical signatures of the base sequences and the sequence-independent DNA shapes [8–10]. The proteins can "feel" the DNA surface shapes and simultaneously recognize the sequences in virtue of electrostatic, van der Waals', and H-bond interactions. However, not all of the interactions give significant contribution to the site specificity. The hydrogen bonds role the specific base readout more significantly than the other interactions, since the transition free energy between best specific binding and nonspecific binding is approximately 16 $k_B T$ below the specific binding energy [11] (experimentally, for instance, 17 $k_B T$ for *Mnt* and ≈16 $k_B T$ for *lac* repressor [12]). As a comparison, the nonspecific binding free energy of the repressor protein CI in λ-virus-infected *E. coli* cells (in vivo) was estimated as only 7 $k_B T$ [11].

In this paper, we present a new model governing the dynamics of the site-specific DNA-protein interaction. We assume a simple, untwisted, ladder-like DNA double strand interacting with a regulatory protein through the hydrogen bonding between functional groups of the protein side chains and the bases in the major groove of the DNA chain. Here the DNA chain contains inhomogeneous base sequences, which would allow the specific binding. Any environmental effects such as the presence of the solution are neglected at this stage.

2. The dynamical model

Our aim is to examine the dynamics of the binding between regulatory proteins and DNA through the mathematical solution and its properties in agreement with biological observations [13, 14]. Here we would like to focus our attention on the major correspondent of the specific DNA-protein interaction, namely the H-bonds. These bonds are linking between protein side chains and groups of bases inside DNA chain and are responsible for the binding specificity. Proteins are composed of many amino acids which linked together by peptide bonds, practically the side chains which interact with DNA are the amino acids or the peptide bonds. This interaction occurs mainly in DNA major groove and affecting the H-bonds inside DNA nucleotides.

In this paper we model the effect of existing protein to the DNA chain. We conjecture that the existence of regulatory proteins near DNA trigger pulse type excitations of the DNA base pairs, this kind of excitations referred as DNA *bubbles* and are well described by the PB solitonic breathers. These propagating solitons carry important genetic information through the DNA and provide an effective communication between two separated proteins binding DNA in vast distances. Our model includes two degrees of freedom, x_m corresponds to the stretching of H-bonds linking a protein side chain and a side chain, and y_n is the stretching of H-bonds that connect two DNA bases in a pair. The indices are labelling locations in DNA. We simplify the complex DNA-protein to a coupled nonlinear oscillator. Nonlinearity arises while we consider the Morse potential to model a group of H-bonds, the potential permits breaking of the bonds when largely stretched. The stacking between base pairs is assumed harmonic at this stage. Thus, our model Hamiltonian is

$$H = H_{\text{DNA}} + H_{\text{prot}} + H_{\text{int}} \tag{1}$$

where

$$H_{\text{DNA}} = \sum_n \frac{p_{yn}^2}{2m} + D_{(\mathbf{z})}(\mathrm{e}^{-\alpha y_n} - 1)^2 + \frac{k}{2}(y_n - y_{n-1})^2 . \tag{2}$$

The base nucleotide mass \$m\$ and momentum p_{yn} are homogenous. The Morse potential depth $D_{(\mathbf{z})}$ is the strength of the bonding between bases (the chain is at **Z**-direction) and α is the inverse width of the well, k is the harmonic coupling of the longitudinal springs connecting the base pairs. H_{prot} contains the Morse potential between a protein and a nearest base pair of DNA,

$$H_{\text{prot}} = \frac{p_{xm}^2}{2M} + E\left(\mathrm{e}^{-\beta x_m} - 1\right)^2, \tag{3}$$

the model is ready to be applied in the case of many proteins, but here we only consider one. Finally, the interaction between these two bonds takes a simple but rather general form,

$$H_{\text{int}} = \sum_n \frac{\chi}{2} x_m^a y_n^b . \tag{4}$$

Here we take a constant potential well depth E as we assume this to be only protein dependent i.e. independent of DNA site; while β is just the inverse width of the well. The coupling constant χ determine the sensitivity and the strength of the interaction, the value is not specified and could be set to fit the reality. In this model the depths $D_{(\mathbf{Z})}$ and E define the specificity of DNA-protein interaction, the values could be determined via experiments (such as in [15]).

To prove that the model is plausible, the qualitative dynamics need to be depicted. This had been done by studying this model stability and phase portraits [16]. The stability was examined with the integer powers $a = 2$ and $b = 1$, the value was taken by assuming the one-way DNA interaction i.e. DNA base pair stretch (y_n) is coupled by protein only (via x_m^2), not by DNA itself. This can be understood by deriving the equations of motion from the Hamiltonian, but this time we will show that we can pick a and b without the assumptions. This will be made clear after we expand the equations for small x and y in the next section.

3. Nonlinear excitations in the DNA-protein interaction

This section will discuss on *how* the interaction occurs, that is, how the protein induces DNA bubbles that precede the total opening of the double helix (i.e. the denaturation process). To explore this aspect, we examine the nonlinear excitations related to the Hamiltonian system.

First we consider the equations of motion derived from the Hamiltonian,

$$\begin{aligned} m\ddot{y}_l &= 2\alpha D_{(\mathbf{Z})}(\mathrm{e}^{-2\alpha y_l} - e^{-\alpha y_l}) + k(y_{l+1} - 2y_l + y_{l-1}) - \tfrac{b}{2}\chi x_m^a y_l^{b-1}\,, \\ M\ddot{x}_m &= 2\beta E(\mathrm{e}^{-2\beta x_m} - e^{-\beta x_m}) - \tfrac{a}{2}\chi x_m^{a-1}\textstyle\sum_l y_l^b\,. \end{aligned} \tag{5}$$

According to the PB approach [17] it is assumed that the oscillations of bases are large enough to be anharmonic, but still insufficient to break the H-bond since the Morse potential plateau is not yet reached. In our model the protein-DNA bond x is also presumed by the same way. Here the base stretching y and the DNA-protein stretching x oscillate around the bottom of the symmetric potential so the following transformation can be safely implemented:

$$Y_n = \epsilon\phi_n \quad \text{and} \quad X_m = \epsilon\psi_m\,, \tag{6}$$

where ϵ is small and $Y \equiv \alpha y,\ X \equiv \beta x$. The continuum approximation is used for now as a first step to construct a solution. We use perturbative approach to obtain a solitonic solution. Inserting this into (5) with continuum limit and expanding in basis ϵ^n, we get (with z_0 replacing m)

$$\begin{aligned} &\phi_{tt} - S\phi_{zz} + V_{(\mathbf{Z})}\left(\phi - \tfrac{3}{2}\epsilon\phi^2 + \tfrac{7}{6}\epsilon^2\phi^3 + O(\epsilon^3)\right) + \tfrac{\mu}{2}\epsilon^{a+b-2}\psi^a|_{(\mathbf{Z}=\mathbf{Z}_0)}\phi^{b-1} = 0 \\ &(\text{at } \mathbf{Z} = \mathbf{Z}_0) \qquad \psi_{tt} + W\left(\psi - \tfrac{3}{2}\epsilon\psi^2 + \tfrac{7}{6}\epsilon^2\psi^3 + O(\epsilon^3)\right) + \eta\epsilon^{a+b-2}\psi^{a-1}\textstyle\int \phi^b \mathrm{d}\mathbf{Z} = 0 \end{aligned} \tag{7}$$

where we have dropped the indices for simplicity and defined

$$V_{(\mathbf{Z})} = \frac{2\alpha^2 D_{(\mathbf{Z})}}{m},\ W = \frac{2\beta^2 E}{M},\ S = \frac{K}{m},\ \mu = \chi\frac{b\alpha^{2-b}}{2m\beta^a},\ \eta = \chi\frac{a\beta^{2-a}}{2M\alpha^b} \tag{8}$$

From here we look straightforward perturbative solution

$$\phi = \phi^{(0)} + \epsilon\phi^{(1)} + \cdots,\quad \psi = \psi^{(0)} + \epsilon\psi^{(1)} + \cdots. \tag{9}$$

The solutions are examined only up to order ϵ to get the nonlinear excitations consisting of second harmonics, which in case of PB model are the breather solitons. For this to be realised, every order of ϵ^n in above equations should contain time or spatial derivatives, thus we use multiple-scale method [18]. Here ψ and ϕ depend on the independent variables $z_0,\ z_1, \ldots$ and $t_0,\ t_1, \ldots$ where $z_m = \epsilon^m z$ and $t_n = \epsilon^n t$ so that the operators $\partial/\partial z$ and $\partial/\partial t$ are expanded as

$$\frac{\partial}{\partial z}=\frac{\partial}{\partial z_0}+\epsilon\frac{\partial}{\partial z_1}+\epsilon^2\frac{\partial}{\partial z_2}+\cdots,\quad \frac{\partial}{\partial t}=\frac{\partial}{\partial t_0}+\epsilon\frac{\partial}{\partial t_1}+\epsilon^2\frac{\partial}{\partial t_2}+\cdots. \tag{10}$$

This method is actually looking for solutions that variate (oscillate) in independent time and spatial scales, it has the advantage that the solution is accurate up to $z=O(1/\epsilon^m)$ and $t=O(1/\epsilon^n)$. For instance, the term $f(t_1)\propto\exp(i\theta(t_1))$ oscillates slower than the term dependent on t_0. In our model the t_1 term acts as the envelope wave for the t_0 term. This kind of behavior had previously been inspected in [19] for single nonlinear oscillator, it is shown that the multiple-scale method gives an envelope wave which the first harmonic amplitude satisfies nonlinear Schrödinger equation (NLS). In this paper we are generalising the method for a coupled equations.

From equations (5) we chose the integer powers $a+b=3$ so that the interaction plays the role in $O(\epsilon)$ terms and is not ruining $\phi^{(0)}$ or $\psi^{(0)}$. It can be chosen in general $a+b>2$, depending on what order ϵ the interaction is wanted to affect. Here we the lowest power state $H_{\text{int}}=\frac{1}{2}\chi x_m^2 y_n$. We use multiple scales up to t_2 and z_1, the asymmetry of this choice will be clear after the equations governing the amplitudes are observed. For simplicity hereafter, we write $z\equiv z_0$, $Z\equiv z_1$, $t\equiv t_0$, $T\equiv t_1$, and $\tau\equiv t_2$. Inserting (9) and (10) into (7), we get

$$O(\epsilon^0):\ \phi_{tt}^{(0)}-S\phi_{zz}^{(0)}+V_{(z)}\phi^{(0)}=0$$

$$\psi_{tt}^{(0)}+W\psi^{(0)}=0 \tag{11}$$

$$O(\epsilon^1):\ \phi_{tt}^{(1)}-S\phi_{zz}^{(1)}+V_{(z)}\phi^{(1)}=-2\left(\phi_{tT}^{(0)}-S\phi_{zZ}^{(0)}\right)+\frac{3}{2}V_{(z)}\left(\phi^{(0)}\right)^2-\frac{\mu}{2}\left(\phi^{(0)}\right)^2$$

$$\psi_{tt}^{(1)}+W\psi^{(1)}=-\psi_{tT}^{(0)}+\frac{3}{2}W\left(\psi^{(0)}\right)^2-\eta\psi^{(0)}\int\int\phi^{((0))}\mathrm{d}z\mathrm{d}Z \tag{12}$$

$$O(\epsilon^2):\ \phi_{tt}^{(2)}-S\phi_{zz}^{(2)}+V_{(z)}\phi^{(2)}=-2\left(\phi_{tT}^{(1)}-S\phi_{zZ}^{(1)}\right)-\left(\phi_{TT}^{(0)}-S\phi_{ZZ}^{(0)}\right)-2\phi_{t\tau}^{(0)}+3V_{(z)}\phi^{(0)}\phi^{(1)}$$

$$-\frac{7}{6}V_{(z)}\left(\phi^{(0)}\right)^3-\mu\psi^{(0)}\psi^{(1)}$$

$$\psi_{tt}^{(2)}+W\psi^{(2)}=-2\psi_{tT}^{(1)}-\psi_{TT}^{(0)}-2\psi_{t\tau}^{(0)}+3W\psi^{(0)}\psi^{(1)}-\frac{7}{6}W\left(\psi^{(0)}\right)^3$$

$$-\eta\left(\psi^{(1)}\int\int\phi^{(0)}\mathrm{d}z\mathrm{d}Z+\psi^{(0)}\int\int\phi^{(1)}dzdZ\right) \tag{13}$$

We have broken down the nonlinear problem into several inhomogeneous linear differential equations where the solutions as follows,

$$\phi^{(0)}=A_1(Z,T,\tau)\mathrm{e}^{\mathrm{i}\theta}+\text{c.c.},\qquad \psi^{(0)}=A_2(Z,T,\tau,z)\mathrm{e}^{\mathrm{i}\varphi}+c.c.,$$

$$\phi^{(1)}=3|A_1|^2-\frac{\mu}{V_{(z)}}|A_2|_{z=z_0}^2-\frac{1}{2}A_1^2\mathrm{e}^{2\mathrm{i}\theta}+\frac{\mu}{6V_{(z)}}A_2|_{z=z_0}^2\mathrm{e}^{2\mathrm{i}\varphi}+\text{c.c.},$$

$$\psi^{(1)}=3|A_2|^2-\frac{1}{2}A_2^2\mathrm{e}^{2\mathrm{i}\varphi}+\text{c.c.}+\eta\left[\frac{A_2\int A_1\mathrm{d}Z}{\omega^2+2\omega\sqrt{W}}\mathrm{e}^{\mathrm{i}(\theta+\varphi)}+\frac{A_2^*\int A_1\mathrm{d}Z}{\omega^2-2\omega\sqrt{W}}\mathrm{e}^{i(\theta-\varphi)}+\text{c.c.}\right]\delta(k).$$

The envelopes A_1 and A_2, by zeroing the secular terms ($\exp(\pm\mathrm{i}\theta)$ and $\exp(\pm\mathrm{i}\varphi)$) in $O(\epsilon^2)$, meet these conditions:

$$\mathrm{i}\frac{\partial A_1}{\partial\tau}+P_1\frac{\partial^2A_1}{\partial\xi^2}+Q_1|A_1|^2A_1=\mu\left[3+\frac{2\eta}{\omega^2-4W}\right]|A_2|^2A_1,$$

$$\mathrm{i}\frac{\partial A_2}{\partial\tau}+P_2\frac{\partial^2A_2}{\partial T^2}+Q_2|A_2|^2A_2=\eta\left[3+\frac{2\eta}{\omega^2-4W}\right]\int|A_1|^2dZ\,A_2, \tag{14}$$

after we introduce a right moving coordinate where $\xi=Z-v_gT$. Here if we use scale ϵ^2z we have to introduce the coordinate moving to left, $\epsilon^2z+v_g\tau$, to form the NLS. This is contrary to our right moving coordinate so we cannot use this. The coefficients are $P_1 = (S - v_g)/2\omega$, $Q_1 = 4V_{(z)}$, $P_2 = -(2\sqrt{W})^{-1}$, and $Q_2 = 4W + 5\mu\eta/(6V_{(z)})$. These are in the form of coupled nonlinear Schrödinger (NLS) equation and the solutions have been widely investigated [20].

We look the solution having the type of bright soliton that is under condition $P_i Q_i > 0$ [20]. The one-soliton solution of A_1 and A_2 when inserted to our full solution will enable a particular breathing mode, this is interpreted as the denaturation bubbles. Further analysis to the solution could bring us to a deeper understanding of the dynamical excitation of DNA bubbles when triggered by a regulatory protein in some specific regions. It is also an interesting question that whether the model can represent the lowering of denaturation free energy as seen from experiments [11, and references therein]. The complete solution and its analysis are in progress.

References

[1] Englander S, Kallenbach N, Heeger A, Krumhansl J and Litwin S 1980 *Proc. Nat. Acad. Sci.* **77** 7222–7226
[2] Peyrard M and Bishop A R 1989 *Physical review letters* **62** 2755
[3] Yakushevich L 1989 *Physics Letters A* **136** 413–417
[4] Sulaiman A, Zen F P, Alatas H and Handoko L T 2012 *Physica D: Nonlinear Phenomena* **241** 1640–1647
[5] Sulaiman A, Zen F P, Alatas H and Handoko L T 2012 *Physica Scripta* **86** 015802
[6] Satarić M and Tuszynski J 2002 *Physical Review E* **65** 051901
[7] Sulaiman A, Zen F P, Alatas H and Handoko L T 2010 *Physical Review E* **81** 061907
[8] Rohs R, Jin X, West S M, Joshi R, Honig B and Mann R S 2010 *Ann. Rev. of biochem.* **79** 233
[9] Halford S E and Marko J F 2004 *Nucleic acids research* **32** 3040–3052
[10] Von Hippel P H and Berg O G 1986 *Proc. Nat. Acad. Sci.* **83** 1608–1612
[11] Williams M C and Maher J L 2010 *Biophysics of DNA-protein interactions: from single molecules to biological systems* (Springer Science & Business Media)
[12] Gerland U, Moroz J D and Hwa T 2002 *Proc. Nat. Acad. Sci.* **99** 12015–12020
[13] Wang J C and Giaever G N 1988 *Science* **240** 300–304
[14] Hogan M, Dattagupta N and Crothers D 1979 *Nature* **278** 521–524
[15] Gao Y, Devi-Prasad K and Prohofsky E 1984 *The Journal of Chemical Physics* **80** 6291–6298
[16] Dwiputra D, Hidayat W, Khairani R and Zen F P 2015 Nonlinear model of specific DNA-protein interactions and its stability (unpublished), submitted at the 6th asian physics symposium
[17] Peyrard M 2004 *Nonlinearity* **17** R1
[18] Jakobsen P 2013 Introduction to the method of multiple scales *Preprint* arXiv:1312.3651
[19] Remoissenet M 1986 *Physical Review B* **33** 2386
[20] Radhakrishnan R and Lakshmanan M 1995 *J. Phys. A: Mathematical and General* **28** 2683

Acknowledgements

The authors would like to thank Ministry of Research Technology and Higher Education of Republic of Indonesia for research funding Desentralisasi 2015 and Desentralisasi 2016. DD thanks the members of Theoretical Physics Laboratory ITB for hospitality.

Investigation of diffusion process inside the cell using Monte Carlo Cell (MCell)

A Sutresno[1,2], F Haryanto[1], S Viridi[1] and I Arif [1]

[1] Department of Physics, Faculty of Mathematics and Natural Sciences, Institut Teknologi Bandung, Jalan Ganesa 10, Bandung 40132, Indonesia
[2] Department of Physics, Faculty of Mathematics and Natural Sciences, Satya Wacana Christian Univeristy, Jalan Diponegoro 52-60, Salatiga 50711, Indonesia

E-mail: adita1204@gmail.com

Abstract. Understanding a structure, function and molecular mechanisms of a cell are not easy. Even though, it is difficult, but it still has a solution to be investigated. The aim of this study is to investigate the compartment model as a structure of a cell and diffusion process inside a cell and nucleus by using Monte Carlo simulation. MCell was used to analyze the compartment model for mechanisms inside a cell and nucleus. Compartment model consist two icosphere with different size, with cell diameter for 0.8 μm and nucleus diameter varies for about 0.4, 0.5, and 0.6 μm, respectively. Another parameter was their width and a model of receptor on the surface of the nucleus. During investigation, an amount of receptor was 120, 150, and 180 grid element. The model was collected and separated. The diffusion process flew from cell to the nucleus and returned to the cell with 2000 molecules on surface. MCell successfully simulated compartment model for a cell and the nucleus. Diffusion process depended on the diameter of the nucleus. The greater the nucleus, more effective the speed of diffusion process would be because the concentration inside the cell was increase and concentration in the nucleus was decrease.

1. Introduction

Cells are the basic unit of life. Understanding structural, functional and molecular mechanisms in a cell [1, 2] is not easy. One of the solutions is by modeling it using a Monte Carlo method to investigate the diffusion processes inside the cell. Several well-identification processes are involved in cell mechanisms such as a molecule diffusion process from cell to the nucleus and vice versa. This process is influenced by diffusion rate, comparison of molecule concentration and amount in a cell and nucleus, and kinds of receptor. The molecule concentration depends on the molecule amount in the cell and nucleus, the comparisons of cell and nucleus dimension or both.

Previous computational studies in biology systems with simulation model of Monte Carlo Cell has shown an effect such as ligand kinetics, density receptor and dimension synapse [3, 4], and urinary system with compartment model [5].

Shape of an icosphere was used as a model of a cell and nucleus compartments in which the icosphere nucleus was inside of the icosphere cell. Simulations were carried out using a variety of nucleus dimension, a number of receptors, and a receptor model. A Variety of parameter and the dimension of compartment gave much information about mechanisms diffusion in a cell, but this model didn't really represent a cell and nucleus in shape. This study is addressed to investigate the compartment model as the structure of a cell and the diffusion process as molecule mechanisms inside a cell and nucleus using Monte Carlo Simulation Cell (MCell).

2. Methods

Blender 2.69 was used to create and visualize the picture and movie for the compartments and added with CellBlender and Monte Carlo Cell (MCell). The simulation was carried out using MCell 3.1 [3, 6] (www.mcell.org) running on a Lenovo ThinkPad 2.2 GHz Intel Core i7 (Ubuntu 14.10).

The Cell in figure 1a represents a model target to create simulations.Fig 1a shows a fluorescent mouse cell that wasvisible under the microscope from Nagai-Group, Osaka University. Model illustrated in Fig 1b shows a result of a model for cell and nucleus. The cell has 0.8 μm diameter and the nucleus has varies diameter each of them were 0.4, 0.5 and 0.6 μm, respectively. Molecules with green color were molecules inside a cell; it called molecules A. Molecules A will be transferred to the nucleus using the diffusion process. Molecules with red color were a non-diffusing receptor on the top of a nucleus. Non-diffusing receptors at a density have two variation models that was collected and separated. Figure 1c shows a model that collected non-diffusion receptor and figure 1d shows a model that separate. The model has variable number of element grid, which represents the wide of area receptor such as120, 150, and 180, respectively.

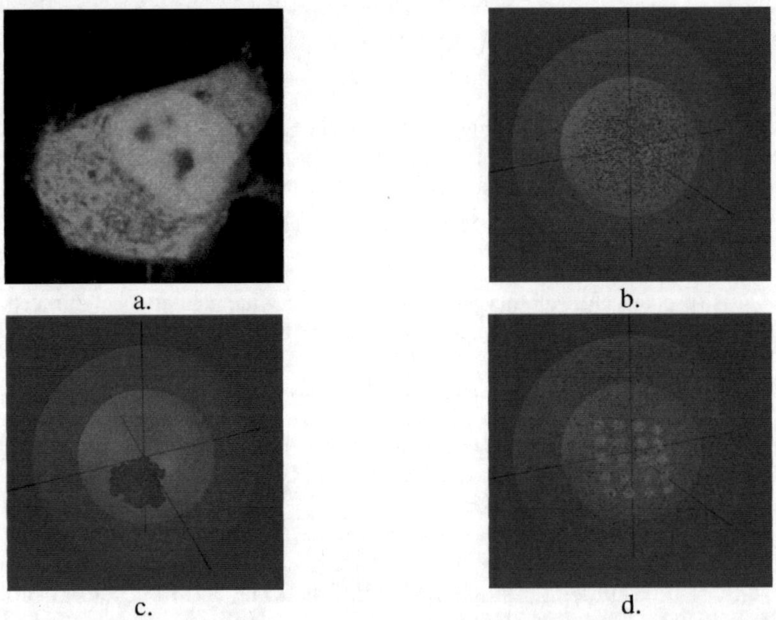

Figure 1. (a) Mouse cell visible under microscope fluorescent from Nagai-Group with permission, (b) Diffusion simulation at 0.01 s, (c) Model of collected non-diffusion receptor at 0.00 s (red), (d) Model of separate non-diffusion receptor at 0.00 s (green)

This study used two steps diffusion to describe the diffusion process in cell and nucleus. The first step from the cell happened as follow: molecules A diffused to the nucleus with reaction to molecules non-diffused t1 on the top of a nucleus with forward rate 2×10^{7} $M^{-1}.s^{-1}$ and the result of diffusion process was molecules B in the nucleus. The second step molecules B diffused back to the cell with forward rate 1×10^{7} $M^{-1}.s^{-1}$. And the reactions of diffusion process were formulated as below.

$$A'+t1' \xrightarrow{fwd1} B,+t' \xrightarrow{fwd2} C'+t1'$$

Simulations were carried out using a set of standard parameter values (table 1). Subsequently, it varied the parameters individually to test, the effect of the parameter on diffusion and a time step of 10 µs was used for simulation with standard parameters.

Table 1. Standard values of the parameters used in the simulations

Parameter	Standard Value
Diffusion constant molecules A, B, C (vol)	$1x10^{-6}$ cm^2s^{-1}
Diffusion constant of surface (Surf)	$1x10^{-8}$ cm^2s^{-1}
Forward rate 1 (Fr1)	$2x10^7$ cm^2s^{-1}
Forward rate 2 (Fr2)	$1x10^7$ cm^2s^{-1}
Number of molecules	2000
Wide of area non-diffusing (receptor)	120, 150, 180 e.g[a]

[a] e.g = element grid

3. Results and discussion

In this study, figure 2a shows the results of a diffusion process that occurs from the cell to the nucleus, then return to the cell again. Molecules that were diffused from the cell into the nucleus called molecule A marked with black which molecules A declined quickly until all molecules A diffused to the nucleus during 0.287s. The red color indicated that molecule B where the molecules presented in the nucleus as a result of diffusion process rose suddenly until its peak at t = 0.04s with 542 number of molecules. Finally, molecule B transferred to cell through diffusion mechanism and called molecules C with a blue line.

In the figure 2b shows half-saturation of molecules A with varying diameter of the nucleus occurs at $t_{0.5}$ = 0.015s, 0.031s and 0.051s that happened for nucleus diameter 0.6 µm, 0.5 µm, and 0.4 µm, respectively. The result shows bigger nucleus diameter influenced to the increasing molecule concentration in a cell because the volume of cell declined in molecules constant number. This condition caused the concentration difference between the cells and changed the nucleus in which the concentration inside a cell increased and inside the nucleus decreased, so the diffusion rate increased. The phenomenon shows the congruence between simulation resulting with a theory in which the diameter of 0.6 µm nucleus had a shorter of diffusion for about 0.015s followed by 0.5 µm and 0.4 µm during 0.031s and 0.051s, respectively. Duration of half-saturation had affected the number of molecules surviving in the nucleus in which the shortest duration 0.6 µm had 906 molecule number and then for 0.5 µm and 0.4 µm had 542 and 288, respectively as shows in figure 2c.

Receptors are gates that connect between the cell and the nucleus. In this investigation, a number of receptors had three wide varieties such as120, 150 and 180 element grids and two model receptors: converged receptor and separate receptor. Figure 2e shows a result of simulation wide varieties in which the receptor with the greater number of element grids had faster diffusion process. Figure 2f shows the receptor model converged and separated, in which the model of receptor which had large number of receptors did not have much effect on the speed of diffusion even though it had a different model, but the small number of receptors, converged model faster than the separated ones.

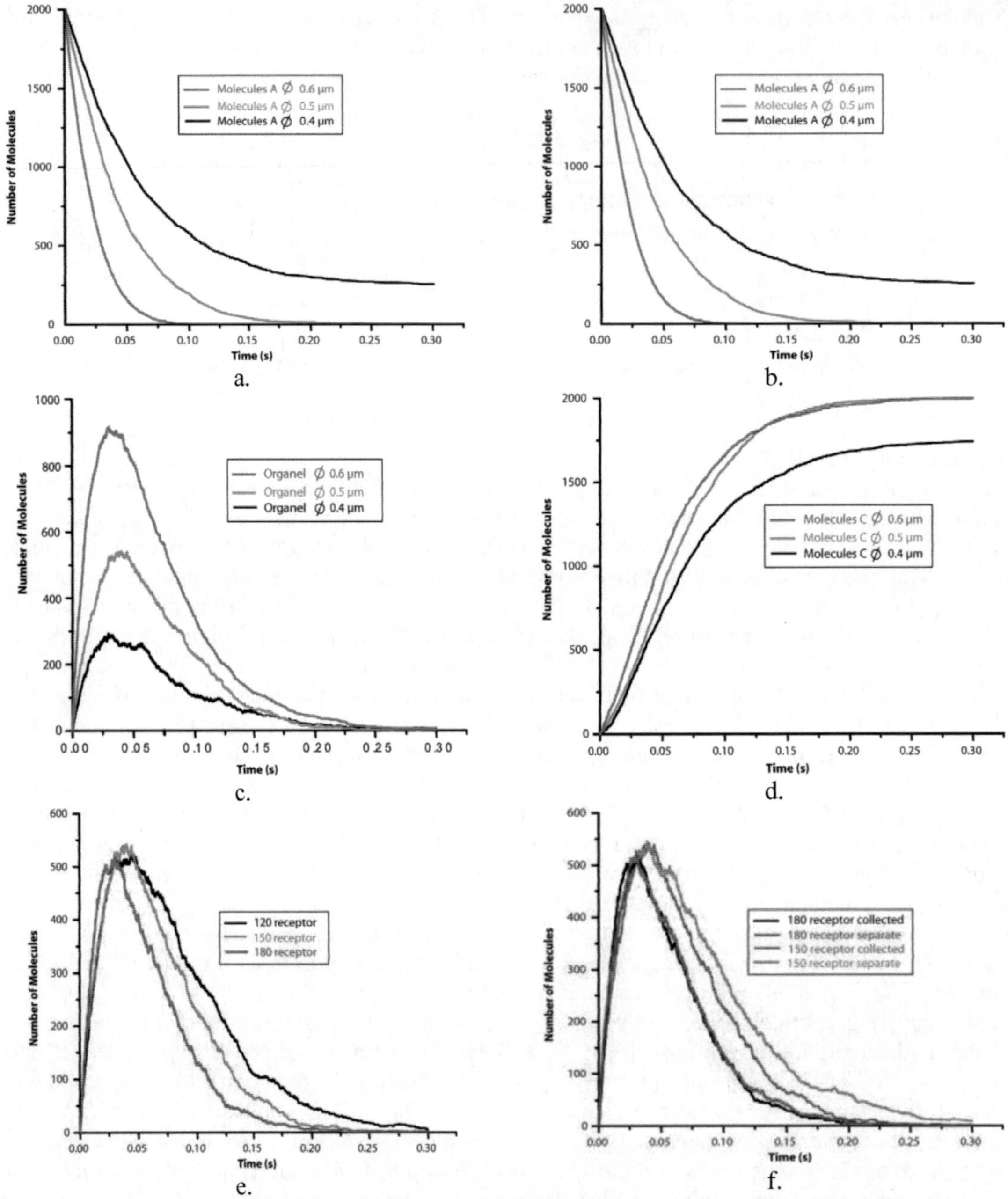

Figure 2. Results of simulation a. diffusion molecules A, B, C with diameter nucleus 0.5 µm, b, c, d diffusion molecules A, B and C with variety diameter nucleus, respectively, e. diffusion molecules B with variety number of receptor (non-diffusion), f. diffusion molecules B with variety model receptor (collected and separate).

4. Conclusions

MCell successfully simulated compartment model for a cell and nucleus. Diffusion process depends on a diameter of the nucleus. The greater the nucleus then the more effective the nucleus will be to speed the diffusion because the concentration inside cell increase and concentration inside the nucleus decreases. The other result, the greater the amount of receptor shows the increasing of the diffusion process and it makes the model collected faster than separated.

References

[1] Sherwood L 2010 *Human physiology from Cell to systems* (Canada: Brooks, Cole)
[2] Mader S S 2012 *Understanding Human Anatomy and Physiology* (English: McGraw-Hill)
[3] Dilger J P 2010 *Biophysical* **98** pp 959-67
[4] Freche D, Pannasch U, Rouach N and Holcman D 2011 *Ploes One* **6(10)** pp 1-15
[5] Sutresno A, Haryanto F, Viridi S and Arif I 2015 *Advance Science, Engineering and Medicine Journal* **7** pp 888-91
[6] Stiles J R and Bartol T M 2001 *Monte Carlo Methods for Simulating Realistic Synaptic Microphysiology Using MCell* in *Computational Neuroscience realisticModeling for experimentalists* (Belgium: CPC Press LLC) Chapter 4

Acknowledgements

This work is partially supported by ITB and JICA 2015 and thanks to Nagai Group at Osaka University for support Mouse cell visible.

Identification and level of organochlorine insecticide contamination in groundwater and iridology analysis for people in Upper Citarum cascade

K Oginawati[1] and M A Pratama[1]

[1] Department of Environmental Engineering, Faculty of Civil and Environmental Engineering, InstitutTeknologi Bandung, Jalan Ganesa 10, Bandung 40132, Indonesia

E-mail: Adhiraga.pratama@gmail.com

Abstract. Organochlorines are the main pollutants in the class of persistent organic pollutants which are types of pollutants that are being questioned worldwide due to chronic persistence, toxicity and bioaccumulation. Human around the Citarum River are still using groundwater as a drinking source. It is very risky for people health that consume groundwater because in 2009 the application of organochlorine still found in the Upper Citarum watershed rice field and had potential to contaminate groundwater. Groundwater was analyzed with nine species belonging to the organochlorine pollutants Organic Peristent types. 7 types of organochlorinesAldrin was detected with an average concentration of 0.09 ppb, dieldrin with an average concentration of 0.24 ppb, heptaklor with an average concentration of 0.51 ppb, with concentrations of endosulfan on average 0.73 ppb, DDT with average concentration of 0.13 ppb, Lindan with an average concentration of 1.2 ppb, endrin with an average concentration of 0.03 ppb. Types with the highest concentration of organochlorine a lindan and endosulfan. Residues of aldrin, dieldrin and heptaklor in groundwater already exceeds the quality standards for drinking water Permenkes 492/2010. Based on the iridology analysis obtained several systems are expected to nervous, immune and reproductive system disorders and toxin deposits under the skin.

1. Background

Pesticides can be defined as a mixture of chemicals used to prevent, eradicate animal/plant with the aim of human welfare [1]. Among various types of pesticides, there are some which use is prohibited because of their toxic and persistent nature. Pesticides are one of the pollutants that come from agricultural activities. One of them is organochlorine insecticide. Some organochlorine insecticides belong to persistent organic pollutants (POP's) which is a type of pollutant at issue all over the world due to its chronic, persistent and bioaccumulativecharacter.Organochlorine pesticides such as DDT, Aldrin, Dieldrin, Endrin, Chlordane, Hexachlorobenzene, MIREX, Toxaphene, Heptachlor are persistent insecticides [2].

The Upper Citarum River Watershed is an area with landuse that dominated by agricultural areas [2]. Furthermore, 80% of the community use groundwater as drinking water for their daily life which have high health effect risk if the groundwater was contaminated by organochlorine. Threfore research about the level and distribution of organochlorines in the groundwater of Upper Citarum River Watershed needs to be conduct based on the habits of the population that use groundwater for drinking source and sanitation.

2. Material and methods

2.1. Groundwater sampling
Thirty groundwater samples were collected from field site. Groundwater is taken from well that surrounded by the agricultural area. Fifteen samples taken from opened well and the other fifteen samples taken from closed well. Sample is taken by water sampler. The sample bottles were rinsed with groundwater sample and then refilled in a softly for minimized aeration. The sample collection bottles were transported in cool box with blue ice and subsequently stored in a refrigerator at 4◦C until further analysis. Sampling points in this research shown in figure 1.

2.2. Laboratory analysis
Laboratory analysis of organochlorine residues in samples divided into two stages, samples preparation of samples and analysis of organochlorine residues. All samples were analyzed using liquid-liquid extraction (LLE) methods. This method was adopted and developed based on Indonesia National Pesticide Commission. At laboratory, 100ml of water samples with 50 ml n-hexane-dichlormethan 85:15 v/v was shaken using separated funnel for three minutes. The lower layer will be extracted twice more with the same procedure clean up the final solution using sodium anhydrate and fluorisil. The filtrate evaporated using rotary evaporator until the volume 1 ml, elute with the acetone until the final volume 10 ml. In the other hand, 25 gram of each sediment sample was shaken with 100ml acetone using vertical shaker for 20 minutes for 3 times. Filter the filtrate using butchner funnel and flask. Evaporate the filtrate by rotary evaporator, clean up using florisil and sodium anhydrate and elute with n-hexane 50ml, evaporate again until the volume 1ml then elute with acetone until final volume 10 ml [3].

Samples were analyzed using Varian 450 Gas Chromatograph fitted with ECD detector with coulomb VF 1701 (30m x 0.25 mm x 0.25 μm) and VF-5 (30m x 0.25 mm x 0.25 μm). Injector and detector temperature were 240°C, column temperature 230°C. Nitrogen gas was used as carrier gas with velocity 40ml/min, with speed recording 5mm/min and sensitivity 102. Recovery analysis was 90-92% for organochlorine pesticide. Water samples resulted in μg/L unit [4]. Organochlorine types that analyzed in this study are aldrin, dieldrin, DDT, heptachlor, endrin and endosulfan.

2.3. Iridology analysis
Iridology analysis will be conducted on 10 people for each hazard class. Those to be analyzed must have the following attributes:

- Not work as pesticide sprayers
- Consume groundwater will be analyzed organochlorine residues as a source of drinking water for at least 10 years for adults and five years for children

Subject iridology analysis can be divided into three groups. Children aged 8-15 years. Adults aged 16-55 years and elderly aged over 55 years but not more than 65 years. Iridology analysis result is in the form of a table which consists of the quality of the organs in the body of the subject. The quality of the organs in the body of the subject is presented in the form of a score with a score of 1-5. The interpretation of these values is

1 = normal 3 = sub chronic 5 = Poor
2 = acute 4 = chronic

3. Results and discussion

3.1. Organochlorine pesticide residue in groundwater
Based on the laboratory analysis of organochlorine in groundwater, there are seven types of organochlorine pesticide that detected include lindan, aldrin, dieldrin, endrin, heptachlor, DDT and endosulfan. The average concentration for those organochlorineinsectisede are 0.09 μg/L for aldrin, 0.51 μg/L for heptachlor, 0.241 μg/L for dieldrin, 0.134 μg/L for DDT, 0.731 μg/L for endosulfan,

1.248 μg/L for lindane and 0.03 μg/L for endrin. Lindan is the highest average concentration of organochlorine that found in groundwater samples.The distribution of aldrin, dieldrin, heptachlor, DDT, endrin and endosulfan in each sampling point and the comparison with water quality standard are shown in figure 1.

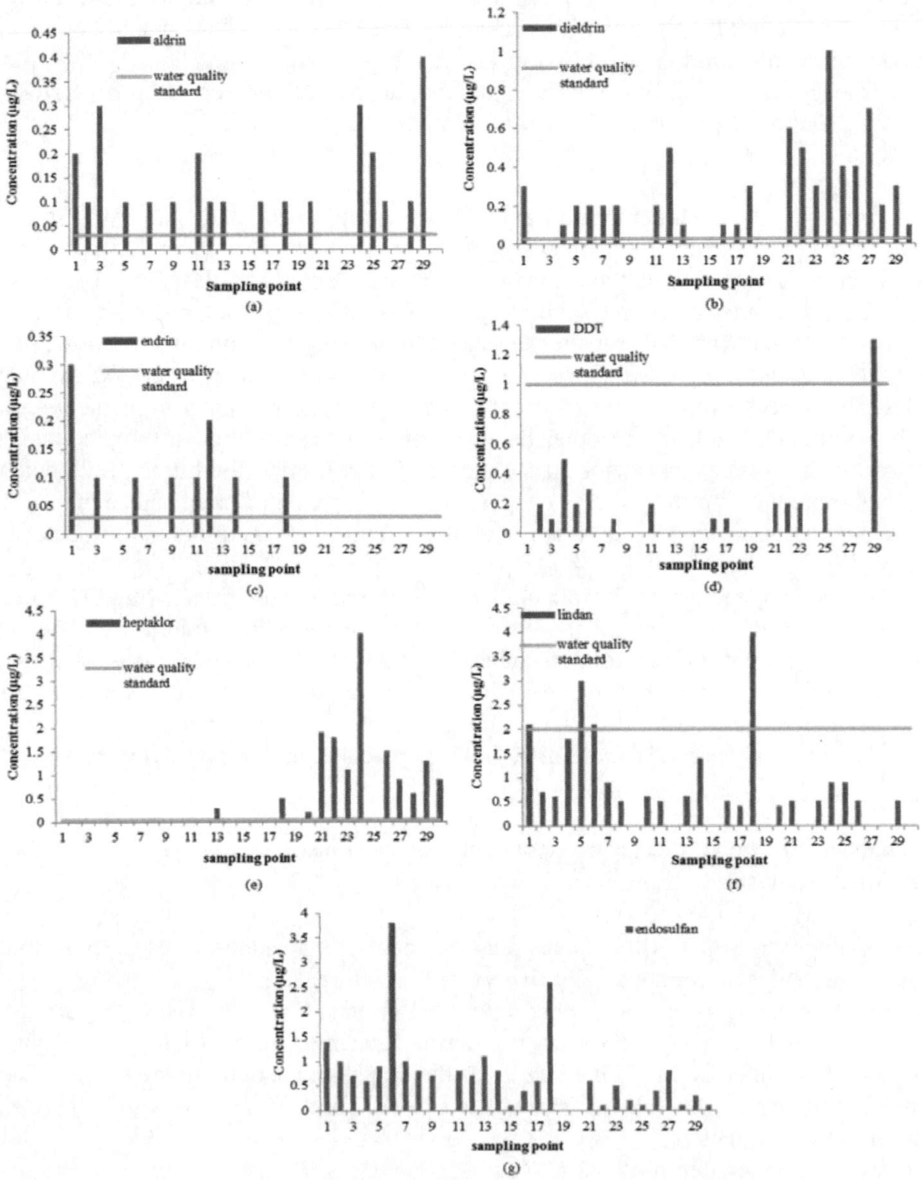

Figure 1. Residues of aldrin (a), dieldrin (b), endrin (c), DDT (d), heptachlor (e), lindane (f) and endosulfan(g) in groundwater at field sit.

3.2. Iridology analysis for people in Upper Citarum Cascade

Iridology analysis results in the nervous system are shown in figure 2.

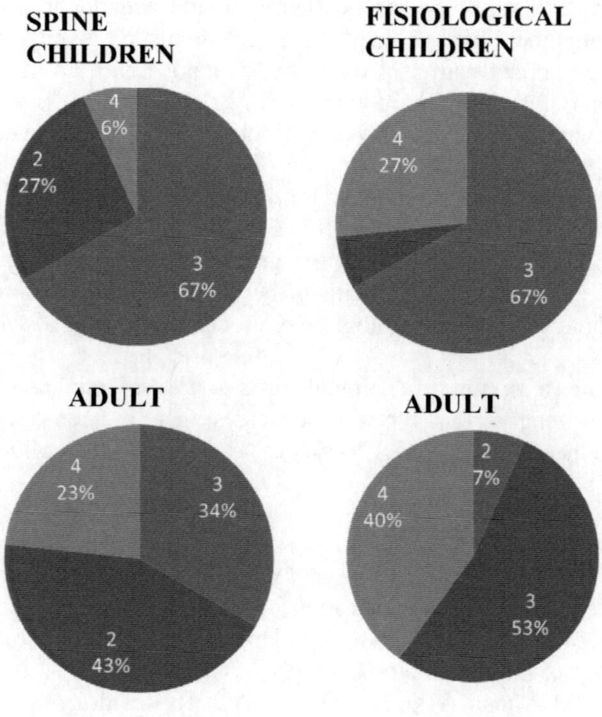

Figure 2. Iridology analysis on the nervous system

Based on the scores obtained from the nervous system iridology analysis, it was found that the physiological brain damage is part of the nervous system that most major disorder. Organochlorines are neurotoxins on the impulse conduction. Impulse conduction is an activity that is regulated by the brain physiologically.

The nervous system is the primary target of DDT after exposure to acute, subchronic and chronic. Exposure to DDT can cause tremors, convulsions in some of the test animals. The effects obtained from an experiment toxicity tests on rats with a dose of 50-600 mg of DDT/kg/day [4]. Mice who received a dose of 160 mg / kg experienced a disruption of tremor, whereas the dose of 200-600 mg/kg/day cause a disruption in the form of convulsions. DDT is also associated with an increased neurotransmitter. An increase in the metabolite 5-HIAA and serotonin degradation were reported to have a correlation with DDT at a dose of 50 mg/ kg/day. DDT can cause increased elepasan acetylcholine and decreased release of norepinephrine in rats exposed to 400 mg / kg of DDT.

Several small groups of humans who consume contaminated meat aldrin and lindane within 6-12 months experienced some nervous system disorders. The form of myoclonic disorders, auditory and visual disturbances, hyperexcitability and irritability. In some cases, the study subjects experienced a loss of the bond and the other subjects are mentally retarded. Hyperexcitability, kehilangatan memory, persistent headaches also reported as a result of exposure to dieldrin at a dose of 120 mg/kg. The effects of endosulfan on the concentration of neurotransmitters in the brain of mice has been done. The study showed that mice exposed to endosulfan with a dose of 6 mg/kg for 8 days led to changes in the level of noradrenaline, dopamine and serotonin in the central nervous system (the olfactory lobe, hippocampus and cerebellum). Dogs that are exposed to concentrations of endosulfan premises 30 ppm- 54 days, 45 days and 52 ppm-60 ppm for 40 days to experience extreme sensitivity to noise, fear terhada optical stimulation and muscle contraction in the face of extreme and legs.

4. Conclusion

Groundwater was analyzed with nine species belonging to the organochlorine pollutants Organic Peristent types. 7 types of organochlorinesAldrin was detected with an average concentration of 0.09 ppb, dieldrin with an average concentration of 0.24 ppb, heptaklor with an average concentration of 0.51 ppb, with concentrations of endosulfan on average 0.73 ppb, DDT with average concentration of 0.13 ppb, Lindan with an average concentration of 1.2 ppb, endrin with an average concentration of 0.03 ppb. Lindan is the highest average concentration of organochlorine that found in groundwater samples. Residues of aldrin, dieldrin and heptachlor in groundwater already exceeds the quality standard for drinking water in Indonesia, PERMENKES 492/2010. The contamination levels of organochlorine pesticide in field site showed that the groundwater remediation needs to recover its quality so that it is safe for consumption. Based on the scores obtained from the nervous system iridology analysis, it was found that the physiological brain damage is part of the nervous system that most major disorder.

Pesticides can be defined as a mixture of chemicals used to prevent, eradicate animal/plant with the aim of human welfare [1]. Among various types of pesticides, there are some which use is prohibited because of their toxic and persistent nature. Pesticides are one of the pollutants that come from agricultural activities.

References

[1] Soemirat and Juli 2003 *Toksikologi Lingkungan* (Yogyakarta: Gadjah MadaUniversity Press)
[2] Rochmanti M 2009 The Use of Pesticides Identification in Vegetable And Paddy Field (Case Study:Agricultural Area in Kertasari Subdistrict, Upper Citarum Watershed, Bandung) *Seminar of Infrastructure and Built Environment, Bandung*
[3] Rahmawati S, Yoneda M, Matsui Y and Nakayama A 2010 Residue of organochlorine pesticide in surface water and sediments in the upper Citarum River, West Java, Indonesia
[4] Singh 2010 *Journal of Environmental Contamination and Toxicology* **67** pp 126-13

Study of budding yeast colony formation and its characterizations by using circular granular cell

D Aprianti[1], F Haryanto[2], A Purqon[2], S N Khotimah[2], S Viridi[2]

[1] Department of Computational Science, Faculty of Mathematics and Natural Sciences, Institut Teknologi Bandung, Jalan Ganesa 10, Bandung 40132, Indonesia
[2] Department of Physics, Faculty of Mathematics and Natural Sciences, Institut Teknologi Bandung, Jalan Ganesa 10, Bandung 40132, Indonesia

E-mail: apriantidev@gmail.com

Abstract. Budding yeast can exhibit colony formation in solid substrate. The colony of pathogenic budding yeast can colonize various surfaces of the human body and medical devices. Furthermore, it can form biofilm that resists drug effective therapy. The formation of the colony is affected by the interaction between cells and with its growth media. The cell budding pattern holds an important role in colony expansion. To study this colony growth, the molecular dynamic method was chosen to simulate the interaction between budding yeast cells. Every cell was modelled by circular granular cells, which can grow and produce buds. Cohesion force, contact force, and Stokes force govern this model to mimic the interaction between cells and with the growth substrate. Characterization was determined by the maximum (L_{max}) and minimum (L_{min}) distances between two cells within the colony and whether two lines that connect the two cells in the maximum and minimum distances intersect each other. Therefore, it can be recognized the colony shape in circular, oval, and irregular shapes. Simulation resulted that colony formation are mostly in oval shape with little branch. It also shows that greater cohesion strength obtains more compact colony formation.

1. Introduction

Yeast is a well-known unicellular organism that becomes a model to study many biological processes, such as cancer and cell cycle in multicellular organism [1, 2]. The yeast cell grows by increasing its volume until reaching a critical size. This growth undergoes in a cycle. Some species of yeast can divide by forming bud at the periphery of the mother cell, called budding yeast such as the popular baker's yeast *Saccharomycess cerevisiae*. The cell division occurs after the DNA replication. Bud of a mother yeast cell can emerge at a place in mother cell circumference. It will stick to its mother cell until it is ready to produce new cells. Cell division of budding yeast will occur until reaching its lifespan limit.

Budding yeast can exhibit some multicellular properties such as colony formation. The colony formation of pathogenic budding yeast, such as *Candida sp*, can colonize in the human body or some medical devices [3]. Moreover, it can form biofilm that considered as a major cause of human infection and can resist drug therapy. This work aims to study the early formation of the budding yeast colony by taking account some parameters, especially cell cohesion and cell budding pattern.

2. Method

Cell growth can be modelled by using several approaches which can be classified into two main categories, continuum approach and discrete approach. Those methods are used based on how the cell is defined. By using the discrete approach, a cell can be defined by its biophysical or biological parameters and approximated by simpler shape such as sphere [4].

Budding yeast's cell actual shape is ovoid. Due to the assumption that colony formation is seen from above, the cell is modelled and simulated in two-dimension. The cell is modelled by using circular granular cell, neglected its internal properties. Cell growth is defined by increasing in cell diameter. After reaching its maximum size, cell will divide by producing buds at its periphery. The new cell will go through a period of diameter growth (t_G) before it has ability to produce bud while the mother cell can directly produce another bud after a delay time. The time interval between two buds birth in a mother cell is defined by t_B.

The site of new bud emergence is considered to affect colony spreading. In this work, simulation is limited by maximum number of cells (N) and four cell budding patterns (θ) which expected to represent a possible pattern in budding yeast colony growth. New cell that occupy some spaces will interact with existed cells and its growth medium. All the interaction is simulated using molecular dynamic method.

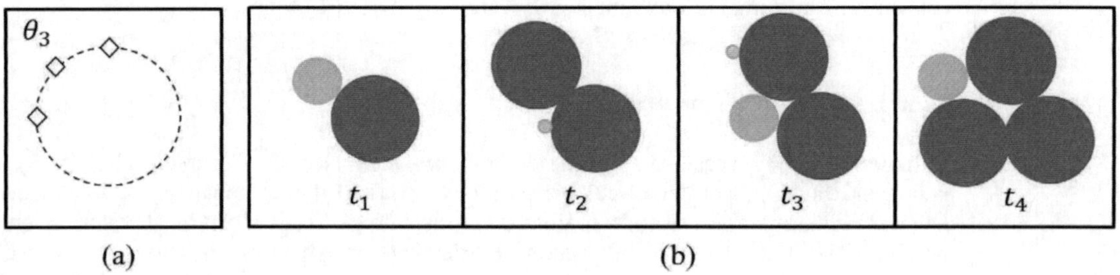

Figure 1. (a) Possible sites of new bud (θ_3); (b) Illustrations of colony formation with budding pattern θ_3.

Cell-cell interactions are represented by two forces, cohesion force and contact force. Medium or growth substrate is assumed to affect the colony formation. Due to this assumption, the movement of cell is also depended by the viscosity of substrate (η). The interaction between cells and growth substrate is represented by Stokes force. Formulation of all forces and their explanation are already provided in previous work. [5].

At the initial condition, there are already three cells in the system. One of them has the ability to produce new cells while others are still growing. Parameter values of simulation are chosen so that it does not need a huge amount of computation time. Most of simulation parameters are provided in table 1.

Table 1. Simulation parameters.

Parameter	Values
η	15.0
m	1.0
Δt	10^{-5}
t_B	15.0
t_G	3.0
N	50

3. Results and discussion

Characterization of the colony was determined by the maximum and minimum distances between two cells within the colony and whether two lines that connect the two cells in the maximum and minimum distances intersect each other. Therefore, it can be recognized the shape of the colony. Colony formation that has the same maximum and minimum distance value will be considered as circular. If the values are different and the intersection line within the colony, then the shape is considered as oval, else the colony is defined as in irregular shape.

Two parameters are varied in this simulation: cohesion strength (G) and budding pattern (θ). The variation of parameter simulation is provided in table 2. Every combination of G and θ values are simulated until five repetitions, since the alteration of the angle in the budding pattern occurs randomly. Figure 2 and 3 show some simulation results.

Table 2. Parameter variations in simulation.

Variation	Cohesion Strength (G)	Budding pattern (θ)
Value	G = 2.0 G = 5.0	$\theta_1 = \frac{\pi}{4}$ $\theta_2 = 0, \frac{\pi}{4}$ $\theta_3 = \frac{\pi}{2}, \frac{5\pi}{4}, \pi$ $\theta_4 = 0, \frac{\pi}{2}, \pi, \frac{3\pi}{2}$

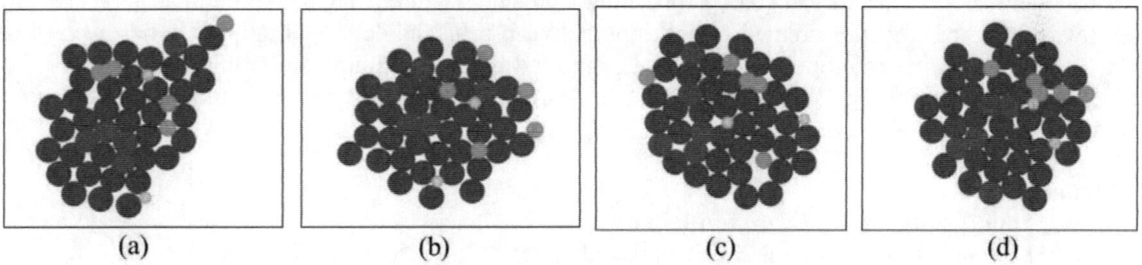

Figure 2. Simulation result of every cell budding pattern (a) θ_1, (b) θ_2, (c) θ_3, (d) θ_4 and cohesion strength G = 5.

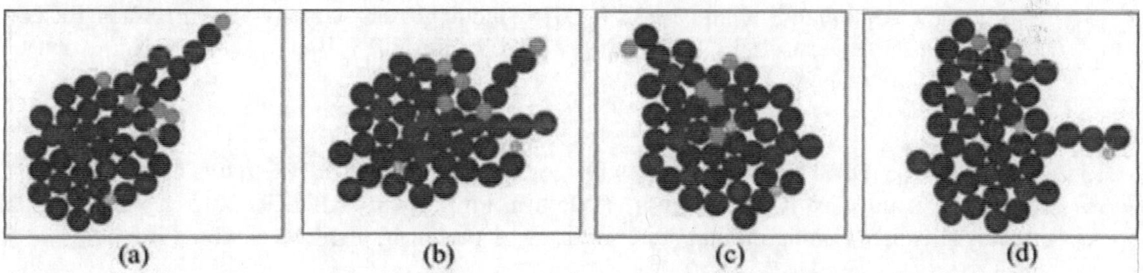

Figure 3. Simulation result of every cell budding pattern (a) θ_1, (b) θ_2, (c) θ_3, (d) θ_4 and cohesion strength G = 2.

Different values of cohesion strength give significant differences in colony formation. Greater cohesion strength result more compact colony formation. It is caused by cells in a colony tend to adhere each other and leaving it birth-site. It is also shown in figure 4, the difference distance between L_{max} and L_{min} for G = 2 are wider that G=5. Besides, branching structure is found in G = 2 for almost all budding patterns. The average graph shows that the average colony formation is in oval shapes.

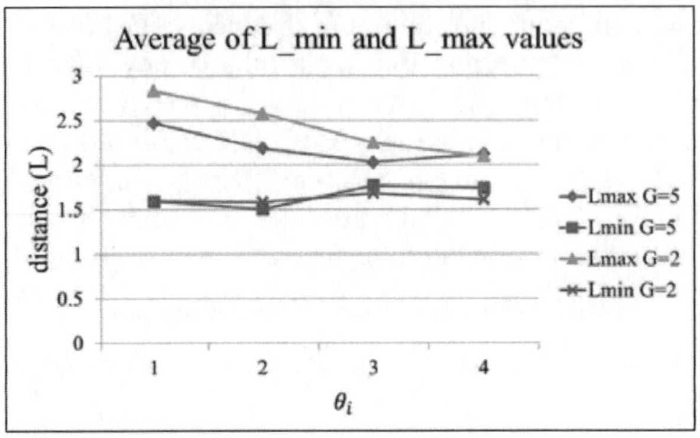

Figure 4. The average of maximum and minimum distances values, L_{min} and L_{max} of a colony for two different values of G = 2, 5.

4. Conclusion

A model for two dimensional colony formation based on the circular granular cell has been presented. It was found out that the strength of cohesion between cells and cell budding pattern has a significant role in determining colony expansion that determined by the maximum and minimum distances of the cell within the colony. The greater cohesion strength value will produce more compact colony formation.

References

[1] Pray L 2008 *Nature Education* **1** 183
[2] Karathia H, Vilaprinyo E, Sorribas A and Alves R 2011 *PLoS ONE* **6** e16015
[3] Chandra J, Kuhn D M, Mukherjee P K, Hoyer L L, McCormick T and Ghannoum M A 2001 *J. Bacteriol* **183** 5385-94
[4] Byrne H and Drasdo D 2009 *J. Math. Biol.* **58** 657–687
[5] Aprianti D, Khotimah S N and Viridi S 2015 Budding Yeast Colony Growth Study Based On Circular Granular Cell, in Asian Physics Symposium-2015, Bandung, Indonesia, August 19-20 2015, *in review*

Acknowledgements

Riset Desentralisasi DIKTI in year 2015 with contract number 310i/I1.C01/PL/2015 supports this research in 13th South East Asia Congress of Medical Physics (SEACOMP) 2015. This work is also supported by Advance Computing Laboratory, Physic department, Institut Teknologi Bandung.

Institute of Physics
Dirac House, Temple Back
Bristol BS1 6BE UK

ISSN: 1742-6588
ISBN 978-1-5108-2236-8